Encyclopaedia of Herbal Antioxidants

Encyclopaedia of Herbal Antioxidants

– *Volume 1* –

Editor
T. Pullaiah
Department of Botany
Sri Krishnadevaraya University
Anantapur – 515 003
Andhra Pradesh

2012
REGENCY PUBLICATIONS
New Delhi - 110 002

ISBN 978-81-892
ISBN 978-81-892
ISBN 978-81-892
ISBN 978-81-892

Published by : **REGENCY PUBLICATIONS**
(*Division of Daya Publishing House*)
4760-61/23, Ansari Road, Darya Ganj,
New Delhi - 110 002
Phone: 23245578, 23244987
Fax: (011) 23260116
e-mail : info@regency-books.com
website : www.dayabooks.com
website : www.regencybooks.com

Laser Typesetting : **Classic Computer Services**
Delhi - 110 035

Printed at : **Chawla Offset Printers**
Delhi - 110 052

PRINTED IN INDIA

Preface

The role of free radicals in many disease conditions has been well established. Several biochemical reactions in our body generate reactive oxygen species and these are capable of damaging crucial bio-molecules. If they are not effectively scavenged by cellular constituents, they lead to disease conditions, *e.g.*, cerebrovascular disease, cancer, Arteriosclerosis, Atherosclerosis, Heart disease, Senelity, Ageing, Behcet's disease, Crohn's disease, Cataracts, Sunburn, Ulcers, Osteoporosis, Rheumatoid Arthritis, Diabetes Mellitus, Emphysema, Stroke, Haemorrhagic Shock, Cardiovascular Disorders, Cystic Fibrosis, Neurodegenerative Diseases (*e.g.*, Parkinsonism, Alzheimer's disease), Gastrointestinal Ulcerogenesis, AIDS and even early Senescence. Free radicals are involved in numerous skin diseases, especially inflammatory reactions and photosenescence. Antioxidants reduce blood clotting and cholesterol level. Antioxidants have antistress, immunomodulatory, cognition-facilitating, anti-inflammatory and anti-aging properties.

The interest on the role of natural antioxidants as a tool to prevent aging and degenerative diseases process is growing due to the toxicity of synthetic antioxidants such as butylated hydroxyl anisole (BHA) and butylated hydroxyl toluene (BHT) at fairly high doses which limits their therapeutic usage. In the last few years there is a flurry of research on antioxidant plants.

The book, *Encyclopaedia of Herbal Antioxidants*, gives information on antioxidant activity of different plant species. All the species are listed in alphabetical sequence. Angiosperms are given first followed by other groups of plants. Plant part used, model system studied, methods of assay and chemical constituents are given for each and every species studied.

I tried to make this book as comprehensive as possible. However for a book of this magnitude there are bound to be omissions. Had I tried for perfection I could have never completed this book. Readers are requested to bring to the notice of the author for inclusion in the revised edition.

T. Pullaiah

Contents

Abbreviations

AAPH: 2, 2 – Azobios-2-amino-propane dihydro Chloride
ABTS: 2,2′-azino-bis-3-ethyl benzothiazoline-6-sulphonic acid
ALP: Alkaline Phosphatase
CAT: Catalase
DMSO: Dimethyl Sulphoxide
DPPH: 1,1,Diphenyl-2-Picryl-hydrazyl
FRAP: Ferric Reducing and Antioxidant Power
GAE: Gallic acid Equivalent
GPx: Glutathione Peroxidase
GSH: Glutathione
IC: Inhibitory Concetration
MDA: Malondialdehyde
ONOO: Peroxynitrite
ORAC: Oxygen Radical Absorbance Capacity
RNS: Reactive Nitrogen Species
ROS: Reactive Oxygen Species
SGOT: Glutamate Oxaloacetate transaminase
SGPG: Glutamate Pyruvate transaminase
SOD: Superoxide Dismutase
TEAC: Trolox Equivalent Antioxidant Capacity

Chapter 1
Introduction

☆ K.S. Nagesh

Antoine Lavoisier, a pioneer oxygen chemist, had pointed out about 150 years ago that animals that respire are true, combustible bodies that burn and consume themselves (Lehninger *et al.*, 1990). The biological combustion produces harmful intermediates called free-radicals. A free radical is simply defined as any species capable of independent existence that contains one or more unpaired electrons, an unpaired electron being one that is alone in an orbital. It may be superoxide ($O_2^{\cdot}$), thiyl (R-S·) or nitric oxide (NO·) in which the unpaired electron delocalized between both atoms. The $O_2^{\cdot-}$hydroxyl radicals (·OH) and other reactive oxygen species (ROS) such as H_2O_2 are continuously produced *in vivo.*

Free radicals are fundamental to any biochemical process and represent an essential part of aerobic life and our metabolism. They are continuously produced by the body's normal uses of oxygen such as respiration and some cell-mediated immune functions. These free radicals are also generated through environmental pollutants, cigarette smoke, automobile exhaust fumes, radiation, air pollutants and pesticides (Tiwari, 2001). These exogenous pollutants generating free-radicals have become part and parcel of our daily inhaling or ingesting life. Continuous interaction of the animal physiological systems with these free radicals generated either indigenously or inhaled from exogenous sources. They lead to excess load of free radicals and cause cumulative damage of protein, lipid, DNA, carbohydrates and membranes, resulting in so-called oxidative stress (Tiwari, 2001). The oxidative stress is thought to contribute to aging (Ashok and Ali, 1999), and a wide range of human diseases (Sies, 1982) including alzheimer's disease (Christen, 2000; Nunomura *et al.*, 2006), Parkinson's disease (Wood-Kaczmar *et al.*, 2006), diabetic pathologies (Davi *et al.*, 2005; Giugliano *et al.*, 1996), rheumatoid arthritis (Hitchon and El-Gabalawy, 2004) and neurodegeneration diseases (Cookson and Shaw, 1999), cancers (Wiseman and Halliwell, 1996), inflammation (Sco. *et al.*, 1995), neurodegeneration (Lebortiz *et al.*,

1996), Parkinson's disease (Jenner, 1994), atherosclerosis (Witztum, 1994) and premature aging (Orr and Sohal, 1994). It also leads to cardiovascular disease in which low density lipoprotein oxidation appears to trigger the process of atherogenesis which results in atherosclerosis and finally cardiovascular disease (Aviram, 2000; Van Gaal *et al.*, 2006).

Living creastures have evolved a highly complicated defense system with antioxidants composed of enzymes and vitamins against oxidative stress in the course of their evolution. Antioxidants are molecules capable of slowing or preventing the oxidation of other molecules. In biological systems the definition for antioxidants has been extended to any substance that when present at low concentrations, compared to those of an oxidizable substrate, significantly delays or prevents oxidation of the substrate like lipids, proteins, DNA and carbohydrates (Halliwell, 1990). Currently, biological antioxidants have further assumed a broad definition to include repair systems such as iron transport proteins, antioxidant enzymes and factors affecting vascular homeostasis, signal transduction and gene expression (Frakel and Meyer, 2000).

The term antioxidant originally was used to refer specifically to a chemical that prevented the consumption of oxygen. In the late 19th and early 20th century, extensive study was devoted to the use of antioxidants in important industrial processes, such as the prevention of metal corrosion, the vulcanization or rubber and the polymerization of fuels in the fouling of internal combustion of engines (Matill, 1947). Early research on the role of antioxidants in biology focused on their views in preventing the oxidation of unsaturated fats which cause rancidity (German, 1999). Identification of vitamins A, C and E as antioxidants has revolutionized the field and led to the realization of importance of antioxidants in living organisms.

Since time immemorial man has been using plant extracts to protect himself against several diseases and also to improve his health and life-style. No doubt, plants are serving several purposes whether heal, nutrition, beauty of medicinal. With the development in techniques and recent researches, it has been proved that certain non-nutritive chemicals in plants such as terpenes and flavonoids which were earlier thought to be of no importance to human diet, possess antioxidant properties.

The plants are susceptible to damage caused by active oxygen and thus develop numerous antioxidant defense system resulting in formation of numerous potent antioxidants.

Till date as such no set of definition of antioxidants exists. Scientists are still striving hard to find out the role of particular dietary supplements in body that have potent health benefits. Since different antioxidant compounds found in diet considerably vary from one another, it is a difficult task to identify the role of a single compound. In simple words, "Antioxidants are a type of complex compounds found in our diet that act as a protective shield for out body against certain disastrous enemies (diseases) such as arterial and cardiac diseases, arthritis, cataracts and also premature ageing along with several chronic diseases".

The above definition gives an idea about what actually an antioxidant is as still a lot of work has to be carried on getting exact information about antioxidants, their exact amount in one's diet and their function. The recent researches on free radicals promise a revolutionary improvement in health and life-style of humans (Mishra *et al.*, 2007).

Reactive Oxygen Species (ROS)

Oxygen is an essential element for life to perform biological functions such as catabolism of fats, proteins and carbohydrates in order to generate energy for growth and other activities. However, a parallel role of oxygen as a toxic agent for living tissues has also been discovered. Oxygen, though not dangerous by itself, is involved in the generation of various kinds of "reactive oxygen species" (ROS). ROS, formed during metabolism or through the action of ionizing radiation, can interact with bio-molecules and ultimately lead to an onset of a variety of diseases. A number of different free radicals and non-radicals are produced during normal aerobic metabolism (Halliwell, 1995; Greenwald, 1991). A collective term, "reactive oxygen species" (ROS), is used for oxygen-derived species including oxygen bearing free radicals, as well as certain non-radicals. Some non-oxygenated radicals are also generated in biological systems, such as carbon-centered free radicals (*e.g.* alkyl radical, R-H_2C) and sulfur-centered radicals (*e.g.*, thiyl radicals, R-S), which are produced by the attack of free radicals on hydrocarbons and the oxidation of glutathione, respectively (Ivanova and Ivanov, 2000; Cheeseman and Slater, 1993).

Free Radicals-Highly Reactive Species

A free radical is a chemical species, which is capable of independent existence and possesses one or more unpaired electrons that bestow it with immense reactivity. This reactivity is inversely related to their stability (Aitken and Fisher, 1994).

Types and Sources of ROS

A number of ROS are continuously generated in living system as a consequence of normal metabolic processes. These can be produced in two ways: by enzymatic reactions involving xanthine oxidase (XO), NADPH oxidases and lipoperoxidases; by non-enzymatic sequence of reactions such as the catalytic action of free transition metals (for example iron and copper), by the toxic action of certain chemicals such as doxorubicin, by the attack of electrons leaked from the mitochondrial electron transport chain and by the effect of radiation including UV light and radon (Ra) gas. Different types of ROS generated include both free radical and non-radical species (Bakonyi and Radak, 2004; Kerr *et al.*, 1996; McCord, 1993). The major ROS have been listed in Table 1.1.

Alkyl, alkoxyl, alkperoxyl radicals and lipid peroxide (R·, RO·, ROO·, and LOOH) are the products of attack of free radicals in lipid and fatty acid molecules. These biomolecules are more susceptible towards free radical attack due to the presence of unsaturation. Superoxide anions ($O_2^{\cdot-}$) are the reduced form of oxygen. Though $O_2^{\cdot-}$ itself is not damaging, it plays an important role in the formation of ROS like H_2O_2 and nitric oxide radicals, and also acts as a reductant for transition metals (Cheesemand and Slater, 1993).

Table 1.1: Types of ROS

ROS	*Types*	*Symbol*
Radicals	Superoxide	$O_2^{\cdot}$
	Hydroxyl	·OH
	Alkoxyl	LO·/RO·
	Peroxyl	LOO·/ROO·
	Nitric oxide	NO·
	Thiyl radical	R-S·
Non-radicals	Hydrogen peroxide	H_2O_2
	Hypochlorous acid	HOCl
	Ozone	O_3
	Singlet oxygen	$^1O_2$
	Peroxynitrite	ONOO-
	Lipid peroxide	LOOH

There are three main sites of $O_2^{\cdot-}$ production in cells: the mitochondria, endoplasmic reticulum, and cell cytoplasm membrances. In mitochondria, besides other enzymes of the electron transfer chain, two enzymatic sites have been clearly identified as major sources for one-electron reduction of oxygen: ubiquinone-cytochrome C reductase, which involves autooxidation of the udisemiquinone (Sun and Trumpower, 2003; Han *et al.*, 2001; Nohi and Jordon, 1986) and NADH dehydrohenase, which involves autoxidation semi-flavin cofactor (Kalinowski and Malinski, 2004; Aust *et al.*, 1972). In the membrance of the endoplasmic reticulum, $O_2^{\cdot-}$ is produced by the oxy complex of cytochrome P-450 and the action of NADH cytochrome P-450 reductase (Henderson and Chappell, 1996). In plasma, NADPH oxidase is involved in the generation of $O_2^{\cdot-}$ radicals by transferring one electron to molecular oxygen by the enzyme electron transfer chain reaction (Curi *et al.*, 2002). Xanthine oxidase has also been proposed to be an important source of $O_2^{\cdot-}$ generation in reperfused tissues. It uses molecular oxygen as its acceptor, producing superoxide anion (Prigmore *et al.*, 1995, McCord and Omar, 1993).

Hydroxyl radical (OH·) is an extremely reactive radical species that can react to every bio-molecule. A number of sources generate these radicals. Hydrogen peroxide, $O_2^{\cdot-}$ and transition metals (Fe^{+2} and Cu^{+}) are generally involved in the generation of OH· Hypochlorous acid is another source of hydroxyl radicals (McCormick *et al.*, 1998; Cheeseman and Slater, 1993).

Hydrogen peroxide (H_2O_2) is an oxidizing agent and a main source of hydroxyl radicals as mentioned earlier. It produces OH· through Haber-Weiss and Fenton reactions in the presence of $O_2^{\cdot-}$ and Fe^{+2}, respectively.

$$O_2^{\cdot-} + H_2O_2 \longrightarrow O_2 + OH^{\cdot} + OH^{-} \text{ (Haber-Weiss reaction)}$$

$$H_2O_2 + Fe^{2+} \longrightarrow OH^{\cdot} + OH^{-} + Fe^{3+} \text{ (Fenton reaction)}$$

H_2O_2 also produces another oxidizing agent, hypochlorous acid (Lin and Kao, 1998; Halliwell and Gutterdge, 1992). Hypochlorous acid (HOCI) is a potent antibactericidal agent produced by the activated polymorphonuclear cells. It is generated by action of myeloperoxidase on chloride ions in the presence of H_2O_2:

$$H_2O_2 + Cl^- \xrightarrow{\text{Myeloperoxidase}} HOCl + OH^-$$

It can cross cell membranes and, in the presences of transitional metal ions, it can generate hydroxyl radicals ($OH^{\cdot}$) by reacting with superoxide anion radicals and ferrous iron (Pullar *et al.*, 2001; Aruoma, 1994):

$$HOCI + Fc^{2+} \longrightarrow OH^{\cdot} + Cl^- + Fe^{3+}$$

$$HOCI + O_2^{\cdot} \longrightarrow OH^{\cdot} + Cl^- + O_2$$

Nitric oxide ($NO^{\cdot}$) is a common gaseous free radical. It plays an important role in vascular physiology and is also known as endothelium derived relaxing factor. Vascular endothelium produces nitric oxide, like neutrophils and macrophages, from arginine, using the enzyme nitric oxide synthase (Mazzetti *et al.*, 2001; Beckman and Crow, 1993; Moncada *et al.*, 1991).

Peroxynitrite (ONOO-) is produced due to the release of significant quantities of $NO^{\cdot}$ and $O_2^{\cdot -}$ from activated macrophages and neutrophils during the inflammatory response (Grace *et al.*, 1998).

$$O_2^{\cdot -} + NO^{\cdot} \longrightarrow ONOO^-$$

Singlet oxygen ($^1O_2$) is an electronically excited and mutagenic form of oxygen. It is generated during exercise, by radiations, by the action of peroxidases or lipoxygenases, and by the reaction of hydrogen peroxide with hypochlorite or peroxynitrite, as well as during the respiratory burst of phagocytes (Mascio *et al.*, 1994; Steinbeck *et al.*, 1993). In mammalian cells, singlet oxygen can be generated during oxidative stress and it is able to attack many cellular molecules such as amino acids, nucleic acid based and membrane lipids. Sunlight contains high energy short-wavelength ultraviolet photons (comprising the UVB spectra, 290-320 nm) which are potentially detrimental because of their destructive interactions with many cellular biomolecules. Chronic exposure to sunlight is thus a significant causative factor in the development of skin cancer (Estevam *et al.*, 2004).

Adverse Effects of Free Radicals

Because of the immense reactivity of free radicals, they can easily react with several bio-molecules including DNA, lipids, proteins and carbohydrates(Figure 1.1). ROS react with the bio-molecules, leading to local injury and eventual organ dysfunction. They also accelerate the aging and related degenerative processes. Morever, ROS are also involved in the promotion of heart diseases, chronic inflammation, and cancers (Ivanova and Ivanov, 2000).

The susceptibility of biological membranes to peroxidation is due to the presence of polyunsaturated fatty acids (PUFA). The presence of double bonds in PUFA weakens the C-H bond of the adjacent carbon atom (allylic carbons) and facilitates the hydrogen

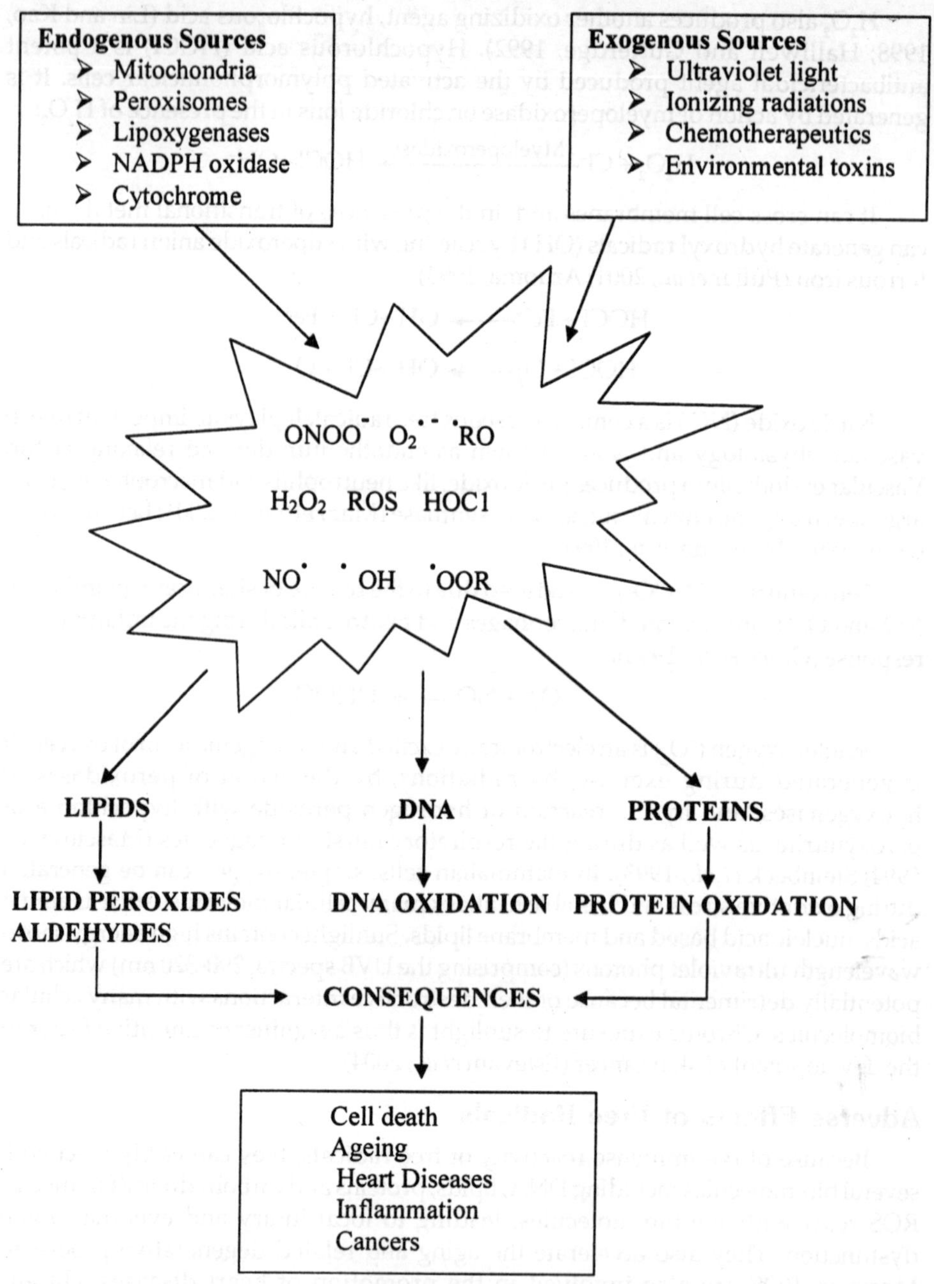

Figure 1.1: Sources and Consequences of Reactive Oxygen Species (ROS): Oxidants are continuously generated as a result of normal intracellular metabolism in mitochondria and peroxisomes as well as from a variety of cytosolic enzyme systems. In addition, number of exogenous sources can generate ROS. The ROS are harmful to cells by damaging biomolecules which results in acceleration in ageing and of certain other age-related diseases.

abstraction step, which initiates the peroxidation reactions (Aitken and Fisher, 1994). Cell membranes contain a variety of PUFA such as linoleic, linolenic and arachidonic acids, mainly in the form of esters with phospholipids, triglycerides, or with cholesterol. Overall attack of one reactive free radical on PUFA molecule can convert multiple fatty acid side chains into lipid peroxides, damage membrane proteins, make the membrane leaky and eventually cause breakdown of the membrane (Schafer *et al.*, 2000; Cheeseman, 1993).

Protein and nucleic acids are generally less susceptible to free radical attack than PUFAs and hence have less possibility to take part in the progression of chain reactions. This can happen only if radicals are allowed to accumulate, or if the damage is focused on a particular site of the protein (Leeuwenburge *et al.*, 1998; Stadman and Oliver, 1991).

Deoxyribonucleric acid (DNA) is a sensitive target for free radicals-mediated damage in a living system. A free radical can damage the specific site of DNA, leading to breaking of strands, or it might delay the repair before replication occurs, leading to mutations (Cheesman and Slater, 1993). It has been reported that free radicals may be involved in cell death or sub-lethal injuries such as mutations, chromosomal aberrations or carcinogenesis by damaging DNA and the DNA repair processes (Aust and Eveleigh, 1999; VanRensburg *et al.*, 1992).

Mechanism of Action of Free Radicals

There are three main stages in free radical mediated reactions: initiation, propagation and termination. Initiation starts with the abstraction of a hydrogen atom from the bio-molecule. For example fatty acid (LH) can be concerted into radicals ($L^{\cdot}$). The hydroxyl ($OH^{\cdot}$), alkperoxyl ($ROO^{\cdot}$), and alkoxyl ($RO^{\cdot}$) radicals are all capable of oxidizing PUFAs. Extremely rapid addition of oxygen to the fatty acid radicals then generates peroxyl radicals ($LOO^{\cdot}$) that propagate the reaction by initiating a new chain of oxidation with the formation of lipid hydroperoxide (LOOH). This chain reaction continues till an antioxidant interrupts it through scavenging the radicals: the termination step (Schafer *et al.*, 2000; Aitken and Fisher, 1994). The mechanism of peroxidation of linoleic acid as a model presented in Figure 1.2

Disease Associated with Free Radicals

Voluminous data in biochemical literature implicate ROS in the causation of diseases whose list is ever growing (Bandopadhya, 1999). Oxidative stress has been implicated in the pathology of many diseases such as inflammatory conditions, cancer, diabetes and aging (Marx, 1987). Free radicals induced by peroxidation have gained much importance because of their involvement in several pathological conditions such as atherosclerosis, ischemia, liver disorder, neural disorder, metal toxicity and pesticide toxicity (Pandey *et al.*, 1994).

Atherosclerosis

Cardiovascular diseases are the main causes of death in USA, Europe and in the different sections of developing countries. Atherosclerosis is the disease of arteries characterized by a local thickening of the intima or inner most part of the vessel. High

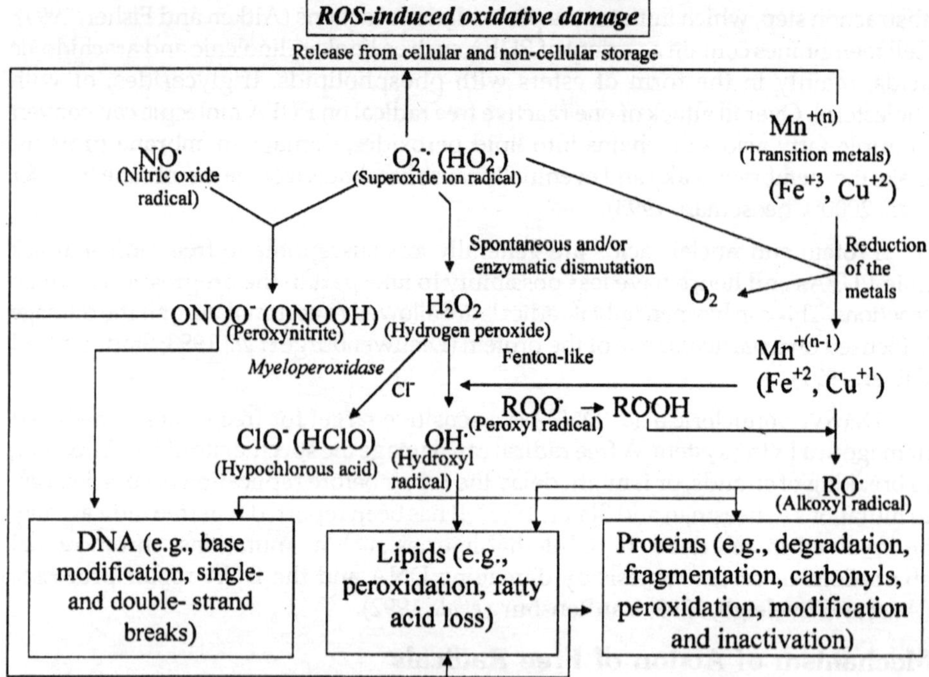

Figure 1.2: Reactive Oxygen Species (ROS)-Induced Oxidative Damage

blood LDL cholesterol is consistently associated with atherosclerosis (Munro and Cotran, 1988).

Cancer

Any substance that damages DNA is potentially carcinogenic. Exposure of DNA to O_2 generating systems causes extensive strand breakage and degradation of deoxyribose (Brawn and Fridovids, 1981) mainly due to formation of OH. Several drugs damage DNA by producing ROS (Lown *et al.*, 1982). The mutagenic effects of O_2^- (Superoxide anion radical) radicals produced during the respiratory burst may promote the development of cancer in chronically inflamed tissues. The radical intermediates of some toxic chemicals form DNA adducts leading to chemical carcinogenesis. Radiation induced cancer is also thought to be due to generation of ROS (Vonsontag, 1987).

Rheumatoid Arthritis

Inflammation is a response to tissue injury which, if prolonged may lead to several diseases. This appears to be the underlying basis for rheumatoid arthritis. The production of ROS (Reactive oxygen speices) and other inflammatory mediators by activated neutrophils is believed to induce damage that occur in the inflamed rheumatoid joint (Halliwell *et al.*, 1988). The subsequent recruitment of more activated

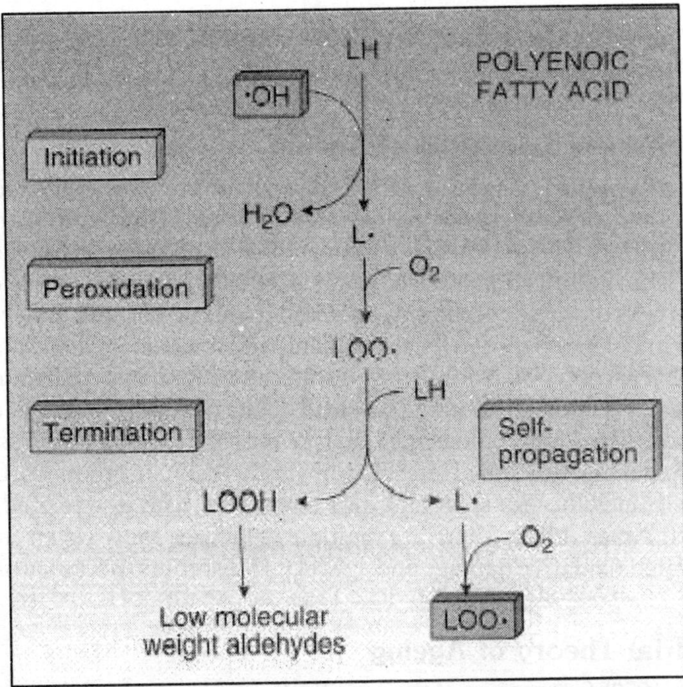

Figure 1.3: Mechanism of Lipid (Linoleic acid) Peroxidation: (A) Hydrogen adstraction from the allylic position of linoleic acid (LH) by an initiator leads to the formation of L˙. (B) Delocalization of the odd electron yields a conjugated diene. (C) Formation of LOO˙ as a result of reaction between oxygen and L˙. (D) Generation of new L˙ with the formation of LOOH as a result of attack of LOO˙ to a new LH.

neutrphils leads to the chronic inflammation and gradual degeneration of the joints was the characteristic of this disease.

Ageing

Rizvi and Jha (2009) gave an excellent review on aging and antioxidants in aging. The idea that free radicals are toxic agents was first suggested by Gerschman (Gerschman *et al.*, 1954). Probably the most influential of the current theories, the free radical/oxidation theory of aging is supported by a large and growing body of evidence. Developed by Drenham Harman in 1956, the theory postulates oxygen free radicals specifically hydroxyl and hydroperoxyl are formed endogenously from normal oxygen utilizing metabolic processes and play an essential role in the aging process (Harman, 1956). It has been shown that accumulation of free radical damages increases with age (Holmes *et al.*, 1992). Support for the free radical theory of aging has increased progressively over the years, and growing numbers of studies implicate free radical reactions in aging and arteriosclerosis (Beckman and Ames, 1998). There are evidences for deleterious changes to the molecular structure of DNA (deoxyguanosine derivatives, proteins, carbonyls, lipids, lipoperoxides, MDA) and prostaglandins (isoprostanes), all markers of oxidative stress, associated with aging

(Harman, 1992, 1993). ROS also play a role in normal signaling processes and that their generation is important to maintain homeostasis and cellular responsiveness (Drodge, 2002).

Oxidative Stress Hypothesis of Ageing

Harman's hypothesis has been refined to encompass not only free radicals but also peroxides and aldehydes in oxidative damage to cells. This modification led to the origin of Oxidative Stress Theory of aging (Sohal and Weindruch, 1996; Yu and Yang, 1996). This modification is based on the fact that a chronic state of oxidative stress exists in cells of aerobic organisms even under normal physiological conditions because of an imbalance between the prooxidants and antioxidants. The antioxidant system is unable to counterbalance all the free radicals continuously generated during the life of a cell. The imbalance leads to a steady state of accumulation of oxidative damage in variety of macromolecules that increases during aging, resulting in progressive loss in the functional efficiency of various cellular processes. Oxidative damage determines the life span and also associated with age related diseases (Beckman and Ames, 1998). There are multiple evidences showing an association between elevated oxidative damage and age related disorders (Beckman and Ames, 1998).

Mitochondrial Theory of Ageing

Harman (1972) first proposed the modification of free radical theory of aging, giving a central role to mitochondria, thereby defining a new theory developed further as Mitochondrial Free Radical theory of aging (Harman, 1972). Mitochondrial dysfunction is heavily implicated in the aging process. The mitochondrial theory of aging proposes that progressive accumulation of somatic mutations in mt DNA during a lifetime leads to an inevitable decline in mitochondrial functions (Harman, 1992). The most important biochemical factors in this process are considered to be reactive oxygen species (ROS) which are generated, during the normal function of the mitochondrial respiratory chain. It has been proposed that ROS can cause somatic mutations in mtDNA (Harman, 1992). These mutations then result in impaired functions of the respiratory chain (Trifunovic and Larsson, 2008) leading to increased ROS production and the subsequent accumulation of more mutations. There is evidence that some pathogenic mutations, such as at 8993T>G, lead to increased ROS production (Geromel *et al.*, 2001); for 8993T>G, free radical-mediated inhibition of OXPHOS can be restored by antioxidants (Mattiazzi *et al.*, 2004). The ROS vicious cycle is believed to account for an exponential increase in oxidative damage during aging, which results in the eventual loss of cellular and tissue functions through a combination of energy insufficiency, signaling defects, apoptosis and replicative senescence. Although the mitochondrial theory of aging was first suggested in the 1950s, the first convincing evidence of a biochemical defect in humans was the detection of COX-deficient cells in aging post-mitotic tissues (Muller-Hocker, 1989). These cells had identical characterstics to those found in patients with inherited mitochondrial disease (Sciacco, 1994).

Oxidative Stress in Aging

Oxidative stress is classically defined as the redox imbalance with an excess of oxidants or a defect in antioxidants (Sies, 1985). It refers to a condition whereby imbalances in the Overview of Mechanism of Aging redox state cause oxidatively modified cellular constituents and dysfunction. The aging process is the accumulation of oxidative damage to cells and tissues, associated with a progressive increase in the chance of morbidity and mortality (Harman, 1956, 1972; Beckman and Ames, 1998). Oxidative stress acts as a major determinant in the diseases that frequently accompany aging (Wick *et al.*, 2000). The antioxidant network counteracts the deleterious action of free radicals on biomolecules. Long lived animal species possess a more efficient antioxidant system and higher liver Cu-Zn superoxide dismutase activity than shorter living species (Cutler, 1991). There are several repair systems which help antioxidant action by the recovery of damaged macromolecules (Cutler and Rodriguez, 2003). These systems altogether play an important role in the ability of the body to respond to the oxidative challenge. There are suggestions that oxidative stress and associated damages are neither parallel with aging alterations nor correlated with maximum life span because there are uncertain repairing processes following up (Sohal *et al.*, 2002; Park and Gerson, 2005). The remaining alterations after repairing are the age related changes, *e.g.*, nondegradable crosslinked amyloids and lipofuscin accumulation with age.

Antioxidants

To protect cells and organs from the oxidants stress induced by ROS, living organisms have evolved with an extremely efficient and highly sophisticated protective system, the so-called "antioxidant defensive system". It involves a variety of components, both endogenous and exogenous in origin (Figure 1.2). These components function interactively and synergistically to neutralize free radicals (Percival, 1998). A broader definition of an antioxidant is "any substance which, when present at low concentrations compared to those of oxidizable substrates, significantly delays or prevents oxidants of those substrates". The term oxidizable substrates includes DNA, lipids, proteins and carbohydrates, which are the essential building blocks of a biological system (Halliwell *et al.*, 1995). Oxidative stress occurs as a result of an increase in oxidative metabolism, which produces a number of ROS. To avoid oxidative stress, antioxidants can play an important role conferring beneficial healthy effects (Figure 1.2) (Vaya and Aviram, 2001). High dietary intake of proven antioxidants can significantly lower the risk of several chronic diseases such as heart diseases, cancers and cataracts.

Classification of Antioxidants

A variety of antioxidants are collectively required for the removal of free radicals to protect the body from adverse effects of ROS (Figure 1.4). Certain enzymes as well as non-enzymatic cellular molecules are involved in the detoxification of ROS. Based on the nature of antioxidants, the human antioxidant system can be categorized into two broader classes: enzymatic and non-enzymatic (Table 1.2) (Jekus, 2000).

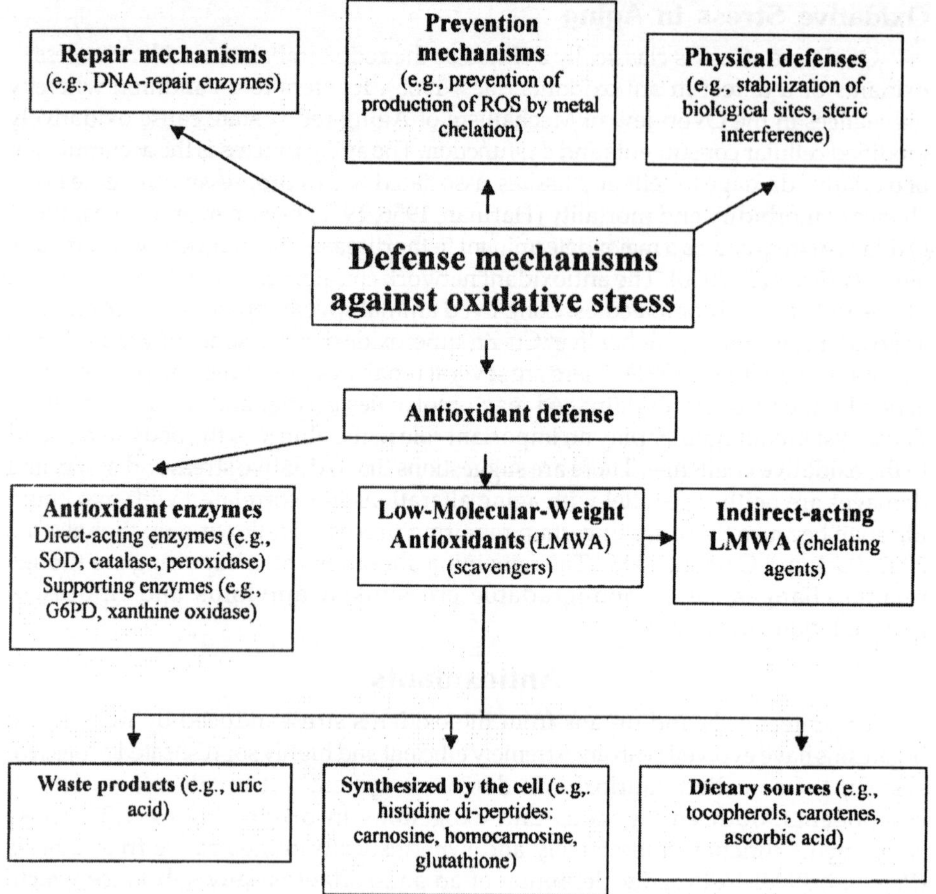

Figure 1.4: Classification of Antioxidant Cellular – Defense Mechanisms

Types of Antioxidants

Basically these are classified into three categories.

Enzymatic and Non-Enzymatic Antioxidants

They are found both in extra cellular as well as intracellular environment. These are tactically arranged within the cell in order to provide maximum protection against free radicals. Some important enzymatic and non-enzymatic physiological antioxidants are shown in Table 1.2.

Antioxidant Derived from Natural and Dietary Sources

Plants develop several antioxidants that aid in antioxidant defense system, protecting plants against damage caused by active O_2 formed due to exposure to ultraviolet radiation. Certain seaweeds also function as antioxidants. Our daily diet contains vegetables, fruits, tea, wine, etc. which possess compounds rich in antioxidative properties.

Table 1.2: Types of Antioxidants

Sl.No.	*Antioxidant*	*Location*	*Function/Property*
(*a*)	**Enzymatic antioxidant**		
1.	Glutathione peroxidase (GSH)	Mitochondria and Cytosol	Removal of H_2O_2 and organic hydro-peroxide
2.	Catalase (CAT)	Mitochondria and Cytosol	Removal of H_2O_2
(*b*)	**Non-Enzymatic antioxidant**		
1.	Carotenoids	Lipid soluble antioxidants in membrane tissue	Removal of reactive oxygen species
2.	Bilirubin	Product of heme metabolism in blood	Extracellular antioxidants
3.	Glutathione	Non-protein thiol in cell	Cellular oxidant defense
4.	α-lipoic acid	Endogenous thiol	Serves as substitute for Glutathione, recycling vitamin C
5.	Vitamin C	Aqueous phase of the cell	Free scavenger, recycle Vitamin E
6.	Vitamin E	Cell	Chain breaking antioxidants
7.	Uric acid	Product of Purine metabolism	Scavenging of OH radical

Antioxidants from Natural Sources

Secondary products of plants which are functioning as antioxidants are:

1. Chlorophyll derivatives
2. Essential oils
3. Carotenoids
4. Alkaloids
5. Phytosterols
6. Phenolics – coumarins, flavonoids
7. Polyphenolics – tannins, proanthocyanidine
8. Nitrogen containing compounds – alkaloids, indoles

Dietary Antioxidants

These reduce the free radical formation as well as oxidative stress and reduce the possibility of cardiovascular diseases. Several compounds such as phenolic diterpenoids –carnosol, rosmanol, carnosoic acid etc. obtained from several aromatic plants possess strong antioxidant properties.

Antioxidants Based on Defense Mechanism

These are four types:

1. *Preventive antioxidants*: These suppress the free radical formation, *e.g.*, enzymes such as peroxidase, catalase, lactoferrin, carotenoids, etc.
2. *Radical scavenging antioxidants*: These suppress the chain initiation reaction, *e.g.*, vitamin – C and Carotenoids

3. *Enzyme inhibitor antioxidants*: These induce production and reaction of free radicals and the transport of appropriate antioxidants to appropriate active site.

Antioxidants are component, which inhibits oxidation, or free radicals induced oxidative damage and therefore are potential quenchers of oxidative stress induced lipid peroxidiation. Antioxidants effectively prevent free radical induced cellular and tissue damage which are solely responsible for several pathogenic condition including cancer.

Antioxidants are naturally present in biological systems such as plant and plant derived products and also in food systems. Such naturally occurring antioxidants provide a protection against oxidation. However, there is a probability that these compounds may be lost during processing steps such as processing of plant and plant derived food products. Addition of exogenous antioxidants therefore appears to be important. Antioxidants help in preventing auto oxidation, increases the shelf life and retards the lipid oxidation.

Numerous synthetic antioxidants have been registered, but only a few species are permitted as food additives by the law because of toxicity effects and other side-effects. Typical antioxidants permitted as the food additives are butylated hydroxy anisole (BHA), butylated hydroxy toluene (BHT), pueraria glycoside (PG), and tertiary-butylated hydroquinone (TBHQ).

Recently, public's concern about the problems of human health caused by food additives has once more evoked food scientists to the enthusiasm of seeking natural antioxidants from the various sources. So far one of the understandings on this issue is that natural antioxidants are primarily plant phenolic compounds occurring in all parts of plant bodies. These antioxidants belong to different classes of compounds, such as carotenoids, polyphenolics, polyamines, gallic acid derivatives, tannins and catechins. Examples include phytic acid, lipoic acid, bilirubin, melatonin, quercetin, carnosol, carnosic acid, hydroxytyrosol, rutin, butylated hydroxyanisole, and butylated hydroxyltoluene. Vitamins E and C are among the most effective antioxidants with preventive effects against heart diseases and cancers.

Enzymatic Antioxidants

The major primary intracellular endogenous antioxidants defenses are the enzyme system. This antioxidant enzymatic system includes superoxide dismutases (SODs), catalase (CAT), and glutathione peroxidase (GSHPx) (Yang *et al.*, 1999; Halliwell and Guttredige, 1990).

Superoxide Dismutase

Superoxide dismutase (SODs) have been found in three isoforms. Manganese containing SOD (MnSOD) is a tetrameric protein that is localized in the mithochondrial matrix. It plays a key role in scavenging $O_2^{\cdot-}$ generated from the electron transport chain. Copper and zinc containing SOD (CuZnSOD) is a dimer protein that is localized in the cell cytoplasm. It is thought to remove $O_2^{\cdot-}$ generated by endoplasmic reticulum and cytosolic oxidases. The extracellular SOD is a tetrameric protein found in the extracellular space (Noor *et al.*, 2002). SODs catalyze the dismutation of superoxide into H_2O_2 (Ivanova and Ivanova, 2000).

$$2O_2^{-} + 2H^{+} \text{ SOD} \xrightarrow{\text{SOD}} H_2O_2 + O_2$$

Catalase (CAT) is located in perxoisomes and mitochondria. It is a large tetrameric protein which removes H_2O_2 by catalyzing its conversion into water (Krinsky, 1992).

$$2H_2O_2 \text{ Catalase} \xrightarrow{\text{Catalase}} 2H_2O_2 + O_2$$

Glutathione Peroxidases

Glutathione peroxidases (GSHPx) are a group of selenium-dependent enzymes. Four isoforms of GSHPx have been described: cytosolic GSHPxl, plasma GSHPx, phospholipids-hydroperoxide PHGSHPx and gastrointestinal GSHPx-Gl. All GSHPx require glutathione (GSH) as a cofactor and secondary enzymes, such as glutathione reductase and glucose-6-phosphate dehydrogenase (G-6-PDH), to function. G-6-PDH generates NADPH to recycle the GSH (Takebe *et al.*, 2002; Ivanova and Ivanova 2000).

$$2GSH + H_2O_2 \xrightarrow{\text{GSHPx}} GSSG + 2H_2O$$

Non-Enzymatic Antioxidants

Non-enzymatic antioxidant may be further classified into two groups: endogenous and exogenous antioxidants. The major extracellular endogenous antioxidants found in human plasma are transition metal binding proteins. These include ceruloplasmin, transferring, hepatoglobin and albumin. They bind with transition metals and hence control the production of metal-catalyzed free radicals. Albumin and ceruloplasmin are the copper ions sequesters. Hepatoglobin binds with haemoglobin while ferritin and transferring bind with free iron (Ivanova and Ivanov, 2000; Halliwell and Gutteridge, 1990). Lipoic and uric acids, bilirubin, ubiquinone and glutathione are non-protein endogenous antioxidants which inhibit the oxidation processes by scavenging free radicals (Shahidi, 1997).

Many effective exogenous antioxidants are generally of dietary origin. The best known are vitamins such as ascorbic acid, vitamin E, carotenoids, quinines, and polyphenols. These molecules can inhibit oxidative reactions by scavenging free radicals, while certain compounds may chelate redox active metals or inhibit particular oxidative enzymes. Vitamin E is a lipid soluble, chain breaking antioxidant which reacts with lipid peroxyl radicals to yield a relatively stable lipid hydroperoxide and thus protects against membrane lipid peroxidation (Maguire *et al.*, 1989). On the other hand, vitamin C has multiple antioxidant properties, including the ability to regenerate α-tocopherol by reducing α-tocopheroyl radicals at the membrane surfaces of membranes. It also scavenges others free radicals and certain non-radicals such as HOCl (Packer and Cadenas, 2002).

Mechanism of Action of Antioxidants

Antioxidants can remove both free radical and non-radical species through different modes of action depending upon what ROS is required for neutralization. On the basis of the nature of ROS, two basic mechanisms have been proposed for the action of Antioxidants: mechanism of removal of ROS initiators and a chain breaking mechanism.

Mechanism of Removal of Initiators of ROS

This process is mainly based on the inhibition of the enzymes involved in the production of ROS. Xanthine oxidase (XO) is one of the major sources of superoxide anions production (Borges *et al.*, 2002). Lipoxygenase, during arachidonic metabolic pathway, produces lipid peroxides (Dailey and Imming, 1999). Antioxidants inhibit these enzymes so that they are not available to harm the cellular system. The other main sources of free radicals are free transition metal ions which, by virtue of their attachment with carrier proteins, act as pro-oxidants and are therefore toxic to the body. There are several metal binding proteins that chelate transition metals, which are capable of reacting with hydroperoxides to produce fiee radicals.

$$HOOH/LOOH + Fe^{2+} \longrightarrow LO^{\cdot}/HO^{\cdot} + OH^{-} + Fe^{3+}$$

The proteins which bind with the transition metal ions include transferring, lactoferrin and ferritin. These proteins function to keep iron-induced oxidant stress in control. Ceruloplasmin and albumin proteins are the copper and iron sequestrants, respectively (Vaya and Aviram, 2001; Krinsky, 1992).

Chain Breaking Mechanism of Action

Antioxidants scavenge free radicals by donating an electron to them while being oxidized themselves during the process. This is known as a "chain breaking antioxidation" (CBA). The chain reactions of free radicals can be stopped with the help of antioxidants, which can neutralize free radicals formed during the overall reaction. Most of the polyphenolics work through the chain breaking (CB) mechanism of action. Vitamins C and E, carotenes, flavonoids and coumarins are example of chain breaking antioxidants (CBA).

Vitamin E (tocopherol, TH_2) is the most widely distributed lipid soluble antioxidant in nature. Its main function is to prevent the peroxidation of membrane phospholipids and to avoid cell membrane damage (Maguire *et al.*, 1989). A detailed study of the effects of benzene ring substituents on the rate of radical scavenging reaction showed that the reaction is accelerated by the presence of a 4-methoxy and C-2 and C-6 methyl groups. The presense of C-1 hydroxyl group contributes to donating $H^{\cdot}$, whereas functional groups at C-4, C-2, and C-6 stabilize the resulting tocopherol radical (Shahidi, 1997; Burton and Ingold, 1981).

Protective Role of Antioxidants on Biological Functions

Substantial research work had been carried out to investigate the preventive role of antioxidants in different diseases. Every antioxidant has some significance and the best protection against oxidative stress comes from the presence of a wide assortment of interrelated antioxidants and their cofactors. The function of each particular antioxidant depends on what type of oxidative stress is imposed (Percival, 1998).

Lipid peroxidation can damage low-density lipoprotein (LDL) particles in several ways. *In vitro* studies have demonstrated that lipoxygenase, superoxide anoin, peroxynitrite and myeloperoxides can oxidize LDL (Yla-Herttuala *et al.*, 2000) which can lead to heart diseases. Studies showed that antioxidants may protect against coronary heart diseases (CHD) (Chow, 2001; Parthasarathy *et al.*, 1999; Diaz *et al.*, 1997). Vitamins have been shown to reduce the susceptibility of protection factors like HDL-cholestrol. Studies suggest that vitamin C may reduce the risk of hypertension (Noguchi and Niki, 2000). In addition, a high intake of vitamin C appears to protect against gastric cancer, probably through scavenging ROS formed in the gastric mucosa (Woodford and Whitehead, 1998). Further investigations on vitamin C have proved its preventive effects on inhibition of tumor promotion (Lee *et al.*, 2002).

Dietary antioxidants are important to maintain good health. Vegetables and fruits contain a variety of important antioxidants that are essential for the normal functioning of the body (Table 1.2).

Most citrus and dried fruits, cruciferous vegetables, garlic, onions, carrots, tomatoes, sweet potatoes, sesame and olive oil are rich sources of antioxidants.

Tomato intake, the main source of lycopene, has been found to be associated with a lower risk of a variety of cancers in several epidemiological studies (Agarwal and Rau, 2000). Epidemiological investigations on vitamin E suggested that it may protect against the occurrence of Parkinson's disease. Some researchers have also shown that vitamin E intake can slow down the progression of Alzhemer's disease (Vatassery *et al.*, 1999). It was reported that supplementations with vitamins C and E, mixed with other antioxidants, can reduce symptoms of oxidative stress during exercise (Clarson and Thompson, 2000).

Flavonids are polyphenols and they are abundantly found in fruits, vegetables, grains, bark, roots, stems, flowers, and tea. They possess anti-inflammatory, antiallergic, antiviral and anticarcinogenic properties due to their antioxidant potential (Nijveldt *et al.*, 2001; Hollman *et al.*, 1996).

Flavonoids

Luteolin

Luteolin is one of the specific and more common flavone. It plays a crucial role as an antioxidant, a free radical scavenger, an agent in the prevention of inflammation, a promoter of carbohydrate metabolism and an immune system modulator. These characteristics of luteolin are also believed to play a vital role in the prevention of cancer (Mann *et al.*, 1992). Luteolin is most often found in leaves, but it is also seen in celery, thyme, dandelion, rinds, barks, clover blossom and ragweed pollen (John *et al.*, 1992). It has also been isolated from *Salvia tomentosa* (Ulubelen *et al.*, 1979). Dietary sources include celery, green pepper, thyme, perilla, chamomile, tea, carrots, olive oil, peppermint, rosemary and oregano (Shimoi *et al.*, 1998; Lopez-Lazaro, 2009).

Table 1.3: Dietary Sources for Antioxidants.

Class/Components	*Source**	*Potential Benefit*
Carotenoids		
Beta-carotene	Carrots, various fruits	Neutralizes free radicals which may damage cells; bolsters cellular antioxidant defenses
Lutein, Zeaxanthin	Kale, collards, spinach, corn, eggs, citrus	May contribute to maintenance of healthy vision
Lycopene	Tomatoes and processed tomato products	May contribute to maintenance of prostate health
Flavonoids		
Anthocyanidins	Berries, cherries, red grapes	Bolster cellular antioxidant defenses; may contribute to maintenance of brain function
Flavanols–Catechins, Epicatechins, Procyanidins	Tea, cocoa, chocolate, apples, grapes	May contribute to maintenance of heart health
Flavanones	Citrus foods	Neutralize free radicals which may damage cells; bolster cellular antioxidant defenses
Flavonols	Onions, apples, tea, broccoli	Neutralize free radicals which may damage cells; bolster cellular antioxidant defenses
Proantho-cyanidins	Cranberries, Cocoa, apples, strawberries, grapes, wine, peanuts, cinnamon	May contribute to maintenance of urinary tract health and heart health
Isothiocyanates		
Sulforaphane	Cauliflower, broccoli, broccoli sprouts, cabbage, kale, horseradish	May enhance detoxification of undesirable compounds and bolster cellular antioxidant defenses
Phenols		
Caffeic acid, Ferulic acid	Apples, pears, citrus fruits, some vegetables	May bolster cellular antioxidant defenses; may contribute to maintenance of healthy vision and heart health
Sulfides/Thiols		
Diallyl sulfide, Allyl methyl trisulfide	Garlic, onions, leeks, scallions	May enhance detoxification of undesirable compounds; may contribute to maintenance of heart health and healthy immune function
Dithiolthiones C	Cruciferous vegetables–broccoli, cabbage, bok choy, collards	Contribute to maintenance of healthy immune function
Whole Grains		
Whole grains	Cereal grains	May reduce risk of coronary heart disease and cancer; may contribute to reduced risk of diabetes

Chart adapted from International Food Information Council Foundation: Media Guide on Food Safety and Nutrition: 2004-2006.

*Not a representation of all sources.

Polyphenols: Natural Antioxidants

Polyphenols are present in a variety of plants utilized as important components of both human and animal diets (Brave, 1998: Chung *et al.*, 1998: Crozier *et al.*, 2000). These include food grains such as sorghum, millet, barley, dry beans, peas, pigeon peas, winged beans, and other legumes; fruits such as apples, blackberries, cranberries, grapes, peaches, pears, plums, raspberries, and strawberries; and vegetables such as cabbage, celery, onion and parsley also contain a large quantity of polyphenols. Phenolic compounds are also present in tea and wine. Forages such as crownvetch, lespedeza, lotus, sainfoin, and trefoil are also reported to contain polyphenolic compounds.

Diets containing an abundance of fruit and vegetables are protective against a variety of diseases, particularly cardiovascular disease and cancer. The primary nutrients thought to provide the protection afforded by fruit and vegetables are the antioxidants (East wood, 1999; Potter, 1997) reviewed 200 epidemiological studies, the majority of which showed a protective effect of increased fruit and vegetable intake. When the role of individual antioxidants, vitamins C and E, and carotenoids, is examined by epidemiological studies or supplementation trials, the results are not as clear-cut as those obtained for fruit and vegetables and are often disappointing. Potter's (1997) conclusion was that fruit and vegetables provide the best polypharmacy against the development of a chronic disease, considering that they contain a vast array of antioxidant components such as polyphenols.

Diets rich in fruits and vegetables, such as vegetarian and Mediterranean diets, contain a large quantity of polyphenols. Dietary habits consistent with protection from coronary heart disease have been considered too restrictive (high in polyunsaturated fats and / or vegetarian); however, the diet in some Mediterranean countries, such as France, Spain and Italy, is varied and characterized by a low consumption of butter and high consumption of bread, vegetables, fruit, cheese, vegetable fat, and wine: the so called Mediterranean type diet. In addition, other foods high in saturated fat are eaten; 14-15 per cent of energy intake corresponds to saturated fat (Renaud and de Lorgeril, 1992; Segasothy and Phillips, 1999). Certainly, a high consumption of vegetables constitutes a healthy habit observed in Mediterranean countries in conjunction with moderate wine consumption. The univariate correlation coefficients between coronary heart disease mortality and the intake of various foodstuffs in a study based on statistics from the 21 most industrialized wine-drinking countries were as follows: vegetables, -0.48 ($P<.05$); vegetable fats, -0.44 ($P<.05$); fruit, -0.28 (NS); dairy products, 0.66 ($P<.001$); and wine, -0.87 ($P<.001$) (Renaud and Ruf, 1994). In this study, the protective effect of wine appears to be much more convincing than that of other vegetable foods. Indeed, wines are rich in polyphenols, especially red wine. There is no accurate information available on the dietary intake of polyphenols because their content in plant foods varies greatly, even among cultivars of the same species. The presence of polyphenols in plant foods is largely influenced by genetic factors and environmental conditions. Other factors, such as germination, degree of ripening, variety, processing, and storage, also influence the content of plant phenolics (Bravo, 1998).

Hertog *et al.* (1993) measured the flavonoid content of fruits, vegetables, and beverages important in the Dutch diet and reported a dietary flavonoid intake of approximately 26 mg per day in this population. The flavonoid content of the diet was defined as the sum of quercetin, kaempferol, myricetin, apigenin, and luteolin. The investigators found that 95 per cent of total dietary flavonoids were from two compounds, 63 per cent from quercetin and 32 per cent from kaempferol, and that the seasonal variation in the flavonoid content of food was low. The major dietary contributor to flavonoid intake was black tea (61 per cent) followed by onions (13 per cent) and apples (10 per cent). Knekt *et al.* (1996) also reported that flavonoid intake ranged from 0 to 41.4 mg per day in the Finnish population, and the main sources of flavonoids were apples and onions. Both groups found an inverse association between intake of dietary flavonoids and cardiovascular disease. These studies, however, consider only the intake of some specific flavonoids and do not look at other phenolic compounds. Thus, an accurate estimation of total polyphenolic intake is not available.

Structure of Plant Polyphenols

Phenolic compounds, or polyphenols, constitute one of the most numerous and widely-distributed groups of substances in the plant kingdom, with more than 8,000 phenolic structures currently known. Polyphenols are products of the secondary metabolism of plants. The expression "phenolic compounds" embraces a considerable range of substances that possess an aromatic ring bearing one or more hydroxyl substituents. Most of the major classes of plant polyphenol are listed in Table 1.4, according to the number of carbon atoms of the basic skeleton. The structure of natural polyphenols varies from simple molecules, such as phenolic acids, to highly polymerized compounds, such as condensed tannins (Harborne, 1980).

Table 1.4: The Major Classes of Phenolic Compounds in Plants.

Number of Carbon Atoms	*Basic Skeleton*	*Class*	*Examples*
6	C6	Simple phenols Benzoquinones	Catechol, hydroquinone 2,6-Dimethoxybenzoquinone
7	C6-C1	Phenolic acids	Gallic, salicylic
8	C6-C2	Acetophenones Tyrosine derivatives Phenylacetic acids	3-Acetyl-6-methoxyben-zaldehyde Tyrosol p-Hydroxyphenylacetic
9	C6-C3	Hydroxycinnamic acids Phenylpropenes Coumarins Isocoumarins Chromones	Caffeic, ferulic Myristicin, eugenol Umbelliferone, aesculetin Bergenon Eugenin
10	C6-C4	Naphthoquinones	Juglone, plumbagin
13	C6-C1-C6	Xanthones	Mangiferin
14	C6-C2-C6	Stilbenes Anthraquinones	Resveratrol Emodin

Contd...

Number of Carbon Atoms	Basic Skeleton	Class	Examples
15	C6-C3-C6	Flavonoids Isoflavonoids	Quercetin, cyanidin Genistein
18	(C6-C3)2	Lignans Neolignans	Pinoresinol Eusiderin
30	(C6-C3-C6)2	Biflavonoids	Amentoflavone
n	(C6-C3)n (C6)n (C6-C3-C6)n	Lignins Catechol melanins Flavolans (Condensed Tannins)	

From Harborne (1980)

Flavonoids represent the most common and widely distributed group of plant phenolics. Their common structure is that of diphenylpropanes (C6-C3-C6) and consists of two aromatic rings linked through three carbons that usually form an oxygenated heterocycle (Harborne, 1980). Figure 1.5 shows the basic structure and the system used for the carbon numbering of the flavonoid nucleus. Structural variations within the rings subdivide the flavonoids into several families: flavonols, flavones, flavanols, isoflavones, antocyanidins and others. These flavonoids often occur as glycosides, glycosylation rendering the molecule more water-soluble and less reactive toward free radicals. The sugar most commonly involved in glycoside formation is glucose, although galactose, rhamnose, xylose and arabinose also occur,

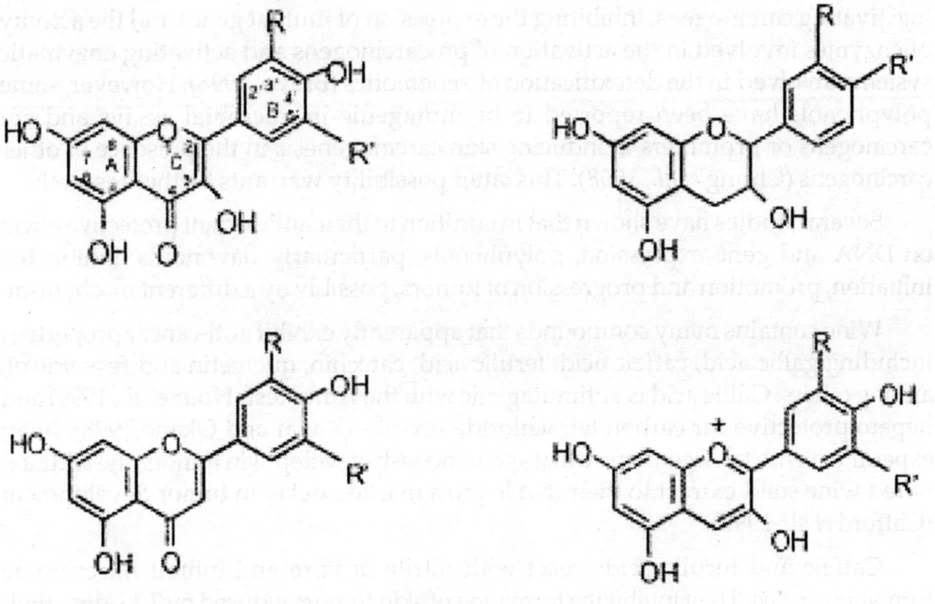

Figure 1.5: Flavonoids (C6-C3-C6) Basic Structure and System Used for Carbon Numbering of the Flavonoid Nucleus. Structural variations within the rings subdivide the flavonoids into several families.

as well as disaccharides such as rutinose. The flavonoid variants are all related by a common biosynthetic pathway, incorporating precursors from both the shikimate and the acetate-malonate pathways (Crozier *et al.*, 2000). Further modification occurs at various stages, resulting in an alteration in the extent of hydroxylation, methylation, isoprenylation, dimerization and glycosylation (producing O-or C-glycosides). Phenolic compounds act as antioxidants with mechanisms involving both free radical scavenging and metal chelation. They have ideal structural chemistry for free radical-scavenging activities, and have been shown to be more effective antioxidants *in vitro* than vitamins E and C on a molar basis (Rice- Evans *et al.*, 1997).

Biological Effects of Polyphenols

Polyphenols exhibit a wide range of biological effects as a consequence of their antioxidant properties. They inhibit LDL oxidation *in vitro* (Frankel *et al.*, 1993). Moreover, LDL isolated from volunteers supplemented with red wine or red wine polyphenols show reduced susceptibility to oxidation (Fuhrman *et al.*, 1995; Nigdikar *et al.*, 1998). Thus, polyphenols probably protect LDL oxidation *in vivo* with significant consequences in atherosclerosis. and also protect DNA from oxidative damage with important consequences in the age-related development of some cancers (Halliwell, 1999). In addition, flavonoids have antithrombotic and anti-inflammatory effects (Gerritsen *et al.*, 1995; Muldoon and Kritchevsky, 1996). The antimicrobial property of polyphenolic compounds has been well documented (Chung *et al.*, 1998).

Several types of polyphenols (phenolic acids, hydrolysable tannins, and flavonoids) show anticarcinogenic and antimutagenic effects. Polyphenols might interfere in several of the steps that lead to the development of malignant tumors, inactivating carcinogens, inhibiting the expression of mutant genes and the activity of enzymes involved in the activation of procarcinogens and activating enzymatic systems involved in the detoxification of xenobiotics (Bravo, 1998). However, some polyphenols have been reported to be mutagenic in microbial assays and co-carcinogens or promoters in inducing skin carcinogenesis in the presence of other carcinogens (Chung *et al.*, 1998). This latter possibility warrants further research.

Several studies have shown that in addition to their antioxidant protective effect on DNA and gene expression, polyphenols, particularly flavonoids, inhibit the initiation, promotion and progression of tumors, possibly by a different mechanism.

Wine contains many compounds that apparently exhibit anti-cancer properties, including gallic acid, caffeic acid, ferulic acid, catechin, quercetin and resveratrol, among others. Gallic acid is antimutagenic with the Ames test (Hour *et al.*, 1999) and hepato protective for carbon tetrachloride toxicity (Kanai and Okano,1998). In an experiment with transgenic mice that spontaneously develop skin tumors, the addition of red wine solid extract to their diet led to a marked delay in tumor development (Clifford *et al.*, 1996).

Caffeic and ferulic acids react with nitrite *in vitro* and inhibit nitrosamine formation *in vivo*. They inhibit the formation of skin tumors induced by 7,12-dimethyl-benz(a) anthracene in mice (Kaul and Khanduja, 1998). They also inhibit tyrosine nitration mediated by peroxynitrite (Pannala *et al.*, 1998).

Resveratrol has been extensively studied. It has been isolated from several sources and shown to inhibit the development of preneoplastic lesions in rat mammary gland tissue in cultures in the presence of carcinogens; it also inhibits skin tumors in mice (*Clifford et al.*, 1996; Jang *et al.*, 1997). Other researchers have shown that the combination of resveratrol and quercetine exerts a synergic effect in the inhibition of growth and proliferation of human oral squamous carcinoma cells (ElAttar and Virji, 1999). In this study, however, the best result was observed with diluted red wine. Since resveratrol and quercetin are present in low concentrations, other polyphenols could also be responsible for this effect and for the potentiation of cell growth inhibition.

Polyphenol Bioavailability and Metabolism

The knowledge of absorption, biodistribution and metabolism of polyphenols is partial and incomplete, yet it is sufficient to state that in general, some polyphenols are bioactive compounds that are absorbed from the gut in their native or modified form. They are subsequently metabolized with products detected in plasma that retain at least part of the antioxidant capacity and then excreted. Experimental studies in animals support the previous general statement (Das and Griffiths, 1969; Das and Sothy, 1971; Griffiths and Smith, 1972; Manach *et al.*, 1995; Manach *et al.*, 1997; Piskula and Terao, 1998; Morand *et al.*, 1998; Okushio *et al.*, 1999a; 1999b). In humans, studies aim at identifying native compounds and their metabolites in plasma and urine after the administration of test meals or drinks. These studies also support the initial general statement. Many of the studies performed with humans are centered on the detection of quercetin after the consumption of onions, tea, and apple juice (Hollman *et al.*, 1996,1997; Aziz *et al.*, 1998; Manach *et al.*, 1998; Lean *et al.*, 1999; McAnlis *et al.*, 1999).

Some of these studies have addressed the question of the biological activity of rutin and quercetin metabolites, such as the ability of quercetin and isorhamnetin to inhibit copper induced LDL oxidation (Manach *et al.*, 1998; Morand *et al.*, 1998). These authors state that the plasma metabolites retain antioxidant activity.

After green tea consumption, epigallocatechin gallate and epicatechin gallate were detected in plasma and urine (Yang *et al.*, 1998). Red wine consumption leads to the accumulation of o-methylcatechin, a catechin metabolic product, in plasma (Donovan *et al.*, 1999). These findings should be considered important initial contributions to the identification of the various bioavailable polyphenols present in tea and wine, as well as the identification of their metabolites. Pietta *et al.* (1998) employed green tea to attempt an overall evaluation of absorption and metabolism. They detected green tea flavanols in plasma and some monohydroxy and dihydroxybenzoic acids in urine, accounting for approximately 15 per cent of the polyphenols administered. These phenolic acids would result from bacterial metabolization of catechin and quercetin in the gut. The intestinal flora has enzymes that cleave the benzopyranosic ring (Das and Griffiths, 1969; Winter *et al.*, 1989). Methylation in one or more phenolic hydroxyls is another possibility in polyphenol metabolism, having been observed for catechin, epicatechin and green tea flavonoids (Piskula and Terao, 1998; Okushio *et al.*, 1999a, 1999b). This reaction is apparently mediated by catechol-O-methyl transferase, an enzyme present in liver and kidney. Epicatechin, methylated and conjugated with glucuronic acid and sulfate, appears as the plasma metabolite with the longer half life, after a single dose of epicatechin to rats (Piskula and Terao, 1998). In rats receiving 0.2 per cent quercetin in their diet for three weeks, the most abundant metabolite was the glucuronic acid and sulfate conjugate of isorhamnetin, the 3' methylation product of quercetin (Morand *et al.*, 1998).

Sulfate and glucuronic acid conjugation, which leads to increased water solubility, is a common strategy for drug metabolism, and in general for xenobiotic metabolism, the products can be more easily eliminated into the urine. Polyphenol glucuronidation occurs in the intestine and in the liver (Sfakianos *et al.*, 1997; Piskula and Terao, 1998; Morand *et al.*, 1998), whereas sulfation apparently occurs only in the liver (Shali *et al.*, 1991; Piskula and Terao, 1998).

Challenges for Research on Polyphenols and their Relationship with Chronic Diseases

There are hundreds of polyphenols with antioxidant activity that are potential contributors to the antioxidant mechanisms in humans and animals in general. These compounds are excellent candidates to explain the health benefits of diets rich in fruits and vegetables, although there is still not enough information on food composition data, bioavailability, interaction with other food components and biological effects (Institute of Medicine, 1998).

Through the number of indexed scientific publications and their distribution over time, it is possible to evaluate the quantity and relative importance of scientific efforts on specific subjects. Thus, we can see that great emphasis has been placed on the subject of chronic disease in recent decades. Vitamins E and C have received sustained attention in the last few decades, perhaps with a particular increment in the last five years. In contrast, the subject of oxidative stress has seen an explosive growth in recent years; 80 per cent of the articles published on the subject have appeared in the last five years.

There is evidence that polyphenols are metabolized by intestinal flora and that they and their metabolites are absorbed. This information is, for the moment, restricted to a few compounds. Similarly, we know that some species are metabolized after absorption. The extent, specificity and localization of polyphenol metabolism in the organism have not been established systematically. In this respect, the known chelating capacity of polyphenols raises the question of their participation in aspects related to metal metabolism and pathology (Morel *et al.*, 1998; Núñez *et al.*, 2000; Opazo *et al.*, 2000; Zago *et al.*, 2000). Another aspect of polyphenol metabolism not yet characterized systematically corresponds to its reaction with other biological antioxidants. Interactions between ascorbate and catechin have been shown (Lotito and Fraga, 2000), leading to the hypothesis that polyphenol antioxidants are part of the antioxidant network of the organism. Indeed, their ability to interact with other antioxidant radicals and peroxyl radicals can be predicted from their reduction potentials (Jovanovic *et al.*, 1998). Attempts have been made to estimate the relative contribution of polyphenols to the total antioxidant capacity in plasma (Perez *et al.*,

2000) but insufficient knowledge on the nature and concentration of circulating polyphenol species renders these results very uncertain. Polyphenol-SH interactions is another subject that remains to be explored systematically. In this respect, Hidalgo *et al.* (2000) describe the effect of redox reagents on the activity of intracellular calcium release channels in muscle and nerve cells, which raises the possibility of another target to explain the biological effects of polyphenol antioxidants.

The interaction of nitric oxide with polyphenol antioxidants is highly relevant in physiological and pathological cellular mechanisms. Atherogenesis is a process markedly dependent of lipid oxidation products that are recognized by specific receptors (*Moriel et al.*, 2000; Rigotti, 2000). Nitric oxide, a free radical itself, participates in the atherogenic process (Rubbo *et al.*, 2000) through membrane lipid and lipoprotein oxidation events (Boveris *et al.*, 2000). Nitric oxide apparently regulates mitochondrial respiration and polyphenol antioxidants are also active at this level (Hodnick and Pardini, 1998; Carreras *et al.*, 2000)

Another rapidly developing aspect of free radical metabolism is its participation in the process of mediating and regulating cellular function. Nitric oxide and superoxide anion are continuously produced in aerobic cells and regulate the mitochondrial function (Valdez *et al.*, 2000) and these and other free radicals can modulate signal transduction pathways and gene expression (Foncea *et al.*, 2000). Thus it seems very likely that dietary polyphenol antioxidants continuously participate in the regulation of cellular function.

Source of Oxidants

Free radicals are produced during normal metabolism in the body. The sources can be divided into endogenous and exogenous sources. The endogenous sources include the mitochondrial sources, which play a major role in the processes of aging and non-mitochondrial sources.

The Endogenous Sources

Mitochondrial Source

Mitochondria seem to be the major source of endogenous oxidants. The rate of respiration is responsible for the rate of generation of oxidants. A small percentage of oxygen in mitochondria inadvertently converses to SOR, which in turn generate H_2O_2, OHR and other FRs (Beckman and Ames, 1998; Ames *et al.*, 1993; Pierrefiche and Laborit, 1995). Mitochondria are also involved in the generation of nitric oxide via the nitric oxide synthase (NOS) reaction. O2- and NO′ react to form another oxidant peroxynitrite (ONOO) which represents a potential source for the more powerful and aggressive hydroxyl radical.

Non Mitochondrial Sources

Fenton Reaction

The degrading H_2O_2 Fenton reaction is catalyzed by free iron bivalent ions and leads to the generation of OH. Recent studies localized the Fenton reaction at endoplasmic reticulum or perinuclear but not at mitochondria or other counterparts

(Liu *et al.*, 2004). H_2O_2 that escapes antioxidant machinery might be converted non-enzymatically in Fenton reaction and act as an RNA or DNA damaging agent.

Microsomal Cytochrome P450 Enzymes

Microsomes contain the cytochrome P450 enzymes which catalyzes univalent oxidation or reduction of xenobiotic compounds (*e.g.*, drugs), simultaneously O_2 is generated (Halliwell and Gutteridga, 1989).

Phagocytic Cells

The respiratory burst of phagocytic cells is the source of via the NADPH oxidase reaction during inflammatory or infectious conditions.

Peroxisomal β Oxidation

Peroxisomal β oxidation of fatty acids generates H_2O_2. As the cell ages the ability to maintain balance between the pro-oxidant and antioxidant gradually compromised and thus, senescent peroxisomes generate increased amount of ROS.

The Exogenous Sources

Polluted urban air, cigarette smoke, iron and copper salts, some phenolic compounds found in many plants and various drugs may all contribute FRs or provoke FRs and constitute the exogenous source of oxidants (Ames *et al.*, 1993; Leibovitz and Siegel, 1980).

Antioxidant Defense System

Free radicals are constantly generated in bioliogical system as a result of internal and external stressors. Antioxidants limit inappropriate exposure ot these stressors. Antioxidants are the small molecules which inhibit or quench FR reactions. An array of antioxidant enzymes persist which are present in both enzymatic and non-enzymatic forms.

The enzymatic and non-enzymatic antioxidant defenses include SOD, Glutathione peroxidase (GPX), catalase (CAT), ascorbic acid, tocopherol, glutathione, β carotene and vitamin A. Ascorbic acid, tocopherol, carotenoids (Chopra and Thurnham, 1993), Ubiquinol-10 and uric acid (Stocker *et al.*, 1991) also possess chain breaking antioxidant properties. Ascorbic acid, the major antioxidant in the extracellular fluids efficiently scavenges O2-, H_2O_2, -OCL, OH, ROO radicals and RNS (Sies *et al.*, 1992), β carotene, lutein, lycopene, β-cryptoxanthin and α carotene, are present in the human blood samples (Cantilena *et al.*, 1992) and the antioxidant nature of these, have been attributed to their unique structure, and extended system of conjugated double bond (Stahl and Sies, 1996).

SOD catalyzes the dismutation of the highly reactive superoxide anion to O_2 and to less reactive H_2O_2. Peroxides are then destroyed by CAT or GPX reactions (Fridovich, 1995: Sandalio *et al.*, 1997; Teixeria *et al.*, 1998). Three types are present in humans: cytosolic Cu/Zn-SOD, mitochondrial Mn-SOD and extracellular SOD (Ee-SOD), in which Cu/Zn-SOD is believed to play an important role in first line of antioxidant defense. Catalase plays an important role in the acquisition of tolerance to oxidative stress in the adaptive responses to cells. The selenium containing glutathione peroxidase caltalyzes the reduction of H_2O_2 using GSH. Small deviation from the physiological activity of antioxidant enzymes may have a dramatic effect on resistance of cells to oxidant-induced damage to the genome and cell killing. Vitamin A contributes to the antioxidant system by limiting the decompartmentalisation of highly catalytic iron (Morrissey and O'Brien, 1998).

Table 1.5: Some Antioxidants of Natural and Synthetic Origin

Sl.No.	*IUPAC/Common Name*	*Source*	*Mechanism of Action*
1.	α-Tocopherol	Vegetables, fruits, seeds, fish oils etc.	Chain breaking
2.	Ascorbic acid	Citrus fruits, vegetable etc.	Chain breaking
3.	β – Carotene	Fruits, vegetables	Singlet oxygen quencher Chain breaking
4.	(+) – Catechin	*Camellia sinensis* L.	Chain breaking
5.	Hypolaetin –8- glucoside	*Siderrities javalambrensis* Pau	Chain breaking
6.	Quercetin	*Polygonum hydropiper* L. *Bridclia ferrunginea*	Chain breaking Inhibits Xanthine oxidase
7.	Aloesin	*Aloe vera*	Chain breaking
8.	A) Carnosic acid B) Carnosol	*Rosnarinus officinalis* L.	Chain breaking
9.	Chrysophanol 8-O-β-D-(6[1]-galloyl)-glucopyranoside	*Rheum undulatum* L.	Chain breaking
10.	5-(3,4-Dihdroxyphenylmethy) oxazolidene-2,4-dione	*Perille frutescens* var. *japonica*	Chain breaking
11.	1β,13,14-Trihydroxy 8,11, 13-podocarpathene-2,7-dione	*Taiapnia cryptonicriodes*	Chain breaking
12.	1,3,4,12-Tetrahydroxy-2-(9-hexadecenoyl am in-0) ocatdecane-1-O glucopyra-noside	*Conyza bonariensis* L.	Inhibits Xanthineoxidase
13.	Reductic acid/2,3-dihy-droxy-2-cyclopentenone	Synthetic	Chain breaking Inhibits Xanthine oxidase
14.	8-11 exylthio-xantbine	Synthetic	Inhibits Xanthine Oxidase
15.	5-Nitro-3,4-dihydroxy-cinamic acid	Synthetic	Chain breaking Inhibits Xanthine Oxidase
16.	3-[5-Hydro-1*H*-indolyl] methy1-4-amino-4,5-dihydro-1*H*-1,2,4-triazole-5-thione	Synthetic	Chain breaking

In addition to the above defense system the presence of metal ion storage and transport protein also serves to maintain cellular integrity. Examples are Transferrin, Lactoferrin, Haptoglobin, Ceruloplasm, Metallothionein (Thurnham,1990) and Carnosine (Chan and Decker, 1994).

Repairing mechanism exists in biological systems depending on the nature of oxidized targets (Pacifici and Davies, 1991). These are based either on regenerating the slightly oxidized macromolecules keeping critical chemical groups in their reduced forms or on degrading defective highly oxidized macromolecules into low molecular mass compounds. Many of the repair systems become deficient in the senescent cells, thus a high amount of biological garbage is accumulated (Terman *et al.*, 2006; Brunk *et al.*, 1992).

Chapter 2
Antioxidant Activity Assays

☆ K.S. Nagesh

Methods of Evaluation

Methods available for the measurement of antioxidant capacity were reviewed by Prior *et al.* (2005), presenting the general chemistry underlysing the assays, the types of molecules detected, and the most important advantages and shortcomings of each method. This overview provided a basis and rationale for developing standardized antioxidant capacity methods for the food, nutraceutical, and dietary supplement industries. From the evaluation of data presented at the First International Congress on Antioxidant Methods in 2004 and in the literature, as well as consideration of potential end uses of antoxidants, it is proposed that procedures and applications for three assays be considered for standardization: the oxygen radical absorbance capacity (ORAC) assay, the Folin-Ciocalteu method, and possibly Trolox equivalent antioxidant capacity (TEAC) assay. ORAC represent a hydrogen atom transfer (HAT) reaction mechanism, which is most relevant to human biology. The Folin-Ciocalteu method is an electron transfer (ET) based assay and gives reducing capacity, which has normally been expressed as phenolic contents. The TEAC assay represents a second ET-based method. Other assays may need to be considered in the future as more is learned about some of the other radical sources and their importance to human bology. Prior *et al.* (2000) reviewed the analysis of botanicals and dietary supplements for antioxidant capacity.

Determination of Reducing Power

The reducing power of the compounds is evaluated according to Oyaizu (1986). Different amounts of the extract are perched in distilled water and diverse with 2.5 ml of 0.2M phosphate buffer (pH 6.6), and 2.5 ml of 1 per cent $K_3Fe(CN)_6$. This mixture is incubated at 50 °C for 20 min, 2.5 ml of 10 per cent TCA is added to the blend and centrifuged at 3000 rpm for 10 min. The upper layer of the solution (2.5 ml) is assorted with distilled water (2.5 ml) and $FeCl_3$ (0.5 ml, 0.1 per cent), and the absorbance are measured at 700 nm. Increase in absorbance of the reaction mixture indicated reducing power.

Free Radical Scavenging Assays

DPPH Assay

The hydrogen atom or electron donating abilities of the resultant compounds and some untainted compounds is measured from the bleaching of the purple-coloured methanol solution of 2,2-diphenyl-1-picrylhydrazyl (DPPH). This spectrophotometric assay uses the stable radical, DPPH as a reagent (Burits and Bucar, 2000; Cuendet *et al.*, 1997). One thousand microlitre of diverse concentrations of the extracts in ethanol were added to 4 ml of 0.004 per cent methanol solution of DPPH. After a 30 min incubation period at room temperature, the absorbance are read against a blank at 517 nm. Inhibition of free radical by DPPH in percent (I per cent) is calculated in following way:

$$1 \text{ per cent } (A_{blank} - A_{sample}/A_{blank}) \times 100$$

where, A blank is the absorbance of the control reaction (containing all reagents except the test compound) and A sample is the absorbance of the test compound. The values of inhibition are calculated for the various concentrations of extract. Tests are conceded out in triplicate. Determination of super oxide anion (O2•") radical scavenging activity (Yen and Chen, 1995). The reaction mixture consist of 1 ml sample solution in distilled water, 1 ml of phenazine methosulphate (60 μM) in phosphate buffer (0.1 M, pH 7.4), 1 ml NADH (450 μM) in phosphate buffer, was incubated at 25 °C for 5 min. The absorbance was then examined at 560 nm against blank samples. The superoxide radical scavenging activity is calculated as where A blank is the absorbance of control and A sample is the absorbance of test. Determination of peroxide (H_2O_2) radical scavenging activity The scavenging ability of H_2O_2 radical by the sample extract was dogged by the scheme of Ruch *et al.* (1989). One millilitre of extract solution [prepared in phosphate buffered saline (PBS)] is incubated with 0.6 ml of 4mM H_2O_2 solution (prepared in PBS) for 10 min. The absorbance of the solution is considered at 230 nm against a blank solution containing the extract without H_2O_2. The concentration of H_2O_2 was spectrophotometrically stubborn from absorption at 230 nm using the molar absorbtivity of 81 M^{-1} cm^{-1}. The H_2O_2 radical scavenging activity is calculated as

$$1 \text{ per cent } (A_{blank} - A_{sample}/A_{blank}) \times 100$$

Where A blank is the absorbance of control and A sample is the absorbance of test.

Determination of Nitric Oxide (NO) Radical Scavenging Activity

This procedure is based on the principle that sodium nitropruside in aqueous solution at physiological pH instinctively generates nitric oxide which interacts with oxygen to produce nitrite ions that can be anticipated using Greiss reagent. Scavengers of nitric oxide contend with oxygen leading to reduced production of nitrite ions. For the experimentation sodium nitropruside (10 mM) in phosphate buffered saline was diverse with different concentrations of extract dissolved in suitable solvent and incubated at room temperature for 150 min. The same reaction mixture without the methanolic extract of sample but with equivalent amount of methanol served as control. After the incubation period, 0.5 ml of Griess reagent (1 per cent sulfanilamide, 2 per cent H_3PO_4 and 0.1 per cent N-(1-naphthyl) ethylenediamine dihydrochloride was added. The absorbance of the chromophore formed was read at 546 nm. Curcumin was used as positive control (Sreejayan, 1997). The NO radical scavenging activity is calculated as

$$1 \text{ per cent } (A_{blank} - A_{sample}/A_{blank}) \times 100$$

Where A blank is the absorbance of control and A sample is the absorbance of test.

Determination of Hydroxyl Radical (OH^-) Scavenging Activity

The scavenging ability for hydroxyl radical was measured by Halliwell *et al.* (1987). Stock solution of EDTA (1mM), $FeCl_3$ (10mM), ascorbic acid (1mM), H_2O_2 (10mM) and deoxyribose (10mM) is prepared in distilled de-ionized water. The attempt is performed by adding up 0.1ml EDTA, 0.01ml of $FeCl_3$, 0.1ml H_2O_2, 0.36 ml deoxyribose, 1ml of sample extract (10-100cg/ml) dissolved in distilled water, 0.33ml of phosphate buffer (50mM, pH 7.4) and 0.1ml of ascorbic acid is added. The mixture is incubated at 37°C for 1 hour. A 1.0 ml of incubated mixture is mixed with 1.0 ml of 10 per cent trichloro acetic acid and 1.0ml of 0.5 per cent thiobarbituric acid (in 0.025M NaOH containing 0.025 per cent BHA) to urbanized the pink color measured at 532nm. The hydroxyl radical scavenging activity is reported as percent inhibition of deoxyribose sugar dilapidation and is calculated as

$$1 \text{ per cent } (A_{blank} - A_{sample}/A_{blank}) \times 100$$

Where A blank is the absorbance of control and A sample is the absorbance of test.

Determination of Protein Oxidation

The protein (BSA) was oxidized by a Fenton-type reaction. BSA (1 mg/ml) is incubated at 25°C in solution with 2.5mM H_2O_2, 1.0mM $FeCl_3$, 1.0mM ascorbate and 3.0mM EDTA in presence or absence of extract. After incubation for 45 min, protein is precipitated with 10 per cent TCA, centrifuged (5000 rpm, 4°C, 10 min) and the supernatants decanted. Protein pellets are dissolved in 1ml of 50mM potassium phosphate buffer, pH 7.5. Total sulphydryl ("SH) group determinations are performed according to the method of Sedlak and Lintion (1968) of metal ion chelating activity using Ellman's reagent.

Determination of Peroxyl Radical Scavengers

The chelation of ferrous ions by the extract was anticipated by the method of Dinis *et al.* (1994). Briefly, 50 µl of 2mM $FeCl_2$ is added to the extract (1 ml). The reaction is initiated by the addition of 0.2 ml of 5mM ferrozine solution. The mixture is enthusiastically shaken and left to rest at room temperature for 10 min. The absorbance of the solution are subsequently measured spectrophotometrically at 562 nm.

Determination of Total Flavonoid Concentration

Flavonoid concentration was determined as follows: Drug extract solution (1 ml) is diluted with 4.3 ml of 80 per cent aqueous ethanol containing 0.1 ml of 10 per cent aluminium nitrate and 0.1 ml of 1 M aqueous potassium acetate. After 40 min at room temperature, the absorbance are determined spectrophotometrically at 415 nm. Total flavonoid concentration is calculated using quercetin as standard (Park *et al.*, 1997):

$$\text{Absorbance} = 0.002108\ \mu\text{g quercetin} - 0.01089 \times (R^2 : 0.9999)$$

Oxygen Radical Absorbance Capacity (ORAC) Assay

The ORAC assay used fluorescein as fluorescent probe and was an amendment of the protocols premeditated (Huang *et al.*, 2002). Fluorescein sodium salt (16.7 mg) is dissolved in 10 ml of phosphate buffer solution (PBS) (75 mM, pH 7.0) to get a stash solution. The operational solution (60 nM) are obtained by ensuing dilution in PBS. A 10 ml solution of AAPH (2,2-azinobis (2-amidinopropan) dihydrochloride) was prepared each day at a concentration of 153 mM and maintained in ice earlier than automatic injection. Trolox (6-hydroxy-2,5,7,8- tetramethylchroman-2-carboxylic acid) solutions used for the calibration curve (8, 16, 24, 32 and 40 M) were prepared each day by dilution in PBS of a 1 mM stock solution made in the same solvent and stored at 80°C. The scrutiny is performed using microplates (96-well, opaque white) and a fluorimeter. Sample (25 µl) is mixed with 250 cl of fluorescein (60 nM) and incubated for 10 min at 37°C in the microplate. AAPH (25 µl 1) solution is then robotically injected and the microplate is traumatized. The fluorescence (kexcitation = 485 nm, kemission = 520 nm) was registered each minute over 50 min. All samples are analyzed at three dilutions and the mean value are taken for ORAC. The quantification of the antioxidant activity is based on the calculation of the area under the curve, as proposed by Cao and Prior (1999). The antioxidant activity by ORAC was articulated as l mol of trolox equivalents (TE) per gram of FW.

Trolox Equivalent Antioxidant Capacity (TEAC) Assay

The antioxidant activity is determined by the TEAC assay via the radical cation $ABTS^+$ (Re *et al.*, 1999). The analysis is adapted to be conducted in a microwell plate (96-well, transparent) with a spectrophotometer k Max 190. The $ABTS^+$ stock solution (7 mM) is prepared using $K_2S_2O_8$ as the oxidant agent. The working solution of $ABTS^+$ is obtained by diluting the stock solution in ethanol to confer an absorption of 0.70 ± 0.02 at k = 734 nm. Extract or trolox (10 µl) diluted in ethanol in the well and the reaction begin with the addition of 290 µl of the $ABTS^+$ working solution. The trolox solution (50 mM) is prepared in ethanol and is stored at 20°C under an atmosphere of nitrogen. For each measurement, a standard curve with trolox is plotted. The TEAC value is calculated by the dimension of the area under the curve, consequential from plotting the percentage reticence of the absorbance as a function of time. The absorbance is recorded every 10 s for 6 min, to allow a close monitoring of the reaction rate, which is important for those antioxidants that show a time dependency when submitted to this assay. The calculation of the area under the curve is performed for one sample dilution, which has a final percentage inhibition between 20 per cent and

80 per cent. Each extract is analysed in triplicate. The antioxidant activity for the plant extracts is articulated as l mol of TE per litre and is calculated by the equation (Vanden *et al.*, 1999).

$$\text{TEAC (TE, μmol/L)} = 30 \times k \times \frac{AUC_{sample}}{r.c._{trolox}}$$

where, k is the dilution of the sample, 30 is a dilution factor originated from the protocol, r.c.trolox is the regression coefficient calculated from the calibration curve:

$$AUC_{trolox} = r.c._{trolox} \times [trolox]$$

AUC sample and AUC trolox are the areas produced correspondingly by the sample and trolox and are calculated by the following equation:

$$AUC = \left(\%Inh_{(t=0)} \times 0.5 + \sum_{i=1}^{36} \%Inh_{(t=10 \times i)} \right) \times 10$$

where, %Inh(t) is the inhibition percentage at t seconds.

Lipid Peroxidation (LPO) Assay

LPO is induced and assayed in rat liver microsomes as per method of Wright *et al.* (1981). The reaction muddle, in a total volume of 1.0 ml, contained 0.58 ml phosphate buffer (0.1 M, pH 7.4), 0.2 ml of hepatic microsome (10 per cent, w/v), 0.2 ml ascorbic acid (100 mM) and 0.02 ml ferric chloride (100 mM) and is incubated at 37°C in a shaking water bath for 1 h. The reaction is clogged by the addition of 1.0 ml TCA (10 per cent, w/v), subsequent which of 1.0 ml TBA (0.67 per cent, w/v) is added and all the tubes are to be found in a boiling water bath for 20 min. At the end, the tubes are shifted to an ice-bath and centrifuged at 2500×g for 10 min. The amount of malonaldehyde formed in each of the samples is assessed by measuring the optical density of the supernatant at 535 nm allied with a reagent blank. The molar extinction coefficient for MDA is taken to be 1.56×105 M^{-1} cm^{-1}.

Animals and Treatment Protocol

Male or female rats/mice are used to evaluate this study. All the experimental procedures are carried out in peace with internationally accepted guiding principle for the care and use of laboratory animals. Rats are divided into five groups containing six rats each. Group 1 served as control, group 2 is administered CCl_4 (negative control), and groups 3–5 are administered sample extract for 7 days. The rats of the groups 1 and 2 are simultaneously administered saline until the seventh day. On the seventh day, the rats of the groups 2– 5 are given a single oral dose of CCl_4 (1:1) in olive oil at 2.0 g/kg of body weight 6 h after the last dose of extract/saline. After 24 h of CCl_4 administration, rats are sacrificed, the livers are secluded and hepatic PMS prepared.

Determination of Aspartate Aminotransferase (AST) and Alanine Aminotransferase (ALT) Activity

AST and ALT are assayed by the means adapted from Reitman and Frankel (1957). Briefly, 1 ml of the substrate (2mM Z-ketoglutarate, 0.2M d,l-aspartate for ALT and 2mM Z -ketoglutarate, 0.2Md, l-alanine for AST) is incubated with 0.2 ml serum for 30 min (for AST) or 60 min (for ALT) in a water bath at 40°C. Subsequent to incubation, the reaction is clogged by addition of 1ml of 1mM DNPH. After 20 min, 10 ml 0.4N NaOH is added to each tube. The absorbance of the solution is measured at 505 nm after 30 min in opponent to water as the blank.

Determination of Glutathione (GSH) Activity

GSH is assayed by the method of Jollow *et al.* (1974). An aliquot of 1.0 ml of hepatic PMS (10 per cent, w/v) is precipitated with 1.0 ml of sulphosalicylic acid (4 per cent, w/v). The samples are detained in reserve at 4°C for 1 h and then centrifuged at 3500 rpm for 15 min. The analyze mixture contain 0.1 ml filtered aliquot, 2.7 ml phosphate buffer (0.1 M, pH 7.4) and 0.2 ml DTNB (40 mg/10 ml of phosphate buffer 0.1 M, pH 7.4) in a total volume of 3.0 ml. The yellow color developed is comprehended at 412 nm on a spectrophotometer.

Determination of Superoxide Dismutase (SOD) Activity

SOD activity is estimated by Beauchamp and Fridovich (1971) method. The reaction mixture consists of 0.5 ml of hepatic PMS, 1ml 50mM sodium carbonate, 0.4 ml of 25cM NBT and 0.2 ml 0.1mM EDTA. The reaction is initiated by addition of 0.4 ml of 1mM hydroxylamine–hydrochloride. The change in absorbance is recorded at 560 nm. The control is simultaneously run without liver homogenate. Units of SOD activity are expressed as the amount of enzyme required inhibiting the reduction of NBT by 50 per cent.

Determination of Catalase (CAT) Activity

CAT activity is assayed by the method of Claiborne (1985). The assay mixture consists of 1.95 ml phosphate buffer (0.05 M, pH 7.0), 1.0 ml H_2O_2 (0.019 M), 0.05 ml of hepatic PMS (10 per cent, w/v). Changes in absorbance are recorded at 240 nm. CAT activity is calculated in terms of nmol H_2O_2 consumed/min/mg of protein.

Determination of Glutathione Peroxidase (GPX) Activity

GPX activity is measured according to course of action of Mohandas *et al.* (1984). The reaction mixture consists of 1.44 ml 0.05M phosphate buffer, pH 7.0, 0.1 ml 1mM EDTA, 0.1mM sodium azide, 0.05 ml 1 U/ml glutathione reductase, 0.10 ml 1mM GSH, 0.1 ml 2mM NADPH, 0.01 ml 0.25mM H_2O_2 and 0.1 ml 10 per cent PMS in a total volume of 2 ml. Disappearance of NADPH at 340 nm is recorded at 25°C. Enzyme activity is calculated as nmol NADP reduced/min/mg protein using a molar extinction coefficient of 6.22×103 M^{-1} cm^{-1}.

Measurement of Effect on Lipid Peroxidation in Rat Liver Homogenate

Rat liver homogenate (10 per cent w/v) method was described by Tripathi *et al.* (1996). In liver tissue peroxidation is induced by Iron-ADP complex in the presence of ascorbic acid. The incubation medium constituted 0.5 ml of liver homogenate (10 per cent w/v), 100 μM $FeCl_3$, 1.7 μM ADP, 500 μM of ascorbate and different concentrations of extract in 2 ml of total incubation medium. The medium is incubated for 20 min at 37°C. Extent of lipid peroxidation is measured by assessment of malondialdehyde (MDA) content (Vani *et al.*, 1997). Results are articulated in terms of decrease in MDA formation by the sample extract. α-tocopherol acetate is used as positive control.

Determination of Total Phenolic Content

Total soluble phenolics in the ethanolic extracts are determined with Folin–Ciocalteu reagent, according to the method of Slinkard and Singleton (1997) using pyrocatechol as a standard. Briefly, 1 ml of extract solution (contains 2000 lg) in a volumetric flask is diluted with glass-distilled water (46 ml). Folin–Ciocalteu reagent (1 ml) is added and the contents of the flask are mixed thoroughly. After 3 min, 3 ml of Na_2CO_3 (2 per cent) is added, then the mixture is allowed to stand for 2 h with intermittent shaking. The absorbance is measured at 760 nm. The concentration of total phenolic compounds in the ethanolic extract is determined as microgrammes of pyrocatechol equivalents, by using an equation that was obtained from the standard pyrocatechol graph, is given as:

$$\text{Absorbance} = 0.00246\ \mu\text{g pyrocatechol} + 0.00325 \times (R^2 : 0.9996)$$

Solid-Phase Spectrometry

A new test method for measuring the antioxidant power of herbal products based on solid-phase spectrometry using tetrabenzo-b,f,j,n,1,5,9,13-tetraazacyclohexadecine-Cu(II) complex immobilized on silica gel, was proposed by Zaporozhets *et al.* (2004). The method represents an alternative to the mostly used radical scavenging capacity assays. The test method is simple ("drop and measure" procedure is applied), rapid (10min/sample), require only the monitoring of time and absorbance, and provides good statistical parameters (Sr less than equal to 0.2). The method was approved in the analysis of the most popular herbal beverages (black and green teas) and drugs *Echinacea* products. The results obtained by Zaporozhets *et al.* (2004) were compared with the content of total flavonoids and tannins of teas and total caffeic acid derivatives of *Echinacea* determined spectrophotometrically.

Chapter 3
Herbal Antioxidants

☆ *T. Pullaiah*

In the following pages the antioxidant activities of different plants are given.

Angiosperms

***Abelmoschus esculentus* (L.) Moench. (Syn.: *Hibiscus esculentus* L.) Family: Malvaceae**

English Name: Okra; Japanese: Maruokura.

Ansari *et al.* (2005) evaluated the antioxidant activity of five vegetables traditionally consumed by South-Asian migrants in Bradford, Yorkshire, UK in the DPPH scavenging assay, and *in vitro* non-enzymatic inhibition of bovine brain lipid peroxidation. In both antioxidant assays a strong activity was shown by extracts derived from *Abelmoschus esculetus*.

Maeda *et al.* (2006) examined the phenolic contents of fruit of okra and its radical scavenging activities for DPPH radical. The polyphenol content was 43.9 mg-gallic acid equivalent/100 g and DPPH scavenging activity was 533.3 µmol-Trolox eq./ 100 g.

Nagashree *et al.* (2009) evaluated the antiperoxidative effect of *Hibiscus esculentus* fruit on various tissues, blood cell counts and haemoglobin of arsenite exposed albino rats. A significant reduction in arsenic induced LPO in brain, heart lungs and liver was observed after the administration of *H.esculentus* fruit extract. The results indicate that people exposed to higher levels of arsenite can increase their intake of *H. esculentus* as the quercetin 3-O-glucoside which is the major antioxidant of *H. esculentus* has a possible protective role of LPO status of certain tissues and blood cell counts of arsenic exposed rats.

Adelakun *et al.* (2009a,b) evaluated the antioxidant activity of Okra seed flour. The antioxidant activity was significantly increased by roasting, while *in vitro* digestibility showed that most antioxidative activities were available in the intestinal phase of gastrointestinal tracts.

Antioxidant potential of selected vegetables commonly used in diet in Asian subcontinent, including *Abelmoschus esculentus* fruits has been evaluated by Gacche *et al.* (2010). The ferrous chelating activity was noted to be 26.5 per cent in the fruits. The amount of vitamin C was found to be 25.60 mg/100g while total phenols was 5.35 mg/g. The DPPH radical scavenging activity was 43.8 per cent, inhibition of lipid peroxidation was 13.8 per cent and FICA was 26.5 per cent.

Antioxidative activities and phenolic content of extracts from Okra were investigated by Khomsug *et al.* (2010) by DPPH radical scavenging and ABTS radical cation decolorization assays. The TPC of pulped and seeds of okra extracts were 10.75±0.02 mg GAE/100 g extract and 142.48±0.02 mg GAE/100 g extract which corresponding with scavenging activities. The predominant phenolic compound was procycanidin B2, following procycanidin B1 and rutin in seeds but in pulped okra they were identified as catechin, procycanidin B2, epicatechin and rutin.

Abelmoschus manihot (L.) Medikus Family: Malvaceae

Phytochemical flavonols, carotenoids and the antioxidant capacity of leaves of *Abelmoschus manihot* was evaluated by Lako *et al.* (2006). *Abelmoschus manihot* leaves have moderate Total antioxidant capacity (100 mg/100g) and Total polyphenol content (TPP) (81mg/100g) and β-carotene (16mg/100g).

Abroma angusta L.f. Family: Sterculiaceae

The administration of aqueous extract of turmeric (300mg/kg) and abromine powder resulted in a significant reduction in blood glucose and an increase in total haemoglobin. Extract also reduced free radical formation in tissue, thiobarbituric reactive substances. These results indicate that mixture of two plants have antidiabetic activity and also reduce oxidative stress in diabetes (Eshrat and Hussain, 2002).

Abutilon bidentatum Hochst. Family: Malvaceae

Antioxidant activity of different extracts of *Abutilon bidentatum* were evaluated by Shahwar *et al.* (2010) through DPPH radical scavenging activity, phosphomolybdate and ferric thiocyanate methods. The DPPH radical scavenging activity of ethyl acetate extract was 88.5 per cent, total antioxidant activity was 0.483 and inhibition of lipid peroxidation was 78.4 per cent while total phenols was 532.6 mg GAE/g of crude extract.

Abutilon indicum (L.) Sweet Family: Malvaceae

Ahmad *et al.* (2009) evaluated the antioxidant activity of seeds of *Abutilon indicum.* Total phenolic compounds were determined according to Folin-Ciocalteu method as 13.770 mg/g oil. For radical scavenging and antioxidant studied ABTS, DPPH, FRAP and linoliec acid peroxidation methods were followed. The seed oil of *A. indicum* exhibited strong antioxidant activities. TEAC values of *A. indicum* seed oils were 14.006 and 21.375 by ABTS and FRAP methods respectively. *A. indicum* exhibited inhibition to peoxidation of linoleic acid. According the results seeds of *Abutilon indicum* have powerful antradical and antioxidant components.

Chakraborthy and Ghorpade (2010) determined the Total Phenolic Content, quantified by Folin-Ciocalteu method and free radical scavenging by DPPH method in methanolic, Hydro-alcoholic and aqueous extracts of the *Abutilon indicum* stem. They reported that *Abutilon indicum* aqueous extract showed a potent total phenolic content and possessed a significant activity. Methanolic, Hydroalcoholic and Aqueous extracts showed TPC 20.94, 27.77 and 35.45 mg GAE/g, respectively. Methanolic, Hydroalcoholic and Aqueous extracts showed DPPH radical scavenging activity 1343.89, 2387.14 and 1154.20 µg/ml IC_{50} values, respectively.

Kaushik *et al.* (2010) evaluated the antioxidant activities of chloroform fraction of alcoholic extract of whole plant of *Abutilon indicum*. The extract was screened for antioxidant and free radical scavenging effects at various concentrations (100, 300 and 500µg/ml) by reducing power assay and DPPH free radical scavenging method. All these antioxidant activities were concentration dependent which were compared with standard antioxidants such as BHA and ascorbic acid.

Abutilon muticum (Del. ex DC.) Sweet Family: Malvaceae

Ahmad *et al.* (2009) evaluated the antioxidant activity of seeds of *Abutilon muticum*. Total phenolic compounds were determined according to Folin-Ciocalteu method as 38.815 mg/g oil. For radical scavenging and antioxidant studied ABTS, DPPH, FRAP and linoliec acid peroxidation methods were followed. The seed oil of *A. indicum* exhibited strong antioxidant activities. TEAC values of *A. muticum* seed oils were 16.369 and 18.08 by ABTS and FRAP methods respectively. *A. muticum* exhibited inhibition to peroxidation of linoleic acid. According to the results seeds of *Abutilon muticum* have powerful antiradical and antioxidant components.

Acacia asak (Forsk.) Willd. Family: Mimosaceae

Methanol extract of *Acacia asak* bark showed effective free radical scavenging activity in the DPPH assay (Al-Fatimi *et al.*, 2007).

Ninety crude extracts, including dichloromethane, methanol and aqueous extracts from 30 medicinal plants used in the Yemeni ethnomedine to treat common infections, were screened *in vitro* for antimicrobial and antioxidant activities (Mohammed *et al.*, 2007). Ten methanol extracts, especially those of *Acacia asak* and *Solanum nigrum* fruits, showed effective free radical scavenging activities in the DPPH assay.

Acacia auriculiformis A. Cunn. Family: Mimosaceae

Singh *et al.* (2007a,b,c) estimated free radical scavenging activity of the methanol, acetone and ethyl acetate extract/fractions of *Acacia auriculiformis* in different assays, *viz.*, DPPH, deoxyribose (site specific and non-site specific), relative reducing power and lipid peroxidation. The bark powder of the plant was extracted with different solvents by maceration method in order increasing and decreasing polarity. The scavenging activity of fractions was found to be more as compared to the crude extract. The per cent inhibition with water fraction of ethyl acetate extract was observed to be 71.2 per cent, 73.66 per cent, 83.37 per cent, 75.63 per cent and 72.92 per cent in DPPH, chelating power, lipid peroxidation, site specific and non-site specific deoxyribose scavenging assays respectively at maximum concentration tested. The water fraction of methanol extract exhibited good scavenging responses of 72.0 per cent, 1.76, 88 per cent and 93.6 per cent in DPPH, reducing power, site specific and non-site specific hydroxyl radical scavenging assay in increasing and (decreasing) order of solvent polarity at maximum concentration, respectively. Maximum inhibitory activity of acetone extract were 72.3 per cent, 91.7 per cent, 1.63, 83.3. per cent and 70.9 per cent in DPPH, deoxyribose, reducing power, chelating power and lipid peroxidation assays, respectively.

Acacia catechu (L.f.) Willd. Family: Mimosaceae

Patil *et al.* (2003) investigated the antioxidant potential of *Acacia catechu* to understand their beneficial role in cancer therapy. The aqueous extract of *Acacia catechu* showed a marked antioxidant activity when compared to ascorbic acid. The study revealed that polyphenols present in the polar extracts have a stronger antioxidant activity compared to the non-polar phytoconstituents.

Naik *et al.* (2003) evaluated the antioxidant activity of aqueous extract of *Acaciaa catechu* for its potential as antioxidant. The antioxidant activity of these extracts was tested by studying the inhibition of radiation induced lipid peroxidation in liver microsomes at different doses in the rages of 100-600 Gy as estimated by TBARS.

Cai *et al.* (2004, 2006) evaluated the antioxidant activity and phenolic compounds in traditional Chinese medicinal plants associated with anticancer. The improved ABTS.+ method was used to systematically assess the total antioxidant capacity (Trolox equivalent antioxidant capacity, TEAC) of the medicinal extracts. The TEAC values and total phenolic content for methanolic extract of branch and stem of *Acacia catechu* were 11197.1 µmol Trolox equivalent/100g dry weight (DW), and 45.70 g of gallic acid equivalent/100g DW respectively. A positive significant linear relationship between antioxidant activity and total phenolic content showed that phenolic compounds were the dominant antioxidant components in the tested herbs. Major types of phenolic compounds in *Acacia catechu* include hydrolysable and condensed tannins, flavan-3-ols ((+)-catechin, epicatechin).

Hazra *et al.* (2010) evaluated the antioxidant and iron chelating property of 70 per cent methanolic extract of 'katha' (heartwood extract of *Acacia catechu*). The extract was found to be an antioxidant with TEAC value of 0.72. The extract has shown its scavenging activity for different radical and 17.8 µg/ml, 39.55 µg/ml, 55.31µg/ml, 746.85µg/ml, 44.2mg/ml, 57.5 µg/ml, 155.48µg/ml, were determined as IC_{50} value for DPPH, superoxide, nitric oxide, peroxinitrite, hydrogen peroxide, singlet oxygen and hypochlorous acid radicals, respectively. The plant was found to inhibit lipid peroxidation with an IC_{50} of 12.35µg/ml. It was also observed that the plant extract (100mg) has 97.13 mg/ml gallic acid equivalent phenolic and 383.66 mg/ml quercetin equivalent flavonoid content.

Acacia confusa Merr. Family: Mimosaceae

Chang *et al.* (2001) evaluatd the antioxidant activity of *Acacia confusa*. The results indicate that both bark and heartwood extract clearly have strong antioxidant effects. Similar inhibitory activities for each test sample were found for both DPPH radical generation and lipid peroxidation. As for the superoxide radical scavenging activity, the heartwood extract was more effective than the bark extract. Further more, the heartwood extract protected PhiX174 supercoiled DNA against strand scission induced by ultraviolet photolysis of H_2O_2 and it reduced the amounts of intracellular hydrogen peroxide, when it was co-incubated with human promyelolytic leukemia (HL-60) cells under oxidative stress.

Tung *et al.* (2007) reported that the ethanolic extracts from the bark of *Acacia confusa* exhibited strong antioxidant activity. Among all the fractions from ethanolic extracts of bark, the EtOAc soluble fraction exhibited the best antioxidant performance. Tung *et al.* (2009) evaluated the antioxidant activities of bark extract of *Acacia confusa* and some of the isolated constituents from its ethyl acetate (EtOAc) fraction in various *in vitro* systems to gain mechanistic insights. Results from antioxidant assays together with authentic antioxidant standards revealed that EtOAc fraction showed strong superoxide radical scavenging activity, reducing power and ferrous ion-chelating ability.

Tung and Chang (2010) evaluated the antioxidant activity of extracts of *Acacia confusa* of different ages. Results indicate that the ethanolic extracts of 5 cm-diameter branch bark and bark of *A. confusa* exhibited stronger antioxidant activity and higher phenolic and proanthocyanidin contents than the other plant parts. Further more twig, 2 cm-diameter branch wood, 2 cm-diameter branch bark and 5 cm-diameter branch wood extracts of 35-year old specimens showed better antioxidant activities and higher phenolic contents, followed by 25 and 15 year-old specimens tested. The total phenolic content of *A. confusa* ethanolic extracts has a high positive correlation with DPPH free-radical scavenging activity, superoxide radical-scavenging activity, ferrous ion-chelating effect and reducing power.

Acacia dealbata Link Family: Mimosaceae

Acetone and methanol extracts of *Acacia dealbata* barks were evaluated for their total phenolics, flavonoid contents and antioxidative activities. The acetone extracts of bark exhibited wide range of total phenolic and flavonoid contents and showed significantly higher DPPH radical scavenging activity than the standards BHA (IC_{50} 33.9µg/ml) and α-tocopherol (IC_{50} 75.5µg/ml). They also contained relatively higher level of FRAP, ABTS radical cation and OH scavenging activity, metal chelation, phosphomolybdenum reduction and peroxidation inhibition (Sowndhararajan *et al.*, 2010).

Acacia nilotica (L.) Willd. Family: Mimosaceae

Saleem *et al.* (2001) extracted fruits of *Acacia nilotica* with 70 per cent acetone and analyzed for their total phenolic concentration and antioxidant potential. The IC_{50} results indicate that the extract is stronger antioxidant than α-tocopherol. Total phenolics concentration, expressed as gallic acid equivalents, showed close correlation with the antioxidant activity. HPLC analysis with diode array detection at 280 nm of the extract indicated the presence of hydroxybenzoic acid derivatives, hydroxycinnamic acid derivatives, flavonol aglycones and their glycosides as main phenolics compounds.

Kaempferol, isolated from the methanol extract of *Acacia nilotica* showed antioxidant potential in several *in vitro* assays: DPPH scavenging assay, hydroxyl radical scavenging activity, metal chelating activity, reducing power and inhibition of lipid peroxidation. The effect of kaempferol was strogly dose-dependent up to the concentrations 1-50µg/ml in DPPH assay and 1-100µg/ml in deoxyribose degradation assay but did not show further change above the highest concentrations (Singh *et al.*, 2007).

Singh and Arora (2007) estimated the free radical scavenging activity of the acetone extract/fractions of *Acacia nilotica* in DPPH free radical scavenging, reducing power and lipid peroxidation assays. The scavenging activity of crude extract was found to be less than the fractions isolated. The per cent of inhibition with water fraction of acetone extract was 71.2 per cent, 73.66 per cent and 83.37 per cent in DPPH, chelating power and lipid peroxidation assays respectively at maximum concentration.

The antioxidant activities of eight extracts were studied by Singh *et al.* (2009) in green pods of *Acacia nilotica*.

Gowri *et al.* (2011) evaluated the free radical scavenging capacity and antioxidant activity of young leaves and barks of *Acacia nilotica*. Though all the extracts exhibited reducing activity, the acetone extracts of both leaf and bark found to be registered highest activity and comparable to that of tannic acid. Like reducing power activity, the similar line of dose dependent trend has also been maintained in the DPPH radical scavenging activity. This might be due to the presence of high molecular compounds like tannins.

***Acacia pennata* (L.) Willd. Family: Mimosaceae**

Chanwitheesuk *et al.* (2005) investigated the antioxidant activity of *Acacia pennata* using a β-carotene bleaching method. The contents of plant chemicals such as vitamin C, Vitamin E, carotenoids, tannin and total phenolics, were also determined. Methanolic extract showed antioxidant activity. *Acacia pennata* has antioxidant index of 1.46, Vitamin C 22.43 mg per cent, Vitamin E 0.0015mg per cent, Total carotenes 1.27 mg per cent, Total xanthophylls 1.59mg per cent, Tannins 11.1mg per cent and Total phenolics of 121mg per cent.

Acetone and methanol extracts of *Acacia pennata* barks were evaluated for their total phenolics, flavonoid contents and antioxidative activities. The acetone extracts of bark exhibited wide range of total phenolic and flavonoid contents and showed significantly higher DPPH radical scavenging activity than the standards BHA (IC_{50} 33.9µg/ml) and α-tocopherol ((IC_{50} 75.5µg/ml). They also contained relatively higher level of FRAP, ABTS radical cation and OH scavenging activity, metal chelation, phosphomolybdenum reduction and peroxidation inhibition (Sowndhararajan *et al.*, 2010).

Acacia pennineruis Sieber ex DC. Family: Mimosaceae

Mothana *et al.* (2009) evaluated the antioxidant activity of 26 Yemeni medicinal plants using DPPH assay. Methanolic extracts of *Acacia penninervis* showed good antioxidant potential at low concentrations (more than 80 per cent at 50g/ml).

Acacia senegal (L.) Willd. Family: Mimosaceae

Marwah *et al.* (2007) investigated the antioxidant activity of wound-healing herbs using *in vitro* DPPH and phosphomolybdenum assay methods. Out of the 19 plants screened, the aqueous ethanol extract of *Acacia senegal* showed the best DPPH scavenging activities with per cent IP value 89.2 per cent, after 15 min of incubation at an effective test concentration of 50µg/ml. The Total Antioxidant Capacity as gallic acid equivalent of 472mg/g of ethanol extract was obtained in the phosphomolybdenum assay. Total phenolic content was 75.5 mg GAE/g of ethanol extract.

Acalypha diversifolia Jacq. Family: Euphorbiaceae

Mosquera *et al.* (2007, 2009) investigated the antioxidant activity of the crude n-hexane, dichloromethane and methanol extracts of *Acalypha diversifolia* collected from Columbia. The strongest antioxidant activity was observed with methanol and dichloromethane extracts. This species had stronger DPPH radical scavenging activity than hydroquinone, the positive control.

Acalypha fruticosa Forsk. Family: Euphorbiaceae

The methanol extract of leaves of *Acalypha fruticosa* was evaluated by Gupta *et al.* (2003) for its antioxidant and anti-inflammatory properties. The antioxidant property of the methanol extract of *Acalypha fruticosa* (MEAF) was evaluated by DPPH radical scavenging assay and *in vitro* lipid peroxidation induced by Fe^{2+} ascorbate system.

Methanolic extracts of *Acalypha fruticosa* showed high free radical scavenging activity (Mothana *et al.*, 2010). Antioxidant potential of methanolic and aqueous extracts of *Acalypha fruticosa* were evaluated by Rajkumar *et al.* (2010b) employing FRAP assay, radical scavenging assays (DPPH and OH), and TBA assay for testing inhibition of lipid peroxidation. Both extracts exhibited promising antioxidant potentials.

Acalypha hispida Burm.f. Family: Euphorbiaceae

The hexane extract of *Acalypha hispida* was screened by Onocha *et al.* (2011) for phytochemical constituents and was found to contain flavonoids, carbohydrates, phenols and alkaloids. Fractions were screened for free radical scavenging activities using DPPH and hydrogen peroxide. Out of the sixteen pooled fractions (S_1-S_{16}) screened, compounds S_{10} (91.8 per cent), S_{11} (93.8 per cent), S_{14} (92.5 per cent) and S_{15}(91.4 per cent) at a concentration of 0.1 mg/ml had significant activities when compared to the known antioxidant ascorbic acid (90.9 per cent). However, in the analysis using hydrogen peroxides S_1 (99.5 per cent), S_9 (99.2 per cent), S_{10} (95.4 per cent), S_{11} (95.8 per cent) and S_{15} (95.6 per cent gave better activity than ascorbic acid (94.8 per cent) while only S_1 and S_9 were effective than BHA (98.9 per cent) and α-Tocopherol (99.1 per cent) at the same concentration).

Acalypha indica L. Family: Euphorbiaceae

Marwah *et al.* (2007) investigated the antioxidant activity of wound-healing herbs using *in vitro* DPPH and phosphomolybdenum assay methods. Out of the 19 plants screened, the aqueous ethanol extract of *Acalypha indica* showed the best DPPH scavenging activities with per cent IP value 81.6 per cent, after 15 min of incubation at an effective test concentration of 50µg/ml. The Total Antioxidant Capacity as gallic acid equivalent of 279mg/g of ethanol extract was obtained in the phosphomolybdenum assay. Total phenolic content was 72.4 mg GAE/g of ethanol extract.

The ethanol and aqueous extracts of root of *Acalypha indica* were screened by Balakrishnan *et al.* (2009) for antioxidant activity using nitric oxide scavenging activity method, which showed significant percentage of inhibition in dose dependent manner.

Acalypha marginata J.J. Sm. Family: Euphorbiaceae

The chloroform and methanolic extracts of 124 Egyptian plant species belonging to 56 families were investigated by Moussa *et al.* (2011) and compared their antioxidant activity by DPPH scavenging assay. Among the 124 plant species tested 18 exhibited extremely high antiradical activity (>80 per cent inhibition). The IC_{50} value of DPPH radical scavenging of *Acalypha marginata* was 24.57µg/ml, while total phenolic and flavonoid contents were 71.36 mg Tannic acid equivalent/g extract and 177.16 mg Rutin equivalent/g extract, respectively. Correlation coefficient between DPPH radical scavenging activity and the total phenolic and flavonoid contents suggest that phenolics and flavonoids in the extracts were partly responsible for the antiradical activities.

Acalypha platyphylla Mull.-Arg. Family: Euphorbiaceae

Mosquera *et al.* (2007) investigated the antioxidant activity of the crude n-hexane, dichloromethane and methanol extracts of *Acalypha platyphylla* collected from Columbia. The strongest antioxidant activity was observed with methanol and dichloromethane extracts. This species had stronger DPPH radical scavenging activity than hydroquinone, the positive control.

Acanthopanax gracilistylus W.W.Smith Family: Araliaceae

Gan *et al.* (2010) evaluated the antioxidant activity of *Acanthopanax gracilistylus* using the FRAP and TEAC assays, and its total phenolic content was measured by the Folin-Ciocalteu method. For FRAP assay the value was 170.19 µmol Fe(II)/g dry weight. For TEAC assay, the value was 125.08µmol Trolox/g dry weight. Total phenolic content was 10.23 mg gallic acid equivalent/g dry weight of plant material.

Acanthopanax senticosus (Rupr. et Maxim.) Harms (Syn.: *Eleutherococcus senticosus* Rupr. et Maxim.) Family: Araliaceae

Lin and Huang (2002) evaluated the antioxidant activity of the crude extract of *Acanthopanax senticosus* and hepatoprotective activities in CCl_4- and acetaminophen-treated rats. Their results suggest that *Acanthopanax senticosus* exerts some antioxidant effects.

The antioxidant activities of *Acanthopanax senticosus* stems were evaluated by Lee *et al.* (2004) in CCl_4-intoxicated research. The n-butanol fraction from the water extract of the stems, when pretreated orally at 200mg/kg/day for 7 consecutive days in rats, was demonstrated to exhibit significant increase in antioxidant enzyme activities such as hepatic cytosolic superoxide dismutase, catalse and glutathione peroxidase. Eleutheroside B, a lignan component, isolated from the n-butanol fraction was found to cause a moderate free radical scavenging effect on DPPH. Lee *et al.* (2004) have suggested that stems of *A. senticosus* possess not only antioxidant but also hepatoprotective activities.

Choi *et al.* (2006) reported that 30kDa glycoprotein, isolated from the stem bark of *Acanthopanax senticosus* may alleviate alcohol-induced hepatotoxicity through increasing ethanol and lipid metabolism, as well as antioxidant defense systems in livers injured by acute and chronic treatment.

Xi *et al.* (2008) has demonstrated that the antidiabetic activity of the extract of root bark of *Acanthopanax senticosus* (EAT) was correlated with its combined antioxidant and antiglycation properties.

Acanthopanax sessiliflorus (Rupr. et Maxim.) Seem. Family: Araliaceae

A compound with proven antioxidant activities, AE-7-II, was isolated and purified by Jun *et al.* (2007) from the ethanol extract of *Acanthopanax sessiliflorus*. Profound inhibitory effects against the generation of superoxide anion radicals and the formation of hydrogen peroxide were observed in differentiated HL-60 cells. The inhibitory effects of the compound were shown to occur in a concentration-dependent manner, with an IC_{50} value of 12.5µg/ml.

Acanthopanax trifoliatum (L.) Voss Family: Araliaceae

Chanwitheesuk *et al.* (2005) investigated the antioxidant activity of *Acanthopanax trifoliatum* using a β-carotene bleaching method. The contents of plant chemicals such as vitamin C, Vitamin E, carotenoids, tannin and total phenolics, were also determined. Methanolic extract showed antioxidant activity. *Acanthopanax trifoliatum* has antioxidant index of 4.37, Vitamin C 15.3 mg per cent, Vitamin E 0.0055mg per cent, Total carotenes 2.54 mg per cent, Total xanthophylls 3.17mg per cent, Tannins 57.3mg per cent and Total phenolics of 275 mg per cent.

Decoction and 75 per cent ethanolic extracts from leaves of *Acanthopanax trifoliatus* were examined by Sithisarn and Jarikem (2010) for antioxidant activity and phenolics content. Both the tested samples showed strong DPPH scavenging activity (EC_{50}<50µg/ml). The leaf decoction extract significantly exhibited the highest *in vitro* antioxidant activity determined by DPPH scavenging assay and the TBARS method. Total phenolic content was 14.21g per cent CAE (decoctionextract) and 10.14 (ethanolic extract), flavonoid content was 0.99g per cent Rutin equivalent (Decoction extract) and 1.75 g per cent Rutin equivalent (Ethanolic extract).

Acanthospermum australe (Loebl.) Kuntze Family: Asteraceae (Compositae)

The antioxidant and prooxidant activities of lyophilized extracts of 19 Argentine plants were studied using the hydroperoxide-initied chemilumiscence assay in liver homogenates. Six extracts were found to have an effective antioxidative activity in the range of 10-100µg dry weight/ml and were further evaluated using TBARS assay. *Acanthospermum australe* showed the higher antioxidant activity with both assay systems (Desmarchelier *et al.*, 1997).

Acanthus ilicifolius L. Family: Acanthaceae

The alcoholic extract of *Acanthus ilicifolius* leaves inhibited the formation of oxygen derived free radicals (ODFR) *in vitro* with IC_{50} of 550µg/ml, 2750µg/ml, 670µg/ml, and 600µg/ml, (Fe2+/ascorbate system), 980µg/ml, (Fe3+/ADP/ascorbate system) for superoxide radical production, hydroxyl radical generation, nitric oxide radical formation and lipid peroxide formation, respectively (Babu *et al.*, 2001).

Acer albopurpurascens Hayata Family: Aceraceae

Lee *et al.* (2006) investigated the antioxidant activity of ethanol extract of leaves of *Acer albopurpurascens* using DPPH, hydroxyl and superoxide radical scavenging and reducing power activities as well as the induction of heme oxygenase-1 (HO-1). The IC_{50} values against DPPH radical was 1.7µg/ml, OH radical 0.32µg, superoxide radical 28.5µg/ml and total phenolics was 28.9 mg of GAE/g.

Acer buerferianum Miq. var. *formosanum* (Hayata) Sasaki Family: Aceraceae

Hou *et al.* (2003) evaluated the antioxidant activity of 70 per cent aqueous acetone extract of *Acer buerferianum* var. *formosanum* by various assays including DPPH, hydroxyl radicals and reducing power assay. The extract exhibited stronger activity against DPPH radicals (IC_{50} = 8.3 µg/ml) and inhibited formation of OH generated in the Fenton reaction system (0.6 µg/ml). The total phenolics were 27.8 mg of GAE/g.

Acer okamotoanum Nakai Family: Aceraceae

Phytochemical analysis of *Acer okamotoanum* leaf and twig led to the isolation of two coumarinolignans, cleomiscosin A and cleomiscosin C. Cleomiscosin C was found to inhibit LDL oxidation mediatedby either catalytic copper ions or free radicals generated with the azo compound AAPH with IC_{50}s of 29.5 and 11.9µM, respectively (Jin *et al.*, 2007). Furthermore, fluorescence analysis clearly indicated that both A and C protect against the oxidative modification of apoB-100 induced by either Cu ions or HOCl (A IC_{50}s of 13.4 and 8.1 µM, respectively). These findings suggest that cleomiscosin A and C could be beneficial in preventing LDL oxidation in atherosclerotic lesions (Jin *et al.*, 2007).

Acer platanoides L. Family: Aceraceae

English Name: Mapple.

The antioxidative activity of phenolic extracts from leaf of *Acer platanoides* was examined by autoxidation of methyl linoleate by Kahkonen *et al.* (1999). Remarkable high antioxidant activity and high total phenolic content were found in mapple. The total phenolics were 31.7 mg/g GAE. The antioxidant activity was 65 per cent at 500 ppm and 97 per cent at 5000 ppm level.

Achillea alexandri-regis Bornm. et Rudski Family: Asteraceae (Compositae)

The antioxidative potential of *Achillea alexandri-regis* has been studied by Kundakovic *et al.* (2005) by measuring the influence of herb extract on hydroxyl and superoxide radicals' quantity in different *in vitro* systems. The ethyl acetate extract exhibited hydroxyl radical scavenging activity in all tested biological systems. Both extracts showed superoxide radical scavenging effect only in hemolyzed blood.

Achillea biebersteinii Afan. Family: Asteraceae (Compositae)

The essential oil and methanol extracts from *Achillea biebersteini* were evaluated by Sokmen *et al.* (2004) for their antioxidant activity using DPPH assay, inhibition of superoxide and hydroxyl radicals and inhibition of the lipid peroxidation assays. Particularly, polar subfraction of the methanol extract showed antioxidant activity.

Antioxidant and DPPH radical scavenging activities, reducing powers and the amount of total phenolic compounds of some medicinal Asteraceae species were studied by Özgen *et al.* (2004). Water extract of *Achillea biebersteinii* showed good antioxidant activity with 62.4 per cent inhibition at 500µg/ml concentration. The DPPH radical scavenging activity was 87.1 per cent at 500µg/ml concentration and reducing power was 17.7 µg/ml Ascorbic acid equivalent. Total phenolic amount was 19.8µg/ml Gallic acid equivalent.

Konyalioglu and Karamanderes (2004) investigated the antioxidant activity of *Achillea bieberstinii*. Total antioxidant capacity was 8.419mM α-tocopherol equivalent/100ml, DPPH radical scavenging capacity of 33.456 per cent, OH radical scavenging ability of 46.478 per cent and H_2O_2 reducing power of 55.820 per cent. Total flavonoid content was 0.232 per cent and Total phenol content was 145.757 mg/l.

Achillea clypeolata Sm. Asteraceae (Compositae)

Konyalioglu and Karamanderes (2004) investigated the antioxidant activity of *Achillea clypeolata*. Total antioxidant capacity was 7.056mM α-tocopherol equivalent/100ml, DPPH radical scavenging capacity of 32.624 per cent, OH radical scavenging ability of 40.879 per cent and H_2O_2 reducing power of 50.150 per cent. Total flavonoid content was 0.131 per cent and Total phenol content was 109.090 mg/l.

Achillea coarctata Sm. Asteraceae (Compositae)

Konyalioglu and Karamanderes (2004) investigated the antioxidant activity of *Achillea coarctata*. Total antioxidant capacity was 4.671mM α-tocopherol equivalent/100ml, DPPH radical scavenging capacity of 23.886 per cent, OH radical scavenging ability of 35.0129 per cent and H_2O_2 reducing power of 40.915 per cent. Total flavonoid content was 0.075 per cent and Total phenol content was 120.605 mg/l.

Achillea collina Becker ex Rchb. Family: Asteraceae (Compositae)

Giorgi *et al.* (2009) investigated the antioxidant capacity and cytoprotective activity against oxidative stress of infusions obtained from the leaves and inflorescences of *Achillea collina* assessed by chemical (free radical scavenging activity by DPPH and Folin Ciocalteau assay) and biological assays (*in vitro* model of cytotoxicity and lipid peroxidation in PC12 cells line). Infusions of leaves had the highest antioxidant properties and cytoprotective activity. The antioxidant activity was significantly correlated with the total phenolic content but not with the cytoprotective profile.

Effects of environmental growth conditions on the antioxidant capacity, total phenolic content and composition of *Achillea collina* were investigated by Giorgi *et al.* (2010). Methanol extracts and infusions obtained from leaves and inflorescences of plants cultivated in the Italian alps at two different altitudes (600 and 1050 msl) were evaluated. Infusions exhibited the highest antioxidant capacity (IC_{50} values from 4.35 to 8.90), total phenolic content (from 31.39 to 49.36 mg GAE/gDW), chlorogenic acid (from 9.21 to 31.27 mg/g DW), 3,5-di-O-caffeoylquinic acid (from 12.28 to 25.13 mg/g DW0 and 4,5-di-O-caffeoylquinic acid (from 7.38 to 12.78 mg/g DW) content.

Achillea crithmifolia Frivald ex Hampe Family: Asteraceae (Compositae)

Konyalioglu and Karamanderes (2004) investigated the antioxidant activity of *Achillea crithmifolia*. Total antioxidant capacity was 6.485mM α-tocopherol equivalent/100ml, DPPH radical scavenging capacity of 26.941 per cent, OH radical scavenging ability of 38.954 per cent and H_2O_2 reducing power of 45.257 per cent. Total flavonoid content was 0.233 per cent and Total phenol content was 158.182 mg/l.

The protective effects of infusions prepared from 15 *Achillea* species was investigated by Konyalioglu and Karamenderes (2005) against H_2O_2-induced oxidative damage in human erythrocytes and leucocytes used in traditional Turkish medicine. The results indicated that all infusions of *Achillea* species were effective on antioxidant enzyme systems of erythrocytes and leucocytes when compared with H_2O group. *Achillea crithmifolia* showed highest activities on CAT.

Achillea falcata L. Family: Asteraceae (Compositae)

Konyalioglu and Karamanderes (2004) investigated the antioxidant activity of *Achillea falcata*. Total antioxidant capacity was 6.742mM α-tocopherol equivalent/100ml, DPPH radical scavenging capacity of 23.739 per cent, OH radical scavenging ability of 39.447 per cent and H_2O_2 reducing power of 46.300 per cent. Total flavonoid content was 0.250 per cent and Total phenol content was 157.787 mg/l.

The protective effects of infusions prepared from 15 *Achillea* species was investigated by Konyalioglu and Karamenderes (2005) against H_2O_2-induced oxidative damage in human erythrocytes and leucocytes used in traditional Turkish medicine. The results indicated that all infusions of *Achillea* species were effective on antioxidant enzyme systems of erythrocytes and leucocytes when compared with H_2O group. *Achillea falcata* was the most effective one on CAT, GPx and SOD enzyme systems of erythrocytes.

Achillea fragrantissima (Forssk.) Sch. Family: Asteraceae (Compositae)

Al-Mustafa and Al-Thunibat (2008) evaluated the antioxidant activity of Jordanian medicinal plant *Achillea fragrantissima* shoot. The level of antioxidant activity was determined by DPPH and ABTS assay in relation to total phenolic contents of the medicinally used part. They concluded that *Achillea fragrantissima* shoot possess low antioxidant capacity (DPPH-TEAC < 20mg/g).

Achillea grandifolia Friv. Family: Asteraceae (Compositae)

Konyalioglu and Karamanderes (2004) investigated the antioxidant activity of *Achillea grandifolia.* Total antioxidant capacity was 8.742mM α-tocopherol equivalent/100ml, DPPH radical scavenging capacity of 34.654 per cent, OH radical scavenging ability of 46.456 per cent and H_2O_2 reducing power of 54.841 per cent. Total flavonoid content was 0.179 per cent and Total phenol content was 119.395s mg/l.

Achillea kotschyi Boiss. ssp. _kotschyi_ Family: Asteraceae (Compositae)

Konyalioglu and Karamanderes (2004) investigated the antioxidant activity of *Achillea kotschyi.* Total antioxidant capacity was 5.599mM α-tocopherol equivalent/100ml, DPPH radical scavenging capacity of 27.381 per cent, OH radical scavenging ability of 37.895 per cent and H_2O_2 reducing power of 44.150 per cent. Total flavonoid content was 0.108 per cent and Total phenol content was 134.242 mg/l.

Achillea ligusticha All. Family: Asteraceae (Compositae)

The antioxidant activity of the methanolic extract and its fractions of *Achillea ligustica* was examined using two different *in vitro* assays, DPPH test and lipid peroxidation of liposomes assay by Confort *et al.* (2005). Methanolic extract showed higher DPPH radical scavenging activity (IC_{50} of 50µg/ml). This activity is probably due to the phenolic fraction with an IC_{50} value of 22µg/ml. A different result was obtained from the methanolic extract on the lipid peroxidation of liposomes (IC_{50} of 416 µg/ml).

Chemical composition and antioxidant activity of the essential oil of *Achillea ligustica* was studied by Tuberosao *et al.* (2005). Antioxidant activity as DPPH radical scavenging activity was expressed in TEAC and ranged between 0.40 and 0.88 mmol/L.

Ethanol extracts of *Achillea ligustica* flowering tops were evaluated for flavonoid contents and antioxidant activity. 6-Hydroxykaempferol-3,6-4′-trimethyl ether, apigenin-6-C-glucoside-8-C-arbinoside, luteolin and apigenin were the most abundant flavonoids. The radical scavenging activity of the extracts was determined by DPPH test and ranged between 4.18 and 12.3 mM. The ability of these extracts to inhibit non-enzymatic lipid peroxidation was studied using the simple *in vitro* system of linoleic acid oxidation: five of the nine extracts exerted a protective effect at the lower amount tested (5 µg). Protection on $CaCo^2$ intestinal cells against TBH-induced toxicity was also investigated: the results showed that two of the extracts tested in this cell system had the ability to protect against oxidative stress induced by TBH starting from concentrations as low as 10µg/ml (Tuberoso *et al.*, 2009).

Achillea millefolium L. Family: Asteraceae (Compositae)

English Name: Yarrow

The antioxidative activity of phenolic extracts from fruits of *Achillea millefolium* was examined by autoxidation of methyl linoleate by Kahkonen *et al.* (1999). Total phenolics were very low, 5.3 mg/g GAE. The antioxidant activity was also very low, 16 per cent at 500 ppm level and 17 per cent at 5000 ppm level.

Burk *et al.* (2010) reported that aqueous extract of *Achillea millefolium* inflorescences suppresses lipopolysaccharide-induced inflammatory responses in RAW 264.7 murine macrophages. Aqueous extract at higher concentrations tested demonstrated good free radical scavenging activity and ability to decrease the levels of intracellular ROS.

An extract of *Achillea millefolium* herb (YE) was investigated for antioxidant activity using chemical and biological assays (Trumbeckaite *et al.*, 2011). An onlie HPLC-DPPH assay showed YE possesses significant antiradical activity which is due to the presence of active components amongst phenolic compounds.

Vitalini *et al.* (2011) assessed *Achillea millefolium* methanolic extract and its isolated components for the free radical scavenging activity against DPPH, total antioxidant capacity based on the reduction of Cu++ to Cu+, and ability to inhibit lipid peroxidation. Methanolic extract, as well as its flavonol glycosides and chlorogenic acids exhibited significant antioxidant properties.

Achillea millefolium L. subsp. *millefolium* Family: Asteraceae (Compositae)

Eucalyptol, camphor, α-terpinol, β-pinene, and borneol were the principal components comprising 60.7 per cent of the essential oil of *Achillea millefolium* subsp. *millefolium* (Candan *et al.*, 2003). The oil strongly reuced the DPPH radical (IC_{50} = 1.56 μg/ml) and exhibited hydroxyl radical scavenging effect in the $Fe^{3+}EDTA-H_2O_2$ deoxyribose system (IC_{50} = 2.7 μg/ml). It also inhibited the nonenzymatic lipid peroxidation of rat liver homogenate (IC_{50} = 13.5 μg/ml). The polar phase of the extract showed antioxidant activity.

The *in vitro* antimictobial and antioxidant activities of the essential oil and methanol extracts of *Achillea millefolium* subsp. *millefolium* were investigated. The oil strongly reduced the DPPH radical (IC(50) = 1.56μg/ml) and exhibited hydroxyl radical scavenging effect in the Fe(3+)-EDTA-H(2))(2) deoxyribose sysetm (IC(50)2.7μg/ml). It also inhibited the nonenzymatic lipid peroxidation of rat liver homogenate (IC(50) = 13.5μg/ml). The polar phase of the extract showed antioxidant activity (Candan *et al.*, 2003).

Achillea millefolium L. subsp. *pannonica* Family: Asteraceae (Compositae)

Konyalioglu and Karamanderes (2004) investigated the antioxidant activity of *Achillea millefolium* ssp. *pannonica*. Total antioxidant capacity was 10.128mM α-tocopherol equivalent/100ml, DPPH radical scavenging capacity of 41.574 per cent, OH radical scavenging ability of 56.471 per cent and H_2O_2 reducing power of 70.300 per cent. Total flavonoid content was 0.329 and Total phenol content was 181.212 mg/l.

The protective effects of infusions prepared from 15 *Achillea* species was investigated by Konyalioglu and Karamenderes (2005) against H_2O_2-induced oxidative damage in human erythrocytes and leucocytes used in traditional Turkish medicine. The results indicated that all infusions of *Achillea* species were effective on antioxidant enzyme systems of erythrocytes and leucocytes when compared with H_2O group. *Achillea millefolium* subsp. *pannonica* showed highest activities on SOD enzyme systems of leucocytes.

Wojcikowski *et al.* (2007) evaluated the antioxidant activity of aerial parts of *Achillea millefolium* var. *pannonica*. Oxygen radical absorbance capacity of the ethyl extract, methanol extract and aqueous extracts were 22.72, 61.60 and 174.92μmol Trolox equivalent per g of dried starting material, respectively.

Achillea multifida (DC.) Boiss. Family: Asteraceae (Compositae)

Konyalioglu and Karamanderes (2004) investigated the antioxidant activity of *Achillea multifida*. Total antioxidant capacity was 6.285mM α-tocopherol equivalent/100ml, DPPH radical scavenging capacity of 32.579 per cent, OH radical scavenging ability of 39.590 per cent and H_2O_2 reducing power of 46.259 per cent. Total flavonoid content was 0.218 and Total phenol content was 151.212 mg/l.

Achillea nobilis L. subsp. _neilrechii_ (Kerner) Formanek Family: Asteraceae (Compositae)

Konyalioglu and Karamanderes (2004) investigated the antioxidant activity of *Achillea nobilis* ssp. *neilrechii*. Total antioxidant capacity was 8.086mM α-tocopherol equivalent/100ml, DPPH radical scavenging capacity of 36.427 per cent, OH radical scavenging ability of 47.789 per cent and H_2O_2 reducing power of 55.758 per cent. Total flavonoid content was 0.136 and Total phenol content was 119.395 mg/l.

Achillea nobilis L. subsp. _nobilis_ Family: Asteraceae (Compositae)

The protective effects of infusions prepared from 15 *Achillea* species was investigated by Konyalioglu and Karamenderes (2005) against H_2O_2-induced oxidative damage in human erythrocytes and leucocytes used in traditional Turkish medicine. The results indicated that all infusions of *Achillea* species were effective on antioxidant enzyme systems of erythrocytes and leucocytes when compared with H_2O group. *Achillea nobilis* subsp. *nobilis* showed highest activities on CAT enzyme systems of leucocytes.

Achillea nobilis L. subsp. _sipylaea_ (Kerner) Formanek Family: Asteraceae (Compositae)

Konyalioglu and Karamanderes (2004) investigated the antioxidant activity of *Achillea nobilis* ssp. *sipylea*. Total antioxidant capacity was 5.485mM α-tocopherol equivalent/100ml, DPPH radical scavenging capacity of 26.691 per cent, OH radical scavenging ability of 36.236 per cent and H_2O_2 reducing power of 45.138 per cent. Total flavonoid content was 0.255 and Total phenol content was 147.877 mg/l

The protective effects of infusions prepared from 15 *Achillea* species was investigated by Konyalioglu and Karamenderes (2005) against H_2O_2-induced oxidative damage in human erythrocytes and leucocytes used in traditional Turkish medicine. The results indicated that all infusions of *Achillea* species were effective on antioxidant enzyme systems of erythrocytes and leucocytes when compared with H_2O group. *Achillea nobilis* subsp. *sipylaea* showed highest activities on LPO enzyme systems of leucocytes.

Achillea phyrigia Boiss. et Bal. Family: Asteraceae (Compositae)

Konyalioglu and Karamanderes (2004) investigated the antioxidant activity of *Achillea phyrigia*. Total antioxidant capacity was 7.242mM α-tocopherol equivalent/

100ml, DPPH radical scavenging capacity of 34.809 per cent, OH radical scavenging ability of 39.709 per cent and H_2O_2 reducing power of 49.350 per cent. Total flavonoid content was 0.2366 and Total phenol content was 110.908 mg/l.

Achillea schischkinii Sosn. Family: Asteraceae (Compositae)

Konyalioglu and Karamanderes (2004) investigated the antioxidant activity of *Achillea schischkinii*. Total antioxidant capacity was 8.514mM α-tocopherol equivalent/100ml, DPPH radical scavenging capacity of 33.586 per cent, OH radical scavenging ability of 45.453 per cent and H_2O_2 reducing power of 55.652 per cent. Total flavonoid content was 0.248 and Total phenol content was 135.757 mg/l.

Achillea setacea Waldst. et Kit. Family: Asteraceae (Compositae)

Konyalioglu and Karamanderes (2004) investigated the antioxidant activity of *Achillea setacea*. Total antioxidant capacity was 6.999mM α-tocopherol equivalent/100ml, DPPH radical scavenging capacity of 27.429 per cent, OH radical scavenging ability of 38.148 per cent and H_2O_2 reducing power of 44.280 per cent. Total flavonoid content was 0.096 and Total phenol content was 121.818 mg/l.

Achillea tenuifolia Lam. Family: Asteraceae (Compositae)

English Name: Yarrow.

Souri *et al.* (2004a) evaluated the antioxidative activity of *Achillea tenuifolia* by linoleic acid peroxidation test using 1,3-diethyl-2-thiobarbituric acid as the reagent. Methanolic extract showed 88.06 per cent inhibition at 40μg concentration with IC_{50} value of 1.88μg which is approximately in the range of α-tocopherol (IC_{50} = 0.60μg).

Souri *et al.* (2008) investigated the antioxidant activity and radical scavenging activity of methanolic extract of flowers. The antioxidant activity of this plant extract against linoleic acid peroxidation expressed as IC_{50} was 39.71ng/ml and radical scavenging activity against DPPH expressed as IC_{50} was 1.89 μg/ml.

Asgarirad *et al.* (2010) investigated the antioxidant activity of different extracts of *Achillea tenuifolia* and total flavonoid and phenol levels in the extracts. Phenol contents were 43.97, 74.16 and 106 mg/g in the ethyl acetate, methanol and methanol-water extracts, respectively. Flavonoid amount obtained in the ethylacetate, methanol and methanol-water extracts were 10.6, 23.1 and 190 mg/g respectively. The percentage of DPPH radical scavenged by the most active extract (methanol-water) was 92 per cent at a concentration of 1mg/ml greater than 95 per cent of BHT at 2 mg/ml. IC_{50} of methanol-water extract and BHT were 0.015 and 0.053 mg/ml respectively. Lipid peroxidation inhibition was observed by the most polar extract (Asgarirad *et al.*, 2010).

Achillea teretifolia Willd. Family: Asteraceae (Compositae)

Konyalioglu and Karamanderes (2004) investigated the antioxidant activity of *Achillea teretifolia*. Total antioxidant capacity was 6.928mM α-tocopherol equivalent/100ml, DPPH radical scavenging capacity of 28.700 per cent, OH radical scavenging ability of 39.036 per cent and H_2O_2 reducing power of 47.100 per cent. Total flavonoid content was 0.151 and Total phenol content was 100.000 mg/l.

Serteser *et al.* (2009) studied the antioxidant activity of 50 per cent aqueous methanol extracts of *Achillea teretifolia* including free radical scavenging, hydrogen peroxide scavenging and metal (Fe^{2+}) chelating activities.The DPPH radical scavenging effects of plant extract was 0.567. Fe^{2+} chelating activity (per cent) of plant extracts was 32.96 and Hydrogen peroxide inhibition activity of plant extracts was 29.61.

The protective effects of infusions prepared from 15 *Achillea* species was investigated by Konyalioglu and Karamenderes (2005) against H_2O_2-induced oxidative damage in human erythrocytes and leucocytes used in traditional Turkish medicine. The results indicated that all infusions of *Achillea* species were effective on antioxidant enzyme systems of erythrocytes and leucocytes when compared with H_2O group. *Achillea teretifolia* showed highest activities on GPx enzyme systems of leucocytes.

***Achillea wilhelmsii* C. Koch Family: Asteraceae (Compositae)**

Souri *et al.* (2004a) evaluated the antioxidative activity of *Achillea wilhelmsii* by linoleic acid peroxidation test using 1,3-diethyl-2-thiobarbituric acid as the reagent. Methanolic extract showed 75.27 per cent inhibition at 40μg concentration with IC_{50} value of 1.41μg which is approximately in the range of α-tocopherol (IC_{50} = 0.60μg).

Antioxidant and DPPH radical scavenging activities, reducing powers and the amount of total phenolic compounds of some medicinal Asteraceae species were studied by Özgen *et al.* (2004). Methanol extract of *Achillea wilhelmsii* showed good antioxidant activity with 66.4 per cent inhibition at 500μg/ml concentration. The DPPH radical scavenging activity was 41.1 per cent at 500μg/ml concentration and reducing power was 13.1 μg/ml Ascorbic acid equivalent. Total phenolic amount was 18.8μg/ml Gallic acid equivalent.

***Achyranthes aspera* L. Family: Amaranthaceae**

Suresh Kumar *et al.* (2008) carried out a study to determine the antioxidant activity of *Achyranthes aspera*. The total alkaloid content was 61.1mg/g while flavonoid content was 58.3 mg/g. The inhibition of lipid peroxidation was low (44 per cent) which was much lower than Butylated hydroxyl toluene (BHT) (93.6 per cent).

Jitendra (2009) reported that *Achyranthes aspera* roots and leaves have been reported to possess DPPH radical scavenging activity with IC_{50} values 241.86μg/ml and 129.91μg/ml, respectively.

Priya *et al.* (2010) screened methanol and aqueous extracts of stem of *Achyranthes aspera* for antioxidant activity by DPPH assay. Methanolic extract exhibited high phenolic content than that of aqueous extract. Antioxidant activity was dose dependent. Patel and Patel (2010) evaluated *in vitro* antioxidant activity of various fractions of roots and seed of *Achyranthes aspera*. The fractions demonstrated different degrees of potency within each assay; however, the observed pattern was not necessarily replicated between assays indicating the importance of the use of more than one screening technique to estimate the antioxidant activity of plant fractions.

Achyranthes bidentata **Blume Family: Amaranthaceae**

Gan *et al.* (2010) evaluated the antioxidant activities of 40 medicinal plants associated with prevention and treatment of cardiovascular and cerebrovascular diseases using FRAP and TEAC assays and their total phenolic contents were measured by the Folin-Ciocalteu methods. The antioxidant activity of *Achyranthes bidenta* was 12.54µmol Fe(II)/g in FRAP assay, 24.79µmol Trolox/g and Total phenolic content was 1.34 mg GAE/g. A high correlation between antioxidant capacity and total phenolic content indicated that phenolic compounds could be the main contributor of the antioxidant activity of these plants.

Achyranthes longifolia **Mak. Family: Amaranthaceae**

Gan *et al.* (2010) evaluated the antioxidant activities of 40 medicinal plants associated with prevention and treatment of cardiovascular and cerebrovascular diseases using FRAP and TEAC assays and their total phenolic contents were measured by the Folin-Ciocalteu methods. The antioxidant activity of *Achyranthes longifolia* was 18.66µmol Fe(II)/g in FRAP assay, 37.05µmol Trolox/g and Total phenolic content was 3.15 mg GAE/g. A high correlation between antioxidant capacity and total phenolic content indicated that phenolic compounds could be the main contributor of the antioxidant activity of these plants.

Achyrocline satureioides **(Lam.) DC. Family: Asteraceae (Compositae)**

The presence of the flavonoid quercetin and its derivatives and of different phenolic acids such as caffeic, chlorogenic and siochlorogenic acids in the aerial parts of *Achyrocline satureioides* has led Desmarchelier *et al.* (1998) to study the antioxidant activity. The aqueous extract reduced hydroperoxide-initiated chemiluminiscence in rat liver homogenates at all concentrations in a dose-dependent manner, with a calculated IC_{50} = 225µg/ml, while the methanolic extract was only effective at higher concentrations (100 and 1000µg/ml). Both aqueous and methanolic extracts were capable of reducing the production of TBARS in rat liver homogenates, with an IC_{50} > 1000µg/ml.

It has been demonstrated that the use of topical antioxidants can contribute to controlling the free radicals excess produced by ultraviolet (UV) irradiation of the skin, being beneficial for processes such as photo aging and eventually cancer. Some plants, particularly because of their polyphenolic constituents, can be beneficial for skin photo damage. Plant extract of *Achyrocline satureioides* (AS) and *Epilobium parviflorum* (EP), potent antioxidant medicinal plants, were mixed with cosmetic base and applied to the back skin of rabbits. Afterwards the skin was exposed to 1 h UV irradiation from a known source. The production of hydroxyl (OH) radical was assessed in the skin after UV by measuring 2,3-dyhydrobenzoic acid (2,3-DHBA) produced by the hydroxylation of sodium salicylate, previously injected intracutaneously in the irradiated areas. The UV provoked a marked increase in 2,3-DHBA that was significantly decreased by AS cosmetic preparation. The EP preparation did not show any effect on 2,3-DHBA production. It is concluded that the cosmetic preparation containing the AS extract was able to scavenge OH production likely to be due to the presence of high concentration of flavonoid aglycones such as quercetin and 3-O-methylquercetin (Desmarchelier *et al.*, 1998).

Arrendondo *et al.* (2004) reported cytoprotection by *Achyrocline saturoides* and some of its main flavonoids against oxidative stress.

Acokanthera oppositifolia Lam. Family: Apocynaceae

Adedapo *et al.* (2008) evaluated the antioxidant activities and phenolic contents of the methanol extracts of the stems of *Acokanthera oppositifolia* using *in vitro* standard methods. The antioxidant activities were determined by ABTS, DPPH and FRAP methods. The results from this study showed that the stem extract possess antioxidant properties. The ABTS activity of the plant extract was comparable to that of BHT.

Aconitum carmichaeli Debx. Family: Ranunculaceae

Cai *et al.* (2004) evaluated the antioxidant activity and phenolic compounds in traditional Chinese medicinal plants associated with anticancer. The improved ABTS.+ method was used to systematically assess the total antioxidant capacity (Trolox equivalent antioxidant capacity, TEAC) of the medicinal extracts. The TEAC values and total phenolic content for methanolic extract of root of *Aconitum carmichaeli* were 67.6 µmol Trolox equivalent/100g dry weight (DW), and 0.50 g of gallic acid equivalent/100g DW respectively. A positive significant linear relationship between antioxidant activity and total phenolic content showed that phenolic compounds were the dominant antioxidant components in the tested herbs. Major types of phenolic compounds in *Aconitum carmichaeli* include phenolic alkaloids (demethylsalsoline).

Aconitum kusnezoffii Reichb. Family: Ranunculaceae

Cai *et al.* (2004) evaluated the antioxidant activity and phenolic compounds in traditional Chinese medicinal plants associated with anticancer. The improved ABTS.+ method was used to systematically assess the total antioxidant capacity (Trolox equivalent antioxidant capacity, TEAC) of the medicinal extracts. The TEAC values and total phenolic content for methanolic extract of root of *Aconitum kusnezoffii* were 149.6 µmol Trolox equivalent/100g dry weight (DW), and 1.29 g of gallic acid equivalent/100g DW respectively. A positive significant linear relationship between antioxidant activity and total phenolic content showed that phenolic compounds were the dominant antioxidant components in the tested herbs.

Aconitum napellus L.ssp. _lusitanicum_ Rory Family: Ranunculaceae

The DPPH radical scavenging activity of two flavonol glycosides obtained from ethanolic extracts of *Aconitum napellus* ssp. *lusitanicum* was studied by Luis *et al.* (2006). The results showed a high DPPH antiradical activity of compound 1 (quercetin 3-O-(6-trans-caffeoyl)-β-glucopyranosyl-(1→2)-β-glucopyranosyl-7-O-α-rhmnopyranoside) when compared with compound 2 (quercetin-3-sophoroside-7-rhamnopyranoside), rutin and ascorbic acid.

Aconitum napellus L. ssp. _neomontanum_ Family: Ranunculaceae

Bioassay-guided fractionation by DPPH radical scavenging test of polar extracts of some Italian *Aconitum* species including *Aconitum napellus* subsp. *neolontanum* by Braca *et al.* (2003) led to the isolation of 13 flavonol glycosides. Their antioxidant activity was determined by measuring the free radical scavenging activity by DPPH test and the coupled oxidation of β-carotene and linoleic acid assay.

***Aconitum napellus* L. ssp. *tauricum* Family: Ranunculaceae**

Bioassay-guided fractionation by DPPH radical scavenging test of polar extracts of some Italian *Aconitum* species including *Aconitum napellus* subsp. *tauricum* by Braca *et al.* (2003) led to the isolation of 13 flavonol glycosides. Their antioxidant activity was determined by measuring the free radical scavenging activity by DPPH test and the coupled oxidation of β-carotene and linoleic acid assay.

***Aconitum paniculatum* Lam. Family: Ranunculaceae**

Bioassay-guided fractionation by DPPH radical scavenging test of polar extracts of some Italian *Aconitum* species including *Aconitum paniculatum* by Braca *et al.* (2003) led to the isolation of 13 flavonol glycosides. Their antioxidant activity was determined by measuring the free radical scavenging activity by DPPH test and the coupled oxidation of β-carotene and linoleic acid assay.

***Aconitum vulparia* Reichb. ex Sprengel (Syn: *Aconitum lycoctyonum* L. subsp. *vulparia*) Family: Ranunculaceae**

Bioassay-guided fractionation by DPPH radical scavenging test of polar extracts of some Italian *Aconitum* species including *Aconitum vulparia* by Braca *et al.* (2003) led to the isolation of 13 flavonol glycosides. Their antioxidant activity was determined by measuring the free radical scavenging activity by DPPH test and the coupled oxidation of β-carotene and linoleic acid assay.

***Acorus calamus* L. Family: Araceae**

The EtOAc extract obtained from *Acorus calamus* was tested in the DPPH free radical assay. High antioxidant activity was obtained from the extract of rhizome (Acuna *et al.*, 2002).

Mahaboob *et al.* (2005) reported acute oral effects of *Acorus calamus* on lipid peroxidaton and glutathione levels in different organs of rat. The results showed that the acute oral treatment of *Acorus calamus* might induce oxidative stress in these organs due to lipid peroxidation at the high dose level.

In order to find their potential sources, the crude extracts of different parts of some selected food and medicinal plants, including *Acorus calamus*, were studied by Bajpai *et al.* (2005) for total phenolic contents (TPC) and antioxidant activity (AOA).TPC in rhizomes and leaves was 11.2 mg/g and 26.2 mg/g (GAE) respectively while antioxidant activity was 85.8 and 74.5 per cent respectively.

Methanolic extracts of *Acorus calamus* rhizome was evaluated for their antioxidant activity by ferric thiocyanate assay and campared with TBA method. The extract showed good antioxidant activity. 20.88 per cent decrease of DPPH standard solution was recorded for the extract at a concentration of 100μg/ml (Zahin *et al.*, 2009).

Rajkumar *et al.* (2010c) investigated the antioxidant potential of methanolic and aqueous extracts of *Acorus calamus* rhizome. Reducing power of the extracts was studied by estimating their efficacy to reduce Fe+3 to Fe+2 by FRAP assay. The abilities of the extracts to scavenge free radicals (DPPH and OH) were also studied. The extracts were tested for potential inhibitory activity against lipid peroxidation with TBA assay. Both extracts were found to possess comprehensive potential for Fe+3 reduction, free radical scavenging and lipid peroxidation inhibition. Significant positive correlation was obtained between total phenolic content and the radical scavenging activities as well as lipid peroxidation inhibition potential.

Narayanaswamy and Balakrishnan (2011) evaluated the antioxidant activity of 13 medicinal plants including leaf of *Acorus calamus*. The aqueous extract showed 80 per cent inhibition of DPPH radical while ethanolic extract showed 40 per cent inhibition of DPPH radical.

Acronychia pedunculata (L.) Miq. Family: Rutaceae

Chemical investigation on the stem and root bark of *Acronychia pedunculata* has resulted in the isolation of five new acetophenones. Among these compounds acrovestone showed marginal antioxidant and antityrosinase activities (Su *et al.*, 2003).

Acroptilon repens (L.) DC. subsp. *repens* Family: Asteraceae (Compositae)

Souri *et al.* (2004) evaluated tha antioxidative activity of *Acroptilon repens* subsp. *repens* by linoleic acid peroxidation test using 1,3-diethyl-2-thiobarbituric acid as the reagent. Methanolic extract showed 85.88 per cent inhibition at 40μg concentration with IC_{50} value of 4.68μg.

Actaea racemosa (Nutt.) L. (Syn: *Cimicifuga racemosa*) Family: Ranunculaceae

English Name: Black cohosh.

Methanol extracts of two plant materials (one collected 85 years back and another collected recently) were tested for their antioxidant activity and both extracts showed similar antioxidant activity (Jiang *et al.*, 2005).

Actinidia chinensis Planch. (Syn.: *Actinidia deliciosa* C.F.Liang et A.R.Ferguson) Family: Actinidiaceae

English Name: Kiwi fruit

The total antioxidant activity of kiwifruit was measured by Wang *et al.* (1996) using automated oxygen radical absorbance capacity (ORAC) assay. On the basis of the wet weight of the fruits (edible portion), kiwifruit had the highest ORAC activity (6.02micromoles of Trolox equivalents per gram). On the basis of dry weight of the fruits kiwifruit had the highest ORAC activity (36.5μmoles/gTrolox equivalent). The contribution of the fruit pulp fraction (extracted with acetone) to the total ORAC activity of a fruit was usually less than 10 per cent.

Cai *et al.* (2004) evaluated the antioxidant activity and phenolic compounds in traditional Chinese medicinal plants and vegetables associated with anticancer. The improved ABTS.+ method was used to systematically assess the total antioxidant capacity (TEAC) of the medicinal extracts. The TEAC values and total phenolic content for methanolic extract of fruit of *Actinidia chinensis* were 48.9 μmol Trolox equivalent/ 100g dry weight (DW), and 0.22 g of gallic acid equivalent/100g DW respectively.

Jung *et al.* (2005) investigated the cardiovascular protective properties of kiwifruit extracts *in vitro*. Aqueous and 70 per cent ethanol extracts from kiwifruit at 50mg/ml showed DPPH radical scavenging activity (72.31 per cent and 70.75 per cent respectively). Total antioxidant activity in linoleic acid emulsion was 85-88 per cent at 10mg/ml and 96-98 per cent at 50mg/ml of kiwifruit extract.

The ethanol-water extract of *A.deliciosa* root (EEAD) was fractionated into n-hexane (EEAD-He), ethyl acetate(EEAD-Ea), n-butanol (EEAD-Bu) and aqueous (EEAD-Aq) fractions according to their different polarity and solubility (Bai *et al.*, 2007). Among the four extracts, it was found that EEAD-Bu was enriched with oleanolic acid (OLA). The antioxidant and hepatoprotective activities of various EEAD fractions and OLA were carefully investigated by the methods of ferric thiocyanate (FTC) and TBA, as well as the model of CCl_4-induced liver toxicity in rats. The results showed that the EEAD-Bu had higher *in vitro* antioxidant and *in vivo* hepatoprotective activities than those of the other types of extracts. Lipid peroxidation (MDA) decreased 42 per cent and glutathione (GSH) increased 114 per cent in the rats liver homogenate, as compared with those of control.

Mohammad (2011) examined the antioxidant activity of kiwifruits employing different *in vitro* assay systems. The extract showed moderate antioxidant activity in some models. In DPPH radical-scavenging model, extract showed potent activity (IC_{50} = 204.0µg/ml). The extract showed good Fe^{2+} chelating activity, IC_{50} was 174.5µg/ml. EDTA showed very strong activity (IC_{50} = 4.2µg/ml). The extract showed weak reducing power between 16 and 256µg/ml but its activity was dose dependent.

Hunter *et al.* (2011) reviewed the antioxidant capacity of kiwifruit. Kiwifruit has high vitamin C and vitamin E content.

Adansonia digitata L. Family: Bombaceae

Lamien-Meda *et al.* (2008) analyzed the phenolic and flavonoid contents of wild edible fruits of 14 species of Burkina Faso and their antioxidant activities using DPPH, FRAP and ABTS methods. The data obtained show that the total phenolic and total flavonoid levels were significantly higher in the acetone than the methanol extracts. *Adansonia digitata* had higher phenolic and flavonoid content. *Adansonia digitata* fruit also showed higher antioxidant activity. There was a strong correlation between total phenolic and flavonoid levels and antioxidant activity.

Oloyede *et al.* (2010) evaluated the antioxidant activity of the hexane, ethyl acetate, butanol and aqueous extracts of the aerial parts of *Adansonia digitata*. Antioxidant activity determined by ferric thiocyanate method showed that the species provided atleast one fraction with highly promising antioxidant activity. Hexane, ethyl acetate, butanol and water extract at various concentrations (50, 100, 200 and 500 µg/ml) showed antioxidant activities in a concentration dependent manner, the butanol extract however at the concentration of 500µg/ml showed 70-78 per cent antioxidant activity which is very close to that of 500µg/ml of α-tocopherol (82 per cent), the reference compound.

The phenolic content and antioxidant capacities of *Parinari curatelifolia*, *Strychnos spinosa* and *Adansonia digitata* were determined and compared to orange juice and baobab nectar, a commercial beverage (Nhukarume *et al.*, 2010). Methanolic extracts were investigated for their ability to scavenge free radicals by the 1, 1-diphenyl-2 picrylhydrazyl radical and superoxide radical scavenging assays whilst the β-Carotene Linoleic Acid Model System and inhibition of phosholipid peroxidation were used as model systems. The reducing power assay was used to determine the reducing potential of the extracts. Results showed that the beverages in this investigation were capable of acting as antioxidant sources as they displayed significant radical scavenging properties. *Adansonia digitata* had the highest and comparable antioxidant activities to *Citrus sinensis* (orange). The total phenolic, ascorbic acid, proantocyanidin and flavonoid contents ranged between 12 and 58 mg GAE/100 ml, 0.00 to 51.26 mg/100 ml, 0.35–1.071 per cent and 18.3–124 mg/100 ml, respectively. There was a positive correlation between antioxidant activities and phenolic compounds content but there was no clear relationship between proanthocyanidin content and antioxidant activity.

Adenia gummifera Harv. Family: Passifloraceae

Adedapo *et al.* (2008) evaluated the antioxidant activities and phenolic contents of the methanol extracts of the stems of *Adenia gummifera* using *in vitro* standard methods. The antioxidant activities were determined by ABTS, DPPH and FRAP methods. The results from this study showed that the stem extract possess antioxidant properties. The ABTS activity of the plant extract was comparable to that of BHT.

Adenia lobata (Jacq.) Engl. Family: Passifloraceae

Mamyrbekova-Bekro *et al.* (2008) evaluated the phytocompounds of the extract and assessed their antioxidant potential of *Adenia lobata* leaves and stem. The extract showed the presence of polyphenols, flavonoids, tannins and coumarins. The extract showd DPPH scavenging activity. This activity is probably due to the presence of the phenolic compounds, flavonoids and tannins.

In vitro antioxidant activities of total flavonoids extracts from leaves and stems of *Adenia lobata* (Jacq.) Engl. were evaluated by Marcel *et al.* (2011). They noticed a relatively high antioxidant potential with ethyl acetate extracts and butanolic extracts, their effective concentration actions were lower.

Adhatoda vasica Nees Family: Acanthaceae

Study was carried out by Jayakumar *et al.* (2009) to evaluate the lipid peroxidation and antioxidant nature of aqueous extract of *Adhatoda vasica* on D-galactosamine induced hepatic damage in wistar albino rats. A significant increase in LPO levels were observed while the levels of enzymic and non-enzymic antioxidant were decreased in D-GalN induced rats.

The aqueous leaf extracts of *Adhatoda vasica* have been analysed by Padmaja *et al.* (2011) for their free radical-scavenging activity, hydroxyl radical-scavenging activity in Fe^{3+}/ascorbate/EDTA/H_2O_2 system, inhibition of lipid peroxidation induced by $FeSO_4$ in egg yolk, metal chelating activity. The results indicated that *Adhatoda vasica* showed significant antioxidant activity *in vitro*.

Roja *et al.* (2011) evaluated the radical-scavenging activities of various crude extracts of *Adhatoda vasica*. Maximum free radical-scavenging activity of DPPH radicals was observed in the water extracts.

Aegiceras corniculatum (L.) Blanco Family: Myrsinaceae

Roome *et al.* (2008) evaluated the antioxidant, free radical scavenging, anti-inflammatory and hepatoprotective actions of *Aegiceras corniculatum* stem extracts. The n-Hexane, ethyl acetate and methanol extracts scavenged superoxide anions and hydroxyl radicals in the nitro blue tetrazolium reduction and deoxyribose degradation assays, respectively. All the extracts inhibited the process of lipid peroxidation at its initiation step. Additionally, in rat liver microsomes n-Hexane and ethyl acetate extracts also caused termination of radical chain reaction supporting their scavenging action towards lipid peroxy radicals.

Vadlapudi and Naidu (2009) evaluated the antioxidant potential of *Aegiceras corniculatum*, a mangrove plant. The SOD was 0.67U/mg, Catalase was 0.76U/mg, Ascorbic acid was 150mg/100g, DPPH radical inhibition was 61 per cent and FRAP units were 186.

Aegiphyla sellowiana Cham. Family: Verbenaceae

Santos *et al.* (2009) evaluated the antioxidant activity of extracts from nine plant species belonging to the Brazilian flora acting on the free radicals DPPH and TEMPOL. Results showed that the extracts of *Aegiphyla sellowiana* display moderate antioxidant activities with DPPH scavenging IC_{50} value of 0.269 mg/ml while Hydroxyl scavenging IC_{50} value was 3.299 mg/ml.

Aegle marmelos (L.) Correa Family: Rutaceae

Sabu and Kuttan (2003) investigated the antioxidant activity of Indian herbal drugs, including *Aegle marmelos*, in rats with alloxan-induced diabetes. The extract of *Aegle marmelos* showed highest activity as an antidiabetic and antioxidant.

Kamalakkannan *et al.* (2003) evaluated the antilipid peroxidative activity of an aqueous extract of *Aegle marmelos* fruits in streptozotocin diabetic rats in heart and pancreas. Oral administration of the extract for 30 days (125 and 250 mg/kg body weight twice daily) produced a significant decrease in the elevated levels of peroxidation products, *viz.*, TBA reactive substances and hydroperoxidees in the tissue of diabetic rats. The depressed activities of SOD, CAT and GPx and lowered glutathione content in the heart and pancreas of diabetic rats was found to increase on treatment with the extract.

Upadhya *et al.* (2004) reported the hypoglycemic and antioxidant activity of *Aegle marmelos* in alloxan induced diabetic rats. Antidiabetic activity of *Aegle marmelos* and its relationship with its antioxidant properties was examined by Sabu and Kuttan (2004). The results indicate that *Aegle marmelos* extract effectively reduced the oxidative stress induced by alloxan and produced a reduction in blood sugar.

The different methanolic extracts of *Aegle marmelos* plant parts like leaves, bark, fruit rind, fruit pulp, floral parts were prepared and screened phytochemically by standard tests. All parts showed the presence of carbohydrates, amino acids, proteins, anthocyanins, glycosides etc. These extracts were evaluated for antioxidant activity by DPPH method. All the parts exhibited significant activity. Among all the extracts, fruit pulp showed the most significant activity (Sharma and Prasad 2008a).

Kruawan and Kangsadalampai (2006) investigated the antioxidant activity and phenolic compound contents of water extract of *Aegle marmelos*. The extract exhibited strong scavenging activity against DPPH radicals (94.07 per cent) and their FRAP value was 1444.43 μmol/g. The content of phenolic compound in the extracts was determined using the Folin-Ciocalteu reagent and calculated as gallic acid equivalent (GAE). The herbal extract gave a high phenolic compound content (GAE 267.22 mg/g).

Dheeba *et al.* (2010) evaluated the antioxidant activity of ethanol, ethyl acetate and aqueous extracts from bark of *Aegle marmelos*. The results showed the maximum DPPH scavenging activity of ethyl acetate extract of *Aegle marmelos* appeared to be as potent as standard vitamin C with a maximum inhibition of 80.16 per cent at 250μg/ml which is comparable to 77.92 per cent for vitamin C at the same concentration, followed by aqueous (70.30 per cent) with a maximum inhibition than vitmin C. Ethanol extract showed least potent effect of (62,03 per cent) when compared to that of standard.

Aerva lanata (L.) Schult. Family: Amaranthaceae

Aerva lanata is endowed with flavonoids, alkaloids, steroids, polysaccharides, tannins, saponins etc. (Afaq *et al.*, 1991; Chandra and Sastry, 1990; Zapesochnaya *et al.*, 1992). The *in vitro* antioxidant studies of the ethanol extract of *A. lanata* exhibited 1,1-diphenyl-2-picrylhydrazyl (DPPH) scavenging properties (Shirwaikar *et al.*, 2004). Flavonoids, which are well-known antioxidants and free radical scavengers, such as kaempferol 3-rhamnoside and kaempferol 3-rhamnogalactoside have been reported to be present in *A. lanata* (Afaq *et al.*, 1991). The significant hepatoprotective, nephroprotective and antioxidant effect of *A. lanata* may be due to presence of alkaloids and its inhibitory effect on microsomal Ctyt 450 enzyme or on lipid peroxidation (Nevin and Vijayammal, 2005).

In vitro antioxidant activity of aerial part of *Aerva lanata* has been studied by Saraswathy *et al.* (2009). The free radical scavenging potential of the alcohol extract of the plant was also evaluated by ferric thiocyanate assay and thiobarbituric acid method. The ethanolic extract of the aerial part was found to exhibit good antioxidant property.

Aerva tomentosa Forsk. Family: Amaranthaceae

Singh *et al.* (2009) evaluated the antioxidant activity of dichloromethane and methanolic extracts of 12 arid zone medicinal plants. *Aerva tomentosa* showed appreciable antioxidant activity and per cent of inhibition of DPPH where RC_{50} (μg/ml) was 19.5 (Dichloromethane extract) and 7 (MeOH extract). per cent of inhibition of DPPH was highest (96.50 per cent) in methanol extract of *Aerva tomentosa*. Key phytochemicals include α-amyrin and chrysin.

Sethi and Sharma (2011) evaluated the antioxidant potential of the different extracts of the whole plant of *Aerva tomentosa* by using DPPH scavenging assay and total phenolic contents was analyzed using Folin-Ciocalteu reagent method. The methanol extract showed significant activity in antioxidant assay compared to other successive extracts such as petether, dichloromethane, ethyl acetate and aqueous. In DPPH scavenging assay the RC_{50} value of the methanol extract was found to be 6.0μg/ml. The highest amount of total phenolic contents was found in methanol soluble fraction (34.5 mg GAE/g extract).

Aesculus indica L. Family: Sapindaceae

Antioxidant activity of chloroform extract of *Aesculus indica* leaves was investigated by Chakraborthy (2009) for its free radical scavenging activity by determining the nitric oxide and superoxide radical found were 27.62 per cent and 48.65 per cent respectively at 250μg/ml concentration.

Aframomum danielli K. Schum. Family: Zingiberaceae

Odukoya *et al.* (2005) evaluated the antioxidant activity of Nigerian dietary spices including *Aframomum danielli*. The antioxidant activity (expressed as per cent inhibition of oxidation) was 46.20 per cent, reducing power was 3.26 per cent and Total phenolic content was 93.78 mg/100g Tannic acid Equivalent. Antioxidant activity correlated significantly and positively with total phenolics while there was no linear correlation between total antioxidant activity and reducing power neither between reducing power and total phenolic content.

Aframomum giganteum K.Schum. Family: Zingiberaceae

The major compounds in the leaf oil of *Aframomum giganteum* were found to be cis-pinocamphone (54 per cent) and pinocarvone (101 per cent), while the essential oil from the rhizomes is dominated by sesquiterpenes, namely β-caryophyllene (10.5 per cent) and their oxygenated derivatives (40 per cent). The antioxidant and antiradical activity of both samples were evaluated by Agnaniet *et al.* (2004).

Aganosma maritima G. Don Family: Apocynceae

Patnibul *et al.* (2008) screened 18 vegetables for antioxidant activity using silica gel Thin Layer chromatography followed by spraying with DPPH. *Aganosma maritima* exhibited strong antioxidant activity.

Agastache rugosa (Fisc. et Mey.) O.Kuntze Family: Lamiaceae (Labiatae)

Oh *et al.* (2006) reported that protein kinase G-dependent heme oxygenase-1 induction by *Agastache rugosa* leaf extract protects RAW264.7 cells from hydrogen peroxide-injury.

Agathosma betulina (Berg.) Pillans Family: Rutaceae

Wojcikowski *et al.* (2007) evaluated the antioxidant activity of aerial parts of *Agathosma betulina*. Oxygen radical absorbance capacity of the ethyl extract, methanol extract and aqueous extracts were 182.54, 127.02 and 78.63μmol Trolox equivalent per g of dried starting material, respectively.

Agave sisalina Perr. ex Engelm. Family: Agavaceae

Chen *et al.* (2009) studied the antioxidant activity of leaves extract towards xanthine oxidase, lipoxygenase and tyrosinase activities. The inhibitory effect of acetone extract on xanthine oxidase activity was less than 20 per cent, tyrosine activity was about 10 per cent and lipoxygenase activity was about 30 per cent.

***Ageratina popaynenses* (Hieron) K. et R. Family: Asteraceae (Compositae)**

Mosquera *et al.* (2009) evaluated the antioxidant activity of forty-six methanol plant extracts using DPPH radical scavenging assay. *Ageratina popayanenses* showed 30.1 per cent antioxidant activity.

***Ageratum conyzoides* L. Family: Asteraceae (Compositae)**

Shirwaikar *et al.* (2003) reported gastroprotective activity of the ethanol extract of *Ageratum conyzoides* and he was of the opinion that this activity could be mediated by its antioxidant activity.

Ethyl acetate extract of *Ageratum conyzoides* and Kaempferol, isolated from the ethylacetate extract, were evaluated for their antioxidant activity. Kaempferol isolated from the extract rapidly scavenged DPPH at a concentration of 130.07 g/kg bw.

Patil *et al.* (2010) have compared two methods (deoxyribose and benzoic acid) on the hydroxyl radical scavenging ability of *Ageratum conyzoides*, mainly constituted by the phenolic chromenes precocene I and precocene II, and they obtained different responses according to the method used. In the benzoic acid hydroxylation method, the essential oil presented better activity than in the deoxyribose method. The authors explained such results to the relative reactivity of hydroxyl radical towards the substrates deoxyribose and benzoic acid. *Ageratum conyzoides* methanolic extracts that had low capacity for preventing lipid oxidation, in the DPPH assay were revealed to be more effective than the essential oil.

***Agrimonia eupatoria* L. Family: Rosaceae**

English Name: Agrimony

Ivanova *et al.* (2005) also reported high phenolics content and antioxidant properties in *Agrimonia eupatoria.*

Wojcikowski *et al.* (2007) evaluated the antioxidant activity of aerial parts of *Agrimonia eupatoria.* Oxygen radical absorbance capacity of the ethyl extract, methanol extract and aqueous extracts were 9.19, 97.72 and 170.90μmol Trolox equivalent per g of dried starting material, respectively.

The antioxidant activity of Agrimony extracts was assessed by Venskutonis *et al.* (2007) measuring in DPPH radical scavenging and ABTS+ radical decolourisation reaction system. Radical scavenging capacity of agrimony extracts varied in a wide range (9.1-91.6 per cent in DPPH reaction and 6.7-79.5 per cent in ABTS reaction) depending on the polarity of the solvent used to obtain the extract.

***Agrimonia pilosa* Ledeb. Family: Rosaceae**

Cai *et al.* (2004, 2006) evaluated the antioxidant activity and phenolic compounds in traditional Chinese medicinal plants associated with anticancer. The improved ABTS.+ method was used to systematically assess the total antioxidant capacity (Trolox equivalent antioxidant capacity, TEAC) of the medicinal extracts. The TEAC values and total phenolic content for methanolic extract of aerial parts of *Agrimonia pilosa* were 698.6 μmol Trolox equivalent/100g dry weight (DW), and 3.66 g of gallic acid equivalent/100g DW respectively. A positive significant linear relationship between antioxidant activity and total phenolic content showed that phenolic compounds were the dominant antioxidant components in the tested herbs. Major types of phenolic compounds in *Agrimonia pilosa* include n-benzo-triphenols (agrimol A, B), flavonoids (luteolin, apigenin, quercetin, rutin), phenolic acids (gallic acid), tannins, coumarins.

Gan *et al.* (2010) evaluated the antioxidant activities of 40 medicinal plants associated with prevention and treatment of cardiovascular and cerebrovascular diseases using FRAP and TEAC assays and their total phenolic contents were measured by the Folin-Ciocalteu methods. The antioxidant activity of *Agrimonia pilosa* was 255.39μmol Fe(II)/g in FRAP assay, 37.05μmol Trolox/g and Total phenolic content was 14.10 mg GAE/g. A high correlation between antioxidant capacity and total phenolic content indicated that phenolic compounds could be the main contributor of the antioxidant activity of these plants.

Agrimonia procera Wallr. Family: Rosaceae

English Name: Fragrant agrimony

The antioxidant activity of Fragrant agrimony extracts was assessed by Venskutonis *et al.* (2007) measuring in DPPH radical scavenging and ABTS+ radical decolourisation reaction system. Radical scavenging capacity of Fragrant agrimony extracts varied in a wide range (26.8-97.5 per cent in DPPH reaction and 23.0-72.6 per cent in ABTS reaction) depending on the polarity of the solvent used to obtain the extract.

Agrostemma githago L. Family: Caryophyllaceae

Avci *et al.* (2006) evaluated the hypercholesterolaemic and antioxidant activities of *Agrostemma githago*. The aqueous extract increased the serum HDL concentration, whereas the ethanolic extracts decreased serum LDL-C concentration.

Ajuga chamaecistus Ging Family: Lamiaceae (Labiatae)

Firuzi *et al.* (2010) evaluated the antioxidant activity and total phenolic content of 24 Lamiaceae species growing in Iran. *Ajuga chamaecistus* showed lowest antioxidant activity with 8.5μM/g DW FRAP value and IC_{50} = 1350μg/ml of DPPH radical scavenging activity. It also showed lowest phenolic content of 3.5 mg CE/g.

Ajuga reptans L. Family: Lamiaceae (Labiatae)

In *Ajuga reptans* Ono *et al.* (2011) reported four new iridoid glucosides, one known diterpenoid glycoside, one known aliphatic alcohol glycoside, and three known ecdysteroids. The diterpeenoid glycoside exhibited DPPH radical (65.77) and H_2O_2 scavenging activity (98.1). Dal Monte *et al.* (2008) reported antioxidant activity in *Ajuga reptans.*

Akebia trifoliata (Thunb.) Koidz. Family: Lardizabalaceae

Gan *et al.* (2010) evaluated the antioxidant activity of *Akebia trifoliata* using the FRAP and TEAC assays, and its total phenolic content was measured by the Folin-

Ciocalteu method. For FRAP assay the value was 102.18 μmol Fe(II)/g dry weight. For TEAC assay, the value was 29.15μmol Trolox/g dry weight. Total phenolic content was 2.38 mg gallic acid equivalent/g dry weight of plant material.

Albizia amara (Roxb.) Boivin Family: Mimosaceae

Suresh Kumar *et al.* (2008) carried out a study to determine the antioxidant activity of *Albizia amara*. The total alkaloid content was 72.6mg/g while flavonoid content was 63.3 mg/g. The inhibition of lipid peroxidation was 96 per cent which was higher than Butylated hydroxyl toluene (BHT) (93.6 per cent).

Albizia julibrissin Durazz. Family: Mimosaceae

The antioxidant activity of stem bark from *Albizia julibrissin* was evaluated by Jung *et al.* (2003) for its potential to scavenge DPPH radicals, to inhibit the generation of the hydroxyl radical (*OH), total reactive oxygen species (ROS) and to scavenge authentic peroxynitrites (ONOO). The methanol extract exhibited strong antioxidant activity in the tested model systems. Therefore, it was further fractionated using several solvent. The antioxidant activity of the individual fractions were in the order of ethyl acetate>n-butanol>dichloromethane> and water. The ethyl acetate soluble fraction, whch exhibited strong antioxidant activity, was further purified by repeated silicagel, Sephadex LH-20 and RP-18 gel column chromatography. Sulfuretin (1) and 3′,4′,7′-trihydroxyflavone (2) were isolated as the active principles. Compounds 1 and 2 exhibited good activity in all tested model systems. Compound 1 exhibited five times more inhibitory activity on the total ROS than rolox. Compound 2 showed six times stronger DPPH radical scavenging activity than L-ascorbic acid (Jung *et al.*, 2003).

Albizia lebbeck (L.) Benth. Family: Mimosaceeae

Resmi *et al.* (2006) evaluated the antioxidant activity of aqueous extract of *Albizia lebbeck* (ALL) in alloxan diabetic rats by estimating the levels of TBARS, conjugated dienes and glutathione (GSH) in liver and kidneys. The antioxidant enzymes registered a decline in activity in diabetic rats thus revealing the damaging effects of free radicals generated due to alloxan exposure. The activities of these enzymes returned to normalacy in ALL-adminstered rats indicating the antioxidant efficacy of the drug in resisting oxidative insult.

Pathak *et al.* (2010) investigated the free radical scavenging activity of *Albizia lebbeck*. The methanolic extract exhibited significant inflammatory and antioxidant activity.

Alchemilla vulgaris L. Family: Rosaceae

Trouillas *et al.* (2003) evaluated the antioxidant activity of *Alchemilla vulgaris* to visualize inhibition of DPPH, superoxide and hydroxyl radicals. They reported enormous antioxidant activity.

Kiselova *et al.* (2006) studied the antioxidant activity of *Alchemilla vulgaris*. Antioxidant activity was measurd by ABTS cation radical decoloration assay and the total polyphenol content was measured according to the Folin-Ciocalteu method. *Alchemilla vulgaris* extract exhibited higher antioxidant activity than mate. A positive correlation between antioxidant activity and polyphenol content was found, suggesting that the antioxidant capacity of the aqueous plant extract is due to a greater extent to their polyphenols.

Alchemilla xanthochlora **Roth Family: Rosaceae**

Effect of *Alchemilla xanthochlora* water extracts on lipid peroxidation and superoxide anion scavenging activity was investigated by Filipek (1994). Ondrojevic *et al.* (2009) isolated antioxidants from *Alchemilla xanthochlora*. TLC-DPPH screening method was followed for the isolation.

Alchornea coelophylla **Pax. et K.Hoffm. Family: Euphorbiaceae**

Mosquera *et al.* (2007) investigated the antioxidant activity of the crude n-hexane, dichloromethane and methanol extracts of *Alchornea coelophylla* collected from Columbia. The strongest antioxidant activity was observed with methanol and dichloromethane extracts. This species had stronger DPPH radical scavenging activity than hydroquinone, the positive control.

Alchornea cordifolia **Schum. et Thonn. Family: Euphorbiaceae**

Kouakou-Siransy *et al.* (2010) evaluated the antioxidant properties of *Alchornea cordifolia* leaves through a scavenger effect on reactive oxygen species (ROS), such as hydrogen peroxide and hypochlorous acid. The plant showed dose-dependent antioxidant activity. Rates of total phenolic, flavonoid and proanthocyaninidin contents were evaluated. The highest rates were to be found in the most active extracts, indicating that antioxidant activity could be influenced by these phytochemical groups.

Alchornea glandulosa **Poepp. Family: Euphorbiaceae**

Mosquera *et al.* (2009) evaluated the antioxidant activity of forty-six methanol plant extracts using DPPH radical scavenging assay. *Alchornea glandulosa* showed 27.0 per cent antioxidant activity.

Alchornea grandiflora **Muell.-Arg. Family: Euphorbiaceae**

Mosquera *et al.* (2009) evaluated the antioxidant activity of forty-six methanol plant extracts using DPPH radical scavenging assay. *Alchornea grandiflora* showed 37.0 per cent antioxidant activity.

Alchornea laxiflora **(Benth.) Pax et Hoffm. Family: Euphorbiaceae**

Oloyede *et al.* (2010) evaluated the antioxidant activity of the hexane, ethyl acetate, butanol and aqueous extracts of the aerial parts of *Alchornea laxiflora*. Antioxidant activity determined by ferric thiocyanate method showed that the species provided atleast one fraction with highly promising antioxidant activity. Hexane, ethyl acetate, butanol and water extract at various concentrations (50, 100, 200 and 500 µg/ml) showed antioxidant activities in a concentration dependent manner, the butanol extract however at the concentration of 500µg/ml showed 70-78 per cent antioxidant activity which is very close to that of 500µg/ml of α-tocopherol (82 per cent), the reference compound.

***Alhagi maurorum* Medik. Family: Fabaceae (Leguminosae)**

Flvanoidisorhamnetin-3-O-[-alpha-1-rhamnopyranosyl-(1→3)]-beta-D-glucopyranoside and 3′-O-methylorobol, isolated from *Alhagi maurorum* exhibited moderate antioxidant activity in the DPPH radical scavenging assay (Ahmad *et al.*, 2010).

***Alisma orientale* (Sam.) Juz. Family: Alismataceae**

Gan *et al.* (2010) evaluated the antioxidant activity of *Alisma orientale* using the FRAP and TEAC assays, and its total phenolic content was measured by the Folin-Ciocalteu method. For FRAP assay the value was 5.54 μmol Fe(II)/g dry weight. For TEAC assay, the value was 25.69μmol Trolox/g dry weight. Total phenolic content was 3.90 mg gallic acid equivalent/g dry weight of plant material.

***Alisma plantago-aquatica* L. var. *orientale* Family: Alismataceae**

Kim *et al.* (2007) investigated the antioxidant effect of methanol extract of *Alisma plantago-aquatica* var. *oreientale* and its main component under conditions of radical generation using allophycocyanin and ferric-thiocyanate assay. Alisol B23-acetate as a main component was isolated from the methanol extract of *Alisma plantaago-aquatica* var. *orientale*.

***Alkanna albugam* Family: Boraginaceae**

Antioxidant activity of the polar extract obtained from *Alkanna albugam* root against LDL oxidation was investigated by Kaliora and Andrikopoulos (2005) and compared with the respective activity of *Mentha piperita*, *Rosmarinus officinalis* and *R. clandestina*, belonging to the Lamiaceae family. Of the extracts tested, only those from *Alkanna albugam* root and *Mentha piperita* exhibited statistically significant activity.

***Alkanna tinctoria* Tausch. Family: Boraginaceae**

Assimopoulou and Papageorgiou (2004) investigated activities of alkannin, shikonin and *Alkannia tinctoria* root extracts in oil substrates. They concluded that Alkannin/shikonin and their derivatives, used as active ingredients in pharmaceutical, cosmetic and food preparations, may simultaneously exert antioxidant effects in oil substrates in addition to wound healing, anti-inflammatory and antimicrobial properties. The antioxidant activity of alkannin was low in corn oil, while monomeric shikonin, combined with citric acid, presents a very moderate antioxidant activity in sunflower oil.

Assimopoulou and Papageorgiou (2005) investigated the radical scavenging activity of *Alkanna tinctoria* root extracts and their main constituents hydroxynaphthaquinones. It was shown that both monomeric and oligomeric alkannin and shikonin and also (A(soI)S esters exhibited extremely high radical scavenging activity. The presence of naphthaquinone moiety seems to be essential for that activity, while the side chain possibly plays a minor role. Esterification of A(soI)S on the side chain hydroxyl group does not affect radical scavenging activity. Organic solvents and olive oil (extracted at room temperature) extracts of *Alkannia tinctoria* roots, which contain as active ingredients A(soI)S esters, exhibited very good antiradical activity (Assimopoulou and Papageorgiou, 2005).

Methanolic extract from *Alkanna tinctoria* was screened by Sengul *et al.* (2009) for its antioxidant and antimicrobial properties. Total phenolic content of extract of this plant was also determined. β-carotene bleaching assay and Folin-Ciocalteu reagent were used to determine total antioxidant activity and total phenols of plant extract. Antioxidant activity was 79.80 per cent and the total phenolic content was 11.57 mgGAE/g DW). There was no correlation between total phenolic content and antioxidant activity.

Allanblackia floribunda **Oliv. Family: Clusiaceae (Guttiferae)**

Methanol extracts of fruits and leves of *Allanblackia floribunda* were screened by Ayoola *et al.* (2008) for phytochemicals and free radical scavenging activities. Alkaloids, anthraquinones, tannins, saponins, steroids, terpenoids, flavonoids and cardiac glycosides were found to be present in both the fruits and leaves. Total phenolic, total flavonoid and proanthocyanidin contents were 65, 0.07 and 2.38 mg/g of powdered plant material for fruits and 12, 51.35, 19.5 mg/g of powdered plant material for leaves as gallic acid, rutin and catechin equivalents respectively.

Alliaria petiolata **(M. Bief.) Cavarra et Grande Family: Brassicaceae (Cruciferae)**

Kumarasamy *et al.* (2004) isolated four flavone-6-C-glycosides: isoorientin, swertiajaponin, swertiasin and isoscoparin-2″-β-D-glucopranoside from the methanolic extract of *Alliaria petiolata*. All these compounds showed prominent free radical scavenging activity (IC_{50} values: 1.25x 10^{-2} to 7.69 x 10^{-3} mg/ml) in DPPH assay.

Allium ascalonicum **L. Family: Liliaceae**

Povichit *et al.* (2010) investigated phenolic content and free radical scavenging effect of *Allium ascalonicum* by DPPH and ABTS assays, anti-lipid peroxidation activity by TBARS and for antiglycation activity. The results revealed that the total phenolic content showed good correlation with free radical scavenging by ABTS and anti-lipid peroxidation by TBARS, but showed no correlation with antiglycation. The extract of *Allium ascalonicum* moderate antioxidant effect and also showed a promising antiglycation effect. The IC_{50} were 1.33mg/ml for the DPPH method, TEAC value of 0.01 mg/Trolox/mg sample for the ABTS method, IC_{50} of 13.74 mg/ml for the TBARS method and IC_{50} of 0.90 for the antiglycation method.

Allium atroviolaceum **Boiss. Family: Liliaceae**

Tepe *et al.* (2005) examined the *in vitro* antioxidant activities of the methanol extracts of five *Allium* species including *Allium atroviolaceum* by two complementary tests; DPPH free radical scavenging and β-carotene/linoleic acid assays. In the first case, non-polar subfractions of the extracts did not show any antioxidant potential, while the polar subfractions exhibited marked activity. Among the polar ones, the most active one was *A. atroviolaceum* with and IC_{50} of 79.0 μg/ml. In the β-carotene/linoleic acid assay, the inhibition ratios of the oxidation of linoleic acid by *A. atroviolaceum* was 71.2 per cent while that of the synthetic antioxidant, BHT, was 96 per cent.

The antioxidative properties of bulb, leaf and stalk of leeks, *Allium atroviolaceum* were investigated by Stajner *et al.* (2008). Activities of the antioxidant enzymes superoxide dismutase, catalase, peroxidase, glutathione peroxidase, the quantities of malondialdehyde, superoxide and hydroxyl radicals and reduced glutathione and the content of flavonoids, chlorophylls, carotenoids, vitamin C and soluble proteins were determined. Results indicated that extracts from all plant organs exhibited antioxidant activity.

Allium cepa L. Family: Liliaceae

English Name: Onion.

Garlic and onion are among the oldest of all cultivated plants. Both have been used as medicinal plants for thousands of years and shown to have applications as antimicrobial, antithrombotic, antitumor, hypolipidaemic, antiarthritic and hypoglycemic agents. These effects have been linked to their influence on eicosanoid metabolism that influences immune functions in various ways (Ali *et al.*, 2000).

Helen *et al.* (1999, 2000) assessed the beneficial effects of onion oil as an antioxidant in nicotine administered rats by studying whether the peroxidative damage caused by nicotine can be effectively combated with the onion oil and the effects compared to vitamin E. Lipid peroxidation products and antioxidant defense system have been studied in liver, lungs and heart. Onion oil supplemented to nicotine treated rats showed increased resistance to lipid peroxidation and the effect was near to that of vitamin E fed rats. On onion oil or vitamin E supplementation the concentration of antioxidants were significantly raised in all the tissues studied, however, a significantly increased concentration of glutathione, vitamin E and retinol was noticed in vitamin E + nicotine treated rats.

The antioxidative properties of bulb, leaf and stalk of *Allium cepa* were investigated by Stajner *et al.* (2003). Activities of the antioxidant enzymes superoxide dismutase, catalase, peroxidase, glutathione peroxidase, the quantities of malondialdehyde, superoxide and hydroxyl radicals and reduced glutathione and the content of flavonoids, chlorophylls, carotenoids, vitamin C and soluble proteins were determined. Results indicated that extracts from all plant organs exhibited antioxidant activity.

Agostini *et al.* (2004) carried out studies to determine the antioxidant capacity of the flavonoids in fresh onions and thermally treated: humid heat (boiled and vapour), dry heat (oven) and high frequency (microwaves). The results indicated that the antioxidant capacity of the flavonoids of onions was of 0.165 equivalent μM Trolox. These values diminished after the thermal treatment at a higher degree by dry heat at 0.048 equivalent μM Trolox and 0.146 equivalent μM Trolox in the boiled onion. The flavonoids showed a higher antioxidant activity in fresh state, and the activity diminished under heat (cooking).

Souri *et al.* (2004) tested the antioxidant activity of both red and white onions and both of them showed antioxidant activity. Agostini *et al.* (2004) carried out studies to determine the antioxidant capacity of the flavonoids in fresh onions and thermally treated: humid heat (boiled and vapour), dry heat (oven) and high frequency (microwaves). The results indicated that the antioxidant capacity of the flavonoids of onions was of 0.165 equivalent μM Trolox. These values diminished after the thermal treatment at a higher degree by dry heat at 0.048 equivalent μM Trolox and 0.146 equivalent μM Trolox.in the boiled onion. The flavonoids showed a higher antioxidant activity in fresh state, and the activity diminished under heat (cooking).

The radical scavenging activity, chain breaking activity, H_2O_2-scavenging, reducing capacity and total phenolics of four types of onions (green onion, yellow, red and purple) were investigated. Total phenolics varied from 30 mg (green onion) to 49 mg/100 g fresh weight. Green onion showed the lowest radical scavenging activity. Chain breaking activity of yellow, red and purple onion extracts was 0.19, 0.048 and 0.032 respectively. However, heating treatment (90°C, 3h) caused an increase in this activity. Low ability of green onion extract to scavenge hydrogen peroxide was noted (35 per cent), whereas high ability was noted in other onion extracts and ranged from 60 to 90 per cent. The lowest reducing capacity was noted in green onion extract (18 per cent). Statistically high significant correlations were observed between total phenolics content and reducing power, scavenging of hydrogen peroxide and chain-breaking activity of extracts (Benkeblia, 2005).

Odukoya *et al.* (2005) evaluated the antioxidant activity of Nigerian dietary spices including *Allium cepa*. The antioxidant activity (expressed as per cent inhibition of oxidation) was 44.23 per cent, reducing power was 1.80 per cent and Total phenolic content was 58.30 mg/100g Tannic acid Equivalent. Antioxidant activity correlated significantly and positively with total phenolics while there was no linear correlation between total antioxidant activity and reducing power neither between reducing power and total phenolic content.

The outer dry layers of red and violet varieties showed better inhibition of lipid peroxidation assayed by ammonium thiocyanate than α-tocopherol. The outer layers were also potential inhibitors of nitroblue tetrazolium chloride (NBT) (Prakash *et al.*, 2007). Red onion scales showed strong antioxidant activity in a β-carotene bleaching method (Velioglu *et al.*, 1998).

Antioxidant activity of aroma extracts isolated from *Allium cepa* were evaluated by Jang *et al.* (2008). The aroma extracts isolated from green pepper inhibited the oxidation of hexanal by 96 per cent for 1 month at a level of 500μg/ml; these extracts inhibited malonaldehyde formation from cod liver oil by 62 per cent, at a level of 500μg/ml. Its antioxidant activities were comparable to that of the natural antioxidant α-tocopherol.

Mariutti *et al.* (2008) evaluated the antioxidant activity of onion employing ABTS and DPPH radical assay. DPPH EC_{50} value was 2224g/kg while DPPH-TEAC value was 0.20mM/g. ABTS-TEAC value was 0.89mM/g.

A study was carried out by Raghu *et al.* (2010) to determine the antioxidant activity of aqueous extracts of ten selected common vegetables. For *Allium cepa* the IC_{50} value for DPPH scavenging was 1.39mg/ml while standard was 2.45μg/ml. Reducing power of the extract was moderate.

Allium dictyoprosum Family: Liliaceae

Tepe *et al.* (2005) examined the *in vitro* antioxidant activities of the methanol extracts of five *Allium* species including *Allium dictyoprosum* by two complementary tests; DPPH free radical scavenging and β-carotene/linoleic acid assays. In the first case, non-polar subfractions of the extracts did not show any antioxidant potential, while the polar subfractions exhibited marked activity. In the β-carotene/linoleic acid assay, the inhibition ratios of the oxidation of linoleic acid by *A. dictyoprosum* was 72.3 per cent while that of the synthetic antioxidant, BHT, was 96 per cent.

Allium fistulosum L. Family: Liliaceae

English Name: Welsh onion, Green onion; Spring onion; Japanese: Wakegi.

Cai *et al.* (2004) evaluated the antioxidant activity and phenolic compounds in traditional Chinese medicinal plants and vegetables associated with anticancer. The improved ABTS.+ method was used to systematically assess the total antioxidant capacity (TEAC) of the medicinal extracts. The TEAC values and total phenolic content for methanolic extract of fruit of *Allium fistulosum* were 43.4 µmol Trolox equivalent/ 100g dry weight (DW), and 0.42 g of gallic acid equivalent/100g DW respectively.

Maeda *et al.* (2006) examined the phenolic contents of leaf of *Allium fistulousum* and its radical scavenging activities for DPPH radical. The polyphenol content was 21.7 mg-gallic acid equivalent/100 g) and DPPH scavenging activity was 98.9 µmol-Trolox eq./100 g).

Wong *et al.* (2006) investigated the antioxidant properties of 25 edible tropical plants including *Allium fistulosum* using DPPH and FRAP assays. Total phenolic content was also estimated. *Allium fistulosum* showed moderate antioxidant activity.

Antioxidant activity and flavonoid content of Welsh onion (green-leafy and white-sheath varieties) and the effect of thermal treatment on them were studied by Aoyama and Yamamoto (2007) by comparing with those of onion (*Allium cepa*) (yellow and red varieties). The order of antioxidant activity was red onion>yellow onion = green Welsh onion > white Welsh onion. The order of total flavonoid content was red onion>yellow onion>green Welsh onion>white Welsh onion. Major flavonoid of yellow and red onions was quercetin and that of green Welsh onion was kaempferol. Antioxidant activity of green Welsh onion was increased, but that of other three vegetables was decreased during boiling for more than 15 minutes. Flavonoids in green Welsh onion were less stable than those in the other three vegetables during boiling procedure.

The antioxidative properties of bulb, leaf and stalk of leeks, *Allium fistulosum* were investigated by Stajner *et al.* (2008). Activities of the antioxidant enzymes superoxide dismutase, catalase, peroxidase, glutathione peroxidase, the quantities of malondialdehyde, superoxide and hydroxyl radicals and reduced glutathione and the content of flavonoids, chlorophylls, carotenoids, vitamin C and soluble proteins were determined. Results indicated that extracts from all plant organs exhibited antioxidant activity.

Antioxidant activity of aroma extracts isolated from *Allium fistulosum* were evaluated by Jang *et al.* (2008). The aroma extracts isolated from green pepper inhibited the oxidation of hexanal by 95 per cent for 1 month at a level of 500μg/ml; these extracts inhibited malonaldehyde formation from cod liver oil by 58 per cent, at a level of 500μg/ml. Its antioxidant activities were comparable to that of the natural antioxidant α-tocopherol.

Allium flavum **L. Family: Liliaceae**

The antioxidative properties of bulb, leaf and stalk of leeks, *Allium flavum* were investigated by Stajner *et al.* (2003). Activities of the antioxidant enzymes superoxide dismutase, catalase, peroxidase, glutathione peroxidase, the quantities of malondialdehyde, superoxide and hydroxyl radicals and reduced glutathione and the content of flavonoids, chlorophylls, carotenoids, vitamin C and soluble proteins were determined. Results indicated that extracts from all plant organs exhibited antioxidant activity.

Cultivated and wild *Allium* species were investigated by Štajner and Popovic (2010) in order to compare the antioxidant capacity of their leaves and bulbs. Leaf and bulb antioxidative enzymes and scavenger activities, along with quantities of non-enzymatic antioxidants, malonyl-dialdehyde and OH radicals were determined. Results obtained suggest that leaves possess higher antioxidant and scavenging activities than bulbs in the majority of cultivated and wild *Allium* species examined.

Allium giganteum **L. Family: Liliaceae**

The antioxidative properties of bulb, leaf and stalk were investigated by Štajner *et al.* (2006). Activities of antioxidant enzymes (superoxide dismutase, catalase, peroxidase, glutathione peroxidase), quantities of malonyldialdehyde, superoxide and hydroxyl radicals and reduced glutathione and also the content of total flavonoids were determined. The highest antioxidant activity was observed in the leaves.

Allium hirtifolium **Boiss. Family: Liliaceae**

English Name: Wild garlic.

Souri *et al.* (2008) investigated the antioxidant activity and radical scavenging activity of methanolic extract of corm. The antioxidant activity of this plant extract against linoleic acid peroxidation expressed as IC_{50} was 30.67ng/ml and radical scavenging activity against DPPH expressed as IC_{50} was 187.88 μg/ml.

Allium neopolitanum **Cyr. Family: Liliaceae**

Nencini *et al.* (2007) evaluated the antioxidant activity of bulbs, leaves and flowers of *Allium neopolitanum*. Flowers showed higher antioxidant power while the antioxidant activity of the bulbs was lower. A correlation between the phenolic contents and the antioxidant activity was discovered.

Allium nevesehirense **Family: Liliaceae**

Tepe *et al.* (2005) examined the *in vitro* antioxidant activities of the methanol extracts of five *Allium* species including *Allium nevesehirense* by two complementary tests; DPPH free radical scavenging and β-carotene/linoleic acid assays. In the first case, non-polar subfractions of the extracts did not show any antioxidant potential, while the polar subfractions exhibited marked activity.

Allium nutans L. Family: Liliaceae

Stajner *et al.* (1999) investigated antioxidant activity of leaf, bulb and root. Biochemical parameters were also determined: activities of antioxidant enzymes (superoxide dismutase, catalase, peroxidase, glutathione peroxidase), quantities of malonydialdehyde, superoxide hydroxyl radicals and reduced glutathione and contents of total flavonoids, chlorophylls a,b, carotenoids, vitamin C and soluble proteins. Their results indicated that *Allium nutans* exhits antioxidant ability in all investigated plant organs. The highest ability was observed in the leaves where all investigated antioxidant enzymes were active and quantities of malonyldialdehyde was low. Reduced glutathione, pigments and carotenoids present in the leaves contribute to the high antioxidant activity.

Allium porrum L. Family: Liliaceae

English Name: Leek.

Souri *et al.* (2004) investigated the antioxidant activity of methanol extracts of Leek, used as vegetable in Iranian diet against linoleic acid peroxidation. The extracts showed good antioxidant activity. The antioxidant activity was lower than dl-α-tocopherol but higher than quercetin.

Aslan *et al.* (2010) studied the antidiabetic and antioxidant activities of the ethanol extracts of bulbs of *Allium porrum*. TBARS and GSH livels in liver, kidney and heart tissues were measured by using spectrophotometric methods for antioxidant assay. Extracts induced significant alleviation on only heart tissue TBARS levels (44.6 per cent), and extracts showed an inhibitory effect on kidney tissue TBARS levels (14.8 per cent). None of the extracts restored GSH levels in kidney, liver and heart tissues of diabetic rats.

Allium psekemense B. Fedtsch. Family: Liliaceae

Štajner *et al.* (2002) investigated the antioxidative properties of the bulbs, leaves and stalks of *Allium psekemense*. The activities of antioxidant enzymes (superoxide dismutase, catalase, peroxidase, glutathione peroxidase), quantities of malondialdehyde, superoxide and hydroxyl radicals and reduced glutathione and also the content of total flavonoids, chlorophylls a and b, carotenoids, vitamin C and soluble proteins were determined. The results indicate that extracts from all plant organs exhibited antioxidant activity. The highest antioxidant activity was observed in the leaves. Furthermore, the ESR signal of DMP-OH radical adducts in the presence of the leaf phosphate buffer (ph7) extract was reduced by 54.3 per cent.

Štajner *et al.* (2008) investigated the antioxidative properties of the bulbs, leaves and stalks of *Allium psekemense*. The activities of antioxidant enzymes (superoxide dismutase, catalase, peroxidase, glutathione peroxidase), quantities of malonyldialdehyde, superoxide and hydroxyl radicals and reduced glutathione and also the content of total flavonoids, chlorophylls a and b, carotenoids, vitamin C and soluble proteins were determined. The results indicate that extracts from all plant organs exhibited antioxidant activity. The highest antioxidant activity was observed in the leaves. Furthermore, the ESR signal of DMPO-OH radical adducts in the presence of the leaf phosphate buffer (pH 7) extract was reduced by 54.3 per cent.

Allium roseum L. Family: Liliaceae

Nencini *et al.* (2007) evaluated the antioxidant activity of bulbs, leaves and flowers of *Allium roseum*. Flowers showed higher antioxidant power while the antioxidant activity of the bulbs was lower. A correlation between the phenolic contents and the antioxidant activity was discovered.

Štajner *et al.* (2008) evaluated the antioxidant and free-radical scavenging activities of bulb, stalk and leaf of *Allium roseum*. The study reports the results concerning antioxidant enzyme activities (superoxide dismutase, catalase, guaiacol peroxidase and glutathione peroxidase), quantities of malondialdehyde, ȮH and O(2)(*-) radicals and total antioxidant capacity determined by the FRAP method. Scavenging activities were determined by ESR method. The total antioxidant capacity was the highest in the leaves of *Allium roseum*. The highest ESR (84.61 per cent) scavenger activity was observed in the leaves of *Allium roseum*.

Allium sativum L. Family: Liliaceae

English Name: Garlic.

The antioxidant properties of three garlic preparations and organosulfur compounds in garlic have been determined by Imai *et al.* (1994). Aged garlic extract inhibited the emission of low level chemiluminiscence and the early formation of TBARS in liver microsomal fraction initiated by t-butyl hydroperoxide. However, the water extracts of raw and heat-treated garlic enhanced the emission of low level chemiluminiscence. Among the variety of organosulfur compounds, S-allylcysteine (SAC) and S-allylmercaptocysteine (SAMC), the major organosulfur compounds found in the aged garlic extract, showed radical scavenging activity in both chemiluminiscence and DPPH assays, indicating that these compounds may play an important role in the antioxidative activity of aged garlic extract.

Garlic and its constituents have been demonstrated to have antioxidant and radical scavenging properties. *In vitro* and *in vivo* antioxidant effects of aged garlic extract and organosulphur compounds derived from aged garlic extract were investigarted in mice by Ide *et al.* (1996). Aged garlic extract and its water-soluble organosulfur compounds demonstrated radical and hydrogen peroxide scavenging activities. Further more, aged garlic extract inhibited *in vivo* lipid peroxidation. The superior bio-availability of SAC (S-allyl cysteine) after oral administration has been reported. Thus both aged garlic extract and SAC would be effective for prevention of diseases induced by oxidative damages through their antioxidant and radical scavenging activities.

A liposome model was used to examine the antioxidant activity of several members of Allium family, including garlic bulb, bakeri garlic bulb, Chinese leek, Chinese Chive, Onion bulb, scallion and shallot bulb (Yin and Cheng, 1998). Allicin formation of these plants was analyzed. The influence of heat, pH, and salt upon the antioxidant activity of these foods was studied. The results showed that part of the

Allium family possessed antioxidant capability. Besides allicin, other compounds could be involved in the antioxidant activity performance because the allicin concentration was not strongly correlated to the antioxidant activity observed in these Allium foods. Heat treatment at 65 to 100°C and acid treatment at pH = 2 reduced the antioxidant activity for most foods.

Garlic is another herb with proclaimed *Rasayana* effect in Ayurveda. Garlic and onion (*Allium cepa*) are among the oldest of all cultivated plants. Both have been used as medicinal plants for thousands of years and shown to have applications as antimicrobial, antithrombotic, antitumor, hypolipidaemic, antiarthritic and hypoglycemic agents. These effects have been linked to their influence on eicosanoid metabolism that influences immune functions in various ways (Ali *et al.*, 2000). Nicotine, a major component of tobacco, is partly responsible for the development of atherosclerosis. The antioxidant effects of oils isolated from onion and garlic on nicotine-induced lipid peroxidation in rat tissues were studied. Lipid peroxidation was significantly increased in the tissues of nicotine treated rats. Both the garlic oil and onion oil supplementation to nicotine-treated rats increased resistance to lipid peroxidation. With garlic oil or onion oil supplementation, nicotine-treated rats showed increased activities of antioxidant enzymes and increased concentrations of glutathione (Helen *et al.*, 1999).

Through a series of investigations, garlic was found to exert a diverse range of health promoting effects. It was found to enhance human immune functions, by stimulating peripheral blood mononuclear cells *in vitro*. Pretreatment with garlic led to a protective effect on heart, liver and pancreas against isoproterenol induced damage. Garlic extract and S-allyl cysteine isolated from garlic were noted to protect pulmonary endothelial cells *in vitro* against hydrogen peroxide induced oxidative damage. Enhanced cell viability and an inhibition were also reported. Diallyl sulfide, a sulfur-containing volatile compound in garlic, exerts anticarcinogenic activity in various rodent tumor models (Singh and Sukla, 1998). Allicin from garlic has been found to induced programmed cell death and therefore, arrest of proliferation in cancer cells (Thatte *et al.*, 2000). Such compounds can thus also exercise immunosuppressive effects.

The antioxidative properties of bulb, leaf and stalk of leeks, *Allium sativum* were investigated by Štajner *et al.* (2003). Activities of the antioxidant enzymes superoxide dismutase, catalase, peroxidase, glutathione peroxidase, the quantities of malondialdehyde, superoxide and hydroxyl radicals and reduced glutathione and the content of flavonoids, chlorophylls, carotenoids, vitamin C and soluble proteins were determined. Results indicated that extracts from all plant organs exhibited antioxidant activity.

Ichikawa *et al.* (2003) evaluated DPPH radical scavenging activity of garlic skin extract. Compounds isolated from the skin extract were also evaluated for the antioxidant activity. A significant decrease in malondialdehyde (MDA) levels and a significant increase in antioxidant activity were observed by Chandra *et al.* (2003) when *Allium sativum* was administered to diabetic rats.

Souri *et al.* (2004) investigated the antioxidant activity of methanol extracts of garlic, used as vegetable in Iranian diet against linoleic acid peroxidation. The extracts showed good antioxidant activity.

Suboh *et al.* (2004) investigated the antioxidant activity effects of various concentrations of methanol extracts of *Allium sativum* and six other medicinal plants. They reported that the extract significantly reduced 10mM hydrogen peroxide-induced protein degradation, lipid peroxidation, loss of deformability and increased osmotic fragility caused by H_2O_2 in human erythrocytes.

Cai *et al.* (2004) evaluated the antioxidant activity and phenolic compounds in traditional Chinese medicinal plants and vegetables associated with anticancer. The improved ABTS.+ method was used to systematically assess the total antioxidant capacity (TEAC) of the medicinal extracts. The TEAC values and total phenolic content for methanolic extract of fruit of garlic were 55 μmol Trolox equivalent/100g dry weight (DW), and 0.20 g of gallic acid equivalent/100g DW respectively.

Influence of garlic extract on temporal characteristics of lipid peroxidation products and antioxidants in tumor bearing rats was investigated by Subramanian *et al.* (2005).

Odukoya *et al.* (2005) evaluated the antioxidant activity of Nigerian dietary spices including *Allium sativum*. The antioxidant activity (expressed as per cent inhibition of oxidation) was 60.06 per cent, reducing power was 1.35 per cent and Total phenolic content was 136.48 mg/100g Tannic acid Equivalent. Antioxidant activity correlated significantly and positively with total phenolics while there was no linear correlation between total antioxidant activity and reducing power, neither between reducing power and total phenolic content.

The radical scavenging activity, chain breaking activity, H_2O_2-scavenging, reducing capacity and total phenolics of garlic were investigated. Total phenolics were 49 mg/100 g fresh weight. High ability of garlic extract to scavenge hydrogen peroxide was noted in garlic extract (90 per cent). The highest reducing capacity of 196 per cent was noted with garlic extract. Statistically high significant correlations were observed between total phenolics content and reducing power, scavenging of hydrogen peroxide and chain-breaking activity of extracts (Benkeblia, 2005).

The leaves and bulb protects against free radical formation, reduced blood clotting and cholesterol level. Best when used with natural antioxidant and vitamins. Alanine, α-tocopherol, ascorbic acid, camphene, eugenol, γ-terpinene and histidine are the main antioxidative constituents present in garlic (Mishra, 2007).

Mariutti *et al.* (2008) evaluated the antioxidant activity of garlic employing ABTS and DPPH radical assay. DPPH EC_{50} value was 41g/kg while DPPH-TEAC value was 6.7mM/g. ABTS-TEAC value was 8.5mM/g.

Koseoglu *et al.* (2009) investigated the effects of acute and subacute garlic supplement administration on serum total antioxidant capacity and lipid parameters in healthy volunteers. The results suggested that garlic used as a dietary supplementation may be beneficial in increasing the antioxidant capacity.

Povichit *et al.* (2010) investigated phenolic content and free radical scavenging effect of *Allium sativum* by DPPH and ABTS assays, anti-lipid peroxidation activity by TBARS and for antiglycation activity. The results revealed that the total phenolic content showed good correlation with free radical scavenging by ABTS and anti-lipid peroxidation by TBARS, but showed no correlation with antiglycation. The extract of *Allium sativum* moderate antioxidant effect and also showed a promising antiglycation effect. The IC_{50} were 5.30mg/ml for the DPPH method, TEAC value of 0.06 mg/Trolox/mg sample for the ABTS method, IC_{50} of 9.75 mg/ml for the TBARS method and IC_{50} of 43.92 for the antiglycation method.

Allium scorodoprasum L. subsp. *rotundum* Family: Liliaceae

The antioxidative properties of bulb, leaf and stalk of leeks, *Allium scorodoprasum* were investigated by Štajner *et al.* (2003). Activities of the antioxidant enzymes superoxide dismutase, catalase, peroxidase, glutathione peroxidase, the quantities of malondialdehyde, superoxide and hydroxyl radicals and reduced glutathione and the content of flavonoids, chlorophylls, carotenoids, vitamin C and soluble proteins were determined. Results indicated that extracts from all plant organs exhibited antioxidant activity.

Antioxidant activity of aroma extracts isolated from *Allium scorodoprasum* were evaluated by Jang *et al.* (2008). The aroma extracts isolated from green pepper inhibited the oxidation of hexanal by 98 per cent for 1 month at a level of 500µg/ml; these extracts inhibited malonaldehyde formation from cod liver oil by 58 per cent, at a level of 500µg/ml. Its antioxidant activities were comparable to that of the natural antioxidant α-tocopherol.

Allium shoenoprasum L. Family: Liliaceae

English Name: Chives.

Souri *et al.* (2004) investigated the antioxidant activity of methanol extracts of chives, used as vegetable in Iranian diet against linoleic acid peroxidation. The extracts showed good antioxidant activity. The antioxidant activity of the aerial part was lower than dl-α-tocopherol but higher than quercetin.

Štajner *et al.* (2004) investigated the antioxidant properties of the bulb, leaf and stalk of *Allium shoenoprasum*. Activities of antioxidant enzymes (superoxide dismutase, catalase, peroxidase, glutathione peroxidase), quantities of malonyldialdehyde, superoxide and hydroxyl radical and reduced glutathione and also the content of total flavonoids, chlorophylls a and b, carotenoids, vitamin C and soluble proteins were determined. The results indicate that extracts from all plant organs exhibited antioxidant activity. The highest antioxidant activity was observed in the leaves (Stajner *et al.*, 2004).

Mariutti *et al.* (2008) evaluated the antioxidant activity of chives employing ABTS radical assay. ABTS-TEAC value was 4.97mM/g.

Allium sivasicum Ozhatay et Kollmann Family: Liliaceae

Tepe *et al.* (2005) examined the *in vitro* antioxidant activities of the methanol extracts of five *Allium* species including *Allium sivasicum* by two complementary tests; DPPH free radical scavenging and β-carotene/linoleic acid assays. In the first case, non-polar subfractions of the extracts did not show any antioxidant potential, while the polar subfractions exhibited marked activity.

Allium sphaerocephalum L. Family: Liliaceae

English Name: Leek.

The antioxidative properties of bulb, leaf and stalk of leeks, *Allium sphaerocephalum* were investigated by Štajner *et al.* (2003). Activities of the antioxidant enzymes superoxide dismutase, catalase, peroxidase, glutathione peroxidase, the quantities of malondialdehyde, superoxide and hydroxyl radicals and reduced glutathione and the content of flavonoids, chlorophylls, carotenoids, vitamin C and soluble proteins were determined. Results indicated that extracts from all plant organs exhibited antioxidant activity. The highest antioxidant activity was observed in the leaves. Furthermore, ESR signal of PBN-OH radical adducts in the presence of leaves phosphate buffer (pH 7) extract was reduced for 64.51 per cent.

Allium subhirsutum L. Family: Liliaceae

Nencini *et al.* (2007) evaluated the antioxidant activity of bulbs, leaves and flowers of *Allium subhirsutum*. Flowers showed higher antioxidant power while the antioxidant activity of the bulbs was lower. A correlation between the phenolic contents and the antioxidant activity was discovered.

Štajner *et al.* (2008) evaluated the antioxidant and free-radical scavenging activities of bulb, stalk and leaf of *Allium subhirsutum*. The study reports the results concerning antioxidant enzyme activities (superoxide dismutase, catalase, guaiacol peroxidase and glutathione peroxidase), quantities of malondialdehyde, OH and O(2)(*-) radicals and total antioxidant capacity determined by the FRAP method. Scavenging activities were determined by ESR method.

Allium tuberosum L. Family: Liliaceae

English Name: Chinese chive; Japanese: Nirai.

Maeda *et al.* (2006) examined the phenolic contents of leaf of *Allium tuberosum* and its radical scavenging activities for DPPH radical. The polyphenol content was 52.8 mg-gallic acid equivalent/100 g) and DPPH scavenging activity was 176.9 µmol-Trolox eq./100 g).

Allium ursinum L. Family: Liliaceae

Stajner *et al.* (2008) investigated the antioxidant properties of bulb, leaf and stalk of *Allium ursinum*. Activities of antioxidant enzymes (superoxide dismutase, catalase, peroxidase, glutathione peroxidase), quantities of malonyldialdehyde, superoxide and hydroxyl radicals and reduced glutathione were determined. The extracts from all organs exhibited antioxidant activity, the highest having been observed in the leaves.

Allium vineale L. Family: Liliaceae

The antioxidative properties of bulb, leaf and stalk of leeks, *Allium vineale* were investigated by Štajner *et al.* (2003). Activities of the antioxidant enzymes superoxide dismutase, catalase, peroxidase, glutathione peroxidase, the quantities of malondialdehyde, superoxide and hydroxyl radicals and reduced glutathione and the content of flavonoids, chlorophylls, carotenoids, vitamin C and soluble proteins were determined. Results indicated that extracts from all plant organs exhibited antioxidant activity.

Allophylus rubifolius **(Hochst. ex A.Rich.) Engl. Family: Sapindaceae**

Marwah *et al.* (2007) investigated the antioxidant activity of wound-healing herbs using *in vitro* DPPH and phosphomolybdenum assay methods. Out of the 19 plants screened, the aqueous ethanol extract of *Allophylus rubifolius* showed the best DPPH scavenging activities with per cent IP value 89.8 per cent, after 15 min of incubation at an effective test concentration of 50µg/ml. The Total Antioxidant Capacity as gallic acid equivalent of 448mg/g of ethanol extract was obtained in the phosphomolybdenum assay. Total phenolic content was 146 mg GAE/g of ethanol extract.

Alnus formosana **(Burk.) Makino Family: Betulaceae**

Lee *et al.* (2006) investigated the antioxidant activity of ethanol extract of leaves using DPPH, hydroxyl and superoxide radical scavenging and reducing power activities as well as the induction of heme oxygenase-1 (HO-1). The IC_{50} values against DPPH, radical were 11.3µg/ml, OH radical 0.36µg, superoxide radical 181.7µg/ml and total phenolics was 15.6 mg of GAE/g.

Alnus glutinosa **(L.) Gaertn. Family: Betulaceae**

English name: Black elder

Middleton *et al.* (2005) evaluated the antioxidant activity of *Alnus glutinosa*. Among the extracts, the methanol extract displayed the highest level of antioxidant activity (RC_{50} = 1.27 X 10^{-1} mg/ml).

Alnus incana **L. ssp. *incana* Family: Betulaceae**

Sevic *et al.* (2010) evaluated antioxidant activity of *Alnus incana* ssp. *incana*. All extracts were found to be strong DPPH radical scavengers, exhibiting IC_{50} value of 3.3-18.9µg/ml and also showed activity in inhibition of lipid peroxidation with IC_{50} values ranging from 38.5 to 157.4µg/ml. The most active extract bark also contained great amounts of total phenolics (316.2 mg GAE/g).

Alnus japonica **(Thunb.) Steud. Family: Betulaceae**

Kim *et al.* (2004) investigated the hepatoprotective and antioxidant effects of *Alnus japonica* extracts on acetaminophen-induced hepatotoxicity in rats. The pretreatment with EtOAc and BuOH fraction led to an increase in free radical scavenging activity and a decrease in inhibition of lipid peroxidation, both superoxide dismutase and catalase prevent the hepatotoxicity by AAP in the treatment of *A. japonica* fraction.

Four new diarylheptanoids along with known diarylheptanoids and flavonoids were isolated from the leaves of *Alnus japonica* by Kuroyanagi *et al.* (2005). The isolated compounds including their main constituent, oregonin, were tested for antioxidant activity against superoxide and DPPH radicals. Some of these compounds having two catechol structure showed potent activity. Compounds having one catechol structure showed moderate activity, but a peracetate of oregonin having no catechol structure exhibited no activity. Thus the catechol structure of the diarylheptanoids is indispensable for antioxidant activity (Kuroyanagi *et al.*, 2005).

Two diarylheptanoids, oreganin and hirsutanone were isolated by Lee *et al.* (2005) from methanol extracts of the leaves of *Alnus japonica.* These compounds exhibited significant LDL antioxidant activity, *e.g.*, TBARS assay (IC_{50} = 3.2μM, 1.5μM respectively) and macrophage-mediated LDL oxidation (inhibition of 1.82 per cent and 2.84 per cent 1μM).

Kang *et al.* (2006) reported that cyclic diaryl-heptanoids isolated from *Alnus japonica* inhibit cell-mediated low-density lipoprotein oxidation. Kim *et al.* (2010) reported antioxidative and hepatoprotective diarylheptanoids from the bark of *Alnus japonica.*

Alnus viridis (Chaix.) DC. ssp. _viridis_ Family: Betulaceae

Sevic *et al.* (2010) evaluated antioxidant activity of *Alnus viridis* ssp. *viridis*. All extracts were found to be strong DPPH radical scavengers, exhibiting IC_{50} value of 3.3-18.9μg/ml and also showed activity in inhibition of lipid peroxidation with IC_{50} values ranging from 38.5 to 157.4μg/ml.

Aloe barbadensis Mill. (Syn.: _Aloe vera_ (L.) Burm.f.). Family: Liliaceae

Antioxidant components in *Aloe vera* were examined for lipid peroxidation using rat liver microsomal and mitochondrial enzymes. Among the aloesin derivatives examined, isorabaichromone showed a potent antioxidative activity. The DPPH radical and superoxide anion scavenging activities were determined. As one of the most potent components, isorabaichromone together with feruloylaloesin and p-coumaroylaloesin showed potent DPPH radical and superoxide anion scavenging activities (Yagi *et al.*, 2002).

This plant is extensively used in skin care preparations and other health products. Shamman *et al.* (1998) have studied the effects of *Aloe vera* gel extract supplementation on hepatocarcinogenesis induced by diethylnitrosamine and 2-acetylamonofluorene in male Sprague-Dawley rats. The severity of the carcinogenesis process was determined by measuring the increase in gamma-glutamyl transpeptidase and the placental form of glutathione S-transferase, histochemically *in situ* and in plasma and liver fractions. Supplementation with *Aloe vera* gel extract in the diseased rats suppressed this increase activity.

In traditional South-Asian medicine the therapeutic value of the parenchymatous leaf-gel of *Aloe vera* for inflammation-based diseases was reported. Extracts from the leaf-gel contains glutathione peroxidase activity (Sabeh *et al.*, 1993). The low molecular constituents of this extract inhibit the release of reactive oxygen species (ROS) by phorbol myristic acetate (PMA)-stimulated human polymorphonuclear leucocytes (PMN). These compounds also inhibit the ROS-dependent extracellular effects of PMN such as lysis of red blood cells. However, the capacity of the PMN to phagocytose and kill microorganisms is not affected. The inhibitory effect of these compounds has been shown to be the indirect result of the diminished availability of intracellular free Ca^{2+} (t-Hart *et al.*, 1990).

The polysaccharide and flavonoid concentrations of two-, three- and four-year old *Aloe vera* were determined and their antioxidant activities were evaluated compared to BHT and α-tocopherol by DPPH radical scavenging method and the linoleic acid system at 100μg of soluble solids per ml of ethanol (Hu *et al.*, 2003). The results showed that three-year-old *Aloe vera* contained significantly higher level of polysaccharides and flavonoids than two and four-year old *Aloe vera*. All the Aloe extracts showed significant antioxidant activity. The antioxidant activity of *Aloe vera* extracts and reference compounds followed the order: three-year-old *Aloe vera*> BHT>four-year-old *Aloe vera*> α-tocopherol>two-year-old *Aloe vera*. Three-year-old extracts exhibited the strongest radical scavenging activity of 72.19 per cent, which is significantly higher than that of BHT at 70.52 per cent and α-tocopherol at 65.20 per cent.

Rajasekaran *et al.* (2005) found that the antioxidant activity of ethanolic extract from the gel of leaf in the plasma and pancreas of diabetic rats was higher than that of the standard used, which supported the ethnopharmacological use of the plant in ameliorating the oxidative stress in diabetes. Kardosova and Machova (2006) investigated the ability of polysaccharides, isolated from leaves, to inhibit peroxidation of soyabean lecithin liposomes by OH radicals. All polysaccharides exhibited antioxidant activity.

The effects of the exudate of *Aloe barbadensis* leaves on oxidative stress and some antioxidant status of streptozotocin induced-diabetic rats were studied by Nwanjo *et al.* (2006). There was significant reduction in scavenging enzymes like SOD activity and significant increase in signs of oxidative damage, such as lipid peroxidation production (plasma MDA) in streptozotocin induced-diabetic rats. Treatment with *Aloe barbadenssis* (150 mg/kg) increased antioxidant enzymes like SOD activities and significantly reduced lipid peroxidation products (Nwanjo, 2006).

The ethanolic extract of *Aloe vera* leaf skin was fractionated by liquid-liquid partition using hexane, ethyl acetate, chloroform-ethanol and butanol. The total phenolic content of the four different fractions were determined by Folin-Ciocalteu method and their antioxidant activity as essayed through some *in vitro* models such as the antioxidant capacity by phosphomolybdenum method, β-carotene bleaching method, radical scavenging activity using DPPH assay and reducing power assay. The chloroform ethanol fraction showed the highest total phenolics (40.50μg GAE/g of extract), the highest scavenging activity and the greatest reducing power, followed by ethyl acetate, butanol and hexane extracts. However, the hexane fraction showed the highest antioxidant capacity (471.30) and the highest antioxidant activity coefficient (AAC) by the β-carotene bleaching method (Miladi and Damak, 2008).

The study by Ozsoy *et al.* (2009) demonstrated that the aqueous extract from *Aloe vera* leaves contained naturally occurring antioxidant components, including total phenols, flavonoids, ascorbic acid, β-carotene and α-tocopherol. The extract exhibited inhibitory capacity against Fe(3+) ascorbic acid induced phosphatidylcholine liposome oxidation, scavenged stable DPPH, ABTS and superoxide anion reductant. In contrast, the leaf inner gel did not show any antioxidant activity.

Twenty-five essential oils, including *Aloe vera* oil, were tested by Wei and Shibamoto (2010) for antioxidant activities using a conjugated diene assay, the aldehyde/carboxylic acid assay, the DPPH free radical scavenging assay, and the malonaldehyde/gas chromatography (MA/GC) assay. They were also tested for lipoxygenase inhibitory activities using the lipoxygenase inhibitor-screening assay. *Aloe vera* oil exhibited the greatest lipoxygenase inhibitory activity (96 per cent) at a concentration of 0.5µg/ml.

The antioxidative properties of extracts of *Aloe vera* gel made in methanol (MEAG), 95 per cent ethanol (EEAG), hexane (HEAG), acetone (AEAG) and chloroform (CEAG) were investigated by Saritha *et al.* (2010) employing various *in vitro* systems *viz.*, DPPH, superoxide anion radicals scavenging, metal ion chelation, reducing power, hydroxyl radical scavenging and total antioxidant activity in linoleic acid emulsion system. The results showed that MEAG and AEAG possessed maximum DPPH free radical and superoxide radical scavenging activities. All the extracts were effective in scavenging the hydroxyl radicals in nonsite specific assay as well as in site specific assay. The formation of the Ferrozine-Fe^{2+} complex was found to be incomplete in the presence of MEAG and AEAG, indicating their capacity in chelating iron. The AEAG was shown more reducing power than MEAG.

Effect of *Aloe vera* gel extract (AGE) on antioxidant enzymes and azoxymethane (AOM)-induced oxidative stress in rats was investigated by Anilakumar *et al.* (2010). It was observed that AOM administration resulted in a significant increase in malodialdehyde and conjugated dienes, with reduction in hepatic glutathione (GSH), vitamin A and uric acid contents. AOM-induced reduction in hepatic GSH and uric acid was brought back to normal by AGE. There was a significant raise in hepatic catalase, superoxidee dismutse and glucose-6-phosphate dehydrogenase activities as a result of feeding the extract.

Aloe dhufarensis Lavranos Family: Liliaceae

Marwah *et al.* (2007) investigated the antioxidant activity of wound-healing herbs using *in vitro* DPPH and phosphomolybdenum assay methods. Out of the 19 plants screened, the aqueous ethanol extract of *Aloe dhufarensis* showed the best DPPH scavenging activities with per cent IP value 58.7 per cent, after 15 min of incubation at an effective test concentration of 50µg/ml. The Total Antioxidant Capacity as gallic acid equivalent of 229mg/g of ethanol extract was obtained in the phosphomolybdenum assay. Total phenolic content was 430 mg GAE/g of ethanol extract.

Aloe ferox Mill. Family: Liliaceae

Due to the presence of the antioxidant polyphenols, indoles and alkaloids, the *Aloe ferox* leaf gel shows antioxidant capacity as confirmed by ORAC and FRAP analyses. Both analytical methods used show the non-flavonoid polyphenols to contribute to the majority of the total polyphenol content (Loots *et al.*, 2007).

Aloe saponaria Haw. Family: Liliaceae

Yoo *et al.* (2008) evaluated the antioxidant and anti-inflammatory activities of ethanol extracts from *Aloe saponaria* using xanthine-xanthine oxidase assay, DPPH assay, lipopolysaccharide (LPS)-induced NO production in RAW 264.7 cell. The ethanol fraction displayed potent antioxidative activities in XO assay. In addition, ethanol fractions showed potent scavenging effects in DPPH assay. Ethanol fractions significantly suppressed NO production from LPS-activaated RAW 264.7 cells.

Aloysia gratissima (Gillies et Hook.) Tronsco Family: Verbenaceae

The aerial parts of 17 Bolivian plants, including *Aloysia gratissima* were screened by Rosas-Romero and Saavedra (2005) to determine antioxidant activity. A methanol extract of each plant was prepared and portioned sequentially with hexane, chloroform, and ethyl acetate, leaving an aqueous solution. All extracts and their 5 fractions for a total of 102 samples, were evaluated using two techniques: an adaptation of the β-carotene bleaching technique using an emulsion of linoleic acid in water as the oxidizable substrate, and the DPPH free radical trapping technique. The results of the β-carotene bleaching technique were more discriminating and better related to the rancidity process under normal conditions; with this assay, 11 species provided at least one fraction with highly promising antioxidant activity. All species gave good results under the DPPH technique, and in most cases they performed better than BHA, which was used as a reference antioxidant (Rosas-Romero and Saavedra, 2005).

Alphonsea hortensis H.Huber Family: Annonaceae

Puvanendran *et al.* (2007) evaluated the antioxidant activity of Dichloromethane (DCM) and MeOH extract of stem, bark, leaves and seeds of *Alphonsea hortensis* using DPPH assay. The extract showed high antioxidant activities. Percentage of inhibition of bark was 35.12 (MeOH extract); leaf extract was 40.62 (DCM extract) and 32.81 (MeOH extract) while that of seed was 30.31 (MeOH extract) and that of stem was 13.69 (DCM extract) and 56.30 (MeOH extract).

Alpinia calcarata Rosc. Family: Zingiberaceae

Narayanaswamy and Balakrishnan (2011) evaluated the antioxidant activity of 13 medicinal plants including leaf of *Alpinia calcarata*. The aqueous extract of *Alpinia calcarata* showed highest 98 per cent inhibition of DPPH radical while ethanolic extract showed 95 per cent inhibition of DPPH radical.

Alpinia galanga (L.) Willd. (Syn: *Languas galanga* L.) Family: Zingiberaceae

English Name: Galangal.

Chanwitheesuk *et al.* (2005) investigated the antioxidant activity of rhizome of *Alpinia galanga* (Syn: *Languas galanga*) using a β-carotene bleaching method. The contents of plant chemicals such as vitamin C, Vitamin E, carotenoids, tannin and total phenolics, were also determined. Methanolic extract showed antioxidant activity. *Languas galanga* has antioxidant index of 3.24, Vitamin C 7.87 mg per cent, Vitamin E 0.0042mg per cent, Total carotenes 0.64 mg per cent, Total xanthophylls 1.05mg per cent, Tannins 17.7mg per cent and Total phenolics of 63.4mg per cent.

Juntachote and Berghofer (2005) and Juntachote *et al.* (2006) assessed the antioxidant activity of ethanolic extracts of Galangal. Galangal extracts exhibited strong superoxide anion scavenging activity, Fe^{2+} chelating activity, and reducing power, in a concentration-dependent manner. Antioxidant activity of extract correlated with reducing power. Further more ethanolic extract of Galangal acted as a radical scavenger and also as lipoxygenase inhibitor. Leelapornpisid *et al.* (2007) reported antioxidant activity of the absolutes (obtained by solvent extraction) of *Alpinia glanaga*.

Gan *et al.* (2010) evaluated the antioxidant activity of *Alpinia galanga* using the FRAP and TEAC assays, and its total phenolic content was measured by the Folin-Ciocalteu method. For FRAP assay the value was 82.21 µmol Fe(II)/g dry weight. For TEAC assay, the value was 45.45µmol Trolox/g dry weight. Total phenolic content was 4.25 mg gallic acid equivalent/g dry weight of plant material.

Alpinia katsumadai Hayata Family: Zingiberaceae

Various aspects of antioxidant activity were evaluated by S.E.Lee *et al.* (2003b) in a total extract derived from *Alpinia katsumadai* seed. Relatively high levels of DPPH radical scavenging activity were detected in the total extract (IC_{50} 1.6µg/ml). The known compounds such as (-)-epigallocatechin-3-gallate and resveratrol showed IC_{50} values of <0.8 and 4.8 1.6µg/ml, respectively. The total extract of *Alpinia katsumadai* also dose-dependently enhanced the activities of superoxide dismutase, catalase and glutathione peroxidase in V79-4 cells.

Gan *et al.* (2010) evaluated the antioxidant activity of *Alpinia katsumadai* using the FRAP and TEAC assays, and its total phenolic content was measured by the Folin-Ciocalteu method. For FRAP assay the value was 42.89 µmol Fe(II)/g dry weight. For TEAC assay, the value was 31.51µmol Trolox/g dry weight. Total phenolic content was 2.52 mg gallic acid equivalent/g dry weight of plant material.

Alpinia nutans Roscoe Family: Zingiberaceae

The methylene chloride extract of *Alpinia nutans* was investigated for its antioxidant constituents by Habsah *et al.* (2003). 5,6-Dehydrokawain, flavokawain-B, 1,7-diphenyl-5-hydroxy-6-hepten-3-one, (-)-pinocembrin and a mixture of stigmasterol and betasitosterol were isolated. The antioxidant activity was measured on isolates using ferric thiocyanate (FTC) and DPPH free radical scavenging techniques.

Alpinia officinarum Hance Family: Zingiberaceae

Methanol aqueous extract of *Alpinia officinarum* was screened by Kim *et al.* (1997) for anti-oxidative activity using Fenton's reagent/ethyl lineolate system. The extract showed antioxidant activity.

The methanol extracts of nine medicinal plants traditionally used in Chinese medicine were screened for antioxidant activity. The extract of *Alpinia officinarum* strongly enhanced viability against H_2O_2-induced oxidative damage in V79-4 cells (S.E.Lee *et al.*, 2003).

Antioxidative compounds were isolated from the methanol extract of fresh rhizome of *Alpinia officinarum* by Tram Ngoc *et al.* (2004). Dry rhizomes of the same species from Tripura (India) also have potent antioxidant property (Kshirsagar and Upadhyay, 2009).

Cai *et al.* (2004) evaluated the antioxidant activity and phenolic compounds in traditional Chinese medicinal plants associated with anticancer. The improved ABTS.+ method was used to systematically assess the total antioxidant capacity (TEAC) of the medicinal extracts. The TEAC values and total phenolic content for methanolic extract of rhizome of *Alpinia officinarum* were 932.8 μmol Trolox equivalent/100g dry weight (DW), and 4.76 g of gallic acid equivalent/100g DW respectively. A positive significant linear relationship between antioxidant activity and total phenolic content showed that phenolic compounds were the dominant antioxidant components in the tested herbs. Major types of phenolic compounds in *Alpinia officinarum* include flavonoids (galangin, kaempferol, kaempferide), tannins, phenolic volatile oils.

The *in vitro* antioxidant activity of the rhizome of *Alpinia officinarum* was investigated by Ramasamy *et al.* (2010) by Ferric Thiocyanate (FTC) and Thio Barbituric Acid (TBA) methods. The ethanol extract of the rhizome possess significant antioxidant activity. The antioxidant activity of the extract is close and identical in magnitude and comparable to that of standard antioxidant compounds used.

Alpinia oxyphylla Miquel Family: Zingiberaceae

Yekuchinone A and Yakuchinone B present in *Alpinia oxyphylla* have inhibitory effects on phorbol ester-induced inflammation and skin carcinogenesis in mice, and oxidative stress *in vitro* (Liu *et al.*, 2003).

An *et al.* (2005) reported antioxidant activity of protocatechuic acid from *Alpinia oxyphylla*. In a study by Guan *et al.* (2006) and Shi *et al.* (2006) the neuroprotetective effects of Alpinia protocatechuic acid (PCA), a phenolic compound isolated from the dried fruits of *Alpinia oxyphylla* was found. The protective effect of Alpinia PCA against H_2O_2 induced oxidative damage on PC12 cells was investigated. The results proved that Alpinia PCA significantly prevented the H_2O_2 induced reduction in cell survival, improved the cognition of aged rats, reduced the content of lipid peroxide, increased the activity of glutathione peroxidase and superoxide dismutase.

Alpinia zerumbet (Pers.) B.L.Burt. et R.M.Sm. Family: Zingiberaceae

Essential oil, total phenolics and antioxidant capacities assayed by DPPH free radical scavenging and β-carotene bleaching methods were evaluated by Elzawely *et al.* (2007) in flowers and seeds of *Alpinia zerumbet*. The ethyl acetate extract of flowers and seeds possessed a high antiradical activity and prevented the bleaching of β-carotene. The HPLC analysis indicated that *p*-hydroxybenzoic acid, ferulic acid and syringic acid were the predominant phenolics in the ethyl acetate extract of flowers, whilst, *p*-hydroxybenzoic acid, syringic acid and vanillin were the major phenolics in seeds.

Chan *et al.* (2009) investigated the effects of different drying methods on the antioxidant properties of leaves of *Alpinia zerumbet*. Freeze-drying resulteα in significant gains in total phenolic content.

Alstonia boonei de Wild. Family: Apocynaceae

Akinmoladun *et al.* (2007, 2010) assessed the phytochemical constituents and antioxidant and free radical scavenging activities of *Alstonia boonei* using seven different antioxidant assay methods. DPPH (2,2-diphenyl-1-picrylhydrazyl) radical scavenging activity, total phenolic content and reducing power were 41.58 per cent, 2.09 mg/g gallic acid equivalent and 0.32 respectively. The results suggest that the methanolic extracts possess significant antioxidant and radical scavenging activity that may be due to the phytochemical content of the plant.

Leaves, roots and stem of *Alstonia boonei* showed moderate antioxidant activity (Oke and Hamburgeer, 2002).

Alstonia scholaris (L.) R.Br. Family: Apocynaceae

Arulmozhi *et al.* (2007, 2010) evaluated antioxidant potential of ethanolic extract and fractions from ethanolic extract of *Alstonia scholaris* using various *in vitro* tests including DPPH free radical scavenging, metal ion chelating, hydrogen peroxide scavenging, superoxide anion radical scavenging and ferric thiocyanate reducing ability. Ethanolic extract had significant free radical scavenging (63 per cent), metal ion chelating (74.88 per cent), hydrogen peroxide scavenging (72.28 per cent), superoxide anion radical scavenging (67.66 per cent) and ferric thiocyanate reducing activities. Dichloromethane and ethyl acetate fractions were found to have significant free radical scavenging and metal ion chelating properties, whereas the petroleum ether and n-butanol fractions lack the *in vitro* antioxidant property.

Leelapornpisid *et al.* (2007) reported antioxidant activity of the absolutes (obtained by solvent extraction) of *Alstonia scholaris*. The methanolic extract of stem bark of *Alstonia scholaris* was found to show moderate antioxidant potential by using DPPH method (Dhobal *et al.*, 2008).

Kulkarni and Juvekar (2009) reported that *Alstonia scholaris* extract at 200 microg/ml showed maximum scavenging of stable radical 1,1-diphenyl, 2-picryl hydrazyl at 90.11 per cent and nitric oxide radical at 62.77 per cent.

Alternanthera bettzickiana (Regel) Nicholsen Family: Amaranthaceae

Chen *et al.* (2009) studied the antioxidant activity of aerial parts extract towards xanthine oxidase, lipoxygenase and tyrosinase activities. The inhibitory effect of acetone extract on xanthine oxidase activity was 40 per cent, tyrosine activity was less than 5 per cent and lipoxygenase activity was about 60 per cent.

Alternanthera brasiliana (L.) Kuntze Family: Amaranthaceae

Costa *et al.* (2009) investigated the antioxidant potential of different extracts from the leaves of *Alternanthera brasiliana. In vivo* and *in vitro* tests were used to detect antioxidant activity of crude methanolic extract (AME), aqueous extract (AAE), ethyl acetate fraction (AEAF), the n-BuOH fraction (ABF) and aqueous residue (AAR). The effects of cellular antioxidant systems on *A. brasiliana* extracts-induced growth inhibition was studied using yeast strains deficient in superoxide dismutases. Scavening of DPPH free radical was measured method. The results showed a significant antioxidant activity with a significant increase of cell survival, for the three strains after treatment with *t*-BOOH for both extracts AME and AAE, and also for the ABF in the case of *sod1D* and *sod1Dsod2D*, mutant strains. On the other hand, the AAR fraction was not able to exert any antioxidant activity. The DPPH free radical scavenging activity of extracts AME and AAE, as well as the three fractions AEAF, ABF and AAR, expressed as inhibition of the DPPH radical were 88.6 per cent, 94.7 per cent, 90.5 per cent, 90.0 per cent and 76.0 per cent, respectively, at an extract concentration of 1000 µg/ml, while it was 55.9 per cent, 76.3 per cent, 71.6 per cent, 56.7 per cent and 19.9 per cent, respectively, at concentration 100 µg/ml of extract.

Alternanthera maritima (Martius) A. St. Hilaire Family: Amaranthaceae

Bioassay-guided fractionation of *Alternanthera maritima* aerial parts by Souza *et al.* (2010) yielded an ethanolic crude extract, its butanolic fraction and seven isolated flavonoids (two aglycones, two O-glycosides and three C-glycosides) with antioxidative activity. The ability of these samples to scavenge enzymatically generated free radicals (luminol-horseradish peroxidase-H^2O^2 reaction) and inhibit reactive oxygen species (ROS) production by opsonized zymosan-stimulated human neutrophils (PMNLs) was evaluated by chemiluminescence methods. In both assays, the butanolic fraction was significantly more active than the ethanolic crude extract, the flavonoid aglycones had high inhibitory activities and the C-glycosylated flavonoids had no significant effect even at the highest concentration tested (50 micromol/L). However, the O-glycosylated flavonoids inhibitory effects on chemiluminescence were strongly dependent on the chemical structure and assay type (cellular or cell-free system). Under the conditions tested, active samples were not toxic to human PMNLs.

Alternanthera philoxeroides (Mart.) Griseb. Family: Amaranthaceae

Cai *et al.* (2011) investigated the influence of ethanol of different concentration on antioxidant activity of polysaccharide *in vitro* from *Alternanthera philoxeroides*. It was evaluated by means by reducing power, DPPH and Total antioxidant capacity. Antioxidant activity varied with ethanol concentration. Antioxidant activity was strongest in 100 per cent ethanol.

Alternanthera sessilis (L.) R. Br. ex DC. Family: Amaranthaceae

Borah *et al.* (2011) evaluated the antioxidant activity of *A.sessilis* plant extracts in different solvents. The antioxidant activity was studied by phosphomolybdate method and DPPH method. In phosphomolybdate method the highest activity was shown by methanolic extract (12.044 mM of ascorbic acid eqvt/gm of sample). The highest radical scavenging activity by DPPH method was found in methanol extracts (IC_{50} 587.093µg/ml). Ferrous chelating activity, superoxide radical scavenging activity, nitric oxide radical scavenging activity was found high in acetone, acetone and methanol extracts respectively. Total Flavonoids and crude phenolics were found to be 0.370 mg/gm dry wt. and 1.529 mg/gm dry wt. respectively. The antioxidant activity increases with increase in the concentration.

Alternanthera tenella Colla Family: Amaranthaceae

Phytochemical analysis of the antioxidant ethanolic extract of *Alternanthera tenella* by Salvador *et al.* (2011) led to the isolation of six flavonoids, acacetin 8-C-[alpha-L-rhamnopyranosyl-(1→2)-β-D-glucopyranoside](1), 2''-O-α-L-rhamnopyranosyl-vitexin (2), 2''-O-beta-D-glucopyranosyl-vitexin (3), vitexin (4), quercetin (5) and kaempferol (6). Antioxidant capacity of extract, fractions and isolated compounds was determined using the oxygen radical absorbance capacity (ORAC) assay and extract, fractions and flavonoids isolated showed antioxidant activity *in vitro*. Moreover, the total soluble phenolic contents of the extract and fractions were measured using the Folin-Ciocalteau reagent and the quantitative analysis of flavone C-glycosides major constituents was performed by HPLC.

Althaea officinalis L. var. _robusta_ Family: Malvaceae

Kardosova and Machova (2006) investigated the ability of polysaccharides, isolated from leaves, to inhibit peroxidation of soyabean lecithin liposomes by OH radicals. All polysaccharides exhibited antioxidant activity. The highest inhibition of lipid peroxidation was observed with glucoronoxylans (30 per cent) from *A. officinalis*.

Alysicarpus vaginalis (L.) DC. Family: Fabaceae (Papilionaceae)

Ethanolic extract of whole plant of *Alysicarpus vaginalis* was evaluated for its nitric oxide radical scavenging activity and reducing power assays using BHT and BHA as standards (Rathi *et al.*, 2010). The extract was studied for its lipid peroxidation inhibition assay using rat liver. The plant extract induced a marked concentration dependent inhibition on lipid peroxidation.

Alyssum homolocarpum (F. et M.) Boiss. Family: Brassicaceae (Cruciferae)

English Name: Alyssum

Souri *et al.* (2008) evaluated the antioxidant activity of *Alyssum homolocarpum* against linoleic acid peroxidation using 1,3-diethyl-2-thiobarbituric acid as reagent. Antioxidant activity (IC_{50}) against peroxidation of linoleic acid (2mg/ml) was 94.25 and phenolic content was 165.68 mg/100g dry weight.

Amaranthus blitum L. Family: Amaranthaceae

Saranya and Guruswamy (2010) assessed the antioxidant activity of ethanolic extract of *Amaranthus blitum*. The ethanolic extract showed inhibition in a dose dependent manner. DPPH radical scavenging activity of *A. blitum* IC_{50} value was 37.5μg/ml when compared to standard ascorbic acid IC_{50} value 25.0μg/ml. FRAP value of *A. blitum* was 33.08μg/ml whereas standard vitamin C was 35.00μg/ml. Superoxide radical scavenging activity of *A. blitum* was 25μg/ml while standard ascorbic acid was IC_{50} = 21.5μg/ml. *A. blitum* (89.89 per cent) showed nitric scavenging activity at the concentration of 100μg/ml, the IC_{50} value was 36μg/ml whereas standard ascorbic acid showed an IC_{50} value of 32μg/ml. In hydroxyl scavenging activity *A. blitum* showed inhibition value of 83.20 per cent at a concentration of 100μg/ml when compared to α-tocopherol with the value of 82.61 per cent. H_2O_2 scavenging was found to be 54μg/ml at a concentration of 100μg/ml. The percentage inhibition of lipid peroxidation was 85.03 at a concentration of 1000μg/ml while standard ascorbic acid was 420μg/ml.

Amaranthus caudatus L. Family: Amaranthaceae

Ethyl acetate extracts of seeds of *Amaranthus caudatus* var. *oscarblanco* and *Amaranthus caudatus* var. *victor* showed antioxidant activity with IC_{50} values of 0.50 mg/ml and 0.62 mg/ml respectively (Conforti *et al.*, 2005).

Odukoya *et al.* (2007) evaluated the antioxidant activity and total phenolic contents of Nigerian green leafy vegetables including *Amaranthus caudatus*. Potential free radical scavenging activity of these vegetables was confirmed by spraying spots of the extracts with DPPH (yellow color on purple background). Antioxidant activity was assayed in linoleic acid model system. Total polyphenols as Tannic Acid Equivalent (TAE) and Ascorbic Acid. There was low correlation between AA/TAE, AA/ASC and TAE/ASC.

The methanolic crude extracts of *Amaranthus caudatus* were screened by Veeru *et al.* (2009) for their free radical scavenging properties using ascorbic acid as standard antioxidant. Free radical scavenging activity was evaluated using DPPH free radical. The overall antioxidant activity of *Amaranthus caudatus* was found to be stronger with IC_{50} value of 0.14 mg/l.The ascorbic acid level was 3.86 mg/100g and the carotenoids content were 15.33 mg/100g in plant extracts. The total phenol content was 3.43 mg/g.

Amaranthus cruentus L. Family: Amaranthaceae

Oboh and Akindahunsi (2004) investigated the effect of preservation of green leafy vegetables of Nigeria on the antioxidant phytoconstituents (Vitamin C and Total phenol) and activity (reducing property and free radical scavenging ability) including *Amaranthus cruentus*. The result of the study revealed that sun-drying of green leafy vegetables cause a significant decrease in the vitamin C content (16.67-64.68 per cent loss). Conversely it leads to a significant increase in the total phenol content (6.45 -223.08 per cent grain), reducing property (16.00-362.50 per cent grain) and free radical scavenging ability (126.00-5757.00 per cent gain) of green leafy vegetables. Adebooye *et al.* (2008) evaluated the antioxidant activity of *Amaranthus cruentus*.

Nsimba *et al.* (2008) evaluated the antioxidant activity of various extracts and fractions of *Amaranthus cruentus* employing DPPH assay and FRAP assay. The antioxidant activity was less correlated to the phenolics content suggesting that non-phenolic compounds might play major free radicals scavenging activity.

Antioxidant activities of eight leafy vegetables of Ghana including *Amaranthus cruentus* were assessed by Morrison and Twumasi (2010). The total antioxidant capacity (TAC) and Total Phenol content (TPC) in the methanol extracts (METE) and hydro-ethanol extracts (HETE) were measured. The TPC of the methanol extract of *Amaranthus cruentus* was 0.238 mg/ml TAE and TAE was 0.228 mg/ml Ascorbic acid equivalent (AAE) while TPC of hydro-ethanol extract was 0.339 mg/ml TAE and TAE was 0.312 mg/ml Ascorbic acid equivalent (AAE). Radical scavenging activity of methanol extract was EC_{50} = 0.1364mg/ml and HETE was 0.1378 mg/ml while Fe^{3+} reducing potential was EC_{50} = 0.1378 (METE) and 0.1378 (HETE).

***Amaranthus gangeticus* L. Family: Amaranthaceae**

Effect of prefeeding dehydrated amaranth (*Amaranthus gangeticus*) leaves at 10 and 20 per cent levels on a chemical toxicant, dimethylhydrazine(DMH)-induced free radical stress in rat liver was evaluated by Anilakumar *et al.* (2004). DMH-induced rise in hepatic malondialdehyde was diminished by Amaranthus leaves. Amaranth leaves at 10 and 20 per cent levels of feeding diminished the hepatic superoxide dismutase and glutathione peroxidase (GSH-Px) activities. DMH influenced adversely the hepatic antioxidant enzyme activities. It was concluded that the consumption of Amaranth leaves at 20 percent level reduced DMH-induced impaired antioxidant status in rat liver.

The antioxidant properties of 25 edible tropical plants including *Amaranthus gangeticus*, expressed as TEAC were studied by Wong *et al.* (2006) using DPPH scavenging and reducing FRAP assays.The value to scavenge DPPH free radicals was about 80μmol Trolox equivalent per g db while FRAP value was about 110 mol Trolox equivalent. A strong correlation between TEAC values obtained for the DPPH assay implied that compounds in the extracts were capable of scavenging the DPPH free radical and reducing ferric ions.

***Amaranthus hybridus* L. Family: Amaranthaceae**

Odukoya *et al.* (2007) evaluated the antioxidant activity and total phenolic contents of Nigerian green leafy vegetables including *Amaranthus hybridus*. Potential free radical scavenging activity of these vegetables was confirmed by spraying spots of the extracts with DPPH (yellow color on purple background). Antioxidant activity was assayed in linoleic acid model system. Total polyphenols as Tannic Acid Equivalent (TAE) and Ascorbic Acid where estimated. Antioxidant activity in *Amaranthus hybridus* was low 3.67 per cent (inhibition of lipid peroxidation). There was low correlation between AA/TAE, AA/ASC and TAE/ASC.

Adefegha and Oboh (2011) reported that cooking decreases the vitamin C contents in the leafy vegetable *Amaranthus hybridus*, while it increased the phenolic content and antioxidant activities.

***Amaranthus hypochondriacus* L. Family: Amaranthaceae**

Nsimba *et al.* (2008) evaluated the antioxidant activity of various extracts and fractions of *Amaranthus hypochondriacus* employing DPPH assay and FRAP assay. The antioxidant activity was less correlated to the phenolics content suggesting that non-phenolic compounds might play major free radicals scavenging activity.

***Amaranthus lividus* L. Family: Amaranthaceae**

Water, methanol and ethyl acetate extracts of stems with leaves and flowers of *Amaranthus lividus*, one of the most popular leafy vegetable consumed in the west Black Sea region of Turkey, were tested *in vitro* for their ability to inhibit peroxidation of phosphatidylcholine liposomes induced with Fe^{3+}/ascorbate, to scavenge $ABTS^{++}$, DPPH and hydroxyl radicals, to reduce Fe (III) to Fe (II) and to chelate Fe(II) ions (Ozsoy *et al.*, 2009). The results showed that amaranth vegetable contained naturally occurring antioxidant activity which may be attributed to its lipid peroxidation inhibitory, radical scavenging and metal chelating activities. The antioxidant activities of the water and ethyl acetate extracts were not concomitant with the development of their reducing power. It was concluded that *A. lividus* might be a potential source of antioxidants (Ozsoy *et al.*, 2009).

Amaranthus paniculatus L. Family.: Amaranthaceae

A. paniculatus has a high content of β-carotene, ascorbic acid (vitamin C) and folate which can be a rich source of antioxidants. Members of Amaranthaceae are good natural sources of carotenoids, vitamin C, nutritionally critical lysine, methionine and proteins (Bhatia and Jain, 2003). The ability of *A. paniculatus* extract to act as a free radical scavenger or hydrogen donor was revealed by DPPH radical scavenging activity assay (Amin *et al.*, 2006).

Amaranthus spinosus L. Family: Amaranthaceae

Ashok Kumar *et al.* (2010) evaluated the antioxidant potential of methanol extract of *Amaranthus spinosus*. Oral administration of the extract (200 and 400 mg/kg) for 15 days showed significant reduction in the elevated blood glucose, MDA and restored GSH, CAT and TT levels as compared with a diabetic control.

Amaranthus viridis L. Family: Amaranthaceae

Amin *et al.* (2006) evaluated the antioxidant activity and phenolic content of raw and blanched *Amaranthus* species. The total antioxidant activity of water soluble components in raw spinach was in the order of *Amaranthus blitum* = *Amaranthus gangeticus* = *Amaranthus viridis* > *Amaranthus paniculatus* whereas free radical scavenging activity was in the order of *Amaranthus blitum* > *A.viridis* > *A. gangeticus* > *A. paniculatus*. All the studied species possessed different antioxidant activities and phenolic contents. Antioxidant activities and phenolic contents of all the spinach were in the order of raw > blanched 10 min > blanched 15 min. Blanching up to 15 min. may affect losses of antioxidant activity and phenolic content, depending on the species.

Ammannia baccifera L. Family: Lythraceae

Ethanol extract of *Ammannia baccifera* was studied by Lavanya *et al.* (2009) for its protective nature against the oxidative damage of lipids, proteins and DNA in carbon tetrachloride-induced toxicity in rats. The study revealed that the administration of the ethanol extract of *A. baccifera* to CCl_4 intoxicated rats could significantly decrease the levels of lipid peroxidation, protein carbonyls and increased the levels of total sulfhydryls in a dose-dependent manner. The results suggest that treatment with the ethanol extract can minimize the deleterious effects caused by CCl_4 through its strong antioxidative and free radical scavenging properties.

Amomum krervanh Pierre ex Gagnep. Family: Zingiberaceae

English Name: Siam cardamom.

Chanwitheesuk *et al.* (2005) investigated the antioxidant activity of fruit of *Amomum krervanh* using a β-carotene bleaching method. The contents of plant chemicals such as vitamin C, Vitamin E, carotenoids, tannin and total phenolics, were also determined. Methanolic extract showed antioxidant activity. *Amomum krervanh* has antioxidant index of 4.00, Vitamin C 2.52 mg per cent, Vitamin E 0.0043mg per cent, Total carotenes 1.29 mg per cent, Total xanthophylls 1.07mg per cent, Tannins 7.77mg per cent and Total phenolics of 46.3mg per cent.

Gan *et al.* (2010) evaluated the antioxidant activity of *Amomum krervanh* using the FRAP and TEAC assays, and its total phenolic content was measured by the Folin-Ciocalteu method. For FRAP assay the value was 43.54 µmol Fe(II)/g dry weight. For TEAC assay, the value was 25.50µmol Trolox/g dry weight. Total phenolic content was 2.77 mg gallic acid equivalent/g dry weight of plant material.

Amomum subulatum Roxb. Family: Zingiberaceae

English Name: Greater cardamom.

In order to gain insight into the antioxidant effect of *Amomum subulatum* hepatic and cardiac antioxidant enzymes, GSH content and lipid conjugated dienes were studied in rats fed high fat diet along with *Amomum subulatum*. The antioxidant enzyme activities were found to be significantly enhanced whereas GSH content was markedly restored in rats fed a fat diet with spices. In addition, this spice partially counteracted increase in lipid conjugated dienes and hydroperoxides, the primary producers of lipid peroxidation (Dhuley, 1999).

Constituents of the fruits of *Amomum subulatum* were fractionated into three fractions, the dichloromethane extract, and the ethyl acetate-soluble and water-soluble fractions of the 70 per cent aqueous acetone extract. The ethyl acetate-soluble fraction showed a high radical-scavenging activity against DPPH. Four compounds were isolated from the ethyl acetate-soluble fraction- protocatechualdehyde (1), protocatechuic acid (2), 1,7-bis (3,4-dihydroxyphenyl) hepta-4E, 6E-dien-3-one (3) and 2,3,7-trihydroxy-5-(3,4-dihydroxy-E-styryl)-6,7,8,9-tetrahydro-5H-benzocycloheptene (4). Compounds 1 and 3 showed stronger activity than such natural antioxidants as α-tocopherol and L-ascorbic acid. Compounds 2 and 4 were comparable to α-tocopherol and L-ascorbic acid (Kikuzaki *et al.*, 2001).

Amomum tsao-ko Crevost et Lemaire Family: Zingiberaceae

Gan *et al.* (2010) evaluated the antioxidant activity of *Amomum tsao-ko* using the FRAP and TEAC assays, and its total phenolic content was measured by the Folin-Ciocalteu method. For FRAP assay the value was 130.16 µmol Fe(II)/g dry weight. For TEAC assay, the value was 100.61µmol Trolox/g dry weight. Total phenolic content was 7.15 mg gallic acid equivalent/g dry weight of plant material.

Yang *et al.* (2010) assayed the antioxidant activity of essential oil of *Amomum tsao-ko* using diverse methods. The lipidic substrate was constituted by liposomes containing lecithin, and the induction of peroxidation was performed with ferrous sulphate. They found weak antioxidant activity due to low phenolic content of oils. *Amomum tsao-ko* oil with low antioxidant activity measured through TBARS method, also had weak ability for scavenging DPPH free radicals and low ferric reducing power, mainly due to the low phenolic content of the oils.

Amomum villosum Lour. Family: Zingiberaceae

Gan *et al.* (2010) evaluated the antioxidant activity of *Amomum villosum* using the FRAP and TEAC assays, and its total phenolic content was measured by the Folin-Ciocalteu method. For FRAP assay the value was 117.57 μmol Fe(II)/g dry weight. For TEAC assay, the value was 80.16μmol Trolox/g dry weight. Total phenolic content was 9.29 mg gallic acid equivalent/g dry weight of plant material.

Amorpha canescens Pursh. Family: Fabaceae

Borchardt *et al.* (2008) evaluated the antioxidant activity of methanol extracts of seeds of *Amorpha canescens* using DPPH antioxidant assay.The antioxidant value reported in μM Trolox/100g (TE) from DPPH radical scavenging activity of crude seeds extract was 58,448.

Amorphophallus campanulatus (Roxb.) Blume ex Decne Family: Araceae

Antioxidant activity of methanolic and aqueous extract of *Amorphophallus campanulatus* tuber was studied by Sahu *et al.* (2009) for its free radical scavenging property on different *in vitro* models, *e.g.*, DPPH method, nitric oxide method and reducing power method. The extracts showed good dose dependent free radical scavenging property in all models. IC_{50} values for water and methanolic extract were found to be 59.91 and 99.40 μg/ml in DPPH method and 77.02 and 70.20 μg/ml in nitric oxide method. In reducing power method, aqueous extracts showed more reducing power as compared to methanolic extract. Ascorbic acid was used as standard.

Jain *et al.* (2009) screened the antioxidant and hepatoprotective activity of ethanolic and aqueous extracts of *Amorphophallus campanulatus* tubers. The antioxidant activity was found positive for both ethanolic and aqueous extracts.

Ampelopsis brevipedunculata (Maxim.) Trautv. Family: Vitaceae

English Name: Porcelain berry.

The stem and root of Porcelain berry was traditionally used as an anti-inflammatory diuretic and anti-hepatotoxic agent in folk medicine. Wu *et al.* (2004) investigated the antioxidant activity of the methanol extract of *Ampelopsis brevipedunculata*. The cell-free system showed that the extract exhibitd dose-dependent antioxidant activities against linoleic acid peroxidation and plasmid DNA oxidation. They also demonstrated that the extract possessed strong reducing power and scavenging effects of hydroxyl radicals and DPPH free radicals. The harmful effects of reactive oxygen metabolites of HepG2 cells and the possible antioxidant effects of the methanol extract was also investigated; pretreatment or cotreatment of HepG2 cells with extract could significantly protct cells from H_2O_2-induced oxidative stress.

Ampelopsis grossodentata Handel-Mazzetti Family: Vitaceae

Gao *et al.* (2009) investigated antioxidant activity of Flavonoid-rich extracts from leaves of *Ampelopsis grossodentata* by DPPH, reducing power, antioxidant activity in linoleic acid system. The tests showed that the antioxidant activity of the flavonoid-rich extracts could be comparable with that of butylhydroquinone. The extracts were rich in dihydromyricetin.

Supercritical carbon dioxide extraction of bioactive compounds including flavonoids and phenolics from *Ampelopsis grossedentata* leaves was carried out by Ying *et al.* (2011). Antioxidant activities of the extracts were assessed by DPPH free radical scavenging assay and ferrous ion chelating (FIC) assay. Maximal yields of total flavonoids and phenolics were 5.99 and 3.26 mg GAE/g dry material, respectively, which were achieved under the optimal conditions of 200 bar, 50°C, 80 min and with a modifier of 95 per cent ethanol. Moreover, results demonstrated that the obtained extract under optimized conditions showed the best DPPH radical scavenging ability, but an insignificant negative correlation was found between FIC and total flavonoid contents (TFC) or total phenolic contents (TPC) by Person correlation analysis. In addition, several unreported flavonoids have been detected in the extracts.

Ampelopsis humulifolia Bunge var. *heterophylla* Family: Vitaceae

The antioxidant activity of *Ampelopsis humulifolia* var. *heterophylla* has been investigated by Chen *et al.* (2005) using both enzymatic and non-enzymatic *in vitro* antioxidant assays. Extracts inhibited xanthine oxidase and lipoxygenase activities and were scavengers of the ABTS radical cation using the Trolox equivalent antioxidant capacity assay.

Ampelopsis sinica (Miq.) W.T.Wang. Family: Vitaceae

The antioxidant activity of *Ampelopsis sinica* has been investigated by Chen *et al.* (2005) using both enzymatic and non-enzymatic *in vitro* antioxidant assays. Extracts inhibited xanthine oxidase and lipoxygenase activities and were scavengers of the ABTS radical cation using the Trolox equivalent antioxidant capacity assay.

Amphimas pterocarpoides Harms Family: Fabaceae (Leguminosae)

Biapa *et al.* (2007) investigated the phytochemical and antioxidant properties of four medicinal plants used in Cameroon, including *Amphimas pterocaroides*. Four extracts (methanol, hydro-ethanolic, aqueous and hydrolysed) from each of the plants were prepared and analysed. Folin Ciocalte, FRAP and DPPH assays were used to determine the antioxidant potential. The highest antioxidant activities were observed in the hydrolysed extracts of each plant, while the aqueous extract showed the least activity irrespective of the method used.

Anacamptis pyramidalis L. Family: Orchidaceae

Stajner *et al.* (2010) investigated the antioxidative properties of flowers and above ground part of *Anacamptis pyramidalis* from Vojvodina. Activities of antioxidant enzymes (superoxide dismutase, catalase, peroxidase and glutathione peroxidase), quantities of malonyldialdehyde, superoxide and hydroxyl radicals and reduced glutathione and also the contents of chlorophyll a and b, carotenoids and soluble proteins were determined. Lipofusion 'plant age pigments', total antioxidant capacity and scavenging activity were also determined. Their results indicated that the above ground part of the plant exhibited higher antioxidant activity due to low MDA and lipofuscin pigment accumulation (2.82 nmol/mg protein; 92.90 per cent), higher scavenging activity (54.16 per cent) and antioxidant capacity (432.00 micromol Fe(2+)/dm(3).

Anacardium occidentale L. Family: Anacardiaceae

English Name: Cashew nut.

Bimla Singh *et al.* (2004) investigated the modulation of antioxidant potential in liver of mice by kernel oil of *Anacardium occidentale* and its lack of tumor promoting ability in DMBA induced skin papillomagenesis. Abbas *et al.* (2006) evaluated the antioxidant activity of *Anacardium occidentale* using DPPH assay. Methanolic extract showed DPPH radical scavenging activity

Schmourlo *et al.* (2007) evaluated the antioxidant activity of the aqueous extract of fruits of *Anacardium occidentale* by DPPH radical scavenging method. The extract showed DPPH radical scavenging activity above 90 per cent at 100µg/ml. The activity was more than 50 per cent even at 10µg/ml.

The total phenolic content and antioxidant activities of methanol, hexane and ethyl acetate extracts of the shoots of *Anacardium occidentale* were measured by Razali *et al.* (2008). Total phenolic content was assessed by the Folin-Ciocalteau assay whereas antioxidant activities were assessed by measuring the ability of the extracts to scavenge the ABTS and DPPH radicals, superoxide anion radicals and nitric oxide radicals as well as their ability to reduce ferric ions. Results indicated that the methanol extract of *Anacardium occidentale* was the most potent reducing agent and radical-scavengers compared to the other two extracts. The methanol extract contained more than 7 fold of total phenolic content compared to the hexane and ethyl acetate extracts indicating the likely possibility that the observed antioxidant activities were partly contributed by the phenolics.

Anacyclus pyrethrum DC. Family: Asteraceae (Compositae)

Kalim *et al.* (2010) evaluated the antioxidant activity of Methanol (50 per cent) extract of *Anacyclus pyrethrum*. IC_{50} values for DPPH, ABTS, NO, OH, O_2 and ONOO were 467.10, 31.76, 41.22, 83.49 and 1.137µg/ml. The extract contained significant amount of polyphenols.

Anadenanthera colubrina (Vell.) Brenan Family: Fabaceae (Leguminosae)

De Melo *et al.* (2010) evaluated the antioxidant capacity and tannin content in *Anadenanthera colubrina*. IC_{50} value of DPPH radical scavenging was 73.24µg/ml while tannin content was 4.41 mg/g.

Anadenanthera macrocarpa Brenan Family: Fabaceae (Leguminosae)

The *in vitro* antioxidant and free radical scavenging properties of bark extract of *Anadenanthera macrocarpa* used as anti-inflammatory agent in the Brazilian state of Bahis, was studied using different bioassays (Desmarchelier *et al.*, 1999). The total reactive antioxidant potential (TRAP) of the aqueous and methanolic extracts was determined by monitoring the intensity of luminal-enhanced chemiluminescence (CL), using 2,2′-azo-bis(2-amidinopropane) as a peroxyl radical source.The extract showed highest activity, measured as Trolox concentration, was observed in the methanolic extract of *A. macrocarpa* (TRAP =3028 µM). Lipid peroxidation was assessed by means of the production of TBRAS and hydroperoxide-initiated CL in rat liver homogenates. As in the case of luminal-enhanced Cl, the extract tested was effective in reducing the production of TBARS. The highest activity was observed in the aqueous extract (IC_{50}=54µg/ml). Extract also reduced hydroperoxide-initiated CL, the methanolic extract showed an IC_{50}=2.0µg/ml (Desmarchelier *et al.*, 1999).

Ananas comosus Merr. Family: Bromeliaceae

The antioxidant capacity and phenol content of honey pineapple (*Ananas comosus*) was studied by Alothman *et al.* (2009). Three solvent systems were used (methanol, ethanol and acetone) at three different concentrations (50, 70 and 90 per cent) and with 100 per cent distilled water. The antioxidant capacity of the fruit extracts was evaluated using a ferric reducing/antioxidant power assay and the free radical-scavenging capacity was evaluated using 2,2-diphenyl-1-picrylhydrazyl radical-scavenging assays. The polyphenol content was 34.7 to 54.7 Gallic Acid Equivalent (GAE)/100g. High phenol content was significantly correlated with high antioxidant capacity.

Kataki (2010) evaluated *in vitro* antioxidant activity of ethanolic extract of *Ananas comosus* tender leaves employing reducing power assay and DPPH radical scavenging activity. The results indicated higher antioxidant activity of the extract than α-tocopherol but lower than BHA. At the concentration of 250µg/ml, the extract showed noticeable DPPH radical scavenging activity.

Anchietea pyrifolia (Mart.) G. Don Family: Violaceae

Menezes *et al.* (2004) investigated the antioxidant activity of hydroalcoholic extracts of leaves of *Anchietea pyrifolia* by DPPH assay and phosphomolybdenum method. The relative activity in phosphomolybdenum complex was 0.35 while scavenging activity against DPPH in IC_{50} (µg/ml) was 391.81.

Andrographis paniculata Nees Family: Acanthaceae

In rats treated with *Andrographis paniculata* the activity of Ca^{2+}-ATPase and $Na^{+}K^{+}$-ATPase increased remarkably and lipid peroxidation decreased significantly (Guo *et al.*, 1995). Zhang and Tan (2000) evaluated antihyperglycemic and antioxidant properties of ethanolic extract of aerial parts of *Andrographis paniculata* in normal and diabetic rats. Results indicate that *A. paniculata* reduced oxidative stress in diabetic rats.

In order to find their potential sources, the crude extracts of different parts of some selected food and medicinal plants, including *Andrographis paniculata,* were studied by Bajpai *et al.* (2005) for total phenolic contents (TPC) and antioxidant activity (AOA). AOA was assayed by auto-oxidation of β-carotene and linoleic acid and expressed as the per cent of inhibition relative to control. TPC in aerial parts was 10.6 mg/g (GAE) while antioxidant activity was 58.2 per cent.

Kruawan and Kangsadalampai (2006) investigated the antioxidant activity and phenolic compound contents of water extract of *Andrographis paniculata*. The extract exhibited low scavenging activity against DPPH radicals (58.02 per cent) and their FRAP value was 486.92 µmol/g. The content of phenolic compound in the extracts was determined using the Folin-Ciocalteu reagent and calculated as gallic acid equivalent (GAE). The herbal extract gave a high phenolic compound content (GAE 38.65 mg/g).

Extract of *Andrographis paniculata* was a potent scavenger of DPPH, ABTS radicals, exemplified by ESR signals, O_2, OH and H_2O_2 displayed excellent reducing power, FRAP potentials to reduce Fe (III) to Fe (II) and had considerable amount of phenolics/ flavonoids contents, an effective antioxidant index (Tripathi and Kamat, 2007). The observed antioxidant effect might be primarily due to its high scavenging ability for ROS. Effect was confirmed *ex vivo* following inhibition in peroxidation, restoration in SOD enzyme, SOD band intensity and protein degradation in *Andrographis paniculata* fed liver homogenate. Phansawan and Poungbangpho (2007) reported antioxidant capacities of *Andrographis paniculata* by ABTS method. Ethanol was the best solvent for extraction to give the highest antioxidant capacity, followed by acetone, methanol, distilled water and acetic acid.

Wojcikowski *et al.* (2007) evaluated the antioxidant activity of aerial parts of *Andrographis paniculata.* Oxygen radical absorbance capacity of the ethyl extract, methanol extract and aqueous extracts were 5.25, 168.10 and 346.61µmol Trolox equivalent per g of dried starting material, respectively. Ojha *et al.* (2009) reported antioxidant, antilipidperoxidative and antiischemic activity of *Adrographis paniculata* hydroalcoholic extract.

Extracts prepared from cultivated *Andrographis paniculata* and their active constituent andrographiolide were evaluated for antioxidant, antioedema and analgesic activities. The results showed that the aqueous extract exhibited a great antioxidant activity, xanthine oxidase inhibition and antilipid peroxidation activities for aqueous extract were 66.8 per cent, 57. per cent and 65.3 per cent respectively. And for ethyl alcohol extract were 57.8 per cent, 52.6 per cent and 34.2 per cent, respectively. Aqueous extract was more potent than ethyl alcohol extract (Lin *et al.*, 2009).

Rafat *et al.* (2009) investigated the antioxidant capacity and total content of phenolic compounds in an ethanolic extract of leaves, stems and fruits of *Andrographis paniculata.* The antioxidant activity of the samples was evaluated by the red cell hemolsis assay, free radical scavenging and superoxide dismutase activities. The Folin-Ciocalteu regent assay was used to estimate the phenolic content of extracts. Leaf extracts showed the highest antioxidant potential followed by stem and fruit extracts with the rabbit erythrocytes hemolysis and superoxide dismutase activity assays. However, *A. paniculata* fruit extracts exhibited the highest DPPH free radical scavenging activity compared to the other extracts. A positive correlation between free radical scavenging capacity and the content of phenolic compounds was found in the fruit, leaf and stem extracts of this plant (Rafat *et al.*, 2009).

Jain and Agrawal (2009) reported antioxidant activity of the methanolic extract of *Andrographis paniculata.* Ojha *et al.* (2009) reported that hydroalcoholic extract of *Andrographis paniculata* prevented isoproterenol induced increase in lipid peroxidation and increased the activities of antioxidant enzymes, *viz.* SOD, CAT, glutathione peroxidase and the levels of reduced glutathione in hearts.

The edaphoclimatic conditions that affect the yield and qualities of *Andrographis paniculata* were studied by Tewari *et al.* (2010). Under seasonal and soil pH variations, andrographiolide content ranged between 2.25 – 3.85 per cent in aerial parts of the plant. Total phenolic content, antioxidant activity and reducing power were 20.3

mg/g GAE, 66.8 per cent and 2.8 ASE/ml. FRSA measured by DPPH showed an IC_{50} of 0.34 mg/ml, EC_{50} of 14.7 mg/ml DPPH, and anti-radical power of 6.76 AOA in different assays expressed as IC_{50} ranged from 0.31 to 1.29 mg/ml. The extract showed significant protective effect against Fenton's reaction on superocoiled pUC 18 DNA.

Saranya *et al.* (2010) evaluated the *in vitro* antioxidant activity of aerial parts of *Andrographis paniculata* by using solvents such as petroleum ether, ethyl acetate, ethanol, hydroalcohol and purified andrographiolide. The antioxidant activity was determined by using DPPH, superoxide, hydroxyl and nitric oxide generating models and reducing power. Among all the fractions studied, hydroalcoholic fractions exhibited significant free radical scavenging activity in all the models. The activities were found to be dose-dependent and this study indicated that both hydroalcoholic extract and andrographiolide exhibited free radical scavenging activity *in vitro.*

Andromeda polifolia L. Family: Ericaceae

English Name: Bog-rosemary.

The antioxidative activity of phenolic extracts from bog-rosemary was examined by autoxidation of methyl linoleate by Kahkonen *et al.* (1999). The inhibition of MeLo-conjugated diene hydroperoxides of the plant extract was very high in Bog-rosemary (97 per cent at 500 ppm level). Total phenolics were 32.8 mg/g GAE. Quercetin, quercetin-monopentosides (quaijaverine and avicularine) and quercetin-dipentoside (polifolioside) were isolated in a methanolic extract of bog-rosemary (Pachaly and Klein, 1987).

Anemarrhena asphodeloides Bge. Family: Liliaceae

Li *et al.* (2008) evaluated the antioxidant properties of 45 medicinal plants including *Anemarrhena asphodeloides* using FRAP and TEAC assays and the total phenolic contents of these plants were measured. *Anemarhena asphodeloides* showed 83.80 µmol Fe(II)/g FRAP value, 63.19µmol Trolox/g of TEAC value and 10.45 mg GAE/g of Phenolic content. A high correlation between antioxidant capacities and their phenolic contents indicated that phenolic compounds were a major contributor of antioxidant activity of these plants.

Anethum graveolens L. Family: Apiaceae (Umbelliferae)

English Name: Dill

Chanwitheesuk *et al.* (2005) investigated the antioxidant activity of seeds of *Anethum graveolens* using a β-carotene bleaching method. The contents of plant chemicals such as vitamin C, Vitamin E, carotenoids, tannin and total phenolics, were also determined. Methanolic extract showed antioxidant activity. *Anethum graveolens* has antioxidant index of 1.09, Vitamin C 11.7 mg per cent, Vitamin E 0.0039mg per cent, Total carotenes 0.64 mg per cent, Total xanthophylls 0.53mg per cent, Tannins 7.78mg per cent and Total phenolics of 66.1mg per cent.

Souri *et al.* (2008) evaluated the antioxidant activity against linoleic acid peroxidation using 1,3-diethyl-2-thiobarbituric acid as reagent. Antioxidant activity (IC_{50}) against peroxidation of linoleic acid (2mg/ml) was 146.75 and phenolic content was 263.59 mg/100g dry weight.

Mariutti *et al.* (2008) evaluated the antioxidant activity of garlic employing ABTS and DPPH radical assay. DPPH EC_{50} value was 75g/kg while DPPH-TEAC value was 4.4mM/g. ABTS-TEAC value was 6.9mM/g.

Antioxidant potential of selected vegetables commonly used in diet in Asian subcontinent, including *Anethum graveolens* leaves has been evaluated by Gacche *et al.* (2010). The DPPH radical scavenging activity was 44.6 per cent, inhibition of lipid peroxidation was 56.9 per cent, Vitamin C content was 10.24 mg/100g and total phenols was 8.70 mg/g.

Angelica archangelica L. Family: Apiaceae (Umbelliferae)

Wojcikowski *et al.* (2007) evaluated the antioxidant activity of aerial parts of *Angelica archangelica*. Oxygen radical absorbance capacity of the ethyl extract, methanol extract and aqueous extracts were 259.51, 42.44 and 46.34µmol Trolox equivalent per g of dried starting material, respectively.

Angelica biserrata Yuan et Shan Family: Apiaceae (Umbelliferae)

Gan *et al.* (2010) evaluated the antioxidant activity of *Angelica biserrata* using the FRAP and TEAC assays, and its total phenolic content was measured by the Folin-Ciocalteu method. For FRAP assay the value was 68.99 µmol Fe(II)/g dry weight. For TEAC assay, the value was 32.67µmol Trolox/g dry weight. Total phenolic content was 7.63 mg gallic acid equivalent/g dry weight of plant material.

Angelica dahurica Benth. et Hook. Family: Apiaceae (Umbelliferae)

Antioxidant capacities of 56 selected Chinese medicinal plants, including *Angelica dahurica* were evaluated by Song *et al.* (2010) using the TEAC and FRAP assays, and their total phenolic content was measured by the Folin-Ciocaleteu method. The TEAC value was 20.79µmol Trolox/g, FRAP value was 27.36µmol Fe^{2+}/g and phenolic content was 2.94mg GAE/g.

Study was conducted by Thuong *et al.* (2010) to evaluate structure-activity relationships of the antioxidant activity of natural coumarins isolated from *Angelica dahurica*. The free radical scavenging and lipid peroxidation assays revealed that phenolic coumarin umbelliferone possessed considerable antioxidant activity.

Angelica gigas Nakai Family: Apiaceae (Umbelliferae)

Ma *et al.* (2009) reported that ESP-102, a combined extract of *Angelica gigas*, *Saurus chinensis* and *Schizandra chinensis*, protects against glutamate-induced toxicity in primary cultures of rat cortical cells. These results support that actual mechanism of neuroprotectivee activity of ESP-102 against gluatamate-induced oxidative stress might be its antioxidative activity.

Angelica keiskei Koidzumi Family: Apiaceae (Umbelliferae)

Cho *et al.* (2008) evaluated the antioxidant activity of *Angelica keiskei*. Water extract increased the induction of IL-6, TNF and nitric oxide on RAW 264.7 cells.

Angelica sinensis (Oliv.) Diels Family: Apiaceae (Umbelliferae)

Wu *et al.* (2005) evaluated the antioxidant activities of *Angelica sinensis*. The results showed that aqueous extracts of these drugs exhibited antioxidant activities in a concentration-dependent manner. The extract displayed an inhibitory effect on $FeCl_2$-ascorbic acid and lipid peroxidation in rat liver homogenate *in vitro.*

Cai *et al.* (2004, 2006) evaluated the antioxidant activity and phenolic compounds in traditional Chinese medicinal plants associated with anticancer. The improved ABTS.+ method was used to systematically assess the total antioxidant capacity (Trolox equivalent antioxidant capacity, TEAC) of the medicinal extracts. The TEAC values and total phenolic content for methanolic extract of root of *Angelica sinensis* were 101.7 µmol Trolox equivalent/100g dry weight (DW), and 0.58 g of gallic acid equivalent/100g DW respectively. A positive significant linear relationship between antioxidant activity and total phenolic content showed that phenolic compounds were the dominant antioxidant components in the tested herbs. Major types of phenolic compounds in *Angelica sinensis* include simple phenols (vanillin, *p*-cresol), phenolic acids (ferulic acid).

Wojcikowski *et al.* (2007) evaluated the antioxidant activity of aerial parts of *Angelica sinensis.* Oxygen radical absorbance capacity of the ethyl extract, methanol extract and aqueous extracts were 47.50, 74.61 and 122.57µmol Trolox equivalent per g of dried starting material, respectively.

Aniba panurensis (Meisn.) Mez. Family: Lauraceae

Alcantara *et al.* (2010) reported antioxidant activity of the essential oil of *Aniba panurensis* from Brazil. Authors followed DPPH method coupled to TLC. The IC_{50} values were generally >1000µg/ml, substantially superior to that of quercetin, the reference used by the authors.

Aniba rosaeodora Ducke Family: Lauraceae

Alcantara *et al.* (2010) reported antioxidant activity of the essential oil of *Aniba rosaeodora* from Brazil. Authors followed DPPH method coupled to TLC. The IC_{50} values were generally >1000µg/ml, substantially superior to that of quercetin, the reference used by the authors.

Anisomeles indica (L.) O.Kuntze. Family: Lamiaceae (Labiatae)

Essential oils obtained from *Anisomeles indica* leaves (AIL EO) and roots (AIR EO) were studied for *in vitro* antioxidant properties. IC_{50} value of AIL EO and AIR EO were found to be 52.63 and 57.77 µg/ml in DPPH radical scavenging activity. AIL EO and AIR EO had powerful dose dependent free radical scavenging property in all antioxidant models. AIR EO showed 56.66, 64/04 and 64.5 per cent inhibition in lipid peroxidation at the 40, 60 and 80 µg/ml of oils respectively, while 40 µg/ml of ascorbic acid, exhibited 43.20 per cent inhibition. AIR EO had effective reducing power, DPPH and nitric oxide scavenging activities at different concentrations. AIL EO also showed significant percent inhibition at 80 µg for all models. All antioxidant activities were compared to standard antioxidants such as BHA and ascorbic acid (Tatiya *et al.*, 2007).

Anisopus mannii N.E.Br. Family: Asclepiadaceae

Aliyu *et al.* (2010) evaluated the antioxidant activity of crude methanol extract, n-butanol and ethyl acetate fractions of *Anisopus mannii* using DPPH radical, total phenolic contents and reducing power assay. The results of the DPPH free radical scavenging at 250 μg/ml indicated maximum antioxidant activity of 94.1, 94.3 and 88.7 per cent for crude methanol extract, ethyl acetate and n-butanol fractions, respectively. The total phenolic content was higher in the ethyl acetate fraction (56.40 mg/g GAE) and crude methanol extract (53.20 mg/g GAE). The reducing power of the n-butanol fraction was 0.112 nm higher than those of ethyl acetate fraction (0.110nm) and crude methanol extract (0.106 nm).

Annona muricata L. Family: Annonaceae

Antioxidant potential of leaves of *Annona muricata* was studied by Baskar *et al.* (2007) by using different *in vitro* models – DPPH, ABTS, nitric oxide, superoxide, hydroxyl radical and lipid peroxidation. The ethanolic extract at 500μg/ml showed maximum scavenging activity (90.05 per cent) of ABTS radical cation followed by the scavenging of hydroxyl radical (85.88) respectively. However, the extract showed only moderate lipid peroxidation inhibition activity.

Annona reticulata L. Family: Annonaceae

Antioxidant potential of leaves of *Annona reticulata* was studied by Baskar *et al.* (2007) using different *in vitro* models – DPPH, ABTS, nitric oxide, superoxide, hydroxyl radical and lipid peroxidation. The ethanolic extract at 500μg/ml showed maximum scavenging activity (89.37 per cent) of DPPH radical followed by (89.05 per cent) ABTS and (80.88 per cent) superoxide, respectively. However, the extract showed only moderate lipid peroxidation inhibition activity.

Annona senegalensis Pers. Family: Annonaceae

Potchoo *et al.* (2008) evaluated the antioxidant activity of leaves of *Annona senegalensis* using the DPPH method. The content of total polyphenol (3.47 per cent) and flavonoid (2.33 per cent) of 70 per cent (v/v) aqueous methanol extract of the specimen from Togo was significantly higher than the one from Burkina Faso (2.66 per cent and 1.64 per cent respectively), whereas the amount of total flavonoid in the ethyl acetate extract of the species from Burkina (40.38 per cent) was triplicated. For the two types of extracts, the species of Burkina Faso showed an improved antioxidant activity than the one of Togo (IC_{50} = 8.51 and 21.08 μg/ml respectively). These flavonoids were rutin and isoquercetin as flavonols (specimen from Togo) of which are added epicatechin and catechin derivatives (flavonols) in the specimen from Burkina.

Annona squamosa L. Family: Annonaceae

English Name: Custard apple.

The ethanolic extract if *Annona squamosa* at 1000 microg/ml showed maximum scavenging of the radical cation ABTS (99.07 per cent) followed by the scavenging stable radical DPPH (89.77 per cent) and nitric oxide radical (73.64 per cent at the same concentration (Shirwaikar *et al.*, 2004). However, the extract showed only moderate scavenging activity of superoxide radicals and antilipid peroxidation potential, which was performed using rat brain homogenate.

Total antioxidant activity of the ethanolic extract of pulp, seed, rind and leaves of *Annona squamosa* was estimated by Padmaja *et al.* (2004) using ABTS decoloration assay. The total antioxidant activity was highest in the seeds (78 per cent) and rind (72 per cent) compared to leaves (61 per cent) and pulp (56 per cent).

Antioxidant potential of leaves of *Annona squamosa* was studied by Baskar *et al.* (2007) by using different *in vitro* models – DPPH, ABTS, nitric oxide, superoxide, hydroxyl radical and lipid peroxidation. The ethanolic extract exhibited inhibition activity in DPPH (88.77 per cent), ABTS (88.06 per cent), Superoxide radical (77.21 per cent) and Hydroxyl radical (79.79 per cent). However, the extract showed only moderate lipid peroxidation inhibition activity.

Ruiz-Terán *et al.* (2008) evaluated the antioxidant activity of aerial parts of *Annona squamosa* using DPPH and β-carotene-linoleic acid bleaching assay. The extract displayed antioxidant activity two times more than the commercial antioxidant BHA.

Anoectochilus formosanus Hay Family: Orchidaceae

Wang *et al.* (2002) evaluated the phytochemical characteristics and antioxidant activities of the crude and fractionated plant extracts of *Anoectochilus formosanus* using five different assay systems. The IC_{50} values of DPPH radicals in the BuOH and acid hydrolyzed BuOH fractions were 0.521 and 0.021 mg/ml, respectively. The acid hydrolyzed BuOH fraction exhibited approximately 5-fold higher activity in scavenging superoxide anion than catechin. The acid-hydrolyzed BuOH fraction also effectively protected ϕ X 174 supercoiled DNA against strand clevage induced by H_2O_2 and reduced oxidative stress in HL-60 cells.

Carbon tetrachloride causes chronic hepatitis, featuring an increase in hepatic hydroxypyroline, spleen weight and serum GPT levels and a decrease in plasma albumin livels. Crude extracts of fresh whole plants of *Anoectochilus formosanus* showed inhibition of chronic hepatitis induced by CCl_4 in mice. Bioactivity-guided fractionation and spectroscopic analysis revealed that kinsenoside was the most active compound. In an *in vitro* study, the LD_{50} values for H_2O_2-induced cytotoxicity in BALB/c normal liver cells were significantly higher after kinsenoside pretreatment than after vehicle along, further confirming that kinsenoside shows significant antihepatotoxic activity (Wu *et al.*, 2006).

Anogeissus dhofarica A.J.Scott Family: Combretaceae

Marwah *et al.* (2007) investigated the antioxidant activity of wound-healing herbs using *in vitro* DPPH and phosphomolybdenum assay methods. Out of the 19 plants screened, the aqueous ethanol extract of *Anogeissus dhofarica* showed the best DPPH scavenging activities with per cent IP value 89.1 per cent, after 15 min of incubation at an effective test concentration of 50μg/ml. The Total Antioxidant Capacity as gallic acid equivalent of 551mg/g of ethanol extract was obtained in the phosphomolybdenum assay. Total phenolic content was 406 mg GAE/g of ethanol extract.

Anogeissus latifolia (Roxb. ex DC.) Wall. ex Guill. et Perr. Family: Combretaceae

Extract of *Anogeissus latifolia* was studied by Govindarajan *et al.* (2004) for total antioxidant capacity, hydrogen-donating ability, nitric oxide, superoxide scavenging activity, hydrogen peroxide decomposition activity along with lipid peroxidation. Integral antioxidative capacity was determined by chemiluminiscence assay by thiobarbituric acid-reactive substances method using rat liver homogenate. The results indicate that the extract has potent antioxidant activity. Also to ascertain the possible season for the potent activity, percentage of gallic acid was estimated and was found to be 0.95 per cent, which could be one of the reasons for potent antioxidant activity. Study was conducted by Govindarajan *et al.* (2005) to isolate antioxidant compounds from bark by activity-guided isolation. Inhibition of DPPH and xanthine oxidase along with photochemiluminiscence assay, were used as bioassay for antioxidant activity. Ethyl acetate and butanol fraction exhibited potent antioxidant activity. Bioassay guided isolation led to the isolation of ellagic acid and dimethyl ellagic acid as the main active compounds.

Status of antioxidant enzymes superoxide dismutase (SOD) and catalase along with lipid peroxidation (LPO) was studied in CRS-induced ulcers. *Anogeissus latifolia* extract decreased LPO and SOD with concomitant increase in catalase activity in CRS-induced ulcers (Govindarajan *et al.*, 2006).

Anthemis tinctoria L. subsp. *tinctoria* var. *pallida* DC.

Family: Asteraceae (Compositae)

From the aerial parts of *Anthemis tinctoria* subsp. *tinctoria* var. *pallida* one glycoside, conduritrol F-1-O-(6′-O-E-p-caffeoyl)-β-D-glucopyranoside (1) and four flavonoids, nicotiflorin (2), isoquercetin (3), rutin (4) and patulitrin (5) were isolated by Papaioannou *et al.* (2007). Methanolic extract and pure isolated compounds were examined for their freee radical scavenging activity using the DPPH free stable radical, and for their inhibitory activity toward soybean lipoxygenase, using linoleic acid as substrate. Compounds 1 and 5 showed strong scavenging effect in the DPPH radical assay. In addition 5 also exhibited high inhibitory activity on soybean lipoxygenase.

Anthocephalus cadamba (Roxb.) Miq. (Syn: *Anthocephalus chinensis* (Lamk.) A.Rich) Family: Rubiaceae

Prahalad *et al.* (2007) investigated the wound healing and antioxidant activities of *Anthocephalus cadamba*. The results indicated that *Anthocephalus cadamba* extract possesses potent superoxide dismutase (SOD) and catalase activities. Sanjay *et al.* (2007) reported that *Anthocephalus cadamba* extract possesses potent antioxidant activity by inhibiting lipid peroxidation and increase in the SOD and catalase activity.

Alam *et al.* (2008) studied the antioxidant property of hydroethanolic extract of the flowering tops of *Anthocephalus cadamba*. The extract induced significant reduction in serum glucose and transminases, *e.g.*, AST, ALT and ALP activities. Significant changes in the TBARS, peroxidase and catalase levels during the experimental period were also observed.

Anthocephalus indicus A. Rich. Family: Rubiaceae

Kumar *et al.* (2009) evaluated the lipid lowering and antioxidant activities in root extract of *Anthocephalus indicus*. The root extract (100-400μg) inhibited the generation of superoxide anions and hydroxyl radicals, in both enzymic and non-enzymic systems, *in vitro*.

Anthurium versicolor **Sodiro Family: Araceae**

Fractionation of a methanolic extract of the leaves of *Anthurium versicolor* has resulted in the isolation of two main fractions, I and II (Aquino *et al.*, 2001). Both the extract and the fractions were assayed for their radical-scavenging activity by means of an *in vitro* test (bleaching of the stable DPPH) and showed a significant radical-scavenging effect.

Anthyllis aurea **Welden Family: Fabaceae (Leguminosae)**

Godevac *et al.* (2008) evaluated the antioxidant activity of nine Fabaceae members including *Anthyllis aurea* using DPPH radical scavenging capacity, TEAC values by ABTS radical cation and inhibition of liposome peroxidation. The plant exhibited strong antioxidant capacity in all the tested methods.

Antidesma thwaitesianum **Muell. Family: Stilaginaceae**

Nuengchamnong and Ingkaninan (2010) investigated the antioxidant activity of phenolic compounds isolated from fruit of *Antidesma thwaitesianum*. The antioxidant compounds were separated and identified as gallic acid, cyanidin-3-sophoroside, monogalloyl glucoside, delphinidin-3-sambubioside, catechin, caffeic acid, and pelargonidin-3-malonyl-glucoside. DPPH assay was used for screening antioxidant activity.

Apama siliquosa **Lamk. Family: Aristolochiaceae**

Thirugnanasampandan *et al.* (2008) investigated the antioxidant activity of five Aristolochiaceae species. *Apama siliquosa* showed good antioxidant, DPPH radical scavenging and reducing power.

Aphanamixis polystachya **(Wall.) Parker Family: Meliaceae**

The ethanolic extract of *Aphanamixis polystachya* was tested in Swiss albino mice transplanted with ehrlich ascites carcinoma and exposed to various doses of gamma-radiation. The treatment with the extract before irradiation elevated lipid peroxidation followed by a reduction in the glutathione contents. Treatment of tumor bearing mice with APE before irradiation further reduced the activities of various antioxidant enzymes at different post last drug administration (Jagetia and Venkatesha, 2005).

Apium graveolens **L. Family: Apiaceae (Umbelliferae)**

English Name: Celery; Japanese: Serurii.

Chanwitheesuk *et al.* (2005) investigated the antioxidant activity of leaves of *Apium graveolens* using a β-carotene bleaching method. The contents of plant chemicals such as vitamin C, Vitamin E, carotenoids, tannin and total phenolics, were also determined. Methanolic extract showed antioxidant activity. *Apium graveolens* has antioxidant index of 6.39, Vitamin C 6.65 mg per cent, Vitamin E 0.0070mg per cent, Total carotenes 1.28 mg per cent, Total xanthophylls 3.72mg per cent, Tannins 4.49mg per cent and Total phenolics of 31.1mg per cent.

Maeda *et al.* (2006) examined the phenolic contents of stem of *Apium graveolens* and its radical scavenging activities for DPPH radical. The polyphenol content was 14.3 mg-gallic acid equivalent/100 g) and DPPH scavenging activity was 15.5 µmol-Trolox eq./100 g).

The antioxidant properties of 25 edible tropical plants including *Apium grveolens,* expressed as TEAC were studied by Wong *et al.* (2006) using DPPH scavenging and reducing FRAP assays. Total phenol content DPPH radical scavenging and FRAP values were very low. A strong correlation between TEAC values obtained for the DPPH assay implied that compounds in the extracts were capable of scavenging the DPPH free radical and reducing ferric ions.

Wojcikowski *et al.* (2007) evaluated the antioxidant activity of aerial parts of *Apium graveolens.* Oxygen radical absorbance capacity of the ethyl extract, methanol extract and aqueous extracts were 60.63, 72.92 and 61.55µmol Trolox equivalent per g of dried starting material, respectively.

Souri *et al.* (2008) evaluated the antioxidant activity of *Apium graveolens* against linoleic acid peroxidation using 1,3-diethyl-2-thiobarbituric acid as reagent. Antioxidant activity (IC_{50}) against peroxidation of linoleic acid (2mg/ml) was 34.75 and phenolic content was 436.05 mg/100g dry weight. The results of this study showed that there is no significant correlation between antioxidant activity and phenolic content of the studied plant materials and phenolic content could not be a good indicator of antioxidant capacity.

Apium graveolens L. var. *dulce* (Miller) Persoon Family: Apiaceae (Umbelliferae)

Souri *et al.* (2004) investigated the antioxidant activity of methanol extracts of *Apium graveolens* var. *dulce,* used as vegetable in Iranian diet against linoleic acid peroxidation. The extracts showed good antioxidant activity.

Aporosa lindleyana (Wt.) Baill. Family: Euphorbiaceae

Four successive and two crude extracts of *Aporosa lindleyana* root were tested for antioxidant activity using standard *in vitro* and *in vivo* models (Badami *et al.*, 2005). The successive methanol and ethyl acetate and crude 50 per cent methanol extracts at 100 and 200 mg/kg body weight showed antioxidant activity with IC_{50} values 3.51, 6.09 and 7.34 respectively, in DPPH method and in the nitric oxide radical inhibition method.

Aquilaria agallocha Roxb. Family: Thymelaeceae

Ethyl acetate extract of heartwood of *Aquilaria agallocha* possess antioxidant activity (Miniyar *et al.*, 2008). Kshirsagar and Upadhyay (2009) also reported antioxidant activity in methanol extract of aerial parts.

Aquilaria malaccensis Lam. Family: Thymelaeaceae

Kshirsagar and Upadhyay (2009) reported antioxidant activity of the methanol extract of stem and aerial part. Their ability of scavenging radicals was measured by DPPH reduction spectrophotometric assay.

Table 3.1: Plants showing Antioxidant Activity

Sl.No.	*Plant Name*	*Family*	*Parts Used*	*References*
1.	*Abelmoschus esculentus*	Malvaceae	Fruits	Shui and Peng, 2004; Ansari *et al.*, 2005; Maeda *et al.*, 2006; Sunilson *et al.*, 2008; Adelaku *et al.*, 2009a,b; Nagashree *et al.*, 2009; Gacche *et al.*, 2010
2.	*Abelmoschus manihot*	Malvaceae	Leaves	Lako *et al.*, 2006
3.	*Abroma angusta*	Malvaceae		Eshrat and Hussain, 2002;
4.	*Abutilon bidentatum*	Malvaceae		Shahwar *et al.*, 2010
5.	*Abutilon indicum*	Malvaceae	Whole plant	Ahmad *et al.*, 2009; Chakraborthy and Ghorpade, 2010; Kaushik *et al.*, 2010
6.	*Abutilon muticum*	Malvaceae		Ahmad *et al.*, 2009
7.	*Acacia asak*	Mimosaceae	Bark	Al-Fatimi *et al.*, 2007; Mohammed *et al.*, 2007
8.	*Acacia auriculiformis*	Mimosaceae	Bark	Singh *et al.*, 2007a, b, c
9.	*Acacia catechu*	Mimosaceae	Bark, stem	Patil *et al.*, 2003; Naik *et al.*, 2003, 2005; Cai *et al.*, 2004, 2006; Hazra *et al.*, 2010
10.	*Acacia confusa*	Mimosaceae	Bark	Chang *et al.*, 2001; Tung *et al.*, 2007, 2009; Tung and Chang, 2010
11.	*Acacia dealbata*	Mimosaceae	Bark	Sowndhararajan *et al.*, 2010
12.	*Acacia nilotica*	Mimosaceae	Fruits	Saleem *et al.*, 2001; Bushra *et al.*, 2007; Singh and Arora, 2007; Singh *et al.*, 2007, 2009; Gowri *et al.*, 2011
13.	*Acacia pennata*	Mimosaceae	Bark	Chanwitheesuk *et al.*, 2005; Sowndhararajan *et al.*, 2010
14.	*Acacia penninivenis*	Mimosaceae		Mothana *et al.*, 2009
15.	*Acacia senegal*	Mimosaceae	Bark	Marwah *et al.*, 2007
16.	*Acalypha diversifolia*	Euphorbiaceae		Mosquera *et al.*, 2007, 2009
17.	*Acalypha fruticosa*	Euphorbiaceae	Leaves	Gupta *et al.*, 2003; Mothana *et al.*, 2010; Rajkumar *et al.*, 2010b
18.	*Acalypha hispida*	Euphorbiaceae	Leaves, twigs	Onocha *et al.*, 2011
19.	*Acalypha indica*	Euphorbiaceae	Whole plant	Marwah *et al.*, 2007; Balakrishnan *et al.*, 2009
20.	*Acalypha platyphylla*	Euphorbiaceae		Mosquera *et al.*, 2007
21.	*Acanthopanax gracilistylus*	Araliaceae		Gan *et al.*, 2010
22.	*Acanthopanax senticosus*	Araliaceae	Stems	Lin and Huang, 2002; Lee *et al.*, 2004; Choi *et al.*, 2006
23.	*Acanthopanax sessiliflorus*	Araliaceae		Jun *et al.*, 2007

Contd...

Table 3.1–*Contd...*

Sl.No.	*Plant Name*	*Family*	*Parts Used*	*References*
24.	*Acanthopanax trifoliatum*	Araliaceae		Chanwitheesuk *et al.*, 2005
25.	*Acanthospermum australe*	Asteraceae		Desmarchelier *et al.*, 1997
26.	*Acanthus ilicifolius*	Acanthaceae	Leaves	Babu *et al.*, 2001
27.	*Acer albopurpurascens*	Aceraceae		Lee *et al.*, 2006
28.	*Acer buerferianum* var. *formosanum*	Aceraceae		Hou *et al.*, 2003
29.	*Acer okamotoanum*	Aceraceae	Leaf, twig	Jin *et al.*, 2007
30.	*Acer platanoides*	Aceraceae	Leaf	Kahkonen *et al.*, 1999
31.	*Achillea alexandri-regis*	Asteraceae	Aerial parts	Kundakovic *et al.* (2005)
32.	*Achillea biebersteini*	Asteraceae	Herb	Sokmen *et al.*, 2004; Özgen *et al.*, 2004; Konyalioglu and Karamanderes, 2004
33.	*Achillea clypeolata*	Asteraceae	Herb	Konyalioglu and Karamanderes, 2004
34.	*Achillea coarctata*	Asteraceae	Herb	Konyalioglu and Karamanderes, 2004
35.	*Achillea collina*	Asteraceae		Giorgi *et al.*, 2009, 2010
36.	*Achillea crithmifolia*	Asteraceae		Konyalioglu and Karamenderes, 2004, 2005
37.	*Achillea falcata*	Asteraceae		Konyalioglu and Karamenderes, 2004, 2005
38.	*Achillea fragrantissima*	Asteraceae	Shoot	Al-Mustafa and Al-Thunibat, 2008
39.	*Achillea grandifolia*	Asteraceae	Herb	Konyalioglu and Karamanderes, 2004
40.	*Achillea kotschyi*	Asteraceae	Herb	Konyalioglu and Karamanderes, 2004
41.	*Achillea ligustica*	Asteraceae	Flowers	Tuberoso *et al.*, 2005, 2009; Conforti *et al.*, 2005
42.	*Achillea millefolium*	Asteraceae	Leaves	Kahkonen *et al.*, 1999; Mantle *et al.*, 2000; Candan *et al.*, 2003; Lado *et al.*, 2004; Konyalioglu and Karamenderes, 2004, 2005; Wojcikowski *et al.*, 2007; Burk *et al.*, 2010; Trumbeckaite *et al.*, 2011; Vitalini *et al.*, 2011
43.	*Achillea multifida*	Asteraceae	Herb	Konyalioglu and Karamanderes, 2004
44.	*Achillea nobilis* subsp. *neilrechii*	Asteraceae	Herb	Konyalioglu and Karamenderes, 2004, 2005
45.	*Achillea phyrigia*	Asteraceae	Herb	Konyalioglu and Karamanderes, 2004

Contd...

Table 3.1–*Contd...*

Sl.No.	Plant Name	Family	Parts Used	References
46.	*Achillea schischkinii*	Asteraceae	Herb	Konyalioglu and Karamanderes, 2004
47.	*Achillea setacea*	Asteraceae	Herb	Konyalioglu and Karamanderes, 2004
48.	*Achillea tenuifolia*	Asteraceae	Flowers	Souri *et al.*, 2004a, 2008; Asgarirad *et al.*, 2010
49.	*Achillea teretifolia*	Asteraceae	Fruit	Konyalioglu and Karamenderes, 2004, 2005; Serteser *et al.*, 2009
50.	*Achillea wilhelmsii*	Asteraceae		Souri *et al.*, 2004a; Özgen *et al.*, 2004
51.	*Achyranthes aspera*	Amaranthaceae	Herb, root, seeds	Suresh Kumar *et al.*, 2008; Jitendra, 2009; Priya *et al.*, 2010; Patel and Patel, 2010
52.	*Achyranthes bidentata*	Amaranthaceae		Gan *et al.*, 2010
53.	*Achyranthes longifolia*	Amaranthaceae		Gan *et al.*, 2010
54.	*Achyrocline satureioides*	Asteraceae	Plant	Desmarchelier *et al.*, 1998; Arrendondo *et al.*, 2004
55.	*Acokanthera oppositifolia*	Apocynaceae	Stems	Adedapo *et al.*, 2008
56.	*Aconitum carmichaeli*	Ranunculaceae	Root	Cai *et al.*, 2004
57.	*Aconitum kusnezoffi*	Ranunculaceae	Root	Cai *et al.*, 2004
58.	*Aconitum napellus* ssp. *lusitanicum*	Ranunculaceae	Herb	Luis *et al.*, 2006
59.	*Aconitum napellus* subsp. *neomontanum*	Ranunculaceae	Herb	Braca *et al.*, 2003
60.	*Aconitum napellus* subsp. *tauricum*	Ranunculaceae	Herb	Braca *et al.*, 2003
61.	*Aconitum paniculatum*	Ranunculaceae	Herb	Braca *et al.*, 2003
62.	*Aconitum vulparia*	Ranunculaceae	Herb	Braca *et al.*, 2003
63.	*Acorus calamus*	Araceae	Rhizome, leaves	Acuna *et al.*, 2002; Govindarajan *et al.*, 2004; Mahaboob *et al.*, 2005; Bajpai *et al.*, 2005; Zahin *et al.*, 2009; Dixit and Ali, 2010; Rajkumar *et al.*, 2010c; Narayanaswamy and Balakrishnan, 2011
64.	*Acronychia pedunculata*	Rutaceae	Stem and root bark	Su *et al.*, 2003
65.	*Acroptilon repens*	Asteraceae		Souri *et al.*, 2004
66.	*Actaea racemosa*	Ranuncuoaceae		Jiang *et al.*, 2005

Contd...

Table 3.1–*Contd...*

Sl.No.	*Plant Name*	*Family*	*Parts Used*	*References*
67.	*Actinidia chinensis* (Syn.: *Actinidia deliciosa*)	Actinidiaceae	Fruit	Wang *et al.*, 1996; Cai *et al.*, 2004; Jung *et al.*, 2005; Bai *et al.*, 2007; Mohammad, 2011
68.	*Adansonia digitata*	Bombacaceae	Fruits	Lamien-Meda *et al.*, 2008; Oloyede *et al.*, 2010; Nhukarume *et al.*, 2010
69.	*Adenia gummifera*	Passifloraceae	Stems	Adedapo *et al.*, 2008
70.	*Adenia lobata*	Passifloraceae	Stems, Leaves	Mamyrbekova-Beka *et al.*, 2008; Marcel *et al.*, 2011
71.	*Adhatoda vasica*	Acanthaceae		Jayakunmar *et al.*, 2009; Padmaja *et al.*, 2011; Roja *et al.*, 2011
72.	*Aegiceras corniculatum*	Myrsinaceae	Stem	Roome *et al.*, 2008; Vadlapudi and Naidu, 2009
73.	*Aegiphyla sellowiana*	Verbenaceae		Santos *et al.*, 2009
74.	*Aegle marmelos*	Rutaceae	Leaves, fruits	Kamalakkannan *et al.*, 2003; Sabu and Kuttan, 2003, 2004; Upadhya *et al.*, 2004; Kalkar and Sambare, 2005; Kruawan and Kangsadalampai, 2006; Sharma and Prasad, 2008a; Dheeba *et al.*, 2010
75.	*Aerva lanata*	Amaranthaceae		Shirwaikar *et al.*, 2004; Nevin and Vijayammal, 2005; Saraswathy *et al.*, 2009
76.	*Aerva tomentosa*	Amaranthaceae	Flowers	Singh *et al.*, 2009; Sethi and Sharma, 2011
77.	*Aesculus indica*	Sapindaceae	Leaves	Chakraborthy, 2009
78.	*Aframomum danielli*	Zingiberaceae	Seed	Odukoya *et al.*, 2005
79.	*Aframomum giganteum*	Zingiberaceae	Leaf, rhizome	Agnaniet *et al.*, 2004
80.	*Aganosma maritima*	Apocynaceae		Patnibul *et al.*, 2008
81.	*Agastache rugosa*	Lamiaceae	Leaf	Oh *et al.*, 2006
82.	*Agathosma betulina*	Rutaceae	Aerial	Wojcikowski *et al.*, 2007
83.	*Agave sisalina*	Agavaceae	Leaves	Chen, C-H. *et al.*, 2009
84.	*Ageratina popayensis*	Asteraceae		Mosquera *et al.*, 2009
85.	*Ageratum conyzoides*	Asteraceae		Shirwaikar *et al.*, 2003; Patil *et al.*, 2010
86.	*Agrimonia eupatoria*	Rosaceae	Herb	Ivanova *et al.*, 2005; Venskutonis *et al.*, 2007; Wojcikowski *et al.*, 2007
87.	*Agrimonia pilosa*	Rosaceae	Aerial parts	Cai *et al.*, 2004, 2006; Gan *et al.*, 2010
88.	*Agrimonia procera*	Rosaceae		Venskutonis *et al.*, 2007
89.	*Agrostemma githago*	Caryophyllaceae	Herb	Avci *et al.*, 2006
90.	*Ajuga chamaecistus*	Lamaceae	Aerial parts	Firuzi *et al.*, 2010

Contd...

Table 3.1–*Contd...*

Sl.No.	Plant Name	Family	Parts Used	References
91.	*Ajuga reptans*	Lamiaceae	Herb	Ono *et al.*, 2000; Dal Monte *et al.*, 2008
92.	*Akebia trifoliata*	Lardizabalaceae		Gan *et al.*, 2010
93.	*Albizia amara*	Momosaceae		Suresh Kumar *et al.*, 2008
94.	*Albizia julibrissin*	Mimosaceae	Stem bark	Jung *et al.*, 2003
95.	*Albizia lebbeck*	Mimosaceae		Resmi *et al.*, 2006; Pathak *et al.*, 2010
96.	*Alchemilla vulgaris*	Rosaceae	Flowers, fruits	Trouillas *et al.*, 2003; Kiselova *et al.*, 2006; Hamd *et al.*, 2007
97.	*Alchemilla xanthochlora*	Rosaceae		Filipek, 1994; Ondrojevic *et al.*, 2009
98.	*Alchornea coelophylla*	Euphorbiaceae		Mosquera *et al.*, 2007
99.	*Alchornea cordifolia*	Euphorbiaceae	Leaves	Kouakou-Siransy *et al.*, 2010
100.	*Alchornea glandulosa*	Euphorbiaceae		Mosquera *et al.*, 2009
101.	*Alchornea grandiflora*	Euphorbiaceae		Mosquera *et al.*, 2009
102.	*Alchornea laxiflora*	Euphorbiaceae		Oloyede *et al.*, 2010
103.	*Alhagi maurorum*	Fabaceae		Ahmad *et al.*, 2010
104.	*Alisma orientale*	Alismataceae		Gan *et al.*, 2010
105.	*Alisma plantago-aquatica*	Alismstaceae	Herb, root	Kim *et al.*, 2007
106.	*Alkanna albugam*	Boraginaceae		Kaliora and Andrikopoulos, 2005
107.	*Alkanna tinctoria*	Boraginaceae	Roots	Assiopoulou and Papageorgiou, 2004, 2005; Sengul *et al.*, 2009
108.	*Allanblackia floribunda*	Clusiaceae	Fruit, leaves	Ayoola *et al.*, 2008
109.	*Alliaria petiolata*	Brassicaceae	Herb	Kumarasamy *et al.*, 2004
110.	*Allium ascalonicum*	Liliaceae	Bulb	Povichit *et al.*, 2010
111.	*Allium atroviolaceum*	Liliaceae	Bulb	Tepe *et al.*, 2005; Stajner *et al.*, 2008
112.	*Allium cepa*	Liliaceae	Bulbs, oils	Velioglu *et al.*, 1998; Helen *et al.*, 1999, 2000; Mcanlis *et al.*, 1999; Kahkonen *et al.*, 1999; Kumari and Augusti, 2002; Nuutila *et al.*, 2003; Stajner *et al.*, 2003; Ichikawa *et al.*, 2003; Agostini *et al.*, 2004; Souri *et al.*, 2004; Odukoya *et al.*, 2005; Benkeblia, 2005; Ichikawa *et al.*, 2006; Prakash *et al.*, 2007; Jang *et al.*, 2008; Stajner *et al.*, 2008; Mariutti *et al.*, 2008; Raghu *et al.*, 2010
113.	*Allium dictyoprosum*	Liliaceae	Bulb	Tepe *et al.*, 2005
114.	*Allium fistulosum*	Liliaceae	Leaf	Cai *et al.*, 2004; Maeda *et al.*, 2006; Wong *et al.*, 2006; Aoyama and Yamamoto, 2007; Stajner *et al.*, 2008

Contd...

Table 3.1–*Contd...*

Sl.No.	*Plant Name*	*Family*	*Parts Used*	*References*
115.	*Allium flavum*	Liliaceae	Bulb	Štajner *et al.*, 2008; Štajner and Popovic, 2010
116.	*Allium giganteum*	Liliaceae	Bulb, leaf and stalk	Štajner *et al.*, 2006
117.	*Allium hirtifolium*	Liliaceae	Corm	Souri *et al.*, 2008
118.	*Allium neopolitanum*	Liliaceae		Nencini *et al.*, 2007
119.	*Allium nevesehirense*	Liliaceae	Bulb	Tepe *et al.*, 2005
120.	*Allium nutans*	Liliaceae		Štajner *et al.*, 1999, 2008
121.	*Allium porrum*	Liliaceae	Aerial part	Souri *et al.*, 2004; Aslan *et al.*, 2010
122.	*Allium psekemense*	Liliaceae	Bulb, stalk, leaves	Štajner *et al.*, 2002; 2008
123.	*Allium roseum*	Liliaceae	Bulb, stalk, leaf	Nencini *et al.*, 2007; Štajner *et al.*, 2008
124.	*Allium sativum*	Liliaceae	Bulbs, oils	Imai *et al.*, 1994; Ide *et al.*, 1996; Yin and Cheng, 1998; Helen *et al.*, 1999; Ali *et al.*, 2000; Chandra *et al.*, 2003; Ichikawa *et al.*, 2003; Nuutila *et al.*, 2003; Stajner *et al.*, 2003; Souri *et al.*, 2004; Suboh *et al.*, 2004; Cai *et al.*, 2004; Odukoya *et al.*, 2005; Benkeblia, 2005; Subramanian *et al.*, 2005; Ichikawa *et al.*, 2006; Mishra, 2007; Mariutti *et al.*, 2008; Štajner *et al.*, 2008; Koseoglu *et al.*, 2009; Povichit *et al.*, 2010
125.	*Allium scorodoprasum*	Liliaceae	Bulb	Štajner *et al.*, 2003; Jang *et al.*, 2008
126.	*Allium shoenoprasum*	Liliaceae	Bulbs	Štajner *et al.*, 2004, 2008; Souri *et al.*, 2004; Cai *et al.*, 2004; Mariutti *et al.*, 2008
127.	*Allium sivasicum*	Liliaceae		Tepe *et al.*, 2005
128.	*Allium sphaerocephalum*	Liliaceae	Bulb, leaf, stock	Štajner *et al.*, 2003, 2008
129.	*Allium subhirsutum*	Liliaceae	Bulb, stalk, leaf	Nencini *et al.*, 2007; Štajner *et al.*, 2008
130.	*Allium tuberosum*	Liliaceae	Leaf	Maeda *et al.*, 2006
131.	*Allium ursinum*	Liliaceae	Flower, bulbs	Štajner *et al.*, 2008
132.	*Allium vineale*	Liliaceae	Bulb	Štajner *et al.*, 2008
133.	*Allophylus rubifolius*	Sapindaceae	Leaves	Marwah *et al.*, 2007
134.	*Alnus formosana*	Betulaceae		Lee *et al.*, 2006
135.	*Alnus glutinosa*	Betulaceae		Middleton *et al.*, 2005
136.	*Alnus incana*	Betulaceae		Stevic *et al.*, 2010

Contd...

Table 3.1–*Contd...*

Sl.No.	*Plant Name*	*Family*	*Parts Used*	*References*
137.	*Alnus japonica*	Betulaceae	Leaves	Kuroyanagi *et al.*, 2005; Lee *et al.*, 2005; Kang *et al.*, 2006; Kim *et al.*, 2004, 2010
138.	*Alnus viridis*	Betulaceeae		Stevic *et al.*, 2010
139.	*Aloe barbadensis* (=*Aloe vera*)	Liliaceae	Leaves	Sabeh *et al.*, 1993; Shamman *et al.*, 1998; Yagi *et al.*, 2002; Hu *et al.*, 2003; Parihar and Hemnani, 2003; Hu *et al.*, 2003; Gajera *et al.*, 2005; Rajasekaran *et al.*, 2005; Kardosova and Machova, 2006; Nwanjo *et al.*, 2006; Ashawat and Saraf, 2007; Miladi and Damak, 2008; Ozsoy *et al.*, 2009; Saritha *et al.*, 2010; Anilakumar *et al.*, 2010; Wei and Shibamoto, 2010
140.	*Aloe dhufarensis*	Liliaceae	Whole plant	Marwah *et al.*, 2007
141.	*Aloe ferox*	Liliaceae	Leaf	Loots *et al.*, 2007
142.	*Aloe saponaria*	Liliaceae		Yoo *et al.*, 2008
143.	*Aloysia gratissima*	Verbenaceae		Rosas-Romero and Saavedra, 2005
144.	*Alphonsea hortensis*	Annonaceae	Leaves, seed, stem, bark	Puvanendran *et al.*, 2011
145.	*Alpinia calcarata*	Zingiberaceae	Leaf	Narayanaswamy and Balakrishnan, 2011
146.	*Alpinia galanga*	Zingiberaceae	Rhizome	Juntachote and Berghofer, 2005; Juntachote *et al.*, 2006; Chanwitheesuk *et al.*, 2005; Leelapornpisid *et al.*, 2007; Gan *et al.*, 2010
147.	*Alpinia katsumadai*	Zingiberaceae	Seed	Lee *et al.*, 2003b; Gan *et al.*, 2010
148.	*Alpinia nutans*	Zingiberaceae		Habsah *et al.*, 2003
149.	*Alpinia officinarum*	Zingiberaceae	Rhizomes	Kim *et al.*, 1997; Lee *et al.*, 2003a; Tram Ngoc *et al.*, 2004; Cai *et al.*, 2004, 2006; Kshirsagar and Upadhyay, 2009; Ramasamy *et al.*, 2010
150.	*Alpinia oxyphylla*	Zingiberaceae	Fruits	Guan *et al.*, 2006; Liu *et al.*, 2006; Shi *et al.*, 2006, An *et al.*, 2005
151.	*Alpinia zerumbet*	Zingiberaceae	Flowers, seeds, leaves	Elzawely *et al.*, 2007;Chan *et al.*, 2009
152.	*Alstonia boonei*	Apocynaceae	Cortex, leaf, root	Oke and Hamburger, 2002; Akinmoladun *et al.*, 2007, 2010

Contd...

Table 3.1–*Contd...*

Sl.No.	*Plant Name*	*Family*	*Parts Used*	*References*
153.	*Alstonia scholaris*	Apocynaceae	Leaves	Arulmozhi *et al.*, 2007, 2010; Leelapornpisid *et al.*, 2007; Dhobal *et al.*, 2008; Kulkarni and Juvekar, 2009
154.	*Alternanthera bettzickiana*	Amaranthaceae	Aerial parts	Chen, C-H. *et al.*, 2009
155.	*Alternanthera brasiliana*	Amaranthaceae	Leavees	Costa *et al.*, 2009
156.	*Alternanthera maritima*	Amaranthaceae		Souza *et al.*, 2010
157.	*Alternanthera philoxeroides*	Amaranthaceae		Cai *et al.*, 2011
158.	*Alternanthera sessilis*	Amaranthaceae	Herb	Borah *et al.*, 2011
159.	*Alternanthera tenella*	Amaranthaceae	Herb	Salvador *et al.*, 2011
160.	*Althaea officinalis*	Malvaceae	Root, flowers	Kardosova and Machova, 2006
161.	*Alysicarpus vaginaalis*	Fabaceae	Whole plant	Rathi *et al.*, 2010
162.	*Alyssum homolocarpum*	Brassicaceae	Seed	Souri *et al.*, 2008
163.	*Amaranthus blitum*	Amaranthaceae		Amin *et al.*, 2006; Saranya and Guruswamy, 2010
164.	*Amaranthus caudatus*	Amaranthaceae	Seeds	Conforti *et al.*, 2005; Odukoya *et al.*, 2007; Veeru *et al.*, 2009
165.	*Amaranthus cruentus*	Amaranthaceae	Leaf	Oboh and Akindahunsi, 2004; Nsimba *et al.*, 2008; Adebooye *et al.*, 2008; Morrison and Twumasi, 2010; Escudero *et al.*,2011
166.	*Amaranthus gangeticus*	Amaranthaceae	Leaves	Anilakumar *et al.*, 2004; Amin *et al.*, 2006; Wong *et al.*, 2006
167.	*Amaranthus hybridus*	Amaranthaceae	Leaf	Odukoya *et al.*, 2007; Adefegha and Oboh, 2011
168.	*Amaranthus hypochondriacus*	Amaranthaceae		Nsimba *et al.*, 2008
169.	*Amaranthus lividus*	Amaranthaceae	Stem, leaves	Ozsoy *et al.*, 2009
170.	*Amaranthus paniculatus*	Amaranthaceae	Leaves	Bhatia and Jain, 2003; Amin *et al.*, 2006
171.	*Amaranthus retroflexus*	Amaranthaceae	Leaf	Özen, 2010
172.	*Amaranthus* sp.	Amaranthaceae	Leaves	Alvarez-Jubete *et al.*, 2010.
173.	*Amaranthus spinosus*	Amaranthaceae	Herb	Ashok Kumar *et al.*, 2010
174.	*Amaranthus viridis*	Amaranthaceae	Leaves	Amin *et al.*2006; Lako *et al.*, 2006
175.	*Ambrosia maritima*	Asteraceae	Herb	Abouzid *et al.*, 2008
176.	*Ammannia baccifera*	Lythraceae	Herb	Lavanya *et al.*, 2009
177.	*Amomum kravanh*	Zingiberaceae	Fruit	Chanwitheesuk *et al.*, 2005; Gan *et al.*, 2010
178.	*Amomum subulatum*	Zingiberaceae	Fruits	Dhuley, 1999; Kikuzaki *et al.*, 2001

Contd...

Table 3.1–*Contd...*

Sl.No.	*Plant Name*	*Family*	*Parts Used*	*References*
179.	*Amomum tsao-ko*	Zingiberaceae	Essential oil	Gan *et al.*, 2010; Yang *et al.*, 2010
180.	*Amomum villosum*	Zingiberaceae		Gan *et al.*, 2010
181.	*Amorpha canescens*	Fabaceae		Borchardt *et al.*, 2008
182.	*Amorphophallus campanulatus*	Araceae	Tuber	Jain *et al.*, 2009; Sahu *et al.*, 2009
183.	*Ampelopsis brevipedunculata*	Vitaceaae	Stem, root	Wu *et al.*, 2004
184.	*Ampelopsis grossodentata*	Vitaceae	Leaves	Gao *et al.*, 2009; Ying *et al.*, 2011
185.	*Ampelopsis humulifolia* var. *heterophylla*	Vitaceae		Chen *et al.*, 2004
186.	*Ampelopsis sinica*	Vitaceae		Chen *et al.*, 2004
187.	*Amphimas pterocaroides*	Fabaceae		Biapa *et al.*, 2007
188.	*Anacamptis pyrimidalis*	Orchidaceae	Flowers, aerial parts	Stajner *et al.*, 2010
189.	*Anacardium occidentale*	Anacardiaceae	Nuts	Bimla Singh *et al.*, 2004; Abbas *et al.*, 2006; Schmourlo *et al.*, 2007; Razali *et al.*, 2008
190.	*Anacyclus pyrethrum*	Asteraceae		Kalim *et al.*, 2010
191.	*Anadenanthera colubrina*	Fabaceae		De Melo *et al.*, 2010
192.	*Anadenanthera macrocarpa*	Fabaceae		Desmarchelier *et al.*, 1999
193.	*Ananas comosus*	Bromeliaceae	Fruit	Huang *et al.*, 2004a; Lako *et al.*, 2006; Alothman *et al.*, 2009; Kataki *et al.*, 2010
194.	*Anchieteea pyrifolia*	Violaceae	Leaves	Menezess *et al.*, 2004
195.	*Andrographis paniculata*	Acanthaceae	Leaves	Guo *et al.*, 1995 Zhang and Tan, 2000; Trivedi and Raawal, 2001; Bajpai *et al.*, 2005; Kruawan and Kangsadalampai, 2006; Tripathi and Kamat, 2007; Phansawan and Poungbangpho, 2007; Wojcikowski *et al.*, 2007; Lin *et al.*, 2009; Ojha *et al.*, 2009; Jain and Agrawal, 2009; Rafat *et al.*, 2010; Tewari *et al.*, 2010; Saranya *et al.*, 2010
196.	*Andromeda polifolia*	Ericaceae		Kahkonen *et al.*, 1999
197.	*Anemarrhena asphodeloides*	Liliaceae		Li *et al.*, 2008
198.	*Anemopsis californica*	Saururaceae	Leaves	Medina *et al.*, 2005
199.	*Anethum graveolens* fruit	Apiaceae	Leaves	Souri *et al.*, 2004, 2008; Chan-witheesuk *et al.*, 2005; Mariutti *et al.*, 2008; Gacche *et al.*, 2010

Contd...

Table 3.1–*Contd...*

Sl.No.	Plant Name	Family	Parts Used	References
200.	*Angelica archangelica*	Apiaceae	Root	Wojcikowski *et al.*, 2007
201.	*Angelica biserrata*	Apiaceae		Gan *et al.*, 2010
202.	*Angelica dahurica*	Apiaceae		Song *et al.*, 2010; Thuong *et al.*, 2010
203.	*Angelica gigas*	Apiaceae		Ma *et al.*, 2009
204.	*Angelica keiskei*	Apiaceae	Leaves	Cho *et al.*, 2008
205.	*Angelica sinensis*	Apiaceae	Root	Wu *et al.*, 2004, 2005; Cai *et al.*, 2004, 2006; Lu *et al.*, 2005; Wojcikowski *et al.*, 2007; Chao *et al.*, 2009
206.	*Angelica sylvestris*	Apiaceae	Seed, root	Sarker *et al.*, 2003
207.	*Aniba panurensis*	Lauraceae	Essential oil	Alcantara *et al.*, 2010
208.	*Aniba rosaeodora*	Lauraceae	Essential oil	Alcantara *et al.*, 2010
209.	*Anisomeles indica*	Lamiaceae	Leaves	Tatiya *et al.*, 2007
210.	*Anisopus mannii*	Asclepiadaceae		Aliyu *et al.*, 2010
211.	*Annona muricata*	Annonaceae	Leaves	Lako *et al.*, 2006; Baskar *et al.*, 2007
212.	*Annona reticulata*	Annonaceae	Leaves	Baskar *et al.*, 2007
213.	*Annona senegalensis*	Annonaceae	Leaves	Potchoo *et al.*, 2008
214.	*Annona squmosa*	Annonaceae	Leaves	Shirwaikar *et al.*, 2004; Padmaja *et al.*, 2004; Punzalan *et al.*, 2006; Kaleem *et al.*, 2006; Baskar *et al.*, 2007; Gupta *et al.*, 2004, 2008; Ruiz-Terán *et al.*, 2008
215.	*Anoectochilus formosanus*	Orchidaceae		Wang *et al.*, 2002; Wu *et al.*, 2006
216.	*Anogeissus dhofarica*	Combretaceae	Tuber	Marwah *et al.*, 2007
217.	*Anogeissus latifolia*	Combretaceae	Bark	Govindarajan *et al.*, 2004, 2005, 2006
218.	*Anthemis tinctoria*	Asteraceae	Aerial parts	Papaioannou *et al.*, 2007
219.	*Anthocephalus cadamba*	Rubiaceae	Flowering tops	Sanjay *et al.*, 2007; Prahalad *et al.*, 2007; Alam *et al.*, 2008
220.	*Anthocephalus indicus*	Rubiaceae		Kumar *et al.*, 2009
221.	*Anthriscus cerifolium*	Apiaceae	Herb, root	Fejes *et al.*, 2000
222.	*Anthriscus sylvestris*	Apiaceae	Herb	Milovanovic *et al.*, 1996
223.	*Anthurium versicolor*	Araceae	Leaves	Aquino *et al.*, 2001
224.	*Anthyllis aurea*	Fabaceae	Herb	Godevac *et al.*, 2008
225.	*Anthyllis vulneraria*	Fabaceae	Herb	Godevac *et al.*, 2008
226.	*Antidesma thwaitesianum*	Stilaginaceae	Fruit	Nuenchamnong and Ingkanian, 2010
227.	*Apama siliquosa*	Aristolochiaceae		Thirugnanasaampandan *et al.*, 2008

Contd...

Table 3.1–*Contd...*

Sl.No.	*Plant Name*	*Family*	*Parts Used*	*References*
228.	*Aphanamixis polystachya*	Meliaceae		Jagetia and Venkatesha, 2005
229.	*Apium graveolens*	Apiaceae	Seeds, fruit	Momin and Nair, 2002; Souri *et al.*, 2004, 2008; Chanwitheesuk *et al.*, 2005; Pendry *et al.*, 2005; Maeda *et al.*, 2006; Wong *et al.*, 2006; Wojcikowski *et al.*, 2007; Souri *et al.*, 2008
230.	*Aporosa lindleyana*	Euphorbiaceae	Root	Badami *et al.*, 2005
231.	*Aquilaria agallocha*	Thymaelaceae		Miniyar *et al.*, 2008
232.	*Aquilaria malaccensis*	Thymaelaceae	Aerial part	Kshirsagar and Upadhyay, 2009
233.	*Aquilegia vulgaris*	Ranunculaceae	Stems, leaves	Murias *et al.*, 2005; Jodynis-Liebert *et al.*, 2005
234.	*Arachis hypogaea*	Fabaceae	Plant parts Hulls, testa	Pratt and Miller, 1984; Green and Danders, 2004;Yen and Duh, 1994; Yen *et al.*, 2005; Bajpai *et al.*, 2005; Duncan *et al.*, 2006; Lee *et al.*, 2006; Wang *et al.*, 2007; Sales and Resurreccion, 2010
235.	*Aralia continentalis*	Araliaceae	Leaves	Kim *et al.*, 1998
236.	*Aralia cordata*	Araliaceae	Fruit	Liu *et al.*, 2010
237.	*Aralia elata*	Araliaceae		Chung and Jung, 2003; Chung *et al.*, 2005; Hu and Wang, 2009
238.	*Aralia taibensis*	Arailaceae		Xi *et al.*, 2010
239.	*Arbutus andrachne*	Ericaceae	Fruit	Kirca and Arslan, 2008; Serce *et al.*, 2010
240.	*Arbutus unedo*	Ericaceae	Leaves	Pabuccuoglu *et al.*, 2003; Palluf *et al.*, 2008; Oliveira *et al.*, 2009
241.	*Arcangelisia flava*	Menispermaceae		Keawpradub *et al.*, 2005
242.	*Archidendron jiringa*	Mimosaceae		Panpipat *et al.*, 2010
243.	*Archontophoenix alexandrae*	Arecaceae	Flowers	Liu *et al.*, 2008
244.	*Arctium lappa*	Asteraceae	Root, flowers	Maruta *et al.*, 1995; Cai *et al.*, 2004, 2006; Pendry *et al.*, 2005; Katalinic *et al.*, 2006; Kardosova and Machova, 2006; Song *et al.*, 2010
245.	*Arctostaphylos uva-ursi*	Ericaceae	Leaves	Amarowicz *et al.*, 2004; Wojcikowski *et al.*, 2007
246.	*Ardisia colorata*	Myrsinaceae		Sumino *et al.*, 2001, 2002
247.	*Ardisia compressa*	Myrsinaceae	Stem bark	Ramirez-Mares *et al.*, 2010
248.	*Ardisia gigantifolia*	Myrsinaceeae		Yang *et al.*, 2008
249.	*Ardisia japonica*	Myrsinaceae		Song *et al.*, 2010

Contd...

Table 3.1–*Contd...*

Sl.No.	*Plant Name*	*Family*	*Parts Used*	*References*
250.	*Areca catechu* var. *dulcissima*	Arecaceae	Fruits	Lee *et al.*, 2003;
251.	*Areca catechu*	Arecaceae	Fruits	Kim *et al.*, 1997; Cai *et al.*, 2004; Naik, *et al.*, 2005
252.	*Argania spinosa*	Sapotaceeae		El Babili *et al.*, 2009
253.	*Argemone mexicana*	Papaveraceae	Whole plant	Lagnika *et al.*, 2011
254.	*Argostemma cf. viscidum*	Rubiaceae		Ahmad *et al.*, 2010
255.	*Argostemma yappi*	Rubiaceae		Ahmad *et al.*, 2010
256.	*Argyreia cymosa*	Rubiaceeae	Bark	Badami *et al.*, 2008
257.	*Argyreia nervosa*	Convolvulaceae		Shreedhara *et al.*, 2009
258.	*Argyreia speciosa*	Convolvulaceae	Root	Habbu *et al.*, 2010
259.	*Arisaema consan-guineum*	Arecaceae		Song *et al.*, 2010
260.	*Aristolochia bracteolata*	Aristolochiaceae	Leaves	Thirugnanasampandan *et al.*, 2008; Shirwaikar *et al.*, 2003
261.	*Aristolochia brasiliensis*	Aristolochiaceae	Leaves	Thirugnanasampandan *et al.*, 2008
262.	*Aristolochia cymbifera*	Aristolochiaceae	Rhizome	Alviano *et al.*, 2008
263.	*Aristolochia gibertii*	Aristolochiaceae		Velazquez *et al.*, 2003
264.	*Aristolochia indica*	Aristolochiaceae	Leaf and roots	Thirugnanasampandan *et al.*, 2008
265.	*Aristolochia mollissima*	Aristolochiaceae	Whole plant	Cai *et al.*, 2004
266.	*Aristolochia tagala*	Aristolochiaceae	Whole plant	Thirugnanasampandan *et al.*, 2008
267.	*Armoracia rusticana*	Brassicaceae	Root	Kahkonen *et al.*, 1999
268.	*Arnica chamissonis* var. *foliosa*	Asteraceae	Seeds	Gawlikdziki *et al.*, 2009
269.	*Arnica montana*	Asteraceae	Seeds	Gawlikdziki *et al.*, 2009
270.	*Aronia melanocarpa*	Rosaceae	Fruit	Kahkonen *et al.*, 1999; Matsumoto *et al.*, 2004; Kedzierska *et al.*, 2009
271.	*Arrabidaea samydoides*	Bignoniaceae	Stem	Pauletti *et al.*, 2003
272.	*Arracacia xanthorrhiza*	Apiaceae		Penarrieeta *et al.*, 2005; Salluca *et al.*, 2008
273.	*Artemesia absinthium*	Asteraceae	Herb	Özgen *et al.*, 2004; Canadanovic-Brunet *et al.*, 2005; Katalinic *et al.*, 2006
274.	*Artemisia abyssinica*	Asteraceae	Essential oils	Burits *et al.*, 2001

Contd...

Table 3.1–*Contd...*

Sl.No.	*Plant Name*	*Family*	*Parts Used*	*References*
275.	*Artemisia afra*	Asteraceae	Essential oils	Graven *et al.*, 1992; Burits *et al.*, 2001
276.	*Artemisia annua*	Asteraceae	Whole plant	Zheng and Wang, 2001; Juteau *et al.*, 2002; Cai *et al.*, 2004
277.	*Artemisia anomala*	Asteraceae		Gan *et al.*, 2010
278.	*Artemisia apiacea*	Asteraceae		Li *et al.*, 2008
279.	*Artemisia arborescens*	Asteraceae		Dessi *et al.*, 2001; Wojcikowski *et al.*, 2007
280.	*Artemisia argyi*	Asteraceae	Leaves	Cai *et al.*, 2004, 2006; Gan *et al.*, 2010
281.	*Artemisia austriaca*	Asteraceae		Özgen *et al.*, 2004
282.	*Artemisia campestris*	Asteraceae	Leaf	Aniya *et al.*, 2000; Maeda *et al.*, 2006; Serteser *et al.*, 2009
283.	*Artemisia capillaris*	Asteraceae		Jung *et al.*, 2004; Cai *et al.*, 2004; Gan *et al.*, 2010
284.	*Artemisia dracunculus*	Asteraceae	Leaves	Mariutti, *et al.*, 2008; Souri *et al.*, 2004
285.	*Artemisia gorgonum*	Asteraceae	Essential oil	Ortet *et al.*, 2010
286.	*Artemisia herba-alba*	Asteraceae	Shoot	Suboh *et al.*, 2004; Al-Mustafa and Al-Thunibat, 2008; Mighri *et al.*, 2010
287.	*Artemisia judaica*	Asteraceae	Whole plant	Liu *et al.*, 2004; Al-Mustafa and Al-Thunobat, 2008
288.	*Artemisia maritima*	Asteraceeae	Budinevi	*et al.*, 1995
289.	*Artemisia montana*	Asteraceae	Leaves	Kim *et al.*, 2000
290.	*Artemisia princeps*	Asteraceae	Leaf	Ohmori *et al.*, 2005; Maeda *et al.*, 2006
291.	*Artemisia santonicum*	Asteraceae	Leaves	Serteser *et al.*, 2009
292.	*Artemisia scoparia*	Asteraceae	Essential oil	Singh *et al.*, 2010
293.	*Artemisia sieversiana*	Asteraceae	Aerial parts	Mangathayaru *et al.*, 2007
294.	*Artemesia vulgaris*	Asteraceae	Herb	Temraj and Tantaway, 2008
295.	*Artocarpus altilis*	Moraceae	Stem	Lako *et al.*, 2006;
296.	*Artocarpus anisophyllus*	Moraceae	Bark	Rahmani *et al.*, 2009
297.	*Artocarpus communis*	Moraceae		Fukai *et al.*, 2003b
298.	*Artocarpus heterophyllus*	Moraceae	Fruit pulp	Jagtap *et al.*, 2010
299.	*Artocarpus lakoocha*	Moraceae	Heartwood	Hasan *et al.*, 2009; Povichit *et al.*, 2010; Singhatong *et al.*, 2010
300.	*Artocarpus nobilis*	Moraceae	Leaves	Jayasinghe *et al.*, 2004 a,b
301.	*Artocarpus obtusus*	Moraceae	Bark	Rahmani *et al.*, 2009
302.	*Aruncus silvester*	Rosaceae	Root	Oszmianski *et al.*, 2007

Contd...

Table 3.1–*Contd...*

Sl.No.	*Plant Name*	*Family*	*Parts Used*	*References*
303.	*Arum dioscoridis*	Araceae	Leaves	Janakat and Al-Thanibat, 2008
304.	*Arum palestinum*	Araceae	Leaves	Al-Mustafa and Al-Thunibat, 2008
305.	*Aruncus silvester*	Rosaceae	Root	Oszmiaski *et al.*, 2007
306.	*Arundo donax*	Poaceae		Menezes *et al.*, 2004
307.	*Aspalathus linearis*	Fabaceae	Herb	Joubert *et al.*, 2005; McKay and Blumberg, 2006
308.	*Asparagus officinalis*	Liliaceae	Young stalk	Maeda *et al.*, 2006
309.	*Asparagus racemosus*	Liliaceae	Leaves, Roots	Kamat *et al.*, 2000; Velavan *et al.*, 2002; Wiboonpun *et al.*, 2004; Visavadiya and Narasimhacharya, 2005; Patnibul *et al.*, 2008; Jain and Agrawal, 2009
310.	*Asperula stylosa*	Rubiaceae		Souri *et al.*, 2004
311.	*Asphodelus tenuifolius*	Liliaceae		Kalim *et al.*, 2010
312.	*Aspila quinquenervis*	Poaceae		Mosquera *et al.*, 2009
313.	*Aster scaber*	Asteraceae		Heo *et al.*, 2009
314.	*Aster tataricus*	Asteraceae	Root	Cai *et al.*, 2004; Song *et al.*, 2010
315.	*Astianthus humilis*	Bignoniaceae		Ruiz-Terán *et al.*, 2008
316.	*Astragalus complanatus*	Fabaceae	Seed	Cai *et al.*, 2004, 2006
317.	*Astragalus hamosus*	Fabaceae	Fruit	Souri *et al.*, 2008
318.	*Astragalus membranaceus*	Fabaceae	Roots	Wojcikowski *et al.*, 2007; Huang *et al.*, 2009; Chen, C-H. *et al.*, 2009
319.	*Astragalus mongolicus*	Fabaceae	Root	Cai *et al.*, 2004, 2006
320.	*Astrodaucus orientalis*	Apiaceae		Souri *et al.*, 2004
321.	*Astronium urundeuva*	Anacardiaceae		Desmarchelier *et al.*, 1999
322.	*Athamanta turbith*	Apiaceae	Tomi	*Tomic et al.*, 2010
323.	*Athrixia phylicoides*	Asteraceae		McGaw *et al.*, 2007
324.	*Atractylodes lancea*	Asteraceae	Rhizome	Cai *et al.*, 2004; Gan *et al.*, 2010
325.	*Avena sativa*	Poaceae	Fruit	Kahkonen *et al.*, 1999; Emmons and Peterson, 1999; Bajpai *et al.*, 2005;
326.	*Averrhoa bilimbi*	Averrhoaceae		Abbas *et al.*, 2006
327.	*Avicennia alba*	Avicenniaceae	Leaves, flowers	Vadlapudi and Naidu, 2009
328.	*Avicennia marina*	Avicenniaceae	Leaves, flowers	Vadlapudi and Naidu, 2009
329.	*Avicennia officinalis*	Avicenniaceae	Leaves, flowers	Vadlapudi and Naidu, 2009
330.	*Azadirachta indica*	Meliaceae	Leaves, bark	Rao *et al.*, 1998; Chandra *et al.*, 2003; Sithisarn *et al.*, 2006; Prakash *et al.*, 2007; Bushra *et al.*, 2007

Contd...

Table 3.1–*Contd...*

Sl.No.	*Plant Name*	*Family*	*Parts Used*	*References*
331.	*Azima tetracantha*	Salvadoraceae	Leaves	Ekbote *et al.*, 2010
332.	*Baccaurea ramiflora*	Euphorbiaceae	Stems	Yang *et al.*, 2007, 2010; Hasan *et al.*, 2009
333.	*Baccharis cordifolia*	Asteraceae		Desmarchelier *et al.*, 1997; Mongelli *et al.*, 1997;
334.	*Baccharis crispa*	Asteraceae	Aerial parts	Desmarchelier *et al.*, 1997; Simoes-Pires *et al.*, 2005
335.	*Baccharis grisebachii*	Asteraceae		Desmarchelier *et al.*, 1997; Tapia *et al.*, 2004; Rosas-Romero and Saavedra, 2005
336.	*Baccharis salicifolia*	Asteraceae		Rosas-Romero and Saavedra, 2005
337.	*Baccharis sessiliflora*	Asteraceae		Borneo *et al.*, 2008
338.	*Baccharis trimera*	Asteraceae	Aerial parts	Simoes-Pires *et al.*, 2005
339.	*Baccharis trinervis*	Asteraceae		De las Heras *et al.*, 1998
340.	*Baccharis usterii*	Asteraceae	Aerial parts	Simoes-Pires *et al.*, 2005
341.	*Backhousia citriodora*	Myraceae		Konczak *et al.*, 2010
342.	*Bacopa monnieri*	Scrophulariaceae	Whole plant	Tripathi *et al.*, 1996; Russo *et al.*, 2003; Rohini *et al.*, 2004; Ghosh *et al.*, 2007;
343.	*Balanites aegyptiaca*	Balanitaceae	Fruits/seed	Cook *et al.*, 1998
344.	*Balanophora laxiflora*	Balanophoraceae	Herb	She *et al.*, 2009
345.	*Balanophora polyandra*	Balanophora-ceae		Wang *et al.*, 2006
346.	*Baliospermum montanum*	Euphorbiaceae	Roots	Desai *et al.*, 2008
347.	*Ballota acetabulosa*	Lamiaceae	Herb	Couladis *et al.*, 2003
348.	*Ballota aucheri*	Lamiaceae		Souri *et al.*, 2004; Firuzi *et al.*, 2010
349.	*Ballota nigra*	Lamiaceae	Herb	Seidel *et al.*, 2000; Vrchovska *et al.*, 2007; Matkowski *et al.*, 2008
350.	*Ballota pseudodictamnus*	Lamiaceae	Aerial parts	Couladis *et al.*, 2003
351.	*Ballota rotundifolia*	Lamiaceae		Gursoy and Tepe, 2009
352.	*Ballota* sp.	Lamiaceae		Citoglu *et al.*, 2004
353.	*Balsamita major*	Asteraceae	Leaves	Ena *et al.*, 2009
354.	*Bambusa breviflora*	Poaceae		Song *et al.*, 2010
355.	*Bambusa vulgaris*	Poaceae		Goyal *et al.*, 2010
356.	*Baphia nitida*	Fabaceae		Kouakou-Siransy *et al.*, 2010
357.	*Baphicanthus cusia*	Acanthaceae		Li *et al.*, 2008
358.	*Barringtonia acutangula*	Lecythidaceae	Root	Babre *et al.*, 2010

Contd...

Table 3.1–*Contd...*

Sl.No.	*Plant Name*	*Family*	*Parts Used*	*References*
359.	*Barringtonia racemosa*	Lecythidaceae		Behbahani *et al.*, 2007; Shikha *et al.*, 2010
360.	*Basella alba*	Basellaceae	Leaf	Oboh and Akindahunsi, 2004; Adefegha and Oboh, 2011
361.	*Bauhinia forficata*	Caesalpiniaceae	Leaves	De Sousa *et al.*, 2004
362.	*Bauhinia galpinii*	Caesalpiniaceae		Aderogba *et al.*, 2007
363.	*Bauhinia microstachya*	Caesalpiniaceae	Fruits	Menezes *et al.*, 2004; Silva *et al.*, 2007
364.	*Bauhinia monandra*	Caesalpiniaceae		Argolo *et al.*, 2004; Aderogba *et al.*, 2006
365.	*Bauhinia purpurea*	Caesalpiniaceae		Chanwitheesuk *et al.*, 2005; Zakaria 2007; Jatwa and Kar, 2009
366.	*Bauhinia racemosa*	Caesalpiniaceae	Stem bark	Kumar *et al.*, 2005
367.	*Bauhinia tarapotensis*	Caesalpiniaceae	Leaves	Braca *et al.*, 2001
368.	*Bauhinia variegata*	Caesalpiniaceae	Stem bark, root	Rajkapoor *et al.*, 2006; Rajani and Ashok, 2009
369.	*Beaumontia grandiflora*	Apocynaceae	Aerial parts	Abdelshaffek *et al.*, 2010
370.	*Becium dhofarense*	Lamiaceae	Shoot	Marwah *et al.*, 2007
371.	*Beilschmedia manni*	Lauraceae	Leaf	Odukoya *et al.*, 2007;
372.	*Belamcanda chinensis*	Iridaceae	Rhizome	Jung *et al.*, 2004; Li *et al.*, 2008
373.	*Bellis perennis*	Asteraceae	Herb	Kavalcioglu *et al.*, 2010; Siatka and Kasparovaa, 2010
374.	*Benincasa hispida*	Cucurbitaceae	Flesh	Huang *et al.*, 2004a,b; Maeda *et al.*, 2006; Gan *et al.*, 2010; Raghu *et al.*, 2010
375.	*Berberis aristata*	Berberidaceae	Root	Singh and Kakkar, 2009
376.	*Berberis crataegina*	Berberidaceae	Fruit, leaf	Souri *et al.*, 2004; Serteser *et al.*, 2009
377.	*Berberis integerrima*	Berberidaceae	Fruit, leaf	Serteser *et al.*, 2009
378.	*Berberis vulgaris*	Berberidaceae	Fruits, leaf	Hanachi, 2009; Montalleb *et al.*, 2005; Serteser *et al.*, 2009
379.	*Bergenia ciliata*	Saxifragaceae		Bagul *et al.*, 2003; Rajkumar *et al.*, 2010d
380.	*Bergenia crassifolia*	Saxifragaceae	Leaves	Shilova *et al.*, 2006; Pozharitskaya *et al.*, 2007b; Ivanov *et al.*, 2010
381.	*Bergenia ligulata*	Saxifragaceae	Rhizome	Bashir and Gilani, 2009
382.	*Beta vulgaris*	Chenopodiaceae	Root	Kahkonen *et al.*, 1999; Pyo *et al.*, 2004; Özen, 2010
383.	*Beta vulgaris* var. *conditiva*	Chenopodiaceae	Root	Jiratanam and Liu, 2004

Contd...

Table 3.1–*Contd...*

Sl.No.	*Plant Name*	*Family*	*Parts Used*	*References*
384.	*Betula pendula*	Betulaceae	Leaf, bark, phloem, flowers	Kahkonen *et al.*, 1999; Trouillas *et al.*, 2003
385.	*Betula platyphylla* var. *japonica*	Betulaceae		Ju *et al.*, 2004
386.	*Betula pubescens*	Betulaceae	Leaf	Kahkonen *et al.*, 1999
387.	*Betula utilis*	Betulaceae		Kumaraswamy *et al.*, 2008
388.	*Bidens andicola*	Asteraceae		Borneo *et al.*, 2008
389.	*Bidens tripartita*	Asteraceae	Herb	Wolniak *et al.*, 2007
390.	*Bieberstena multifida*	Berberidaceae	Root	Souri *et al.*, 2008
391.	*Biophytum reinwardtii*	Oxalidaceae	Whole plant	Srinivasan *et al.*, 2008
392.	*Biophytum sensitivum*	Oxalidaceae	Herb	Guruvaayoorappan *et al.*, 2006
393.	*Bletilla striata*	Orchidaceae		Gan *et al.*, 2010
394.	*Blumea balsamifera*	Asteraceae	Leaves	Fazilatun *et al.*, 2004
395.	*Blumea napifolia*	Asteraceae		Patnibul *et al.*, 2008
396.	*Boehmeria nivea*	Urticaceae		Gan *et al.*, 2010
397.	*Boerhavia diffusa*	Nyctaginaceae	Leaves	Satheesh and Pari, 2004; Bajpai *et al.*, 2005; Kouakou-Siransy *et al.*, 2010; Olaleye *et al.*, 2010
398.	*Boesenbergia pandurata*	Zingiberaceae	Rhizome	Chanwitheesuk *et al.*, 2005
399.	*Bolusanthus speciosus*	Fabaceae	Root wood	Erasto *et al.*, 2004
400.	*Bombax costatum*	Bombacaceae	Fruit	Cook *et al.*, 1998
401.	*Bombax malabaricum*	Bombaceae		Moussa *et al.*, 2011
402.	*Borago officinalis*	Boraginaceae	Herb	Wettasinghe and Shahidi 1999, 2000; Andoniene and Murkovic, 2002; Bandoniene and Murkovic, 2002
403.	*Boscia senegalensis*	Capparaceae	Leaf	Cook *et al.*, 1998
404.	*Boswellia carteri*	Burseraceae		Yang *et al.*, 2010
405.	*Boswellia dioscorides*	Burseraceae		Mothana *et al.*, 2009
406.	*Boswellia serrata*	Burseraceae	Resin	Gajera *et al.*, 2005; Assimopoulou *et al.*, 2005; Sharma *et al.*, 2011
407.	*Boswellia socotrana*	Burseraceae		Mothana *et al.*, 2009
408.	*Bougainvillea glabra*	Nyctaginaceae	Leaves	Adib *et al.*, 2010
409.	*Brassica alba*	Brassicaceae		Kim *et al.*, 1997; Song *et al.*, 2010
410.	*Brassica campestris*	Brassicaceae	Leaf	Maeda *et al.*, 2006
411.	*Brassica campestris* var. *japonica*	Brassicaceae	Oil cake	Nagatsu *et al.*, 2004

Contd...

Table 3.1–*Contd...*

Sl.No.	*Plant Name*	*Family*	*Parts Used*	*References*
412.	*Brassica chinensis*	Brassicaceae	Leaves	Lako *et al.*, 2006
413.	*Brassica juncea*	Brassicaceae	Leaf	Khan *et al.*, 1997; Choi *et al.*, 2002; Yokozawa *et al.*, 2002; Kim *et al.*, 2003; Maeda *et al.*, 2006; Jung *et al.*, 2009
414.	*Brassica napus*	Brassicaceae	Root peel	Kahkonen *et al.*, 1999; Brettonnet *et al.*, 2010
415.	*Brassica nigra*	Brassicaceae	Seeds, leaves	Chung *et al.*, 1997; Lako *et al.*, 2006
416.	*Brassica oleracea*	Brassicaceae	Aerial parts	Souri *et al.*, 2004
417.	*Brassica oleracea* var. *acephala*	Brassicaceae	Leaf	Özen, 2010
418.	*Brassica oleracea* var. *botrytis*	Brassicaceae	Inflore-scence	Llorach *et al.*, 2003; Souri *et al.*, 2004; Koksal and Gulcin, 2008; Gacche *et al.*, 2010
419.	*Brassica oleracea* var. *bullata*	Brassicaceae	Leaf	Lako *et al.*, 2006
420.	*Brassica oleracea* var. *capitata*	Brassicaceae	Leaf	Maeda *et al.*, 2006; Gacche *et al.*, 2010; Raghu *et al.*, 2010
421.	*Brassica oleracea* var.*caulocarpa*	Brassicaceae	Stem tuber	Andallu *et al.*, 2003
422.	*Brassica oleracea* var. *costata*	Brassicaceae	Leaves	Llorach *et al.*, 2003; Ferreres *et al.*, 2007
423.	*Brassica oleracea* var. *gemmifera*	Brassicaceae	Aerial part	Souri *et al.*, 2004; Pellegrini *et al.*, 2010
424.	*Brassica oleracea* var. *gongylodes*	Brassicaceae		Raghu *et al.*, 2010
425.	*Brassica oleracea* var. *italica*	Brassicaceae	Bud	Cai *et al.*, 2004; Maeda *et al.*, 2006
426.	*Brassica rapa* var. *rapa*	Brassicaceae	Seeds	Fernandes *et al.*, 2007
427.	*Brickellia veronicaefolia*	Asteraceae		Perez *et al.*, 2004
428.	*Bridelia airy-shawii*	Euphorbiaceae	Stem bark	Tatiya *et al.*, 2008
429.	*Broussonetia papyrifera*	Moraceae	Fruits	Zhou *et al.*, 2010
430.	*Brucea javanica*	Simaroubaceae	Fruit	Cai *et al.*, 2004
431.	*Brucea amarissima*	Simaroubaceae		Peteros and Uy, 2010
432.	*Bruguira cylindrica*	Rhizophoraceae	Leaves, flowers	Vadlapudi and Naidu, 2009
433.	*Bruguira gymnorrhiza*	Rhizophoraceae	Leaves, flowers	Vadlapudi and Naidu, 2009; Homhual *et al.*, 2006
434.	*Bryonia alba*	Cucurbitaceae	Root	Karagezian *et al.*, 1981
435.	*Bryophyllum calycinum*	Crassulaceae	Leaf	Jain *et al.*, 2010

Contd...

Table 3.1–*Contd...*

Sl.No.	*Plant Name*	*Family*	*Parts Used*	*References*
436.	*Buddleja asiatica*	Buddlejaceae	Leaves	El-Sayed *et al.*, 2008
437.	*Buddleja globosa*	Buddlejaceae		Houghton *et al.*, 2005; Backhouse *et al.*, 2008
438.	*Buddleja officinalis*	Buddlejaceae		Piao *et al.*, 2003; Houghton *et al.*, 2005; Li *et al.*, 2008; Pan *et al.*, 2010
439.	*Bunchosia canesens*	Malpighiaceae		Ruiz-Terán *et al.*, 2008
440.	*Bunium persicum*	Apiaceae	Fruit	Souri *et al.*, 2008; Nickavar and Abolhasani, 2009
441.	*Bupleurum chinense*	Apiaceae		Song *et al.*, 2010
442.	*Bupleurum falcatum*	Apiaceae		Matsumoto *et al.*, 2005; Wojcikowski *et al.*, 2007
443.	*Bupleurum marginatum*	Apiaceae	Essential oil	Ashour *et al.*, 2009
444.	*Bupleurum scorzoneri-folium*	Apiaceae	Root	Cai *et al.*, 2004; Lu *et al.*, 2005; Wu *et al.*, 2005; Chao *et al.*, 2009;
445.	*Burkea africana*	Leguminosae	Stem bark	Mathisen *et al.*, 2002
446.	*Bursera grandifolia*	Burseraceae		Ruiz-Terán *et al.*, 2008
447.	*Butea monosperma*	Fabaceae	Bark, leaves	Panda *et al.*, 2009; Hasan *et al.*, 2009; Sharma and Garg, 2009
448.	*Butea superba*	Fabaceae	Flowers	Nuengchamnong *et al.*, 2005
449.	*Buxus hyrcana*	Buxaceae		Souri *et al.*, 2004
450.	*Byrsocarpus coccineus*	Connaraceae	Leaf	Oke and Hamburger, 2002
451.	*Caesalpinia benthamiana*	Caesalpiniaceae	Root bark	Dickson *et al.*, 2007
452.	*Caesalpinia bonduc*	Caesalpiniaceae	Seeds, leaves	Gupta *et al.*, 2005; Shukla *et al.*, 2009
453.	*Caesalpinia digyna*	Caesalpiniaceae	Roots	Sundaram and Mitra, 2006; Srinivasan *et al.*, 2007
454.	*Caesalpinia mimosoides*	Caesalpiniaceae		Chanwitheesuk *et al.*, 2005
455.	*Caesalpinia pulcherrima*	Caesalpiniaceae	Wood	Padma *et al.*, 2000; Liu *et al.*, 2008; Pawar *et al.*, 2009; Hasan *et al.*, 2009
456.	*Caesalpinia pyramidalis*	Caesalpiniaceae	Leaves	Alviano *et al.*, 2008
457.	*Caesalpinia sappan*	Caesalpiniaceae	Heartwood	Choi *et al.*, 1997; Badami *et al.*, 2003; Safitri *et al.*, 2003; Cai *et al.*, 2004, 2006; Wetwitayaklung *et al.*, 2005; Sarumathy *et al.*, 2010; Gan *et al.*, 2010
458.	*Cajanus cajan*	Fabaceae	Seeds, leaves	Bajpai *et al.*, 2005; Oboh, 2006a; Ghosh and Sil, 2007; Wu *et al.*, 2009; Sreeramulu *et al.*, 2009; Ghosh *et al.*, 2009

Contd...

Table 3.1–*Contd...*

Sl.No.	Plant Name	Family	Parts Used	References
459.	*Cakile maritima*	Brassicaceae		Shams *et al.*, 2010
460.	*Calamintha glandulosa*	Lamiaceae	Aerial parts	Couladis *et al.*, 2003
461.	*Calceolaria chelido-nioides*	Scrophularia-ceae	Aerial parts	Falcao *et al.*, 2006
462.	*Calea angosturana*	Asteraceae		Mosquera *et al.*, 2007
463.	*Calendula officinalis*	Asteraceae	Flowers	Cordova *et al.*, 2002; Pendry *et al.*, 2005; Mariassyova 2006; Khalil *et al.*, 2007; Braga *et al.*, 2009
464.	*Calliandra haemato-cephala*	Mimosaceae	Flowers	Liu *et al.*, 2008
465.	*Calligonum leucocladum*	Polygonaceae	Aerial parts	Okasaka *et al.*, 2004
466.	*Calluna vulgaris*	Ericaceae	Herb	Kahkonen *et al.*, 1999; Deliorman-orhan *et al.*, 2009
467.	*Calotropis gigantea*	Asclepiadaceae		Rathod *et al.*, 2009
468.	*Calotropis procera*	Asclepiadaceae	Latex	Padhy *et al.*, 2007; Abouzid *et al.*, 2008; Joshi *et al.*, 2010
469.	*Calpurnia aurea*	Fabaceae	Leaves, stems	Adedapo *et al.*, 2008
470.	*Calycotome villosa*	Fabaceae		Dessi *et al.*, 2001
471.	*Camellia sinensis*	Theaceae	Leaves	Tanizawa *et al.*, 1984; Ruch *et al.*, 1989; Wang *et al.*, 1989; Zhao *et al.*, 1989; Xu *et al.*, 1992; Miura *et al.*, 1994; Yen and Chen 1995; Miller *et al.*, 1996; Lin *et al.*, 1996; Luo *et al.*, 1997; Wiseman *et al.*, 1997; Zhang and Shen, 1997; Pietta *et al.*, 1998; Benzie and Szeto, 1999; Yildirim *et al.*, 2000; Sunitha *et al.*, 2000; Higashi-Okai *et al.*, 2001, Cai *et al.*, 2004, 2006; Katsube *et al.*, 2004; Rababah *et al.*, 2004; Ohmori *et al.*, 2005; Abeywickrama *et al.*, 2006; Thuong *et al.*, 2006; Morteza-Semnani *et al.*, 2007; Saito *et al.*, 2007; Borse *et al.*, 2007; Chen *et al.*, 2007; Htay *et al.*, 2007; Wojcikowski *et al.*, 2007; Celic and Tuluce 2007; Khalaf *et al.*, 2008; Chen Haixia *et al.*, 2008; Chen, C-H. *et al.*, 2009; Raihan *et al.*, 2009; Divya *et al.*, 2009; Povichit *et al.*, 2010
472.	*Campsis grandiflora*	Bignoniaceae	Flower	Cui *et al.*, 2006; Gan *et al.*, 2010
473.	*Campsis radicans*	Bignoniaceae	Flower	Cai *et al.*, 2004

Contd...

Table 3.1–*Contd...*

Sl.No.	Plant Name	Family	Parts Used	References
474.	*Camptotheca acuminata*	Nyssaceae	Leaves	Chen, C-H. *et al.*, 2009
475.	*Cananga odorata*	Annonaceae	Leaves oil	Sacchetti *et al.*, 2005; Leelapornpisid *et al.*, 2007;
476.	*Canarium* sp.	Burseraceae		Oke and Hamburger, 2002
477.	*Cannabis sativa*	Cannabinaceae		Kim *et al.*, 1997;
478.	*Cansjera rheedii*	Opiliaceae		Mounnissamy *et al.*, 2008; Shalini *et al.*, 2010
479.	*Canthium parviflorum*	Rubiaceae	Leaves	Sathish Kumar *et al.*, 2008
480.	*Canthium setosum*	Rubiaceae	Aerial part	Lagnika *et al.*, 2011
481.	*Capparis atamisguea*	Capparaceae		Borneo *et al.*, 2009
482.	*Capparis cartilaginea*	Capparaceae	Leaf, stem	Marwah *et al.*, 2007
483.	*Capparis ovata* var. *canescens*	Capparaceae	Buds, leaves	El-Ghorab *et al.*, 2007
484.	*Capparis spinosa*	Capparaceae	Buds	Bonind *et al.*, 2002; Souri *et al.*, 2004; Tesoriere *et al.*, 2007; Kulisic-Bilusic *et al.*, 2010
485.	*Capsella bursa-pastoris*	Brassicaceae	Herb	Katalinic *et al.*, 2006; Gan *et al.*, 2010
486.	*Capsicum annuum*	Solanaceae	Fruits	Mateos *et al.*, 2003; Ochi *et al.*, 2003; Lee *et al.*, 2006 : Howard *et al.*, 2000; Maeda *et al.*, 2006; Jang *et al.*, 2008; Chuah *et al.*, 2008; Serrano *et al.*, 2010
487.	*Capsicum annuum* var. *grossam*	Solanaceae	Fruits	Gacche *et al.*, 2010
488.	*Capsicum frutescens*	Solanaceae	Friuts	Howard *et al.*, 2000; Odukoya *et al.*, 2005; Maeda *et al.*, 2006
489.	*Caragana tibetica*	Fabaceae	Stems	Xiang *et al.*, 2005
490.	*Caralluma adscendens*	Asclepiadaceae	Herb	Tatiya *et al.*, 2010
491.	*Caralluma edulis*	Asclepiadaceae	Aerial part	Ansari *et al.*, 2005
492.	*Caralluma flava*	Asclepiadaceae	Whole plant	Marwah *et al.*, 2007
493.	*Caralluma quadrangula*	Asclepiadaceae	Whole plant	Marwah *et al.*, 2007
494.	*Cardiospermum halicacabum*	Sapindaceae		Kumaran and Karunakaran, 2006b; Sheeba and Asha, 2006; Huang *et al.*, 2011
495.	*Carduus nutans*	Asteraceae	Aerial parts	Deliorman Orhan *et al.*, 2004
496.	*Careya arborea*	Barringtoniaceae	Bark	Sambathkumar *et al.*, 2006; Senthilkumar *et al.*, 2007, 2008
497.	*Careya sphaerica*	Barringtoniaceae	Leaves	Maisuthisakul, 2004; Maisuthisakul *et al.*, 2008

Contd...

Table 3.1–*Contd...*

Sl.No.	*Plant Name*	*Family*	*Parts Used*	*References*
498.	*Carica papaya*	Caricaceae	Leaves/ Fruits	Lako *et al.*, 2006; Mehdipour *et al.*, 2006; Ayoola *et al.*, 2008; Maeda *et al.*, 2006; Alfredo, 2007; Omonkhua and Onoagbe, 2008; Faller and Fialho, 2010; Isabelle *et al.*, 2010
499.	*Carissa carandas*	Apocynaceae	Fruits	Hegde and Joshi, 2009; Sulaiman *et al.*, 2010
500.	*Carissa opaca*	Apocynaceae	Fruits	Sahreen *et al.*, 2010; Khan *et al.*, 2010
501.	*Carlina acanthifolia*	Asteraceae	Essential oil	Dordevic *et al.*, 2007
502.	*Carlina acaulis*	Asteraceae	Root	Benzel *et al.*, 1988
503.	*Carpinus laxiflora*	Carpinaceae	Herb	Jung *et al.*, 2006;
504.	*Carthamus oxyacantha*	Asteraceae		Souri *et al.*, 2004
505.	*Carthamus tinctorius*	Asteraceae	Oil cake, dried flowers	Zhang *et al.*, 1997; Lee *et al.*, 2002; Cai *et al.*, 2004, 2006; Kruawan and Kangsadalampai, 2006; Gan *et al.*, 2010
506.	*Carum carvi*	Apiaceae	Fruits	Kahkonen *et al.*, 1999; Lado *et al.*, 2004; Hinneburg *et al.*, 2005, 2006; Mariutti *et al.*, 2008; Samijlik *et al.*, 2010
507.	*Carum nigrum*	Apiaceae	Seeds	Singh *et al.*, 2006
508.	*Caryopteris incana*	Verbenaceae		Gao *et al.*, 1999, 2000
509.	*Caryota mitis*	Arecaceae	Fruit	Fu *et al.*, 2010
510.	*Casearia grayi*	Flacourtiaceae	Mature stem	Mosaddik *et al.*, 2004
511.	*Casearia sylvestris*	Flacourtiaceae		Menezes *et al.*, 2004
512.	*Casimiroa edulis*	Rutaceae	Leaves	Awaad *et al.*, 2006
513.	*Cassia alata*	Caesalpiniaceae		Akinmoladun *et al.*, 2010
514.	*Cassia aurcualata*	Caesalpiniaceae	Flowers, leaves	Pari and Latha, 2002; Kumaran and Karunakaran, 2007; Suresh Kumar *et al.*, 2008; Jain, 2009
515.	*Cassia fistula*	Caesalpiniaceae	Flowers, fruits	Munasinghe *et al.*, 2001; Chaminda *et al.*, 2001; Siddhuraju *et al.*, 2002; Luximon-Ramma *et al.*, 2002; Manonmani *et al.*, 2005; Bahorun *et al.*, 2005; Ilavarasan *et al.*, 2005; Prakash *et al.*, 2007; Suresh kumar *et al.*, 2008
516.	*Cassia javanica*	Caesalpiniaceae	Bark, leaves	Kaur and Arora, 2011
517.	*Cassia occidentalis*	Caesalpiniaceae		Kouakou-Siransy *et al.*, 2010; Mehta *et al.*, 2010
518.	*Cassia siamea*	Caesalpiniaceae	Flowers	Chanwitheesuk *et al.*, 2005; Kaur *et al.*, 2006; Kaur and Arora, 2011

Contd...

Table 3.1–*Contd...*

Sl.No.	Plant Name	Family	Parts Used	References
519.	*Cassia sophera*	Caesalpiniaceae	Leaves	Rahman *et al.*, 2008; Yele *et al.*, 2008
520.	*Cassia surattensis*	Caesalpinaceae		Liu *et al.*, 2008
521.	*Cassia tora*	Caesalpiniaceae	Fruits	Cai *et al.*, 2004; Zhenbao *et al.*2007; Ashawat and Saraf, 2007; Patnibul *et al.*, 2008; Abu Backer *et al.*, 2009; Lobo *et al.*, 2010; Arya and Yadav, 2010
522.	*Castanopsis cuspidata*	Fagaceae	Herb	Jung *et al.*, 2006
523.	*Casuarina equisetifolia*	Casuarinaceae	Bark	Prakash *et al.*, 2007
524.	*Catalpa bignonioides*	Bignoniaceae	Leaves	Dvorska *et al.*, 2007
525.	*Catharanthus roseus* (=*Vinca rosea*)	Apocynaceae	Leaves	Kaleem *et al.*, 2005; Ferreres *et al.*, 2008; Zheng *et al.*, 2001; Zheng and Wang, 2001; Pareira *et al.*, 2010; Biswas *et al.*, 2010
526.	*Cecropia pachystachya*	Cecropiaceae		Velazquez *et al.*, 2003
527.	*Celastrus hindsii*	Celastraceae	Leaves	Ly *et al.*, 2006
528.	*Celastrus orbiculatus*	Celastraceae	Aerial parts	Hwang *et al.*, 2001
529.	*Celastrus paniculatus*	Celastraceae	Seeds, seed oil	Kumar and Gupta, 2002, 2006; Godkar *et al.*, 2003; Makari *et al.*, 2000; Ramadan *et al.*, 2009; Zohera *et al.*, 2010
530.	*Celosia argentea*	Amaranthaceae	Leaf	Li *et al.*, 2008
531.	*Celosia argentea* var. *cristata*	Amaranthaceae		Odukoya *et al.*, 2007; Gan *et al.*, 2010
532.	*Centaurea americana*	Asteraceae	Seeds	Shoeb *et al.*, 2006
533.	*Centaurea bruguieriana* subsp. *belangerana*	Asteraceae		Souri *et al.*, 2004
534.	*Centaurea calcitrapa*	Asteraceae	Young whorls	Pieroni *et al.*, 2002
535.	*Centaurea depressa*	Asteraceae		Hosseinimehr *et al.*, 2007
536.	*Centaurea ensiformis*	Asteraceae	Aerial parts	Ugur *et al.*, 2009
537.	*Centaurea mucronifera*	Asteraceae		Tepe *et al.*, 2006
538.	*Centaurium erythraea*	Gentianaceae	Herb	Valentão *et al.*, 2001, 2003; Katalinic *et al.*, 2006
539.	*Centaurium spicatum*	Gentianaceae		Shahat *et al.*, 2003
540.	*Centella asiatica*	Apiaceae	Leaves	Jadhav and Bhutani, 2002; Jayashree *et al.*, 2003; Chanwitheesuk *et al.*, 2005; Gajera *et al.*, 2005; Bajpai *et al.*, 2005; Wong *et al.*, 2006; Wojcikowski *et al.*, 2007; Huda-Faujan *et al.*, 2007, 2009; Jain and Agrawal, 2009; Huda-Faujan *et al.*, 2009; Subhasree *et al.*, 2009

Contd...

Table 3.1–*Contd..*

Sl.No.	*Plant Name*	*Family*	*Parts Used*	*References*
541.	*Centipeda minima*	Asteraceae	Aerial parts	Cai *et al.*, 2004; Song *et al.*, 2010
542.	*Cephalonoplos segetum*	Asteraceae		Gan *et al.*, 2010
543.	*Cerasus mahaleb*	Rosaceae	Fruit	Serteser *et al.*, 2009
544.	*Ceratonia siliqua*	Leguminosae	Fruits	Kumazawa *et al.*, 2002; Papagiannopoulos *et al.*, 2004; El Hajaji *et al.*, 2011
545.	*Ceriops decandra*	Rhizophoraceae	Leaves, flowers	Vadlapudi and Naidu, 2009
546.	*Chaenomeles speciosa*	Rosaceae		Gan *et al.*, 2010
547.	*Chaerophyllum hirsutum*	Apiaceae	Roots, aerial parts	Dall'Acqua and Innocenti, 2004
548.	*Chaerophyllum byzantinum*	Apiaceae	Leaf	Özen, 2010
549.	*Chaerophyllum libanoticum*	Apiaceae	Fruits	Demirci *et al.*, 2007
550.	*Chaerophyllum macrophyllum*	Apiaceae		Coruh *et al.*, 2007
551.	*Chamaerion angustifolium*	Onagraceae	Herb	Myagmar and Aniya, 2000; Katalinic *et al.*, 2006
552.	*Chamomilla recutita*	Asteraceae	Herb	Nissen *et al.*, 1996
553.	*Changium smyrnoides*	Apiaceeae		Song *et al.*, 2010
554.	*Chassalia chartacea*	Rubiaceae		Saha *et al.*, 2004
555.	*Chassalia minor*	Rubiaceae		Ahmad *et al.*, 2010
556.	*Chassalia pubescens*	Rubiaceae		Ahmad *et al.*, 2010
557.	*Cheiloclinium cognatum*	Celastraceae	Root bark	Jeller *et al.*, 2004
558.	*Chelidonium majus*	Papaveraceae	Aerial parts	Then *et al.*, 2003; Hadaruga and Hadaruga, 2009
559.	*Chenopodium album*	Chenopodiaceae	Herb	Singh *et al.*, 2009
560.	*Chenopodium ambrosioides*	Chenopodiaceae	Herb	Abouzid *et al.*, 2008
561.	*Chenopodium botrys*	Chenopodiaceae		Souri *et al.*, 2004
562.	*Chenopodium foliosum*	Chenopodiaceae	Fruit	Serteser *et al.*, 2009
563.	*Chenopodium pallidecaule*	Chenopodiaceae		Panarrieeta *et al.*, 2005
564.	*Chenopodium quinoa*	Chenopodiaceae	Fruit	Panarrieeta *et al.*, 2005; Nsimba *et al.*, 2008; Alvarez-Jubete *et al.*, 2010
565.	*Chimarrhis turbinata*	Rubiaceae		Cardoso *et al.*, 2004, 2006
566.	*Chionanthus virginicus*	Oleaceae	Root bark	Gülçin, 2009
567.	*Chlorophytum borivilianum*	Liliaceae	Tubers	Govindarajan *et al.*, 2005; Kenjale *et al.*, 2007

Contd...

Table 3.1–*Contd...*

Sl.No.	Plant Name	Family	Parts Used	References
568.	*Chlorophytum tuberosum*	Liliaceae	Tubers	Narasimhan *et al.*, 2005; Bajpai *et al.*, 2005
569.	*Chondrilla juncea*	Asteraceae		Pieroni *et al.*, 2002
570.	*Chromolaena odorata*	Asteraceae	Leaves	Akinmoladun *et al.*, 2010; Povichit *et al.*, 2010; Rao *et al.*, 2010
571.	*Chrozophora tinctoria*	Euphorbiaceae		Delazar *et al.*, 2006; Shahwar *et al.*, 2010
572.	*Chrysanthellum indicum*	Asteraceae		Borneo *et al.*, 2008
573.	*Chrysanthemum cinerariaefolium*	Asteraceae		Souri *et al.*, 2004
574.	*Chrysanthemum coronarium*	Asteraceae	Leaf	Maeda *et al.*, 2006
575.	*Chrysanthemum frutescens*	Asteraceae		Moussa *et al.*, 2011
576.	*Chrysanthemum indicum*	Asteraceae	Inflore-scence	Cai *et al.*, 2004; Kruawan and Kangsadalampai, 2006; Nikolova and Dzhurmanski, 2009; Song *et al.*, 2010
577.	*Chrysanthemum morifolium*	Asteraceae	Flowers	Osawa and Namiki, 1981; Duh *et al.*, 1999; Cai *et al.*, 2004; Kim and Lee, 2005; Song *et al.*, 2010
578.	*Chrysanthemum multiflorum*	Asteraceae		Chen *et al.*, 2005
579.	*Chrysophyllum canito*	Sapotaceae	Fruits	Luo *et al.*, 2002
580.	*Chrysophyllum inornatum*	Sapotaceae		Santos *et al.*, 2009
581.	*Chukrasia tabularis*	Meliaceae	Leaves	Kaur *et al.*, 2008
582.	*Cicer arietinum*	Fabaceae	Seeds	Bajpai *et al.*, 2005; Han *et al.*, 2008; Sreeramulu *et al.*, 2009
583.	*Cichorium intybus*	Asteraceae	Aerial parts	Sultana *et al.*, 1995; Pieroni *et al.*, 2002; Souri *et al.*, 2004; Özgen *et al.*, 2004; Katalinic *et al.*, 2006
584.	*Cienfuegosia digitata*	Malvaceae	Herb	Kiessoun *et al.*, 2010
585.	*Cimcifuga dahurica*	Ranunculaceae	Rhizome	Cai *et al.*, 2004;
586.	*Cimcifuga foetida*	Ranunculaceae		Song *et al.*, 2010
587.	*Cimcifuga racemosa*	Ranunculaceae	Root	Wojcikowski *et al.*, 2007
588.	*Cinchona ledgeriana*	Rubiaceae	Leaves	Al-Mustafa and Al-Thunibat, 2008
589.	*Cinchona pubescens*	Rubiaceae		Mosquera *et al.*, 2007
590.	*Cinnamomum burmanni*	Lauraceae	Leaves	Prasad *et al.*, 2009
591.	*Cinnamomum camphora*	Lauraceae		Lee *et al.*, 2006

Contd...

Table 3.1–*Contd...*

Sl.No.	*Plant Name*	*Family*	*Parts Used*	*References*
592.	*Cinnamomum cassia*	Lauraceae	Bark	Lin *et al.*, 2003; Bozan *et al.*, 2003; Lee *et al.*, 2003; Cai *et al.*, 2004; Kim *et al.*, 2008; Prasad *et al.*, 2009; Liu *et al.*, 2009; Song *et al.*, 2010
593.	*Cinnamomum insulari-montanum*	Lauraceae		Hou *et al.*, 2003
594.	*Cinnamomum kanehirai*	Lauraceae	Leaves	Lee *et al.*, 2006
595.	*Cinnamomum pauciflorum*	Lauraceae	Leaves	Prasad *et al.*, 2009
596.	*Cinnamomum tamala*	Lauraceae	Leaves oil	Bajpai *et al.*, 2005; Devi *et al.*, 2007; Prasad *et al.*, 2009; Kapoor *et al.*, 2009; Shahwar *et al.*, 2010
597.	*Cinnamomum verum*	Lauraceae	Leaves	Dhuley, 1999; Mathew and Abraham 2004; Shahwar *et al.*, 2010
598.	*Cinnamomum zeylanicum*	Lauraceae	Barks	Mancini-Filho *et al.*, 1998; Jaythilakan *et al.*, 2007; Prasad *et al.*, 2009; Wei and Shibamoto, 2010; Shahwar *et al.*, 2010
599.	*Cirsium arvense*	Asteraceae	Flowers	Nazaruk, 2008
600.	*Cirsium japonicum*	Asteraceae		Gan *et al.*, 2010
601.	*Cirsium oleraceum*	Asteraceae	Inflorescence, leaves	Nazaruk, 2008
602.	*Cirsium palustre*	Asteraceae	Inflorescence, leaves	Nazaruk, 2008
603.	*Cirsium rivulare*	Asteraceae	Inflorescence, leaves	Nazaruk, 2008
604.	*Cirsium setosum*	Asteraceae	Root	Cai *et al.*, 2004
605.	*Cirsium vulgare*	Asteraceae	Inflorescence, leaves	Nazaruk, 2008
606.	*Cissampelos owarensis*	Menispermaceae		Habila *et al.*, 2011
607.	*Cissampelos pareira*	Menispermaceae	Roots	Cao *et al.*, 1996; Hwang *et al.*, 2002; Amresh *et al.*, 2007; Bafna and Mishra, 2009
608.	*Cissus quadrangularis*	Vitaceae	Stem	Murthy *et al.*, 2003; Jainu *et al.*, 2006
609.	*Cistanche deserticola*	Orobanchaceae	Stem	Xiong *et al.*, 1996
610.	*Cistanche salsa*	Orobanchaceae	Stem	Yang *et al.*, 2009
611.	*Cistus incanus*	Cistaceae		Attaguile *et al.*, 2000
612.	*Cistus ladanifer*	Cistaceae	Leaves	Barrajon-Catalan *et al.*, 2010
613.	*Cistus laurifolius*	Cistaceae	Fruit, leaves	Serteser *et al.*, 2009
614.	*Cistus monspelianus*	Cistaceae		Attaguile *et al.*, 2000
615.	*Cistus populifolius*	Cistaceae	Leaves	Barrajon-Catalan *et al.*, 2010
616.	*Citrullus colocynthis*	Cucurbitaceae	Fruit	Souri *et al.*, 2008

Contd...

Table 3.1–*Contd...*

Sl.No.	Plant Name	Family	Parts Used	References
617.	*Citrullus lanatus*	Cucurbitaceae	Fruit	Wang *et al.*, 1996; Tlili *et al.*, 2010; Isabelle *et al.*, 2010
618.	*Citrullus vulgaris*	Cucurbitaceae	Fruit	Lako *et al.*, 2006
619.	*Citrus aurantifolia*	Rutaceae	Fruit	Ghafar *et al.*, 2010
620.	*Citrus aurantium*	Rutaceae	Fruit	Kim *et al.*, 1997; Cai *et al.*, 2004, 2006
621.	*Citrus bergamia*	Rutaceae	Fruit	Trovato *et al.*, 2009; Wei and Shibamoto, 2010
622.	*Citrus grandis*	Rutaceae	Fruits	Lim *et al.*, 2006; Mokbel and Suganuma, 2006; Zhang and Yu, 2007; Kim *et al.*, 2009; Jang *et al.*, 2010
623.	*Citrus hystrix*	Rutaceae	Fruit	Wong *et al.*, 2006; Butryee *et al.*, 2009; Ghafar *et al.*, 2010
624.	*Citrus incanus*	Rutaceae	Friut	Attaguile *et al.*, 2000
625.	*Citrus indica*	Rutaceae	Fruit	Povichit *et al.*, 2010
626.	*Citrus limetta*	Rutaceae	Fruit	KunduSen *et al.*, 2010
627.	*Citrus limon*	Rutaceae		Oza and Pandya, 2007; Yang *et al.*, 2010
628.	*Citrus maxima*	Rutaceae	Fruit	Huang *et al.*, 2004a; KunduSen *et al.*, 2010; Mathur *et al.*, 2010; Singh *et al.*, 2010
629.	*Citrus medica*	Rutaceae	Fruit peel	Conforti *et al.*, 2007
630.	*Citrus microcarpa*	Rutaceae	Fruit	Ghafar *et al.*, 2010
631.	*Citrus monspliensis*	Rutaceae	Fruit	Attaguile *et al.*, 2000
632.	*Citrus nobilis*	Rutaceae	Peel oil	Gursoy *et al.*, 2010
633.	*Citrus paradisii*	Rutaceae	Seeds	Wang *et al.*, 1996; Vinson *et al.*, 2002; Giamperi *et al.*, 2004; Mohamed, 2004; Deyhim *et al.*, 2006; Yang *et al.*, 2010
634.	*Citrus ponki*	Rutaceae	Peel	Chen, C-H. *et al.*, 2009
635.	*Citrus reticulata*	Rutaceae	Fruits/ pulps	Tanizawa *et al.*, 1992; Wang *et al.*, 1996; Tosun and Ustun, 2003; Kang *et al.*, 2006; Xu *et al.*, 2008; Faller and Fialho, 2010; Mathur *et al.*, 2010
636.	*Citrus sinensis*	Rutaceae	Fruit	Arena *et al.*, 2001; Vinson *et al.*, 2002; Gorinstein *et al.*, 2003, 2004; Mohamed, 2004; Cai *et al.*, 2004; Lo Scalzo *et al.*, 2004; Odukoya *et al.*, 2005; Anagnostopoulou *et al.*, 2006; Deyhim *et al.*, 2006; Lako *et al.*, 2006; Kelebek *et al.*, 2007; Mathur *et al.*, 2010; Gharfar *et al.*, 2010; Singh *et al.*, 2010; Ahmad *et al.*, 2011

Contd...

Table 3.1–*Contd...*

Sl.No.	*Plant Name*	*Family*	*Parts Used*	*References*
637.	*Citrus x tangerina*	Rutaceae	Fruit	Vinson *et al.*, 2002; Huang *et al.*, 2004a; Faller and Fialho, 2010
638.	*Citrus unshiu*	Rutaceae	Fruits	Suguiura *et al.*, 2006; Kim *et al.*, 2008; Higashi-Okai *et al.*, 2002
639.	*Clausena lansium*	Rutaceae	Stem bark, fruits	Adebajo *et al.*, 2009; Nagendra Prasad *et al.*, 2010
640.	*Clematis chinensis*	Ranunculaceae	Root	Cai *et al.*, 2004; Gan *et al.*, 2010
641.	*Clematis flammula*	Ranunculaceae	Leaves	Nadjet *et al.*, 2008
642.	*Cleome arabica*	Cleomaceae	Leaves	Selloum *et al.*, 1997
643.	*Cleome gynandra* (Syn: *Gynandropsis gynandra)*	Capparaceae	Leaf	Cook *et al.*, 1998; Anbazhagi *et al.*, 2009
644.	*Cleome icosandra*	Capparaceae	Seed, leaves, flower	Kalim *et al.*, 2010
645.	*Cleome viscosa*	Capparaceeae	Leaves	Vadlapudi and Naidu, 2010
646.	*Clerodendron cyrtophyllum*	Verbenaceae		Li *et al.*, 2008
647.	*Clerodendrum infortunatum*	Verbenaceae	Leaves	Gouthamchandra *et al.*, 2010
648.	*Clerodendron trichotomum*	Verbenaceae		Chae *et al.*, 2004, 2005
649.	*Cleyera japonica* var. *mori*	Theaceae		Hou *et al.*, 2003
650.	*Clibadium funzike*	Asteraceae		Mosquera *et al.*, 2007
651.	*Clinopodium vulgare*	Lamiaceae	Herb	Tepe *et al.*, 2007
652.	*Clitoria ternatea*	Fabaceae	Flowers	Kamkaen and Wilkinson, 2009; Vadlapudi and Naidu, 2010; Narayanaswamy and Balakrishnan, 2011
653.	*Cnestis ferruginea*	Connaraceae	Leaf, root	Oke and Hamburger, 2002
654.	*Cnidium officinale*	Apiaceae	Essential oil	Jeong *et al.*, 2009a,b
655.	*Cnidoscolus aconitifolius*	Euphorbiaceae	Leaves	Kuti and Kunuru, 2004; Oloyede *et al.*, 2010; Adefegha and Oboh, 2011
656.	*Cnidoscolus chayamansa*	Euphorbiaceae	Leaves	Kuti and Kunuru, 2004
657.	*Coccinia grandis (=Coccinia indica)*	Cucurbitaceae	Leaves	Kamble *et al.*, 1998; Venkateswaran and Pari, 2003; Chanwitheesuk *et al.*, 2005; Mazumder *et al.*, 2008; Biswas *et al.*, 2010
658.	*Cochlospermum tinctorium*	Cochlospermaceae		Nergard *et al.*, 2005

Contd...

Table 3.1–*Contd...*

Sl.No.	*Plant Name*	*Family*	*Parts Used*	*References*
659.	*Cocos nucifera*	Arecaceae	Mesocarp; Coconut water	Mantena *et al.*, 2003; Lako *et al.*, 2006; Alviano *et al.*, 2008; Chakraborthy and Mitra, 2008; Hasan *et al.*, 2009
660.	*Codonopsis pilosula*	Campanulaceae		Ng *et al.*, 2004
661.	*Coffea arabica*	Rubiaceae	Seeds	Dianzani *et al.*, 1991; Devasagayam *et al.*, 1996; Yanagimoto *et al.*, 2004; Gomez Ruiz *et al.*, 2007; Ramalakshmi *et al.*, 2008: Naidu *et al.*, 2008
662.	*Coffea robusta*	Rubiaceae	Seeds	Naidu *et al.*, 2008
663.	*Coix lacryma-jobi*	Poaceae		Kuo *et al.*, 2002; Gan *et al.*, 2010
664.	*Cola nitida*	Sterculiaceae		Rakotoarison *et al.*, 1996
665.	*Colchicum balansae*	Colchicaceae	Tubers, leaves	Mammadov *et al.*, 2009
666.	*Colchicum speciosum*	Colchicaceae	Root stock	Souri *et al.*, 2008
667.	*Coleus amboinicus*	Lamiaceae	Herb	Chanwitheesuk *et al.*, 2005
668.	*Coleus aromaticus*	Lamiaceae	Leaves	Kumaran and Karunakaran, 2006a, 2007a; Anasuya and Gomathi, 2010
669.	*Coleus blumei*	Lamiaceae	Stems, leaves	Zakaria *et al.*, 2008
670.	*Colocasia esculenta*	Araceae	Tubers, leaves	Lindsey *et al.*, 2002; Maeda *et al.*, 2006; Lako *et al.*, 2006
671.	*Combretum acutifolium*	Combretaceae	Leaves	Masoko and Eloff, 2007
672.	*Combretum apiculatum*	Combretaceae	Leaves	Masoko and Eloff, 2007
673.	*Combretum caffrum*	Combretaceae	Leaves	Masoko and Eloff, 2007
674.	*Combretum celastroides*	Combretaceae	Leaves	Masoko and Eloff, 2007
675.	*Combretum collinum*	Combretaceae	Leaves	Masoko and Eloff, 2007
676.	*Combretum edwardsii*	Combretaceae	Leaves	Masoko and Eloff, 2007
677.	*Combretum erythrophyllum*	Combretaceae	Leaves	Masoko and Eloff, 2007
678.	*Combretum hereroense*	Combretaceae	Leaves	Masoko and Eloff, 2007
679.	*Combretum imberbe*	Combretaceae	Leaves	Masoko and Eloff, 2007
680.	*Combretum kraussii*	Combretaceae	Leaves	Masoko and Eloff, 2007
681.	*Combretum micranthum*	Combretaceae	Leaves	Karou *et al.*, 2005
682.	*Combretum moggi*	Combretaceae	Leaves	Masoko and Eloff, 2007
683.	*Combretum molle*	Combretaceae	Leaves	Masoko and Eloff, 2007
684.	*Combretum nelsonii*	Combretaceae	Leaves	Masoko and Eloff, 2007
685.	*Combretum petrophilum*	Combretaceae	Leaves	Masoko and Eloff, 2007
686.	*Combretum padoides*	Combretaceae	Leaves	Masoko and Eloff, 2007
687.	*Combretum woodii*	Combretaceae	Leaves	Masoko and Eloff, 2007

Contd...

Table 3.1–*Contd...*

Sl.No.	*Plant Name*	*Family*	*Parts Used*	*References*
688.	*Combretum zeyheri*	Combretaceae	Leaves	Masoko and Eloff, 2007
689.	*Commelina benghalensis*	Commelinaceae	Aerial parts	Hasan *et al.*, 2009
690.	*Commelina diffusa*	Commelinaceae		Houghton *et al.*, 2005
691.	*Commiphora berryi*	Burseraceae		Gowri Shankar *et al.*, 2008
692.	*Commipora caudata*	Burseraceae	Leaves	Annu *et al.*, 2010
693.	*Commiphora myrrha*	Burseraceae		Mishra *et al.*, 2007
694.	*Commiphora ornifolia*	Burseraceae	Bark	Mothana *et al.*, 2009, 2010
695.	*Commiphora parvifolia*	Burseraceae	Essential oil	Mothana *et al.*, 2010
696.	*Commiphora wightii* (Syn.: *Commiphora mukul*)	Burseraceae	Bark	Chander *et al.*, 2000; Zhu *et al.*, 2001; Al-Burtamani and Fatope, 2004
697.	*Comocladia engleriana*	Anacardiaceae		Ruiz-Terán *et al.*, 2008
698.	*Convolvulus pluricaulis*	Convolvulaceae		Joshi *et al.*, 2003; Parihar and Hemnani, 2003
699.	*Conyza canadensis*	Asteraceae	Herb	Olas *et al.*, 2006; Hayat *et al.*, 2009
700.	*Conyza dioscorides*	Asteraceae	Herb	Abouzid *et al.*, 2008
701.	*Copaifera langsdorffii*	Caesalpiniaceae		Santos *et al.*, 2009
702.	*Copernicia cerifera*	Arecaceae	Leaves, mesocarp	Silva *et al.*, 2005
703.	*Coptis chinensis*	Ranunculaceae	Root	Schinella *et al.*, 2002; Cai *et al.*, 2004; Huang *et al.*, 2007; Li *et al.*, 2008
704.	*Corchorus capsularis*	Tiliaceae		Zakaria, 2007
705.	*Corchorus olitorius*	Tiliaceae	Leaf	Azuma *et al.*, 1994; Oboh and Akindahunsi, 2004; Maeda *et al.*, 2006; Odukoya *et al.*, 2007; Morrison and Twumasi, 2010; Salawu *et al.*, 2011
706.	*Cordia gilletii*	Boraginaceae	Root bark	Okusa and Penge, 2007
707.	*Cordia multispicata*	Boraginaceae	Leaves	Da Silva *et al.*, 2010
708.	*Cordia myxa*	Boraginaceae	Fruit	Souri *et al.*, 2008
709.	*Cordia perrottettii*	Boraginaceae	Bark	Marwah *et al.*, 2007
710.	*Cordia wallichii*	Boraginaceae		Makari *et al.*, 2008
711.	*Coriandrum sativum*	Apiaceae	Fruits	Chitra and Leelamma, 1999; Sabu and Kuttan, 2003; Wangensteen *et al.*, 2004; Souri *et al.*, 2004; Chanwitheesuk *et al.*, 2005; Hashim *et al.*, 2005; Singh *et al.*, 2006; Wong and Kitts, 2006; Wong *et al.*, 2006; Souri *et al.*, 2008; Mariutti *et al.*, 2008; Nickavar and Abolhasani, 2009; Gacche *et al.*, 2010; Samijlik *et al.*, 2010

Contd...

Table 3.1–*Contd...*

Sl.No.	*Plant Name*	*Family*	*Parts Used*	*References*
712.	*Coridothymus capitatus*	Lamiaceae	Essential oils, herb	Lagouri *et al.*, 1993; Demo *et al.*, 1999; Kosar *et al.*, 2003; Dorman *et al.*, 2004
713.	*Cornus mas*	Cornaceae	Fruits, leaves	Gulcin *et al.*, 2005; Leskovac *et al.*, 2007; Tural and Koca, 2008; Serteser *et al.*, 2009; Hassanpour *et al.*, 2011
714.	*Cornus officinalis*	Cornaceae	Fruit	Kim *et al.*, 1997; Cai *et al.*, 2004
715.	*Cornus sanguinea*	Cornaceae	Fruit, leaves	Serteser *et al.*, 2009
716.	*Cornus stolonifera*	Cornaceae		McCune, 1999
717.	*Coronopus didymus*	Brassicaceae	Whole plant	Prabhakar *et al.*, 2006
718.	*Corylus avellana*	Corylaceae	Fruit, leaves	Alasalvar *et al.*, 2006; Shahidi *et al.*, 2007; Oliveira *et al.*, 2007, 2008
719.	*Coscinium blumeanum*	Menispermaceae	Leaves	Keawpradub *et al.*, 2005
720.	*Coscinium fenestratum*	Menispermaceae	Leaves	Venukumar and Latha, 2002b; Punitha *et al.*, 2005a,b; Keawpradub *et al.*, 2005; Divyasree and Nair, 2008
721.	*Cosmos caudatus*	Asteraceae	Leaves	Abbas *et al.*, 2006; Wong *et al.*, 2006; Zaibunnisa *et al.*, 2008; Huda-Faujan *et al.*, 2009; Andarwulan *et al.*, 2010
722.	*Costus afer*	Costaceae	Stem	Anyasor *et al.*, 2010
723.	*Costus igneus*	Costaceae	Leaves	Vishalakshi Devi and Urooj, 2010
724.	*Costus speciosus*	Costaceae	Whole plant	Vijayalakshmi and Sarada, 2008; Vishalakshi Devi and Urooj, 2010
725.	*Cotinus coggygria*	Anacardiaceae		Ivanova *et al.*, 2005; Niforovi *et al.*, 2010
726.	*Coulutea persica*	Fabaceae		Souri *et al.*, 2004
727.	*Coussarea hydrangeifolia*	Rubiaceae	Leaves	Hamerski *et al.*, 2005
728.	*Craibiodendron henryi*	Ericaceae	Roots	Huang *et al.*, 2005
729.	*Crassocephalum crepidioides*	Asteraceae	Leaf	Aniya *et al.*, 2005; Odukoya *et al.*, 2007
730.	*Crassocephalum rubens*	Asteraceae		Iwalewa *et al.*, 2005
731.	*Crassopteryx febrifuga*	Rubiaceae		Maiga *et al.*, 2006
732.	*Crassula* cv. *himaturi*	Crassulaceae		Lee *et al.*, 2005
733.	*Crataegus aronia*	Rosaceae		Ljubuncic *et al.*, 2005
734.	*Crataegus azarolus*	Rosaceae	Stem, leaves, fruit	Al-Mustafa and Al-Thunibat, 2008
735.	*Crataegus laevigatus*	Rosaceae	Berries, leaves, flowers	Pendry *et al.*, 2005

Contd...

Table 3.1–*Contd...*

Sl.No.	Plant Name	Family	Parts Used	References
736.	*Crataegus monogyna*	Rosaceae	Leaves	Rakotoarison *et al.*, 1996; Ivanova *et al.*, 2005; Kiselova *et al.*, 2006; Leskovac *et al.*, 2007; Demiray *et al.*, 2009
737.	*Crataegus oxyacantha*	Rosaceae		Da Silva *et al.*, 2000
738.	*Crataegus pinnatifida*	Rosaceae	Fruit	Cai *et al.*, 2004, 2006; Nam and Kang, 2004
739.	*Crataeva nurvala*	Cappridaceae		Bhakar *et al.*, 1996; Sunitha *et al.*, 2001; Shirwaikar *et al.*, 2004
740.	*Cratoxylum cochinchinense*	Clusiaceae	Roots	Tang *et al.*, 2004; Mahabusarakam *et al.*, 2006; Phuwapraisirisan *et al.*, 2006
741.	*Cratoxyum formosum*	Clusiaceae		Kukongviriyapan *et al.*, 2007; Maisuthisakul *et al.*, 2006, 2007a,b; 2008
742.	*Cremastra variabilis*	Orchidaceae		Li *et al.*, 2008
743.	*Crepidiastrum lanceolatum*	Asteraceae	Leaf	Maeda *et al.*, 2006
744.	*Crinum ornatum*	Liliaceae		Oloyede and Farombi, 2010
745.	*Crinum purpurascens*	Liliaceae	Bulb	Oke and Hamburger, 2002
746.	*Crithmum maritimum*	Apiaceae	Herb	Ruberto *et al.*, 2000; Kulisic-Bilusic *et al.*, 2010
747.	*Crocus sativus*	Iridaceae	Aerial parts, petals	Li *et al.*, 2004; Assimopoulou *et al.*, 2005; Sengul *et al.*, 2009
748.	*Crossopteryx febrifuga*	Rubiaceae		Maiga *et al.*, 2006
749.	*Crotalaria aegyptiaca*	Fabaceae		Ali *et al.*, 2004
750.	*Crotalaria burhia*	Fabaceae	Roots	Alang *et al.*, 2010
751.	*Crotalaria mucronata*	Fabaceae		Jiang *et al.*, 2010
752.	*Crotalaria sessiliflora*	Fabaceae		Munim *et al.*, 2003; Woo *et al.*, 2005
753.	*Croton andinus*	Euphorbiaceae		Rosas-Romero and Saavedra, 2005
754.	*Croton cajucara*	Euphorbiaceae	Herb	Tieppo *et al.*, 2006
755.	*Croton celtidifolius*	Euphorbiaceae	Bark	Nardi *et al.*, 2003
756.	*Croton emporiorum*	Euphorbiaceae		Rosas-Romero and Saavedra, 2005
757.	*Croton lechlei*	Euphorbiaceae	Latex	Risco *et al.*, 2003
758.	*Croton lobatus*	Euphorbiaceae	Aerial part	Lagnika *et al.*, 2011
759.	*Croton mortibensis*	Euphorbiaceae		David *et al.*, 2007
760.	*Croton sparsiflorus*	Euphorbiaceae		Ahmad *et al.*, 2010
761.	*Cruciata taurica*	Rubiaceae		Mavi *et al.*, 2004

Contd...

Table 3.1–*Contd...*

Sl.No.	Plant Name	Family	Parts Used	References
762.	*Cryptotaenia japonica*	Apiaceae	Leaf	Maeda *et al.*, 2006
763.	*Cucumis melo*	Cucurbitaceae	Fruits/ pulp, leaf	Vouldoukis *et al.*, 2004; Ismail *et al.*, 2010; Isabelle *et al.*, 2010
764.	*Cucumis sativus*	Cucurbitaceae	Leaf, seed, fruit	Kahkonen *et al.*, 1999; Cai *et al.*, 2004; Maeda *et al.*, 2006; Souri *et al.*, 2008; Isabelle *et al.*, 2010; Nema *et al.*, 2011
765.	*Cucurbita maxima*	Cucurbitaceae	Fruits/pulp	Stangeland *et al.*, 2009; Nara *et al.*, 2009; Gacche *et al.*, 2010
766.	*Cucurbita moschata*	Cucurbitaceae	Fruit	Chanwitheesuk *et al.*, 2005; Yang *et al.*, 2007
767.	*Cucurbita pepo*	Cucurbitaceae	Flowers seeds	Tarhan *et al.*, 2007;Nkosi *et al.*, 2006; Xanthopoulou Marianna *et al.*, 2009
768.	*Culcitum reflexum*	Asteraceae	Leaves	Aquino *et al.*, 2002
769.	*Cuminum cyminum*	Apiaceae	Fruits	Gachkar *et al.*, 2007; Souri *et al.*, 2008; Mariutti *et al.*, 2008; Nickavar and Abolhasani, 2009; El-Ghorab *et al.*, 2010
770.	*Cuphea carthagenesis*	Lythraceae	Leaves	Schuldt *et al.*, 2004
771.	*Curculigo orchioides*	Hypoxidaceae	Rhizomes	Venukumar and Latha, 2002a; Wu *et al.*, 2005: Nagesh *et al.*, 2008
772.	*Curcuma alismatifolia*	Zingiberaceae	Leaves	Akter *et al.*, 2008; Hasan *et al.*, 2009
773.	*Curcuma amada*	Zingiberaceae	Rhizome	Policegoudra *et al.*, 2007; Siddaraju and Dharmesh, 2007; Leelapornpisid *et al.*, 2007
774.	*Curcuma aromatica*	Zingiberaceae	Leaves	Kim *et al.*, 1997; Al-Reza *et al.*, 2009; Gan *et al.*, 2010
775.	*Curcuma comosa*	Zingiberaceae		Jariyawat *et al.*, 2009; Povichit *et al.*, 2010
776.	*Curcuma domestica*	Zingiberaceae		Leelapornpisid *et al.*, 2007
777.	*Curcuma kwangsiensis*	Zingiberaceae	Rhizome	Zhao *et al.*, 2010
778.	*Curcuma longa*	Zingiberaceae	Rhizome, leaves	Toda *et al.*, 1985b; Selvam *et al.*, 1995; Ruby *et al.*, 1995; Anto *et al.*, 1995; Sreejayam, 1997; Kim *et al.*, 1997; Singhal *et al.*, 1999; Scartezzini and Speroni, 2000; Jayaprakasha *et al.*, 2002; Eshrat and Hussain, 2002; Braga *et al.*, 2003; Kim *et al.*, 2003; Ramos *et al.*, 2003; Cai *et al.*, 2004, 2006; Tilak *et al.*, 2004; Padmaja and Raju, 2004, 2005; Sacchetti *et al.*, 2005;

Contd...

Table 3.1–*Contd...*

Sl.No.	Plant Name	Family	Parts Used	References
				Odukoya *et al.*, 2005; Lako *et al.*, 2006; Ashawat and Saraf, 2007; Wojcikowski *et al.*, 2007; Phansawan and Poungbangpho, 2007; Chan *et al.*, 2009; Li and Li, 2009; Dai *et al.*, 2009; Aftab and Vieira, 2009; Gan *et al.*, 2010; Zhao *et al.*, 2010
779.	*Curcuma mangga*	Zingiberaceae		Abbas *et al.*, 2006; Leelapornpisid *et al.*, 2007
780.	*Curcuma phaeocaulis*	Zingiberaceae	Rhizome	Zhao *et al.*, 2010
781.	*Curcuma wenyujin*	Zingiberaceae		Li and Li, 2009; Gan *et al.*, 2010; Zhao *et al.*, 2010
782.	*Curcuma xanthorrhiza*	Zingiberaceae	Rhizome	Masuda *et al.*, 1992; Chanwitheesuk *et al.*, 2005
783.	*Curcuma zedoaria*	Zingiberaceae	Rhizome, leaves	Mau *et al.*, 2003; Prakash *et al.*, 2007; Souri *et al.*, 2008; Paramapojn and Gritsanapan, 2009; Gan *et al.*, 2010
784.	*Cuscuta chinensis*	Cuscutaceae	Seeds	Bao *et al.*, 2002; Cai *et al.*, 2004, 2006; Yen *et al.*, 2007
785.	*Cuscuta reflexa*	Cuscutaceae	Stem	Yadav *et al.*, 2001; Yadav and Tripathi, 2004
786.	*Cussona bancoensis*	Araliaceae	Stem bark	Tapondjou *et al.*, 2005
787.	*Cussona barteri*	Araliaceae	Root	Diallo *et al.*, 2001; Biapa *et al.*, 2007
788.	*Cyamopsis tetragonaloba*	Fabaceae	Fruits	Gacche *et al.*, 2010
789.	*Cyanthula officinalis*	Amaranthaceae		Gan *et al.*, 2010
790.	*Cyclea peltata*	Menispermaceae	Leaf	Fijesh and Padikkala, 2008
791.	*Cyclobalanopsis stenophylla* var. *stenophylloides*	Fagaceae		Hou *et al.*, 2003
792.	*Cyclocarya paliurus*	Juglandaceae		Xie *et al.*, 2010
793.	*Cyclocodon lancifolia*	Campanulaceae	Aerial	Chen, C-H. *et al.*, 2009
794.	*Cyclopia genistoides*	Fabaceae		Joubert *et al.*, 2008
795.	*Cyclopia intermedia*	Fabaceae	Herb	McKay and Blumberg, 2006
796.	*Cyclotrichium niveum*	Lamiaceae	Aerial parts	Cetinus *et al.*, 2007; Orhan *et al.*, 2009; Gursoy *et al.*, 2009
797.	*Cyclotrichium origanofolium*	Lamiaceae	Herb	Tepe *et al.*, 2005
798.	*Cydonia oblonga*	Rosaceae	Fruit	Silva *et al.*, 2004; Aslan *et al.*, 2010
799.	*Cydonia vulgaris*	Rosaceae	Peela	Yildirin *et al.*, 2001; Alesiani *et al.*, 2010

Contd...

Table 3.1–*Contd...*

Sl.No.	*Plant Name*	*Family*	*Parts Used*	*References*
800.	*Cymbopogon citratus*	Poaceae	Rhizome, stems	Cheel *et al.*, 2005; Kruawan and Kangsadalampai, 2006; Leelapornpisid *et al.*, 2007
801.	*Cymbopogon flexuosus*	Poaceae		Mishra *et al.*, 2007
802.	*Cymbopogon schoenanthus*	Poaceae	Rhizome, shoots	Khadri *et al.*, 2008, 2010
803.	*Cynanchum atraum*	Asclepiadaceae		Li *et al.*, 2008
804.	*Cynanchum paniculatum*	Asclepiadaceae		Gan *et al.*, 2010
805.	*Cynanchum stauntonii*	Asclepiadaceae		Song *et al.*, 2010
806.	*Cynanthillium cinerum*	Asteraceae	Whole plant	Guha *et al.*, 2009b
807.	*Cynara cardunculus*	Asteraceae	Leaf	Valentão *et al.*, 2002 b; Kusku-Kiraz *et al.*, 2009
808.	*Cynara scolymus*	Asteraceae	Leaves	Gebhardt, 1996, 1997; Fintelman, 1996; Jimenez-Escrig *et al.*, 2003
809.	*Cynodon dactylon*	Poaceae	Grass	Auddy *et al.*, 2003; Valamarthi and Harathi, 2009
810.	*Cynomorium songaricum*	Balanophoraceae	Stem	Lu *et al.*, 2008
811.	*Cyperus rotundus*	Cyperaceae	Roots, rhizomes	Kilani *et al.*, 2005; Pal and Dutta, 2006; Nagulendran *et al.*, 2007; Yzdanparast and Ardestani, 2007
812.	*Cyperus scariosus*	Cyperaceae		Kalim *et al.*, 2010
813.	*Cytisus scoparius*	Fabaceae		Sundararajan *et al.*, 2006; Raja *et al.*, 2007
814.	*Dalbergia odorifera*	Fabaceae		Cheng *et al.*, 1998; Gan *et al.*, 2010
815.	*Dalbergia saxatilis*	Fabaceae	Leaf	Sofidya *et al.*, 2006
816.	*Dalbergia sissoo*	Fabaceae	Leaves, flowers	Vadlapudi and Naidu, 2010
817.	*Daniella oliveri*	Fabaceae	Bark	Muanda *et al.*, 2011
818.	*Daphne giraldii*	Thymeleaceae	Roots	Zhou *et al.*, 2007
819.	*Daphne gnidium*	Thymeleaceae		Dessi *et al.*, 2001
820.	*Daphne oleoides*	Thymeleaceae	Leaves	Demo *et al.*, 1999
821.	*Daphniphyllum calycinum*	Daphniphyllaceae	Leaves	Gamez *et al.*, 1998
822.	*Datura metel*	Solanaceae	Leaves, flowers	Vadlapudi and Naidu, 2010
823.	*Datura stramonium*	Solanaceae		Suresh Kumar *et al.*, 2008

Contd...

Table 3.1–*Contd...*

Sl.No.	*Plant Name*	*Family*	*Parts Used*	*References*
824.	*Daucus carota*	Apiaceae	Root tuber, leaf	Kahkonen *et al.*, 1999; Cai *et al.*, 2004; Chantaro *et al.*, 2008; Arabshahi *et al.*, 2005; Mishra *et al.*, 2007; Sun *et al.*, 2009; Raghu *et al.*, 2010
825.	*Decalepis hamiltonii*	Asclepiadaceae	Roots	Murthy *et al.*, 2006; Srivastava *et al.*, 2006; Naik *et al.*, 2007; Srivastava and Sivanandappa, 2010; Surveswaran *et al.*, 2010
826.	*Delonix elata*	Caesalpiniaceae	Leaf	Wijayasiriwardena *et al.*, 2009
827.	*Delonix regia*	Caesalpiniaceae	Flowers	Bade *et al.*, 2009
828.	*Delphinium linearilobum*	Ranunculaceae	Roots	Kolak *et al.*, 2006
829.	*Dendrobium candidum*	Orchidaceae	Stems	Li *et al.*, 2009a,b
830.	*Dendrobium denneanum*	Orchidaceae		Fan *et al.*, 2009; Luo *et al.*, 2011
831.	*Dendrobium linawianum*	Orchidaceae	Herb	Lo *et al.*, 2004b
832.	*Dendrobium moniliforme*	Orchidaceae	Herb	Lo *et al.*, 2004
833.	*Dendrobium nobile*	Orchidaceae	Stem	Cai *et al.*, 2004; Luo *et al.*, 2009, 2010
834.	*Dendrobium officinale*	Orchidaceae	Aerial	Chen, C-H. *et al.*, 2009
835.	*Dendrobium tosaense*	Orchidaceae	Herb	Lo *et al.*, 2004
836.	*Dendrobium denneanum*	Orchidaceae	Stem	Fan *et al.*, 2009; Luo *et al.*, 2011
837.	*Dendrobium linawianum*	Orchidaceae	Herb	Lo *et al.*, 2004
838.	*Dendrobium moniliforme*	Orchidaceae	Herb	Lo *et al.*, 2004
839.	*Dendrobium nobile*	Orchidaceae		Cai *et al.*, 2004; Luo *et al.*, 2009, 2010
840.	*Dendrobium sonia*	Orchidaceae	Flower	Shafazila *et al.*, 2010
841.	*Dendrobium tosaense*	Orchidaceae	Herb	Lo *et al.*, 2004a
842.	*Dendrocnide sinuate*	Urticaceae	Herb, Leaves	Sanchez-Moreno *et al.*, 1998
843.	*Dendrocnide sinuata*	Urticaceae	Leaves	Tanti *et al.*, 2010.
844.	*Dendrophthoe falcata*	Loranthaceae	Leaves	Jadhav and Bhutani, 2002; Hasan *et al.*, 2006; Pattanayak and Sunitha, 2008; Patil *et al.*, 2011
845.	*Dendrophthoe trigona*	Loranthaceae	Stem bark	Singh *et al.*, 2004
846.	*Derris scandens*	Fabaceae	Stems	Laupattarksem *et al.*, 2004
847.	*Descureania sophia*	Brassicaceae	Seed	Souri *et al.*, 2008
848.	*Desmodium ascendes*	Fabaceae	Stems, leaves	Mamyrbekova-Beka *et al.*, 2008
849.	*Desmodium gangeticum*	Fabaceae	Root and aerial parts	Govindarajan *et al.*, 2003; Niranjan and Tewari, 2008; Kurian *et al.*, 2009; Veeru *et al.*, 2009

Contd...

Table 3.1–*Contd...*

Sl.No.	*Plant Name*	*Family*	*Parts Used*	*References*
850.	*Desmodium triflorum*	Fabaceae		Lai *et al.*, 2010
851.	*Desmos zeylanica*	Annonaceae	Stem, leaves	Puvanendran *et al.*, 2011
852.	*Detarium microcarpum*	Caesalpiniaceae	Fruit flesh	Lamien-Meda *et al.*, 2008
853.	*Dialium guineense*	Caesalpiniaceae	Fruit flesh	Lamien-Meda *et al.*, 2008
854.	*Dianthus superbus*	Caryophyllaceae	Aerial parts	Cai *et al.*, 2004; Gan *et al.*, 2010
855.	*Dicentra eximia*	Fumariaceae	Seeds	Borchardt *et al.*, 2008
856.	*Dichrostachys cinerea*	Mimosaceae	Roots	Sreedevi *et al.*, 2009
857.	*Dictamnus dasycarpus*	Rutaceae		Li *et al.*, 2008
858.	*Didymopanax venosum*	Araliaceae		Santos *et al.*, 2009
859.	*Dimocarpus longan*	Sapindaceae	Seed	Soong and Barlow, 2006; Hsieh *et al.*, 2008; Pan *et al.*, 2008
860.	*Dioclea grandiflora*	Fabaceae		Vianna *et al.*, 2006
861.	*Dioclea lasiophylla*	Fabaceae	Stems	David *et al.*, 2004
862.	*Diodia apiculata*	Rubiaceae		David *et al.*, 2007
863.	*Dioscorea alata*	Dioscoreaceae	Tubers	Mendoza *et al.*, 1994; Chang *et al.*, 2004; Chen *et al.*, 2004; Lako *et al.*, 2006; Lubag Jr. *et al.*, 2008; Raghu *et al.*, 2010
864.	*Dioscorea batatas*	Dioscoreaceae	Tubers	Hou *et al.*, 2001, 2002
865.	*Dioscorea bulbifera*	Dioscoreaceae	Tuber	Song *et al.*, 2010
866.	*Dioscorea collettii*	Dioscoreaceae		Gan *et al.*, 2010
867.	*Dioscorea esculenta*	Dioscoreaceae	Tuber	Lako *et al.*, 2006
868.	*Dioscorea glandulosa*	Dioscoreaceae	Leaves	Menezes *et al.*, 2004
869.	*Dioscorea nipponica*	Dioscoreaceae		Gan *et al.*, 2010
870.	*Dioscorea nummularia*	Dioscoreaceae	Tuber	Lako *et al.*, 2006
871.	*Dioscorea villosa*	Dioscoreaceae	Root, seeds	Wojcikowski *et al.*, 2007; Borchardt *et al.*, 2008
872.	*Diospyros abyssinica*	Ebenaceae		Maiga *et al.*, 2006
873.	*Diospyros discolor*	Ebenaceae	Leaves	Lee *et al.*, 2006
874.	*Diospyros kaki*	Ebenaceae		Katsube *et al.*, 2004
875.	*Diospyros malabarica*	Ebenaceae	Bark	Mondal *et al.*, 2006
876.	*Diospyros mespiliformis*	Ebenaceae	Fruit	Lamien-Meda *et al.*, 2008
877.	*Diospyros nipponica*	Ebenaceae	Bark	Luo *et al.*, 2008
878.	*Dipsacus asper*	Dipsacaceae		Huang *et al.*, 2006
879.	*Dodecadenia grandiflora*	Lauraceae	Essential oil	Joshi *et al.*, 2010
880.	*Dodonaea viscosa*	Sapindaceae		Pawar *et al.*, 2009; Mothana *et al.*, 2010; Moussa *et al.*, 2011

Contd...

Table 3.1–*Contd...*

Sl.No.	Plant Name	Family	Parts Used	References
881.	*Dorstenia ciliata*	Moraceae		Kansci *et al.*, 2003
882.	*Dorstenia foetida*	Moraceae	Tuber	Marwah *et al.*, 2007
883.	*Dorstenia manni*	Moraceae	Leaves	Dufall *et al.*, 2003
884.	*Dorstenia multiformis*	Moraceae		Balestrin *et al.*, 2008
885.	*Dorstenia psilurus*	Moraceae		Kansci *et al.*, 2003
886.	*Dorycnium hirsutum*	Fabaceae	Aerial parts	Pistelli *et al.*, 2006
887.	*Dorystoechas hastata*	Lamiaceae	Leaves	Karagozler *et al.*, 2008
888.	*Draba nemerosa*	Brassicaceae		Rahman *et al.*, 2007
889.	*Dracocephalum moldavica*	Lamiaceae	Leaves	Povilaitye *et al.*, 2001; Khalil *et al.*, 2007
890.	*Dracocephalum surmandicum*	Lamiaceae		Sonboli *et al.*, 2010
891.	*Dregea volubilis*	Asclepiadaceae		Chanwitheesuk *et al.*, 2005
892.	*Droogmansia gofroyana*	Fabaceae		Patnibul *et al.*, 2008
893.	*Drosera burmanni*	Droseraceae		Gan *et al.*, 2010
894.	*Drymaria diandra*	Caryophyllaceae	Leaves, stems	Mandal *et al.*, 2009
895.	*Durio zibethinus*	Bombacaceae	Fruit	Isabelle *et al.*, 2010
896.	*Ebenus pinnata*	Fabaceae	Aerial parts	Abru *et al.*, 2007
897.	*Echinacea angustifolia*	Asteraceae	Roots	Hu and Kitts, 2000; Sloley *et al.*, 2001; Amarowicz *et al.*, 2004; Pellati *et al.*, 2004, 2005
898.	*Echinacea atrorubens*	Asteraceae	Root	Pellati *et al.*, 2005
899.	*Echinacea laevigata*	Asteraceae	Root	Pellati *et al.*, 2005
900.	*Echinacea pallida*	Asteraceae	Herb, roots	Hu and Kitts, 2000; Sloley *et al.*, 2001; Pellati *et al.*, 2004, 2005; Orhan *et al.*, 2009
901.	*Echinacea paradoxa*	Asteraceae	Root	Pellati *et al.*, 2005
902.	*Echinacea purpurea*	Asteraceae	Roots	Pietta *et al.*, 1998; Hu and Kitts, 2000; Sloley *et al.*, 2001; Miliuskas *et al.*, 2004; Mishima *et al.*, 2004; Pellati *et al.*, 2004, 2005; Dalby-Brown *et al.*, 2005; Masteikova *et al.*, 2007; Orhan *et al.*, 2009
903.	*Echinacea sanguinea*	Asteraceae	Root	Pellati *et al.*, 2005
904.	*Echinacea simulata*	Asteraceae	Root	Pellati *et al.*, 2005
905.	*Echinacea tennesseensis*	Asteraceae	Root	Pellati *et al.*, 2005
906.	*Echium amoenum*	Boraginaceae		Ranjbar *et al.*, 2006
907.	*Echium rubrum*	Boraginaceeae	Root	Niciforovic *et al.*, 2010

Contd...

Table 3.1–*Contd...*

Sl.No.	Plant Name	Family	Parts Used	References
908.	*Echium vulgare*	Boraginaceae	Root	Niciforovic *et al.*, 2010
909.	*Eclipta prostrata* (=*E.alba*)	Asteraceae	Aerial parts	Sabu and Kuttan, 2003; Cai *et al.*, 2004; Sharma *et al.*, 2007; Veeru *et al.*, 2009
910.	*Ekebergia senegalensis*	Meliaceae	Leaf	Sofidya *et al.*, 2006
911.	*Elaegnus angustifolia*	Elaegnaceae	Flowers, branch	Bucur *et al.*, 2008
912.	*Elais guineensis*	Arecaceae	Fruit	Balasundram *et al.*, 2005; Neo *et al.*, 2010
913.	*Elettaria cardamomum*	Zingiberaceae	Fruits	Badei *et al.*, 2002; Hinneburg *et al.*, 2005; Singh *et al.*, 2007; Omankutty, 2007; Mariutti, *et al.*, 2008; Khalaf *et al.*, 2008; Souri *et al.*, 2008; Bhatti *et al.*, 2010
914.	*Eleusine coracana*	Poaceae	Fruit	Bajpai *et al.*, 2005; Rao and Murali-krishna, 2006; Sreeramulu *et al.*, 2009
915.	*Eleutherococcus gracilistylus*	Araliaceae	Root	Wojcikowski *et al.*, 2007
916.	*Elsholtzia splendens*	Lamiaceae		Song *et al.*, 2010
917.	*Elymus repens*	Poaceae	Rhizome	Wojcikowski *et al.*, 2007; Petrova *et al.*, 2009
918.	*Embelia ribes*	Myrsinaceae		Joshi *et al.*, 2007
919.	*Emilia sonchifolia*	Asteraceae	Leaves	Shylesh and Padikkala, 1999; Patnibul *et al.*, 2008
920.	*Eminium spiculatum*	Araceae	Leaves	Janakat and Al-Thnaibat, 2008
921.	*Empetrum nigrum*	Empetraceae	Fruit	Kahkonen *et al.*, 1999.
922.	*Enicosanthum acuminata*	Annonaceae	Stem, bark, leaves	Puvanendran *et al.*, 2011
923.	*Enicostemma axillare*	Gentianaceae	Leaves	Vasu *et al.*, 2005; Deore *et al.*, 2008; Jaishree *et al.*, 2008, Jaishree and Badami, 2009, 2010
924.	*Enicostemma littorale*	Gentianaceae	Leaves	Maroo *et al.*, 2003; Vasu *et al.*, 2005; Vihas *et al.*, 2005
925.	*Entada africana*	Mimosaceae	Leaf	Cook *et al.*, 1998; Tibri *et al.*, 2010
926.	*Entada pursaetha*	Mimosaceae	Seed	Pakutharivu and Suriyavadhana, 2010
927.	*Epilobium angustifolium*	Onagraceae	Herb, seeds	Hiermann *et al.*, 1991; Kahkonen *et al.*, 1999; Shikov *et al.*, 2006; Stajner *et al.*, 2007; Borchardt *et al.*, 2008; Hevesi Toth *et al.*, 2009a,b
928.	*Epilobium hirsutum*	Onagraceae	Herb	Wojdylo *et al.*, 2007

Contd...

Table 3.1–*Contd...*

Sl.No.	Plant Name	Family	Parts Used	References
929.	*Epilobium montanum*	Onagraceae	Herb	Hevesi Toth *et al.*, 2009a,b
930.	*Epilobium parviflorum*	Onagraceae	Herb	Balazs *et al.*, 2003; Arredondo *et al.*, 2004; Steenkamp *et al.*, 2006; Hevesi Toth *et al.*, 2009
931.	*Epilobium roseum*	Onagraceae		Hevesi Toth *et al.*, 2009a,b
932.	*Epilobium tetragonum*	Onagraceae		Hevesi Toth *et al.*, 2009a,b
933.	*Epimedium brevicornum*	Berberidaceae	Aerial parts	Cai *et al.*, 2004
934.	*Epipremnum pinnatum*	Araceae	Aerial parts	Linnert *et al.*, 2010
935.	*Erechtites valerianaefolia*	Asteraceae		Rosas-Romero and Saavedra, 2005
936.	*Eremostachys adenantha*	Lamiaceae	Aerial parts	Firuzi *et al.*, 2010
937.	*Eremostachys laciniata*	Lamiaceae		Erdemoglu *et al.*, 2006
938.	*Erica arborea*	Ericaceae	Leaves, Flowers	Ay *et al.*, 2007; Nazemiyeh *et al.*, 2008
939.	*Erigeron acris*	Asteraceae		Nalewajo-Sielwoniuk *et al.*, 2008
940.	*Erigeron annuus*	Asteraceae		Jeong *et al.*, 2011
941.	*Erigeron breviscapus*	Asteraceae		Liu *et al.*, 2005
942.	*Eriobotrya deflexa*	Rosaceae		Hou *et al.*, 2003
943.	*Eriobotrya japonica*	Rosaceae	Seeds	Hamada *et al.*, 2004; Yokota *et al.*, 2006; Song *et al.*, 2010
944.	*Erodium cicutarium*	Geraniaceae	Herb	Sroka *et al.*, 1994; Fecka *et al.*, 1997
945.	*Erodium oxyrrhynchym* subsp. *oxyrhynchum*	Geraniaceae		Souri *et al.*, 2004
946.	*Eruca sativa*	Brassicaceae		Ozturk *et al.*, 2002; Kim *et al.*, 2004; Alam *et al.*, 2007
947.	*Erucaria microcarpa*	Brassicaceae		Hashim, 2007
948.	*Erycibe obtusifolia*	Convolvulaceae		Hsu *et al.*, 1999.
949.	*Eryngium campestre*	Apiaceae	Herb	Kupeli *et al.*, 2006
950.	*Eryngium cucasicum*	Apiaceae		Ebrahimzadeh *et al.*, 2009
951.	*Eryngium foetidum*	Apiaceae		Chanwitheesuk *et al.*, 2005; Wong *et al.*, 2006
952.	*Erythrina abyssinica*	Fabaceae	Stem bark	Habtemariam and Dagne, 2010
953.	*Erythrina crista*	Fabaceae		Maisuthisakul *et al.*, 2008
954.	*Erythrina latissima*	Fabaceae	Stem wood	Chacha *et al.*, 2005
955.	*Erythrina mildbraedii*	Fabaceae		Najmen *et al.*, 2003
956.	*Erythrina sigmoidea*	Fabaceae		Njamen *et al.*, 2004
957.	*Erythrina stricta*	Fabaceae	Leaves	Ashok Kumar *et al.*, 2008
958.	*Erythrina variegata*	Fabaceae		Gan *et al.*, 2010

Contd...

Table 3.1–*Contd...*

Sl.No.	Plant Name	Family	Parts Used	References
959.	*Etlingera elatior*	Zingiberaceae	Leaves	Mohamed *et al.*, 2005; Chan *et al.*, 2009; Andarwulan *et al.*, 2010
960.	*Eucalyptus camaldulensis*	Myrtaceae	Leaves/ oils	El-Ghorab *et al.*, 2003; Souri *et al.*, 2004; El-Ghorab *et al.*, 2009; Barra *et al.*, 2010
961.	*Eucalyptus globulus*	Myrtaceae	Bark/oil, leaves	Dessi *et al.*, 2001; Sacchetti *et al.*, 2005 ; Bajpai *et al.*, 2005; Nazquez *et al.*, 2008; Amakura *et al.*, 2009
962.	*Eucalyptus grandis*	Myrtaceae	Wood	Kulkarni *et al.*, 2007
963.	*Eucalyptus robusta*	Myrtaceae	Fruit	Fu *et al.*, 2010
964.	*Eucalyptus rostrata*	Myrtaceae		Okamura *et al.*, 1993; Moussa *et al.*, 2011
965.	*Eucalyptus tereticornis*	Myrtaceae	Essentil oil	Singh *et al.*, 2009
966.	*Eucommia ulmoides*	Eucommiaceae	Bark	Hsieh and Yen, 2000; Ohmori *et al.*, 2005; Luo *et al.*, 2009; Xu *et al.*, 2010
967.	*Eugenia aggregata*	Myrtaceae	Fruits	Reynertson *et al.*, 2008
968.	*Eugenia brasiliensis*	Myrtaceae	Fruits	Reynertson *et al.*, 2008
969.	*Eugenia caryophyllata*	Myrtaceae	Buds	Kim *et al.*, 1997; Glelcin *et al.*, 2004; Jayathilakan *et al.*, 2007; Khalaf *et al.*, 2008; Wei and Shibomoto, 2010; Wei *et al.*, 2010; Viuda-Martos *et al.*, 2010; Politeo *et al.*, 2010
970.	*Eugenia jambolana*	Myrtaceae	Seed	Ravi *et al.*, 2004
971.	*Eugenia luschnathiana*	Myrtaceae	Fruits	Reynertson *et al.*, 2008
972.	*Eugenia polyantha*	Myrtaceae		Wong *et al.*, 2006; Leleno *et al.*, 2009
973.	*Eugenia reinwardtiana*	Myrtaceae	Fruits	Reynertson *et al.*, 2008
974.	*Eugenia siamensis*	Myrtaceae		Maisuthisakul *et al.*, 2008
975.	*Eugenia uniflora*	Myrtaceae		Velazquez *et al.*, 2003
976.	*Euonymus alatus*	Celastraceae		Kim *et al.*, 2008; Chon *et al.*, 2009
977.	*Euonymus europaeus*	Celastraceae	Seeds	Kumaraswamy *et al.*, 2003
978.	*Euonymus alatus*	Celastraceae		Kim *et al.*, 2008; Chon *et al.*, 2009
979.	*Euonymus latifolius*	Celastraceae	Fruit, leaves	Serteser *et al.*, 2009
980.	*Eupatorium articulatum*	Asteraceae		De las Heras *et al.*, 1998
981.	*Eupatorium bunifolium*	Asteraceaae		Borneo *et al.*, 2008
982.	*Eupatorium fortunei*	Asteraceae		Gan *et al.*, 2010
983.	*Eupatorium odoratum*	Asteraceae		Chomnawang *et al.*, 2007
984.	*Euphorbia acanthothamnos*	Euphorbiaceae		Barla *et al.*, 2007

Contd...

Table 3.1–*Contd...*

Sl.No.	*Plant Name*	*Family*	*Parts Used*	*References*
985.	*Euphorbia antiquorum*	Euphorbiaceae		Veeresh *et al.*, 2009
986.	*Euphorbia formosana*	Euphorbiaceae	Aerial	Chen, C-H. *et al.*, 2009
987.	*Euphorbia geniculata*	Euphorbiaceae	Latex	Chougale *et al.*, 2011
988.	*Euphorbia hirta*	Euphorbiaceae	Leaves, root, stem, flower	Sharma *et al.*, 2007, 2008b; Kumar *et al.*, 2010; Basma *et al.*, 2011; Habila *et al.*, 2011
989.	*Euphorbia macroclada*	Euphorbiaceae		Barla *et al.*, 2007
990.	*Euphorbia rigida*	Euphorbiaceae		Barla *et al.*, 2007
991.	*Euphorbia royleana*	Euphorbiaceae		Shahwar *et al.*, 2010
992.	*Euphorbia thymifolia*	Euphorbiaceae		Lin *et al.*, 2002; Thuong *et al.*, 2006; Habila *et al.*, 2011
993.	*Euphrasia officinalis*	Orobanchaceae	Herb	Katalinic *et al.*, 2006
994.	*Eurya nitida*	Theaceae	Fruits	Fu *et al.*, 2010
995.	*Euryale ferox*	Nymphaeaceae	Seed	Lee *et al.*, 2002
996.	*Evodia daniellii*	Rutaceae		Thuong *et al.*, 2010
997.	*Evodia officinalis*	Rutaceae		Kim *et al.*, 1997;
998.	*Evolvulus alsinoides*	Convolvulaceae		Auddy *et al.*, 2003; Achliya *et al.*, 2003; Jain and Agrawal, 2009
999.	*Excoeria agallocha*	Euphorbiaceae	Leaves, flowers	Subhan *et al.*, 2008; Vadlapudi and Naidu, 2009; Patra *et al.*, 2009
1000.	*Excoecaria kawakamii*	Euphorbiaceae	Leaves	Lee *et al.*, 2006
1001.	*Eysenhardtia platycarpa*	Fabaceae		Narvaez-Mastache *et al.*, 2007
1002.	*Eysenhardtia punctata*	Fabaceae		Narvaez-Mastache *et al.*, 2007
1003.	*Eysenhardtia subcoriacea*	Fabaceae		Narvaez-Mastache *et al.*, 2007
1004.	*Fagaropsis angolensis*	Rutaceae		Kuglerova *et al.*, 2008
1005.	*Fagonia cretica*	Zygophyllaceae		Rawal *et al.*, 2004
1006.	*Fagopyrum cymosum*	Polygonaceae	Rhizome	Cai *et al.*, 2004
1007.	*Fagopyrum esculentum*	Polygonaceae	Seed	Bajpai *et al.*, 2005; Wojcicki *et al.*, 2006, 2007; Liu *et al.*, 2007, 2008; Alvarez-Jubete *et al.*, 2010; Metzger *et al.*, 2010
1008.	*Fagopyrum tataricum*	Polygonaceae	Sprouts	Liu *et al.*, 2008
1009.	*Fagraea blumei*	Potaliaceae	Stem bark	Cuendet *et al.*, 1997
1010.	*Falcaria vulgaris*	Apiaceae	Essential oil	Shafaghat, 2010
1011.	*Fallopia convolvulus*	Polygonaceae	Herb	Kahkonen *et al.*, 1999;
1012.	*Feronia elephantum*	Rutaceae	Fruit	Nithya and Saraswathi, 2010
1013.	*Feronia limonia*	Rutaceae	Leaves	Hasan *et al.*, 2009
1014.	*Ferula hermonis*	Apiaceae		Subon *et al.*, 2004

Contd...

Table 3.1–*Contd...*

Sl.No.	Plant Name	Family	Parts Used	References
1015.	*Ferula orientalis*	Apiaceae		Kartal *et al.*, 2005
1016.	*Ferulago campestris*	Apiaceae		Basile *et al.*, 2009
1017.	*Fibraurea tinctoria*	Menispermaceae		Keawpradub *et al.*, 2005
1018.	*Ficus asperifolia*	Moraceae		Annan and Houghton, 2008
1019.	*Ficus benghalensis*	Moraceae	Bark, aerial root, leaf	Daniel *et al.*, 1998; D'Mello *et al.*, 2000; Shukla *et al.*, 2004; Manian *et al.*, 2008; Thakare *et al.*, 2010; Patel *et al.*, 2010; Nidhiya *et al.*, 2010
1020.	*Ficus capensis*	Moraceae		Muanda *et al.*, 2011
1021.	*Ficus carica*	Moraceae	Fruit, leaves	Souri *et al.*, 2004; Serteser *et al.*, 2009
1022.	*Ficus cotinifolia*	Moraceae		Ruiz-Terán *et al.*, 2008
1023.	*Ficus deltoidea*	Moraceae	Leaf	Hakiman and Maziah, 2009
1024.	*Ficus involuta*	Moraceae		Ruiz-Terán *et al.*, 2008
1025.	*Ficus lutea*	Moraceae	Whole plant	Marwah *et al.*, 2007
1026.	*Ficus microcarpa*	Moraceae		Ao *et al.*, 2008
1027.	*Ficus racemosa*	Moraceae	Stem bark, leaf	Veerapur *et al.*, 2007; Channabasavaraj *et al.*, 2008; Manian *et al.*, 2008; Ahmed and Urooj, 2010; Ara Jahan *et al.*, 2009; Patel *et al.*, 2010
1028.	*Ficus religiosa*	Moraceae	Fruit	Rathee *et al.*, 2010
1029.	*Ficus sycomorus*	Moraceae	Fruit peel	Lamien-Meda *et al.*, 2008
1030.	*Filipendula hexapetala*	Rosaceae	Flowers	Maksimovi *et al.*, 2007
1031.	*Filipendula ulmaria*	Rosaceae	Fruits	Kahkonen *et al.*, 1999; Trouillas *et al.*, 2003; Nikolova and Dzhurmanski, 2009
1032.	*Filipendula vulgaris*	Rosaceae	Root	Oszmianski *et al.*, 2007;
1033.	*Flemingia strobilifera*	Fabaceae	Root, leaf	Madan *et al.*, 2010
1034.	*Flueggea leucopyrus*	Euphorbiaceae	Fruit	Karuppusamy *et al.*, 2011
1035.	*Foeniculum vulgare*	Apiaceae	Seeds	Budinevi *et al.*, 1995; Ruberto *et al.*, 2000; Oktay *et al.*, 2003; Cai *et al.*, 2004; Choi and Hwang, 2004; Hinneburg *et al.*, 2005; Maeda *et al.*, 2006; Khalil *et al.*, 2007; De Marino *et al.*, 2007; Souri *et al.*, 2008; Nickavar and Abolhasani, 2009; Miguel *et al.*, 2010
1036.	*Forsythia suspensa*	Oleaceae	Fruit	Schinella *et al.*, 2002; Cai *et al.*, 2004; Li *et al.*, 2008

Contd...

Table 3.1–*Contd...*

Sl.No.	Plant Name	Family	Parts Used	References
1037.	*Fragaria ananassa*	Rosaceae	Fruit flesh	Wang *et al.*, 1996; Kahkonen *et al.*, 1999; Wang and Lin, 2000, 2003; Meyers *et al.*, 2003; Abay *et al.*, 2005
1038.	*Fragaria vesca*	Rosaceae	Fruit	Joseph *et al.*, 1999; Agostini *et al.*, 2004; Katalinic *et al.*, 2006; Kiselova *et al.*, 2006
1039.	*Fraxinus excelsior*	Oleaceae	Flowers	Myer *et al.*, 1995; Middleton *et al.*, 2005
1040.	*Fraxinus formosana*	Oleaceae	Leaves	Lee *et al.*, 2006
1041.	*Fraxinus ornus*	Oleaceae	Cortex	Marinova *et al.*, 1994; Kostova *et al.*, 2001
1042.	*Fraxinus rhynchophylla*	Oleaceae		Cai *et al.*, 2006; Li *et al.*, 2008; Thuong *et al.*, 2009; 2010
1043.	*Fritillaria cirrhosa*	Liliaceae		Song *et al.*, 2010
1044.	*Fritillaria verticillata*	Liliaceae		Song *et al.*, 2010
1045.	*Fumaria officinalis*	Fumariaceae	Aerial parts	Sengul *et al.*, 2009
1046.	*Funturnea elastica*	Apocynaceae	Leaf	Oke and Hamburger, 2002
1047.	*Gaillardia megapotamica*	Asteraceae		Borneo *et al.*, 2008
1048.	*Galanthus reginae-olgae* subsp. *vernalis*	Amaryllidaceae	Aerial parts, bulbs	Conforti *et al.*, 2010
1049.	*Galeopsis speciosa*	Lamiaceae	Plant	Matkowski and Piotrowska, 2006; Matkovski *et al.*, 2008
1050.	*Galeopsis tetrahit*	Lamiaceae	Herb	Matkovski *et al.*, 2008
1051.	*Galium aparine*	Rubiaceae	Herb	Lindsey *et al.*, 2002
1052.	*Galium verum*	Rubiaceae	Herb	Mavi *et al.*, 2004; Katalinic *et al.*, 2006
1053.	*Garcinia afzelii*	Clusiaceae	Stem bark	Kamden Waffo *et al.*, 2006
1054.	*Garcinia atroviridis*	Clusiaceae	Fruits	Kruawan and Kangsadalampai, 2006; Povichit *et al.*, 2010
1055.	*Garcinia hombroniana*	Clusiaceae	Bark	Rema *et al.*, 2007
1056.	*Garcinia indica*	Clusiaceae	Leaves	Yamaguchi *et al.*, 2000a, b; Mishra *et al.*, 2006
1057.	*Garcinia kola*	Clusiaceae	Seed	Eddy *et al.*, 1997; Nwankwo *et al.*, 2000; Terahima *et al.*, 2002; Farombi and Nwaokeafor, 2005; Farombi, 2009; Farombi and Owoeye, 2011
1058.	*Garcinia mangostana*	Clusiaceae	Fruit	Yamaguchi *et al.*, 2000a, b; Chomnawang *et al.*, 2007; Isabelle *et al.*, 2010; Nainwal *et al.*, 2010; Povichit *et al.*, 2010

Contd...

Table 3.1–*Contd...*

Sl.No.	*Plant Name*	*Family*	*Parts Used*	*References*
1059.	*Garcinia multiflora*	Clusiaceae	Stems	Chiang *et al.*, 2003
1060.	*Garcinia subelliptica*	Clusiaceae	Fruit	Weng *et al.*, 2004; Chen, C-H. *et al.*, 2009
1061.	*Garcinia vieillardii*	Clusiaceae	Stem bark	Hay *et al.*, 2004
1062.	*Garcinia xanthochymus*	Clusiaceae	Fruits	Baggett *et al.*, 2005
1063.	*Gardenia angusta*	Rubiaceae		Leelapornpisid *et al.*, 2007
1064.	*Gardenia erubescens*	Rubiaceae	Fruit	Lamien-Meda *et al.*, 2008
1065.	*Gardenia gummifera*	Rubiaceae		Kalim *et al.*, 2010
1066.	*Gardenia jasminoides*	Rubiaceae	Fruit	Cai *et al.*, 2004; Kim *et al.*, 2006; Li *et al.*, 2008
1067.	*Gastrodia elata*	Orchidaceae		Hsieh *et al.*, 2001; Liu and Mori, 1992
1068.	*Gaultheria fragrantissima*	Ericaceae	Fruit	Karuppusamy *et al.*, 2011
1069.	*Gaultheria shallon*	Ericaceae	Fruit	Acuna *et al.*, 2002
1070.	*Gaura biennis*	Onagraceae	Seeds	Borchardt *et al.*, 2008
1071.	*Gazania splendens*	Asteraceae		Moussa *et al.*, 2011
1072.	*Genista cadasonensis*	Fabaceae		Serrelli *et al.*, 2010
1073.	*Genista sandrasica*	Fabaceae		Orhan *et al.*, 2011
1074.	*Genista tenera*	Fabaceae		Rauter *et al.*, 2009
1075.	*Genista tinctoria*	Fabaceae	Herb	Koelzer *et al.*, 2009
1076.	*Genista vuralii*	Fabaceae		Orhan *et al.*, 2011
1077.	*Gentiana decumbens*	Gentianaceae	Herb	Myagmar and Aniya, 2000
1078.	*Gentiana lutea*	Gentianaceae	Leaves, root	Kusar *et al.*, 2006
1079.	*Gentiana macrophylla*	Gentianaceae		Gan *et al.*, 2010
1080.	*Gentiana piasezkii*	Gentianaceae		Wu *et al.*, 2006
1081.	*Gentiana scabra*	Gentianaceae		Li *et al.*, 2008
1082.	*Gentianella alborosea*	Gentianaceae	Plant	Acero *et al.*, 2006
1083.	*Gentianella austriaca*	Gentianaceae	Aerial parts	Leskovac *et al.*, 2007
1084.	*Gentianella nitida*	Gentianaceae		Lock *et al.*, 2005
1085.	*Geranium macrorrhizum*	Geraniaceae	Aerial parts	Miliauskas *et al.*, 2004; Venskutonis *et al.*, 2004
1086.	*Geranium niveum*	Geraniaceae	Roots	Maldonado *et al.*, 2005
1087.	*Geranium pratense*	Geraniaceae	Herb	Myagmar and Aniya, 2000;
1088.	*Geranium purpurem*	Geraniaceae	Leaves	Sabuncuoglu *et al.*, 2008
1089.	*Geranium robertianum*	Geraniaceae	Herb	Katalinic *et al.*, 2006
1090.	*Geranium sanguineum*	Geraniaceae		Murzakhmetova *et al.*, 2008
1091.	*Geranium sibiricum*	Geraniaceae	-	Shim and Lim, 2009; Wu *et al.*, 2010

Contd...

Table 3.1–*Contd...*

Sl.No.	*Plant Name*	*Family*	*Parts Used*	*References*
1092.	*Geranium wilfordii*	Geraniaceae		Gan *et al.*, 2010
1093.	*Geum quellyon*	Rosaceae		Russo *et al.*, 2005
1094.	*Geum rivale*	Rosaceae	Root	Oszmianski *et al.*, 2007
1095.	*Geum triflorum*	Rosaceae	Seeds	Borchardt *et al.*, 2008
1096.	*Geum urbanum*	Rosaceae	Rhizome	Mantle *et al.*, 2000
1097.	*Gevuina avellana*	Proteaceae	Hulls	Moure *et al.*, 2000, 2001
1098.	*Gisekia pharnaceoides*	Aizoaceae	Herb	Singh *et al.*, 2009
1099.	*Glaucium contortup-licatum*	Papaveraceae		Souri *et al.*, 2004
1100.	*Glaucium elegans*	Papaveraceae		Souri *et al.*, 2004
1101.	*Glaucium fimbrilligerum*	Papaveraceae		Souri *et al.*, 2004
1102.	*Glechoma hederacea*	Lamiaceae	Herb	Kumarasamy *et al.*, 2002
1103.	*Gleditsia japonica*	Caesalpiniaceae		Kim *et al.*, 1997; Nam and Kang, 2004
1104.	*Gleditsia sinensis*	Caesalpiniaceae	Thorns	Ha *et al.*, 2008; Song *et al.*, 2010
1105.	*Glehnia littoralis*	Apiaceae		Ng *et al.*, 2004
1106.	*Glinus oppositifolius*	Molluginaceae	Leaves	Diallo *et al.*, 2001; Asok Kumar *et al.*, 2009; Behera *et al.*, 2010
1107.	*Gliricidia sepium*	Fabaceae		Ruiz-Terán *et al.*, 2008
1108.	*Globimetula cupulatum*	Loranthaceae		Akinmoladun *et al.*, 2010
1109.	*Globularia alypum*	Globulariaceae	Aerial parts	Khlifi *et al.*, 2005; El-Safi *et al.*, 2005, 2007; Harzallah *et al.*, 2010
1110.	*Glochidion wallichianum*	Euphorbiaceae		Panpipat *et al.*, 2010; Wu *et al.*, 2010
1111.	*Glossocardia bosvallea*	Asteraceae		Prasanna Anjaneya Reddy *et al.*, 2011
1112.	*Glyceria grandis*	Poaceae	Seeds	Borchardt *et al.*, 2008
1113.	*Glyceria striata*	Poaceae	Seeds	Borchardt *et al.*, 2008
1114.	*Glycine max*	Fabaceae	Beans	Lin *et al.*, 2001; Lee *et al.*, 2004; Bajpai *et al.*, 2005; Maleneic *et al.*, 2007, 2008; Han *et al.*, 2008: Malencic *et al.*, 2008; Cho *et al.*, 2008; Sreeramulu *et al.*, 2009
1115.	*Glycyrrhiza glabra*	Fabaceae	Root, leaves	Kittagawa *et al.*, 1994; Vaya *et al.*, 1997; Belinky *et al.*, 1998a,b; Kishore *et al.*, 2001; Naik, *et al.*, 2003, 2005; Guo *et al.*, 2007; Fukai *et al.*, 2003; Naik *et al.*, 2003; Souri *et al.*, 2004; Biondi *et al.*, 2004; Gajera *et al.*, 2005; Ashawat and Saraf, 2007; Chin *et al.*, 2007; Wojcikowski *et al.*, 2007; Jain and Agarwal, 2009

Contd...

Table 3.1–*Contd...*

Sl.No.	*Plant Name*	*Family*	*Parts Used*	*References*
1116.	*Glycyrrhiza inflata*	Fabaceae		Haraguchi *et al.*, 1998; Fukai *et al.*, 2003b
1117.	*Glycyrrhiza lepidota*	Fabaceae	Root	Amarowicz *et al.*, 2004
1118.	*Glycyrriza raddi*	Fabaceae	Leaves	Guo *et al.*, 2007
1119.	*Glycyrrhiza uralensis*	Fabaceae	Root	Cai *et al.*, 2004; Fukai *et al.*, 2003b; Lee *et al.*, 2003
1120.	*Glyphaea brevis*	Tiliaceae	Roots, leaves	Mamyrbekova-Beka *et al.*, 2008
1121.	*Gmelina arborea*	Verbenaceae		Pandey *et al.*, 2010
1122.	*Gmelina hystrix*	Verbenaceae	Leaves	Adib *et al.*, 2010
1123.	*Gnaphalium obtusifolium*	Asteraceae	Seeds	Borchardt *et al.*, 2008
1124.	*Gongronema latifolium*	Asclepiadaceae	Leaf	Ugochukwu and Babady, 2002; Odukoya *et al.*, 2007
1125.	*Goniothalamus hookeri*	Anonaceae	Leaves, stem bark	Puvanendran *et al.*, 2011
1126.	*Gontscharovia popovi*	Lamiaceae	Herb	Firuzi *et al.*, 2010
1127.	*Gordonia axillaris*	Theaceae	Fruit	Fu *et al.*, 2010
1128.	*Gossypium arboreum*	Malvaceae		Annan and Houghton, 2008
1129.	*Graptopetalum paraguayense*	Crassulaceae	Stem	Chung *et al.*, 2005; Chen *et al.*, 2008
1130.	*Grewia avellana*	Tiliaceae		Moure *et al.*, 2000
1131.	*Grewia nervosa*	Tiliaceae	Twig	Kshirsagar and Upadhyay, 2009
1132.	*Grewia sapida*	Tiliaceae	Twig, stem	Kshirsagar and Upadhyay, 2009
1133.	*Grewia tiliaefolia*	Tiliaceae	Fruit	Khadeer Ahmed *et al.*, 2010; Karuppusamy *et al.*, 2011
1134.	*Grindelia robusta*	Asteraceae	Aerial parts	Fraternale *et al.*, 2007
1135.	*Guiera senegalensis*	Combretaceae	Different parts	Bouchet *et al.*, 1998
1136.	*Guizotia abyssinica*	Asteraceae	Seed	Ramadan *et al.*, 2003; Shahidi *et al.*, 2003
1137.	*Gundelia tournefortii*	Asteraceae	Aerial parts	Coruh *et al.*, 2007
1138.	*Gymnema inodorum*	Asclepiadaceae		Chanwitheesuk *et al.*, 2005
1139.	*Gymnema montanum*	Asclepiadaceae	Leaves	Anantha *et al.*, 2004
1140.	*Gymnema sylvestre*	Asclepiadaceae	Leaves	Ohmori *et al.*, 2005; Bajpai *et al.*, 2005; Gajera *et al.*, 2005; Biswas *et al.*, 2010
1141.	*Gynostemma pentaphyllum*	Cucurbitaceae	Whole plant	Cai *et al.*, 2004; Wang and Luo, 2007; Yang *et al.*, 2008; Wang *et al.*, 2010

Contd...

Table 3.1–*Contd...*

Sl.No.	*Plant Name*	*Family*	*Parts Used*	*References*
1142.	*Gynoxys psilophylla*	Asteraceae		Rosas-Romero and Saavedra, 2005
1143.	*Gynura bicolor*	Asteraceae	Leaf	Maeda *et al.*, 2006; Chen, C-H. *et al.*, 2009
1144.	*Gynura formosana*	Asteraceae	Herb	Hou *et al.*, 2005
1145.	*Gynura procumbens*	Asteraceae	Leaf	Akowuah *et al.*, 2009
1146.	*Gypsophila eriocalyx*	Caryophyllaceae	Leaves	Serteser *et al.*, 2009
1147.	*Gypsophila parva*	Caryophyllaceae	Leaves	Serteser *et al.*, 2009
1148.	*Gypsophila perfoliata*	Caryophyllaceae	Leaves	Serteser *et al.*, 2009
1149.	*Gypsophila pilosa*	Caryophyllaceae	Leaves	Serteser *et al.*, 2009
1150.	*Gypsophila tubulosa*	Caryophyllaceae	Leaves	Serteser *et al.*, 2009
1151.	*Gyrocarpus jathrophifolius*	Hernandiaceae		Ruiz-Terán *et al.*, 2008
1152.	*Habenaria repens*	Orchidaceae		Johnson *et al.*, 1999
1153.	*Halacsya sendtneri*	Boraginaceae		Niciforovic *et al.*, 2010
1154.	*Halenia elliptica*	Gentianaceae		Huang *et al.*, 2010
1155.	*Hamamelis virginiana*	Hamamelidaceae		Da Silva *et al.*, 2000; Tourino *et al.*, 2008
1156.	*Hamelia patens*	Rubiaceae		Ruiz-Terán *et al.*, 2008
1157.	*Haplopappus multifolius*	Asteraceae		Torres *et al.*, 2006
1158.	*Harpagophytum procumbens*	Pedaliaceae	Callus	Georgiev *et al.*, 2010
1159.	*Harpephyllum caffrum*	Anacardiaceae		Moyo *et al.*, 2010
1160.	*Harungana madagascariensis*	Clusiaceae		Biapa *et al.*, 2007
1161.	*Hedeoma drummondii*	Lamiaceae	Aerial parts	Viveros-Valdez *et al.*, 2008
1162.	*Hedera colchica*	Araliaceae		Gulcin *et al.*, 2004
1163.	*Hedera helix*	Araliaceae		Gulcin *et al.*, 2004
1164.	*Hedranthera batteri*	Apocynaceae	Leaf	Oke and Hamburger, 2002
1165.	*Hedychium spicatum*	Zingiberaceae		Rawat *et al.*, 2011
1166.	*Hedychium thyrsiforme*	Zingiberaceae	Rhizome	Mooi *et al.*, 2003
1167.	*Hedyosmum angustifolium*	Chloranthaceae		Rosas-Romero and Saavedra, 2005
1168.	*Hedyotis capitellata*	Rubiaceae		Ahmad *et al.*, 2005
1169.	*Hedyotis corymbosa*	Rubiaceae	Aerial parts	Sasikumar *et al.*, 2010
1170.	*Hedyotis dichotoma*	Rubiaceae		Ahmad *et al.*, 2005
1171.	*Hedyotis diffusa*	Rubiaceae	Whole plant	Cai *et al.*, 2004
1172.	*Hedyotis havilandi*	Rubiaceae		Ahmad *et al.*, 2005

Contd...

Table 3.1–*Contd...*

Sl.No.	*Plant Name*	*Family*	*Parts Used*	*References*
1173.	*Hedyotis herbacea*	Rubiaceae		Ahmad *et al.*, 2005
1174.	*Hedyotis philippensis*	Rubiaceae		Ahmad *et al.*, 2005
1175.	*Hedyotis verticillata*	Rubiaceae		Saha *et al.*, 2004
1176.	*Heinsia crinita*	Rubiaceae	Leaf	Odukoya *et al.*, 2007;
1177.	*Helianthus tuberosus*	Asteraceae	Root	Souri *et al.*, 2004; Aslan *et al.*, 2010
1178.	*Helichrysum arenarium*	Asteraceae		Tepe *et al.*, 2004; Özgen *et al.*, 2004; Albayrak *et al.*, 2010
1179.	*Helichrysum artvinense*	Asteraceae		Albayrak *et al.*, 2010
1180.	*Helichrysum chasmolycium*	Asteraceae		Albayrak *et al.*, 2010
1181.	*Helichrysum chionophilum*	Asteraceae		Tepe *et al.*, 2004; Albayrak *et al.*, 2010
1182.	*Helichrysum compactum*	Asteraceae	Capitula	Suzec *et al.*, 2005; Albayrak *et al.*, 2010
1183.	*Helichrysum dasyanthum*	Asteraceae		Lourens *et al.*, 2004
1184.	*Helichrysum excisum*	Asteraceae		Lourens *et al.*, 2004; Albayrak *et al.*, 2010
1185.	*Helichrysum felinum*	Asteraceae		Lourens *et al.*, 2004; Albayrak *et al.*, 2010
1186.	*Helichrysum goulandriorum*	Asteraceae		Albayrak *et al.*, 2010
1187.	*Helichrysum graveolens*	Asteraceae		Lourens *et al.*, 2004; Aslan *et al.*, 2007; Albayrak *et al.*, 2010
1188.	*Helichrysum heywoodianum*	Asteraceae		Albayrak *et al.*, 2010
1189.	*Helichrysum italicum*	Asteraceae		Sala *et al.*, 2002; Schinella *et al.*, 2002; Poli *et al.*, 2003; Albayrak *et al.*, 2010
1190.	*Helichrysum kitianum*	Asteraceae		Albayrak *et al.*, 2010
1191.	*Helichrysum longifolium*	Asteraceae		Aiyegoro and Okoh, 2010
1192.	*Helichrysum noeanum*	Asteraceae		Tepe *et al.*, 2004; Albayrak *et al.*, 2010
1193.	*Helichrysum orientale*	Asteraceae		Albayrak *et al.*, 2010
1194.	*Helichrysum pallasii*	Asteraceae		Albayrak *et al.*, 2010
1195.	*Helichrysum pamphylicum*	Asteraceae		Albayrak *et al.*, 2010
1196.	*Helichrysum peshmenianum*	Asteraceae		Albayrak *et al.*, 2010
1197.	*Helichrysum petiolare*	Asteraceae		Lourens *et al.*, 2004; Albayrak *et al.*, 2010
1198.	*Helichrysum plicatum*	Asteraceae		Tepe *et al.*, 2004; Aslan *et al.*, 2007; Albayrak *et al.*, 2010

Contd...

Table 3.1–*Contd...*

Sl.No.	*Plant Name*	*Family*	*Parts Used*	*References*
1199.	*Helichrysum sanguineum*	Asteraceae		Albayrak *et al.*, 2010
1200.	*Helichrysum stoechas* subsp. *Barellieri*	Asteraceae		Albayrak *et al.*, 2010
1201.	*Helicia formosana*	Proteaceae	Seed	Chen, C-H. *et al.*, 2009
1202.	*Helicteres isora*	Sterculiaceae	Fruits	Kumar *et al.*, 2007, 2008; Basniwal *et al.*, 2009; Suthar *et al.*, 2009
1203.	*Heliotropium glutinosum*	Boraginaceae	Resin	Modak *et al.*, 2007
1204.	*Heliotropium indicum*	Boraginaceae	Root	Souza *et al.*, 2005
1205.	*Heliotropium marifolium*	Boraginaceae	Herb	Singh *et al.*, 2009
1206.	*Heliotropium sinuatum*	Boraginaceae		Modak *et al.*, 2003
1207.	*Heliotropium strigosum*	Boraginaceae	Herb	Hussain *et al.*, 2010
1208.	*Helleborus hercegovinus*	Ranunculaceae	Herb	Caker *et al.*, 2011
1209.	*Helleborus multifidus*	Ranunculaceae	Herb	Caker *et al.*, 2011
1210.	*Helleborus odorus*	Ranunculaceae	Herb	Caker *et al.*, 2011
1211.	*Helixanthera parasitica*	Loranthaceae	Plant	Lirdprapamongkol *et al.*, 2003
1212.	*Hemerocallis fulva*	Liliaceae	Leaf	Maeda *et al.*, 2006; Que *et al.*, 2007; Fu *et al.*, 2009; Liu *et al.*, 2010
1213.	*Hemidesmus indicus*	Asclepiadaceae	Roots	Alam *et al.*, 1998; Prabhakan *et al.*, 2000; Ravishakara *et al.*, 2002; Mary *et al.*, 2003; Shetty *et al.*, 2005; Mohana Rao *et al.*, 2005; Zahin *et al.*, 2009; Surveswaran *et al.*, 2010; Murali *et al.*, 2010
1214.	*Henophyton deserti*	Brassicaceae	Seeds, leaf	Derbel *et al.*, 2010
1215.	*Heracleum nepalense*	Apiaceae		Dash *et al.*, 2005
1216.	*Heracleum pastinacifolium*	Apiaceae		Firuzi *et al.*, 2010a
1217.	*Heracleum persicum*	Apiaceae	Fruit	Souri *et al.*, 2004; Coruh *et al.*, 2007; Nickavar and Abolhasani, 2009; Firuzi *et al.*, 2010a
1218.	*Heracleum rechingeri*	Apiaceae		Firuzi *et al.*, 2010a
1219.	*Heracleum transcaucasicum*	Apiaceae		Firuzi *et al.*, 2010a
1220.	*Heritiera fomes*	Sterculiaceae	Stem bark	Wangensteen *et al.*, 2009
1221.	*Hernandia nymphaefolia*	Hernandiaceae		Chen *et al.*, 2001
1222.	*Heterosmilax erythrantha*	Smilacaceae	Root	Thuong *et al.*, 2006
1223.	*Heterothalamus alienus*	Asteraceae		Borneo *et al.*, 2008
1224.	*Heterotheca inuloides*	Asteraceae		Haraguchi *et al.*, 1996, 1997; Kubo *et al.*, 1996
1225.	*Hibiscus aculeatus*	Malvaceae	Roots	Sunilson *et al.*, 2008

Contd...

Table 3.1–*Contd...*

Sl.No.	Plant Name	Family	Parts Used	References
1226.	*Hibiscus calyphyllus*	Malvaceae	Leaf	Odukoya *et al.*, 2007;
1227.	*Hibiscus cannabinus*	Malvaceae	Leaf, flowers	Agbor *et al.*, 2003; Mukherjee *et al.*, 2010; Mukherjee *et al.*, 2010
1228.	*Hibiscus rosa-sinensis*	Malvaceae	Flower	Neeli *et al.*, 2008; Liu *et al.*, 2008
1229.	*Hibiscus sabdariffa*	Malvaceae	Leaves, seeds, flowers, calyx	Osawa and Namiki, 1981; Tseng *et al.*, 1997; Suboh *et al.*, 2004; Farombi and Fakoya, 2005; Usoh *et al.*, 2005; Hirunpanich *et al.*, 2006; Wong *et al.*, 2006; Kruawan and Kangsadalampai, 2006; Aderogba *et al.*, 2007; Farombi and Ige, 2007; Ochani and D'Mello, 2009; Morrison and Twumasi, 2010; Mohd-Esa *et al.*, 2010
1230.	*Hibiscus tiliaceous*	Malvaceae	Leaves, flowers	Ajaikumar *et al.*, 2002; Vadlapudi and Naidu, 2009
1231.	*Hieracium cappadocicum*	Moraceae		Tepe *et al.*, 2006
1232.	*Hieracium pilosella*	Asteraceae	Herb	Trouillas *et al.*, 2003; Stanojevi *et al.*, 2008, 2009
1233.	*Himatanthus lancifolius*	Apocynaceae	Bark	Baggio *et al.*, 2005
1234.	*Hippomarathrum microcarpum*	Apiaceae	Essential oil	Ozer *et al.*, 2007
1235.	*Hippophae rhamnoides*	Elaegnaceae	Seeds, fruits	Geetha *et al.*, 2002; Manea *et al.*, 2006; Ercisli *et al.*, 2007; Chauhan *et al.*, 2007; Zhang *et al.*, 2010
1236.	*Holarrhena antidysenterica*	Apocynaaceae	Bark	Zahin *et al.*, 2009; Kalim *et al.*, 2010
1237.	*Holoptelea integrifolia*	Ulmaceae	Leaves	Reddy *et al.*, 2008; Saraswathy *et al.*, 2008
1238.	*Homalomena occulta*	Araceae		Gan *et al.*, 2010
1239.	*Hopea hainanensis*	Dipterocarpaceae		Ge *et al.*, 2009
1240.	*Hopea odorata*	Dipterocarpaceae	Leaves	Hasan *et al.*, 2009
1241.	*Hopea ponga*	Dipterocarpaceae	Seed wings	Sukesh *et al.*, 2011
1242.	*Hordeum sativum*	Poaceae	Fruit	Kahkonen *et al.*, 1999
1243.	*Hordeum vulgare*	Poaceae	Corns,	Osawa and Namiki, 1981; Kahkonen *et al.*, 1999; Bonoli *et al.*, 2004; Bajpai *et al.*, 2005; Ohmori *et al.*, 2005; Zhao *et al.*, 2006; Kim *et al.*, 2007; Siebenhandl *et al.*, 2007; Korycinska *et al.*, 2009; Qingming *et al.*, 2010

Contd...

Table 3.1–*Contd...*

Sl.No.	*Plant Name*	*Family*	*Parts Used*	*References*
1244.	*Houttuynia cordata*	Saururaceae	Aerial parts	Cai *et al.*, 2004; Chomnawang *et al.*, 2007; Li *et al.*, 2008; Chou *et al.*, 2009
1245.	*Hovenia dulcis*	Rhamnaceae	Vinegar	Li *et al.*, 2011
1246.	*Humulus lupulus*	Cannabinaceae	Glandulae	Kahkonen *et al.*, 1999; Trouillas *et al.*, 2003; Yamaguchi *et al.*, 2009
1247.	*Hunteria umbellata*	Apocynaceae	Seed	Adejuwon *et al.*, 2011
1248.	*Hybanthus enneaspermus*	Violaceae	Herb	Vuda *et al.*, 2011
1249.	*Hydnocarpus annamensis*	Flacourtiaceae	Bark	Shi *et al.*, 2006
1250.	*Hydnocarpus wightiana*	Flacourtiaceae	Seed hulls	Reddy *et al.*, 2005
1251.	*Hydnophytum formicarum*	Rubiaceae		Ahmad *et al.*, 2010
1252.	*Hydrastis canadensis*	Hydrastidaceae		Da Silva *et al.*, 2000
1253.	*Hydrocotyle batrachium*	Apiaceae	Herb	Huang *et al.*, 2008
1254.	*Hydrocotyle nepalensis*	Apiaceae	Herb	Huang *et al.*, 2008
1255.	*Hydrocotyle setulosa*	Apiaceae	Herb	Huang *et al.*, 2008
1256.	*Hydrocotyle sibthorpioides*	Apiaceae	Herb	Huang *et al.*, 2008
1257.	*Hyeronima antioquiensis*	Euphorbiaceae		Mosquera *et al.*, 2007
1258.	*Hyeronima macrocarpa*	Euphorbiaceae		Mosquera *et al.*, 2007, 2009
1259.	*Hygrophila auriculata* (=*Asteracantha longifolia*)	Acanthaceae	Root	Ahmed *et al.*, 2001; Vijayakumar *et al.*, 2005, 2006; Shanmugasundaram and Venkataraman, 2006
1260.	*Hylocereus polyrhizus*	Cactaceae	Fruit	Vaillant *et al.*, 2005; Wu *et al.*, 2006; Rebecca *et al.*, 2009, 2010; Kim *et al.*, 2011
1261.	*Hymenocardia acida*	Hymenocardiaceae	Leaves	Sofidya *et al.*, 2006, 2009; Ogbunugafor *et al.*, 2010
1262.	*Hymenocrater longiflorus*	Lamiaceae		Ahmadi *et al.*, 2010
1263.	*Hyoscyamus niger*	Solanaceae	Seed	Cai *et al.*, 2004; Souri *et al.*, 2004
1264.	*Hypericum androsaemum*	Hypericaceae		Valentão *et al.*, 2002 c; Souri *et al.*, 2004
1265.	*Hypericum ascyron*	Hypericaceae		Chon *et al.*, 2009
1266.	*Hypericum calycinum*	Hypericaceae		Kirmizibekmez *et al.*, 2009
1267.	*Hypericum densiflorus*	Hypericaceae	Aerial parts	Henry *et al.*, 2009
1268.	*Hypericum empetrifolium*	Hypericaceae	Aerial parts	Meral and Konyalioglu, 2003
1269.	*Hypericum hookerianum*	Hypericaceae		Wahile *et al.*, 2007
1270.	*Hypericum hyssopifolium*	Hypericaceae	Aerial parts	Cakir *et al.*, 2003
1271.	*Hypericum jovis*	Hypericaceae	Herb	Athanasas *et al.*, 2004

Contd...

Table 3.1–*Contd...*

Sl.No.	*Plant Name*	*Family*	*Parts Used*	*References*
1272.	*Hypericum papuanum*	Hypericaceae		Heilmann *et al.*, 2003
1273.	*Hypericum perforatum*	Hypericaceae	Leaves	Budinevi *et al.*, 1995; Chatterjee *et al.*, 1998; Tripathi *et al.*, 1999; Zheng and Wang, 2001; Meral and Konyalioglu, 2003; Heilmann *et al.*, 2003; Benedi *et al.*, 2004; Zou *et al.*, 2004, 2010; Ivanova *et al.*, 2005; Silva *et al.*, 2005; Katalinic *et al.*, 2006; Kirca and Arslan, 2008; Gioti *et al.*, 2009; Zou *et al.*, 2010; Spiridon *et al.*, 2011
1274.	*Hypericum styphelioides*	Hypericaceae	Leaves	Gamiotea-Turro *et al.*, 2004.
1275.	*Hypericum triquetrifolium*	Hypericaceae	Aerial Parts	Conforti *et al.*, 2002; Meral and Konyalioglu, 2003
1276.	*Hyphaene thebeica*	Arecaceae	Shell, fruit, leaves	Cook *et al.*, 1998; Hsu *et al.*, 2006; Eldahshan *et al.*, 2008
1277.	*Hypoxis hemerocallidea*	Hypoxidaceae		Steenkamp *et al.*, 2006; Nair *et al.*, 2007; Katerere and Eloff, 2008
1278.	*Hyptis fasciculata*	Lamiaceae	Aerial parts	Silva *et al.*, 2005, 2009
1279.	*Hyptis fruticosa*	Lamiaceae	Leaves	Andrade *et al.*, 2010
1280.	*Hyptis suaveolens*	Lamiaceae		Narayanaswamy and Balakrishnan, 2011
1281.	*Hyptis tetracephala*	Lamiaceae		Mensor *et al.*, 2001
1282.	*Hyssopus officinalis*	Lamiaceae	Essential oil	Lugasi *et al.*, 2006; Kizil *et al.*, 2010
1283.	*Icacina tricantha*	Icacinaceae	Leaf, shoot	Oke and Hamburger, 2002; Sofidiya *et al.*, 2006
1284.	*Ilex cornuta*	Aquifoliaceae		Zhu *et al.*, 2009
1285.	*Ilex kudingcha*	Aquifoliaceae		Zhu *et al.*, 2009
1286.	*Ilex paraguariensis*	Aquifoliaceae	Tubers	Vander Jagt *et al.*, 2002; Bixbu *et al.*, 2005; Bravo *et al.*, 2007, Filip, 2007; Paglioso *et al.*, 2010
1287.	*Ilex pubescens*	Aquifoliaceae		Gan *et al.*, 2010
1288.	*Illicium anisatum*	Illiciaceae	Essential oil	Kim *et al.*, 2009
1289.	*Impatiens balsamina*	Balsaminaceae		Gan *et al.*, 2010
1290.	*Impatiens bicolor*	Balsaminaceae		Shahwar *et al.*, 2010
1291.	*Impatiens capensis*	Balsaminaceae	Seeds	Borchardt *et al.*, 2008
1292.	*Impatiens edgeworthii*	Balsaminaceae		Shahwar *et al.*, 2010
1293.	*Imperata cylindrica* var. *major*	Poaceae	Root	Cai *et al.*, 2004; Gan *et al.*, 2010
1294.	*Indigofera coulutea*	Fabaceae		Bakasso *et al.*, 2008
1295.	*Indigofera lineata*	Fabaceae	Aerial parts	Dhenge *et al.*, 2008

Contd...

Table 3.1–*Contd...*

Sl.No.	Plant Name	Family	Parts Used	References
1296.	*Indigofera macrocalyx*	Fabaceae		Bakasso *et al.*, 2008
1297.	*Indigofera nigritana*	Fabaceae		Bakasso *et al.*, 2008
1298.	*Indigofera pulchra*	Fabaceae		Bakasso *et al.*, 2008
1299.	*Indigofera tinctoria*	Fabaceae		Bakasso *et al.*, 2008
1300.	*Indigofera trifoliata*	Fabaceae		Dhenge and Itankar, 2008
1301.	*Indigofera truxillensis*	Fabaceae	Leaves	Farias-Silva *et al.*, 2007
1302.	*Inga verna* ssp. *affinis*	Leguminosae		Vivot *et al.*, 2001
1303.	*Intsia bijuga*	Fabaceae		Peteros and Uy, 2010
1304.	*Inula aucheriana*	Asteraceae	Aerial parts	Sengul *et al.*, 2009
1305.	*Inula britannica*	Asteraceae	Inflore-scence	Cai *et al.*, 2004, 2006; Song *et al.*, 2010
1306.	*Inula crithmoides*	Asteraceae	Aerial parts	Giamperi *et al.*, 2010
1307.	*Inula helenium*	Asteraceae	Root	Wojcikowski *et al.*, 2007; Spiridon *et al.*, 2011
1308.	*Inula racemosa*	Asteraceae	Roots	Haseena Banu *et al.*, 2010
1309.	*Inula viscosa*	Asteraceae	Shoot	Schinella *et al.*, 2002; Al-Mustafa and Al-Thunibat, 2008
1310.	*Ipomoea aquatica*	Convolvulaceae	Leaf, stalk	Huang *et al.*, 2005; Maeda *et al.*, 2006; Lako *et al.*, 2006
1311.	*Ipomoea batatas*	Convolvulaceae	Flowers, leaf, stalk	Hou *et al.*, 2001; Huang *et al.*, 2005; Lako *et al.*, 2006; Dini *et al.*, 2006; Wong *et al.*, 2006; Maeda *et al.*, 2006; Xu *et al.*, 2010; Morrison and Twumasi, 2010; Adefegha and Oboh, 2011
1312.	*Ipomoea hederacea*	Convolvulaceae	Seeds	Dhobal *et al.*, 2008
1313.	*Ipomoea obscura*	Convolvulaceae		Srinivasan *et al.*, 2007
1314.	*Ipomoea quamoclit*	Convolvulaceae	Aerial parts	Hasan *et al.*, 2009
1315.	*Ipomoea reniformis*	Convolvulaceae	Herb	Sanja *et al.*, 2009
1316.	*Iris versicolor*	Iridaceae	Rhizome	Wojcikowski *et al.*, 2007
1317.	*Irvingia gabonensis*	Ixonanthaceae		Agbor *et al.*, 2005
1318.	*Irvingea grandifolia*	Ixonanthaceae		Omonkhua and Onoagbe, 2008
1319.	*Iryanthera juruensis*	Myrtistichaceae		Santos *et al.*, 2009
1320.	*Iryanthera lancifolia*	Myristichaceae		Lock *et al.*, 2005
1321.	*Iryanthera sagotiana*	Myristichaceae	Leaves	Davino *et al.*, 1998.
1322.	*Isatis indigotica*	Brassicaceae	Root	Cai *et al.*, 2004
1323.	*Isatis tinctoria*	Brassicaceae		Li *et al.*, 2008
1324.	*Ixeridium gracile*	Asteraceae		Ma *et al.*, 2007
1325.	*Ixeris dentata*	Asteraceae		Heo *et al.*, 2007

Contd...

Table 3.1–*Contd...*

Sl.No.	*Plant Name*	*Family*	*Parts Used*	*References*
1326.	*Ixora chinensis*	Rubiaceae		Liu *et al.*, 2008
1327.	*Ixora coccinea*	Rubiaceae	Flowers	Saha *et al.*, 2008
1328.	*Ixora williamsii*	Rubiaceae		Liu *et al.*, 2008
1329.	*Jacaranda micrantha*	Bignoniaceae		Menezes *et al.*, 2004
1330.	*Jacaratia mexicana*	Caricaceae		Ruiz-Terán *et al.*, 2008
1331.	*Jasminum fruticans*	Oleaceae	Fruit, leaves	Serteser *et al.*, 2009
1332.	*Jasminum grandiflorum*	Oleaceae	Flowers	Umamaheshwari *et al.*, 2007
1333.	*Jasonia glutinosa*	Asteraceae		Lopez *et al.*, 2008
1334.	*Jasonia montana*	Asteraceae	Aerial parts	Hussain, 2011
1335.	*Jatropha curcas*	Euphorbiaceae	Seed oil, aerial parts, fruit	Getwichit *et al.*, 2008; Balaji *et al.*, 2009; El Diwani *et al.*, 2009; Gallegos-Tintore *et al.*, 2010; Sawant *et al.*, 2010; Oskoueian *et al.*, 2011; Narayanaswamy and Balakrishnan 2011
1336.	*Jatropha gaumeri*	Euphorbiaceae	Leaves	Can-Ake *et al.*, 2004
1337.	*Jatropha gossypifolia*	Euphorbiaceae	Rhizome	Povichit *et al.*, 2010; Shahwar *et al.*, 2010; Narayanaswamy and Balakrishnan, 2011
1338.	*Jatropha mollissima*	Euphorbiaceae		De Melo *et al.*, 2010
1339.	*Jatropha multifida*	Euphorbiaceae		Narayanaswamy and Balakrishnan, 2011
1340.	*Jatropha panduraefolia*	Euphorbiaceae	Leaves	Adib *et al.*, 2010
1341.	*Juglans regia*	Juglandaceae	Leaves, flowers	Cai *et al.*, 2004; Katalinic *et al.*, 2006; Pereira *et al.*, 2007; Labuckas *et al.*, 2008; Oliveira *et al.*, 2008; Miliauskas *et al.*, 2009
1342.	*Juncus effusus*	Juncaceae		Gan *et al.*, 2010
1343.	*Jussiaea repens*	Onagraceae	Aerial parts	Thuong *et al.*, 2006
1344.	*Justicia flava*	Acanthaceae	Seed	Odukoya *et al.*, 2005
1345.	*Kadsura longipedunculata*	Schisandraceae	Stem bark	Mulyaningsih *et al.*, 2010
1346.	*Kaempferia galanga*	Zingiberaceae	Leaves	Chan *et al.*, 2009; Chanwitheesuk *et al.*, 2005
1347.	*Kaempferia rotunda*	Zingiberaceae	Rhizome	Lotulung *et al.*, 2008
1348.	*Kalanchoe pinnata*	Crassulaceae	Leaf	Harika *et al.* (2007)
1349.	*Kelussia odoratissima*	Apiaceae	Herb	Ahmadi *et al.*, 2007
1350.	*Khaya senegalensis*	Meliaceae	Leaves, bark	Karou *et al.*, 2005; Moussa *et al.*, 2011
1351.	*Kigelia africana*	Bignoniaceae		Olaleye and Rocha, 2007, 2008.

Contd...

Table 3.1–*Contd...*

Sl.No.	Plant Name	Family	Parts Used	References
1352.	*Kochia scoparia*	Chenopodiaceae		Gan *et al.*, 2010
1353.	*Koelreuteria henryi*	Sapindaceae	Leaves	Lee *et al.*, 2006; Chen, C-H. *et al.*, 2009
1354.	*Krameria triandra*	Polygalaceae	Leaves	Carini *et al.*, 2002
1355.	*Lactuca indica*	Asteraceae		Wang *et al.*, 2003
1356.	*Lactuca sativa*	Asteraceae	Leaf, seed	Caldwell 2003; Cai *et al.*, 2004; Souri *et al.*, 2004, 2008; Lako *et al.*, 2006; Maeda *et al.*, 2006
1357.	*Lactuca scariola*	Asteraceae	Aerial parts	Kim, 2001
1358.	*Ladambergia macrocarpa*	Rubiaceae		Mosquera *et al.*, 2007
1359.	*Lagerstroemia indica*	Lythraceae	Fruit	Fu *et al.*, 2010; Moussa *et al.*, 2011
1360.	*Lagerstroemia speciosa*	Lythraceae	Leaves	Unno *et al.*, 1998; Priya *et al.*, 2008; Anil *et al.*, 2010; Fu *et al.*, 2010
1361.	*Laggera pterodonta*	Asteraceae		Wu *et al.*, 2007
1362.	*Lamium album*	Lamiaceae	Flowers, herb	Matkowski and Piotrowska, 2006; Trouillas *et al.*, 2003
1363.	*Lamium maculatum*	Lamiaceae	Herb	Matkovski *et al.*, 2008
1364.	*Lamium purpureum*	Lamiaceae	Flowers	Matkowski and Piotrowska, 2006
1365.	*Landolfia owaricensis*	Apocynaceae	Leaf	Oke and Hamburger, 2002
1366.	*Lannea acida*	Anacardiaceae	Bark	Outtara *et al.*, 2010
1367.	*Lannea microcarpa*	Anacardiaceae	Fruit	Lamien-Meda *et al.*, 2008; Outtara *et al.*, 2010
1368.	*Lannea velutina*	Anacardiaceae	Leaves	Diallo *et al.*, 2001; Maiga *et al.*, 2006; Outtara *et al.*, 2010
1369.	*Lantana camara*	Verbenaceae		Mensor *et al.*, 2001; Tripathi 2003; Rosas-Romero and Saavedra, 2005; Basu and Hazra, 2006
1370.	*Lantana magnibracteata*	Verbenaceae		Rosas-Romero and Saavedra, 2005
1371.	*Lantana trifolia*	Verbenaceae		Mensor *et al.*, 2001; Rosas-Romero and Saavedra, 2005
1372.	*Laportea aestuans*	Urticaceae	Leaf	Morrison and Twumasi, 2010
1373.	*Laportea meyneana*	Utricaceae		Peteros and Uy, 2010
1374.	*Larrea tridentata*	Zygophyllaceae	Leaves	Abou-Gazar *et al.*, 2004
1375.	*Lasianthera africana*	Icacinaceae	Leaf	Odukoya *et al.*, 2007;
1376.	*Lasianthus cyanocarpus*	Rubiaceae		Ahmad *et al.*, 2010
1377.	*Lasianthus maingayi*	Rubiaceae		Ahmad *et al.*, 2010
1378.	*Lasianthus oblongus*	Rubiaceae		Saha *et al.*, 2004
1379.	*Lasianthus pilosus*	Rubiaceae		Ahmad *et al.*, 2010

Contd...

Table 3.1–*Contd...*

Sl.No.	Plant Name	Family	Parts Used	References
1380.	*Lathyrus binatus*	Fabaceae	Herb	Godevac *et al.*, 2007
1381.	*Launaea pinnatifida*	Asteraceae	Leaves	Nagalapur and Paramjyothi, 2010
1382.	*Laurocerasus officinalis*	Rosaceae	Fruit	Kolayli *et al.*, 2003.
1383.	*Laurus nobilis*	Lauraceae	Leaves	Demo *et al.*, 1999; Kang *et al.*, 2002; Simic *et al.*, 2003; Pabuccuoglu *et al.*, 2003; Mariutti *et al.*, 2008; Hinneburg *et al.*, 2005; Moussa *et al.*, 2011
1384.	*Lavandula angustifolia*	Lamiaceae	Aerial parts	Hohman *et al.*, 1999; Yang *et al.*, 2010; Spiridon *et al.*, 2011
1385.	*Lavandula intermedia*	Lamiaceae		Torras-Claveria *et al.*, 2007
1386.	*Lavandula officinalis*	Lamiaceae	Leaves, Cell cultures	Nitzsche *et al.*, 2004; Bouayed *et al.*, 2007
1387.	*Lavandula stoechas*	Lamiaceae	Essential oil	Gulcin *et al.*, 2004
1388.	*Lawsonia inermis* (=*Lawsonia alba*)	Lythraceae	Fruits	Ahmed *et al.*, 2000; Mikhaeil *et al.*, 2004; Prakash *et al.*, 2007; Hosein and Zinab, 2007; Endrini *et al.*, 2007; Guha *et al.*, 2009
1389.	*Ledum groenlandicum*	Ericaceae	Herb	Dufour *et al.*, 2007
1390.	*Leea asiatica*	Leeaceae	Leaves	Dalal *et al.*, 2010
1391.	*Leea guinense*	Leeaceae	Root	Oke and Hamburger, 2002; Op de Beck *et al.*, 2003
1392.	*Leea indica*	Leeaceae		Saha *et al.*, 2004
1393.	*Lens culinaris*	Fabaceae	Pulses	Han *et al.*, 2008; Amarowicz *et al.*, 2010
1394.	*Lens esculenta*	Fabaceae	Seeds	Bajpai *et al.*, 2005; Sreeramulu *et al.*, 2009
1395.	*Leontice smirnowii*	Leonticaceae	Tuber	Gülçin *et al.*, 2006
1396.	*Leonurus cardiaca*	Lamiaceae	Plant	Mantle *et al.*, 2000; Nam and Kang, 2004; Matkowski and Piotrowska, 2006; Matkowski *et al.*, 2008
1397.	*Leonurus heterophyllus*	Lamiaceae		Gan *et al.*, 2010
1398.	*Leopoldia comosa* (=*Muscari comosum*)	Liliaceae	Bulb	Pieroni *et al.*, 2002
1399.	*Lepechinia graveolens*	Lamiaceae		Parejo *et al.*, 2004
1400.	*Lepechinia mayenii*	Lamiaceae	Herb	Lock *et al.*, 2005
1401.	*Lepidagathis cristata*	Acanthaceae		Achliya *et al.*, 2003.
1402.	*Lepidagathis trinervis*	Acanthaceae	Herb	Singh *et al.*, 2009
1403.	*Lepidium apetalum*	Brassicaceae		Song *et al.*, 2010
1404.	*Lepidium perfoliatum*	Brassicaceae	Seed	Souri *et al.*, 2008

Contd...

Table 3.1–*Contd...*

Sl.No.	Plant Name	Family	Parts Used	References
1405.	*Lepidium sativum*	Brassicaceae		Souri *et al.*, 2004
1406.	*Leptadenia hastata*	Asclepiadaceae	Leaf	Cook *et al.*, 1998
1407.	*Lespedeza capitata*	Fabaceae	Seeds	Borchardt *et al.*, 2008
1408.	*Leucaena glauca*	Mimosaceae		Pongsawatmanit and Maisuthisakul, 2004.
1409.	*Leucaena leucocephala*	Mimosaceae		Chanwitheesuk *et al.*, 2005; Panpipat *et al.*, 2010
1410.	*Leucas aspera*	Lamiaceae	Leaves, root	Rahman *et al.*, 2007; Meghashri *et al.*, 2010
1411.	*Leucas cephalotes*	Lamiaceae	Leaves	Mathur *et al.*, 2010
1412.	*Leucas indica*	Lamiaceae	Aerial parts	Mostafa *et al.*, 2007
1413.	*Levisticum officinale*	Apiaceae	Root	Wojcikowski *et al.*, 2007
1414.	*Licania arborea*	Chrysobalanaceae		Ruiz-Terán *et al.*, 2008
1415.	*Licania licaniaeflora*	Chrysobalanaceae	Leaves	Braca *et al.*, 2002
1416.	*Licaria martiniana*	Lauraceae	Essential oil	Alcantara *et al.*, 2010
1417.	*Ligaria cuneifolia*	Loranthaceae		Borneo *et al.*, 2009
1418.	*Ligularia fischeri*	Asteraceae		Choi *et al.*, 2007; Shang *et al.*, 2010
1419.	*Ligusticum chuanxiong*	Apiaceae		Jeong *et al.*, 2009a; Gan *et al.*, 2010
1420.	*Ligusticum sinense*	Apiaceae		Song *et al.*, 2010
1421.	*Ligustrum delavayanum*	Oleaceae	Leaves	Nagy *et al.*, 2006
1422.	*Ligustrum japonicum*	Oleaceae		Katsube *et al.*, 2004
1423.	*Ligustrum lucidum*	Oleaceae	Fruit	Cai *et al.*, 2004; Gao *et al.*, 2009
1424.	*Ligustrum robustum*	Oleaceae		He *et al.*, 2003; Zhu *et al.*, 2009
1425.	*Ligustrum sinense*	Oleaceae		Ouyang *et al.*, 2003
1426.	*Ligustrum vulgare*	Oleaceae	Leaves, flowers	Nagy *et al.*, 2006; Sersen *et al.*, 2005
1427.	*Limnophila geoffrayi*	Scrophulariaceae	Aerial part	Suksamaran *et al.*, 2003
1428.	*Limonium bicolor*	Plumbaginaceae		Zhen *et al.*, 2009; Li *et al.*, 2010
1429.	*Limonium wrightii*	Plumbaginaceae		Aniya *et al.*, 2002.
1430.	*Linaria pyramidata*	Scrophulariaceae		Souri *et al.*, 2004
1431.	*Lindelofia stylosa*	Boraginaceae	Aerial parts	Choudhary *et al.*, 2005
1432.	*Lindera pulcherrima*	Lauraceae	Essential oil	Joshi *et al.*, 2010
1433.	*Lindera strychnifolia*	Lauraceae		Kim *et al.*, 1997
1434.	*Lindernia anagallis*	Scrophulariaceae	Herb	Shyur *et al.*, 2005
1435.	*Linum usitatissimum*	Linaceae	Seed	Kahkonen *et al.*, 1999; Souri *et al.*, 2008; Zanwar *et al.*, 2010

Contd...

Table 3.1–*Contd...*

Sl.No.	*Plant Name*	*Family*	*Parts Used*	*References*
1436.	*Lippia adoensis*	Verbenacae	Leaves	Mikre *et al.*, 2007
1437.	*Lippia alba*	Verbenaceae	Leaves	Ramos *et al.*, 2003; Stashenko *et al.*, 2004; Ara and Nur, 2009
1438.	*Lippia boliviana*	Verbenaceae		Rosas-Romero and Saavedra, 2005
1439.	*Lippia citriodora*	Verbenaceae		Valentão *et al.*, 2002a
1440.	*Lippia dulcis*	Verbenaceae	Aerial parts	Ono *et al.*, 2005
1441.	*Lippia graveolens*	Verbenaceae	Essential oil	Sánchez-Arana *et al.*, 2010
1442.	*Lippia multiflora*	Verbenaceae	Leaf	Agnaniet *et al.*, 2005; Avlessi *et al.*, 2005
1443.	*Lippia sidoides*	Verbenaceae	Leaf	Monteira *et al.*, 2007
1444.	*Liquidamber formosana*	Altingiaceae		Wang *et al.*, 2009; Gan *et al.*, 2010
1445.	*Liriope spicata*	Liliaceae	Leaf	Hou *et al.*, 2004
1446.	*Litchi chinensis*	Sapindaceae	Fruit, pericarp	Alfredo, 2007; Liu *et al.*, 2007, 2009; Yang *et al.*, 2007: Duan *et al.*, 2007;Yang *et al.*, 2007; Sun *et al.*, 2010; Wang *et al.*, 2011
1447.	*Lithocarpus pachyphyllus*	Fagaceae	Leaves	Yang *et al.*, 2004
1448.	*Lithospermum erythrorhizon*	Boraginaceae		Han *et al.*, 2008; Cai *et al.*, 2004, 2006; Li *et al.*, 2008
1449.	*Litsea cubeba*	Lauraceae		Hwang *et al.*, 2005
1450.	*Litsea glutinosa*	Lauraceae	Stem	Kshirsagar and Upadhyay, 2009
1451.	*Lobelia chinensis*	Lobeliaceae	Whole plant	Cai *et al.*, 2004; Li *et al.*, 2008
1452.	*Loiseleuria procumbens*	Ericaceae		Cuendet *et al.*, 2000
1453.	*Lomatogonium carianthiacum*	Gentianaceae	Herb	Myagmar and Aniya, 2000
1454.	*Lonicera japonica*	Caprifoliaceae	Floral bud	Cai *et al.*, 2004, 2006; Choi *et al.*, 2007; Li *et al.*, 2008
1455.	*Lophatherum gracile*	Poaceae		Li *et al.*, 2008
1456.	*Loranthus parasiticus*	Loranthaceae		Gan *et al.*, 2010
1457.	*Lotus corniculatus*	Fabaceae	Herb	Koelzer *et al.*, 2009
1458.	*Ludwigia octovalvis*	Onagraceae	Herb	Shyur *et al.*, 2005
1459.	*Luffa acutangula*	Cucurbitaceae	Fruits	Ansari *et al.*, 2005; Raghu *et al.*, 2010
1460.	*Luffa cylindrica*	Cucurbitaceae	Fruit	Maeda *et al.*, 2006
1461.	*Luffa echinata*	Cucurbitaceae		Kumar *et al.*, 2000
1462.	*Lumnitzera racemosa*	Combretaceae	Leaves flowers	Vadlapudi and Naidu, 2009
1463.	*Lupinus angustifolius*	Fabaceae		Kahkonen *et al.*, 1999

Contd...

Table 3.1–*Contd...*

Sl.No.	*Plant Name*	*Family*	*Parts Used*	*References*
1464.	*Lupinus mutabilis*	Fabaceae		Penarrieeta *et al.*, 2005
1465.	*Lycianthes acutifolia*	Solanaceae		Mosquera *et al.*, 2009
1466.	*Lycianthes radiata*	Solanaceae		Mosquera *et al.*, 2009
1467.	*Lycium barbarum*	Solanaceae	Fruit, Root bark	Ren *et al.*, 1995; Luo *et al.*, 2004; Wu *et al.*, 2004; Cai *et al.*, 2004, 2006; Li *et al.*, 2007
1468.	*Lycium chinense*	Solanaceae		Wong *et al.*, 2006; Li *et al.*, 2008; Zhang *et al.*, 2010
1469.	*Lycopersicon esculentum*	Solanaceae	Fruits	Wang *et al.*, 1996; Abdulnabi *et al.*, 1997; Kahkonen *et al.*, 1999; Cai *et al.*, 2004; Kim *et al.*, 2004; Huang *et al.*, 2004a; Agostini *et al.*, 2004; Andrews *et al.*, 2004; Toor and Savage, 2005; Lana *et al.*, 2006; Lenucci *et al.*, 2006; Odriozola-Serrano *et al.*, 2008; Chang Ching-Hui and Liu Yung-Chuan, 2008; Hsu *et al.*, 2008; Guil-Guerrero *et al.*, 2009; Walia *et al.*, 2010; Raghu *et al.*, 2010
1470.	*Lycopus europaeus*	Lamiaceae	Herb	Lamaison *et al.*, 1991
1471.	*Lycopus lucidus*	Lamiaceae	Aerial parts	Cai *et al.*, 2004; Nam and Kang, 2004; Woo and Piao, 2004; Slusarczyk *et al.*, 2008; Gan *et al.*, 2010
1472.	*Lysimachia christinae*	Primulaceae		Gan *et al.*, 2010
1473.	*Lysimachia vulgaris*	Primulaceae	Herb	Kahkonen *et al.*, 1999
1474.	*Lythrum alatum*	Lythraceae	Seeds	Borchardt *et al.*, 2008
1475.	*Lythrum salicaria*	Lythraceae	Seeds	Kahkonen *et al.*, 1999; Mantle *et al.*, 2000; Tunalier *et al.*, 2007; Borchardt *et al.*,2008; Lopez *et al.*, 2008; Lee *et al.*, 2010
1476.	*Mabea montana*	Euphorbiaceae		Mosquera *et al.*, 2007
1477.	*Macadamia integrifolia*	Proteaceae	Fruit	Wall, 2010
1478.	*Machaerium villosum*	Fabaceae		Santos *et al.*, 2009
1479.	*Machilus zuihoensis*	Lauraceae		Hou *et al.*, 2003
1480.	*Maclura pomifera*	Moraceae	Fruit	Tsao *et al.*, 2003; Vesela *et al.*, 2004; Hamed and Hussain, 2005
1481.	*Maclura tinctoria*	Moraceae	Stem bark	Cioffi *et al.*, 2003
1482.	*Macropanax dispermus*	Araliaceae		Chanwitheesuk *et al.*, 2005
1483.	*Macrotyloma uniflorum*	Fabaceae	Seeds	Siddhuraju and Manian 2007
1484.	*Madhuca indica*	Sapotaceae	Bark	Pawar and Bhutani, 2004; Kuashik *et al.*, 2010

Contd...

Table 3.1–*Contd...*

Sl.No.	*Plant Name*	*Family*	*Parts Used*	*References*
1485.	*Maesa lanceolata*	Myrsinaceae		Muhammad *et al.*, 2003
1486.	*Magnolia liliflora*	Magnoliaceae		Song *et al.*, 2010
1487.	*Magnolia officinalis*	Magnoliaceae	Bark	Cai *et al.*, 2004, 2006: Gan *et al.*, 2010
1488.	*Magonia glabrata*	Sapindaceae	Fruit bark	Lemos *et al.*, 2006
1489.	*Mahonia aquifolium*	Berberidaceae	Seeds	Muller and Ziereis, 1994; Misik *et al.*, 1995;Bezakova *et al.*, 1996; Racková *et al.*, 2004, 2007; Kardasova and Machova, 2007
1490.	*Mahonia leschenaultii*	Berberidaceaee	Fruit	Karuppusamy *et al.*, 2011
1491.	*Majorana hortensis*	Lamiaceae	Aerial parts	Kosar *et al.*, 2003; El-Ghorab *et al.*, 2004; Dorman *et al.*, 2004; Mariutti *et al.*, 2008
1492.	*Majorana syriaca*	Lamiaceae	Herb	Al-Bandak *et al.*, 2009
1493.	*Mallotus japonicus*	Euphorbiaceae		Katsube *et al.*, 2004
1494.	*Mallotus oppositifolium*	Euphorbiaceae	Leaf	Farombi *et al.*, 2001
1495.	*Mallotus repandus*	Euphorbiaceae		Lin *et al.*, 1995b
1496.	*Malpighia emarginata*	Malpighiaceae	Fruit	Mezadri *et al.*, 2008
1497.	*Malpighia glabra*	Malpighiaceae	Fruits	Schmourlo *et al.*, 2007
1498.	*Malus docmeri*	Rosaceae		Hou *et al.*, 2003
1499.	*Malus pumila*	Rosaceae	Fruit	Wang *et al.*, 1996; Kahkonen *et al.*, 1999; Eberhardt *et al.*, 2000; Chinnici *et al.*, 2004; Agostini *et al.*, 2004; Cai *et al.*, 2004; D'Abrosca *et al.*, 2007
1500.	*Malus sylvestris*	Rosaceae	Fruit	Serteser *et al.*, 2009
1501.	*Malva neglecta*	Malvaceae		Mavi *et al.*, 2004
1502.	*Malva sylvestris*	Malvaceae	Flowers, fruits	Asolini *et al.*, 2006; DellaGreca *et al.*, 2009
1503.	*Malva verticillata*	Malvaceae		Gan *et al.*, 2010
1504.	*Mammea africana*	Clusiaceae		Nguelefack-Mbuyo *et al.*, 2010
1505.	*Mammea americana*	Clusiaceae	Seeds	Yang *et al.*, 2005
1506.	*Mammea siamensis*	Clusiaceae		Leelapornpisid *et al.*, 2007
1507.	*Mangifera indica*	Anacardiaceae	Fruits, peel, bark Kernel	Gregorio *et al.*, 2000; Scartezzini and Speroni, 2000; Martinez *et al.*, 2000; Soong and Barlow, 2006; Ribeiro *et al.*, 2008; Ayoola *et al.*, 2008; Maisuthisakul and Pasuk, 2008; Kim *et al.*, 2010; Vaghasiya. and Chanda, 2010; Niwano *et al.*, 2011

Contd...

Table 3.1–*Contd...*

Sl.No.	Plant Name	Family	Parts Used	References
1508.	*Manihot esculenta*	Euphorbiaceae	Roots, leaves	Wong *et al.*, 2006; Lako *et al.*, 2006; Hosseinian *et al.*, 2007; Patnibul *et al.*, 2008; Ndidi and Akeem, 2011; Suresh *et al.*, 2011; Eleazu *et al.*, 2011
1509.	*Manihot utilissima*	Euphorbiaceae	Roots	Zaibunnisa *et al.*, 2005. Zaibunnisa *et al.*, 2008; Salawu *et al.*, 2011
1510.	*Manilkara zapota*	Sapotaceae	Fruit	Isabelle *et al.*, 2010
1511.	*Mansonia gagei*	Sterculiaceae		Tiew *et al.*, 2003
1512.	*Marrubium astracanicum*	Lamiaceae	Aerial parts	Firuzi *et al.*, 2010
1513.	*Marrubium globosum* subsp. *globosum*	Lamiaceae		Sarikurkcu *et al.*, 2008
1514.	*Marrubium vulgare*	Lamiaceae	Plant	Vander Jagt *et al.*, 2002; Souri *et al.*, 2004; Matkowski and Piotrowska, 2006; Katalinic *et al.*, 2006; Matkowski *et al.*, 2008; Firuzi *et al.*, 2010
1515.	*Marsdenia glabra*	Asclepiadaceae		Chanwitheesuk *et al.*, 2005
1516.	*Matricaria aurea*	Asteraceae	Shoot	Al-Mustafa and Al-Thunibat, 2008
1517.	*Matricaria chamomilla*	Asteraceae	Flower	Kahkonen *et al.*, 1999; Bajpai *et al.*, 2005
1518.	*Matricaria matricarioides*	Asteraceae		Kahkonen *et al.*, 1999
1519.	*Matricaria recutita*	Asteraceae	Flowers	Kahkonen *et al.*, 1999; Lado *et al.*, 2004; Miliuskas *et al.*, 2004; Asolini *et al.*, 2006
1520.	*Maytenus aquifolium*	Celastraceae		Corsino *et al.*, 2003
1521.	*Maytenus ilicifolia*	Celastraceae	Root bark	Vellosa *et al.*, 2006
1522.	*Maytenus krukovii*	Celastraceae	Bark	Bruni *et al.*, 2006
1523.	*Medicago sativa*	Fabaceae	Seeds	Al-Mamun *et al.*, 2007; Rana *et al.*, 2010; Bora and Sharma, 2011
1524.	*Melaleuca alternifolia*	Myrtaceae	Herb	Kim *et al.*, 2004
1525.	*Melaleuca leucadendron*	Myrtaceae	Fruit	Fu *et al.*, 2010
1526.	*Melaleuca leucandra*	Myrtaceae	Essential oil	Pino *et al.*, 2010
1527.	*Melampyrum barbatum*	Scrophulariaceae	Flower, leaf	Stajner *et al.*, 2009
1528.	*Melastoma candidum*	Melastomataceae		Lee *et al.*, 2001
1529.	*Melastoma decemfidum*	Melastomataceae	Leaves	Sarju *et al.*, 2011
1530.	*Melastoma malabathricum*	Melastomataceae	Flowers	Deny *et al.*, 2007; Zakaria, 2007

Contd...

Table 3.1–*Contd...*

Sl.No.	*Plant Name*	*Family*	*Parts Used*	*References*
1531.	*Melastoma sanguineum*	Melastomaceae	Fruit	Fu *et al.*, 2010
1532.	*Melicope ptelefolia*	Rutaceae		Abbas *et al.*, 2006
1533.	*Melilotus officinalis*	Fabaceae	Aerial parts	Parejo *et al.*, 2002; Pourmorad *et al.*, 2006; Miliauskas *et al.*, 2004, 2009; Spiridon *et al.*, 2011
1534.	*Melissa officinalis*	Lamiaceae	Leaves	Hohman *et al.*, 1999: Mimica-Duki *et al.*, 2004; Marongin *et al.*, 2004; de Souza *et al.*, 2004; Ivanova *et al.*, 2005; Lugasi *et al.*, 2006; Bouayed *et al.*, 2007; Spiridon *et al.*, 2011
1535.	*Melodorum fruticosum*	Annonaceae		Pripdeeveech and Chukeatirote, 2010
1536.	*Mentha aquatica*	Lamiaceae	Leaves	Dorman *et al.*, 2003; Serteser *et al.*, 2009; Ebrahimzadeh *et al.*, 2010
1537.	*Mentha arvensis*	Lamiaceae	Leaves, stems	Chanwitheesuk *et al.*, 2005; Ivanova *et al.*, 2005; Wong *et al.*, 2006; Zaibunnisa *et al.*, 2008; Zakaria *et al.*, 2008
1538.	*Mentha cordifolia*	Lamiaceae	Leaves	Chanwitheesuk *et al.*, 2005; Spiridon *et al.*, 2011
1539.	*Mentha haplocalyx*	Lamaceae	Leaves	Song *et al.*, 2010
1540.	*Mentha longifolia*	Lamiaceae	Leaves	Souri *et al.*, 2004; Nickavar *et al.*, 2008; Serteser *et al.*, 2009; Mkaddem *et al.*, 2009; Gursoy *et al.*, 2009; Bhatti *et al.*, 2010; Ebrahimzadeh *et al.*, 2010; Motamed and Naghibi, 2010; Spiridon *et al.*, 2011
1541.	*Mentha piperita*	Lamiaceae	Leaves	Yadegarinia *et al.*, 2006; Davod *et al.*, 2006;Yang *et al.*, 2010; Mathur *et al.*, 2010; Ebrahimzadeh *et al.*, 2010
1542.	*Mentha pulegium*	Lamiaceae	Leaves	Souri *et al.*, 2004; Nickavar *et al.*, 2008; Serteser *et al.*, 2009; Kamkar *et al.*, 2010
1543.	*Mentha spicata*	Lamiaceae	Leaves	Souri *et al.*, 2004, 2008; Arabshahi *et al.*, 2005; Kiselova *et al.*, 2006; Hosseinimehr *et al.*, 2007; Serteser *et al.*, 2009; Nikolova and Dzhurmanski, 2009; Ebrahimzadeh *et al.*, 2010
1544.	*Mentha viridis*	Lamiaceae	Leaves	Arumugam *et al.*, 2004; Mkaddem *et al.*, 2009
1545.	*Mesembryanthemum crystallinum*	Aizoaceae	Herb	Hanen *et al.*, 2010

Contd...

Table 3.1–*Contd...*

Sl.No.	Plant Name	Family	Parts Used	References
1546.	*Mesembryanthemum edule*	Aizoaceae	Herb	Hanen *et al.*, 2010
1547.	*Mesembryanthemum nodiflorum*	Aizoaceae	Herb	Hanen *et al.*, 2010
1548.	*Mesona procumbens*	Lamiaceae	Leaf gum	Lai *et al.*, 2001
1549.	*Mezoneuron benthamianum*	Leguminosae		Dickson *et al.*, 2006
1550.	*Michelia alba*	Magnoliaceae	Leaves	Leelapornpisid *et al.*, 2007; Wang *et al.*, 2010
1551.	*Michelia champaca*	Magnoliaceae	Leaf, flower	Hossain *et al.*, 2009; Hasan *et al.*, 2009; Vivek Kumar *et al.*, 2010
1552.	*Miconia lemannii*	Melastomataceae		Mosquera *et al.*, 2009
1553.	*Microdesmis keayana*	Pandaceae	Roots	Zamble *et al.*, 2006, 2009
1554.	*Microglossa pyrifolia*	Asteraceae		Dickson *et al.*, 2006
1555.	*Microlabium candidum*	Asteraceae		Borneo *et al.*, 2008
1556.	*Micromeria graeca*	Lamiaceae	Aerial parts	Couladis *et al.*, 2003
1557.	*Mikania buchtienii*	Asteraceae		Rosas-Romero and Saavedra, 2005
1558.	*Mikania leiostachya*	Asteraceae		Mosquera *et al.*, 2007
1559.	*Mikania psilostachya*	Asteraceae		Parejo *et al.*, 2005
1560.	*Mikania scandens*	Asteraceae	Leaves	Hasan *et al.*, 2009
1561.	*Millettia dielsiana*	Fabaceae		Gan *et al.*, 2010
1562.	*Millettia pulchra* var. *microphylla*	Fabaceae	Leaves	Lee *et al.*, 2006
1563.	*Millingtonia hortensis*	Bignoniaceae		Leelapornpisid *et al.*, 2007
1564.	*Mimosa hamata*	Mimosaceae	Roots, leaf	Jain *et al.*, 2009; Singh *et al.*, 2009
1565.	*Mimosa pudica*	Mimosaceae	Leaves	Muthukumaran *et al.*, 2011; Zhang *et al.*, 2011
1566.	*Mimosa verrucosa*	Mimosaceae		Desmarchelier *et al.*, 1999
1567.	*Mimusops elengi*	Sapotaceae		Saha *et al.*, 2008; Gayathri and Suresh, 2010
1568.	*Minuartia lineata*	Caryophyllaceae		Souri *et al.*, 2004
1569.	*Misodendrum punctulatum*	Myzodendraceae		Desmarchelier *et al.*, 2005
1570.	*Mitragyna rotundifolia*	Rubiaceae	Stem, twig	Kshirsagar and Upadhyay, 2009; Kang *et al.*, 2009
1571.	*Mollugo nudicaulis*	Molluginaceae	Herb	Singh *et al.*, 2009
1572.	*Mollugo pentaphylla*	Molluginaceae	Aerial parts	Maharana *et al.*, 2010

Contd...

Table 3.1–*Contd...*

Sl.No.	Plant Name	Family	Parts Used	References
1573.	*Momordica charantia*	Cucurbitaceae	Fruits, leaves, stem, seeds	Scartezzini and Speroni 2000; Sabu and Kuttan, 2003; Naik *et al.*, 2003, 2005; Chandra *et al.*, 2003; Ansari *et al.*, 2005 ; Bajpai *et al.*, 2005; Chanwitheesuk *et al.*, 2005; Maeda *et al.*, 2006; Kubola and Siriamornpun, 2008; Liu *et al.*, 2010; Xie and Zhang, 2010; Saeed *et al.*, 2010; Biswas *et al.*, 2010; Raghu *et al.*, 2010
1574.	*Momordica cochinensis*	Cucurbitaceae		Gan *et al.*, 2010
1575.	*Momordica dioica*	Cucurbitaceae	Leaves, roots	Shreedhara and Vaidya, 2006; Jain *et al.*, 2008
1576.	*Momordica grosvenori*	Cucurbitaceae	Fruits	Song *et al.*, 2007, 2010
1577.	*Monechma ciliatum*	Acanthaceae	Seed cake	Mariod *et al.*, 2010
1578.	*Moricandia arvensis*	Brassicaceae	Flowers	Braham *et al.*, 2005; Skandrani *et al.*, 2007, 2010
1579.	*Morinda citrifolia*	Rubiaceae	Roots, Fruit	Zin *et al.*, 2002; Singh *et al.*, 2004; Su *et al.*, 2005; Pongaravane *et al.*, 2006; Zin *et al.*, 2002; Basu and Hazra, 2006; Ramamoorthy and Bono, 2007; Wang *et al.*, 2009
1580.	*Morinda citriodora* var. *citriodora*	Rubiaceae		Dorman and Deans, 2004
1581.	*Morinda lucida*	Rubiaceae		Akinmoladun *et al.*, 2010
1582.	*Morinda officinalis*	Rubiaceae		Soon and Tan, 2002; Wu *et al.*, 2006
1583.	*Moringa oleifera* (Syn: *Moringa pterigosperma*)	Moringaceae	Fruits, Leaves, seeds	Siddaraju and Becker, 2003; Kumar *et al.*, 2007Arabshahi *et al.*, 2005; Bajpai *et al.*, 2005; Wong *et al.*, 2006; Lako *et al.*, 2006; Charde *et al.*, 2009
1584.	*Moringa peregrina*	Moringaceae	Bark	Marwah *et al.*, 2007
1585.	*Morus alba*	Moraceae	Leaves, fruits	Cheong *et al.* 1998; Kim *et al.*, 1999; Chang *et al.*, 2002; Fukai *et al.*, 2003b; Cai *et al.*, 2004, 2006; Ercisli and Orhan, 2007; Du *et al.*, 2008; Katsube *et al.*, 2006; Thuong *et al.*, 2006; Jiang *et al.*, 2008; Chao *et al.*, 2009; Serteser *et al.*, 2009; Povichit *et al.*, 2010; Wanyo *et al.*, 2010; Gan *et al.*, 2010; Vadlapudi and Naidu, 2010; Song *et al.*, 2010
1586.	*Morus bombycis*	Moraceae	Roots	Jin *et al.*, 2005, 2006
1587.	*Morus indica*	Moraceae	Leaves	Andallu and Varadacharyulu, 2003
1588.	*Morus macroura*	Moraceae	Stem bark	Dai *et al.*, 2004

Contd...

Table 3.1–*Contd...*

Sl.No.	*Plant Name*	*Family*	*Parts Used*	*References*
1589.	*Morus mongolica*	Moraceae	Stem and root bark	Kang *et al.*, 2006
1590.	*Morus nigra*	Moraceae	Leaves, fruits	Al-Mustafa and Al-Thunibat, 2008; Özgen *et al.*, 2009; Naderi *et al.*, 2004; Ercisli and Orhan, 2007; Serteser *et al.*, 2009; Iqbal *et al.*, 2010; Kumar *et al.*, 2011
1591.	*Morus rubra*	Moraceae	Fruits	Ercisli and Orhan, 2007; Özgen *et al.*, 2009
1592.	*Mosla chinensis*	Lamiaceae		Cao *et al.*, 2009
1593.	*Mouriri pusa*	Melastomataceae		Andreo *et al.*, 2006
1594.	*Mucuna pruriens*	Fabaceae	Fruits	Sidduraju and Becker, 2003; Rajeswar *et al.*, 2005; Misra and Wagner, 2007; Yerra *et al.*, 2005; Tripathi and Upadhya, 2002; Gajera *et al.*, 2005; Dhansekaran *et al.*, 2008
1595.	*Mukia maderaspatana*	Cucurbitaceae	Leaves	Petrus *et al.*, 2011
1596.	*Muntingia calabura*	Elaeocarpaceae		Zakaria, 2007
1597.	*Murraya koenigii*	Rutaceae	Leaves	Patel, 1979; Khan *et al.*, 1997; Sabu and Kuttan, 2003; Wong *et al.*, 2006; Rao *et al.*, 2007; Ningappa *et al.*, 2008; Huda-Faujan *et al.*, 2009; Mathur *et al.*, 2010
1598.	*Musa nana*	Musaceae	Fruit	Lako *et al.*, 2006
1599.	*Musa paradisiaca*	Musaceae	Fruits	Wang *et al.*, 1996; Geol *et al.*, 2001; Vijayakumar *et al.*, 2008; Alothman *et al.*, 2009; Faller and Fialho, 2010
1600.	*Mutisia friesiana*	Asteraceae		Viturro *et al.*, 1999
1601.	*Myracrodendruon urundeuva*	Anacardiaceae	Stem bark	Nobre Junior *et al.*, 2007
1602.	*Myrciaria cauliflora*	Myrtaceae	Fruits	Reynertson *et al.*, 2008
1603.	*Myrciaria dubia*	Myrtaceae	Fruits	Reynertson *et al.*, 2008
1604.	*Myrciaria vexator*	Myrtaceae	Fruits	Reynertson *et al.*, 2008
1605.	*Myrianthus arboreus*	Urticaceae	Leaf	Odukoya *et al.*, 2007; Biapa *et al.*, 2007
1606.	*Myrica gale*	Myricaceae	Fruit	Mathiesen *et al.*, 1995
1607.	*Myrica rubra*	Myricaceae	Fruit	Bao *et al.*, 2005
1608.	*Myriostachya wightiana*	Poaceae	Aerial parts	Vadlapudi and Naidu, 2009
1609.	*Myristica argentea*	Myristicaceae	Fruit	Calliste *et al.*, 2010
1610.	*Myristica communis*	Myristicaceae	Berries	Yadegarinia *et al.*, 2006; Mantora *et al.*, 2006

Contd...

Table 3.1–*Contd...*

Sl.No.	Plant Name	Family	Parts Used	References
1611.	*Myristica fragrans*	Myristicaceae	Fruits	Dorman and Deans, 2004; Singh *et al.*, 2005; Chatterjee *et al.*, 2006; Olaleye *et al.*, 2006; Mishra *et al.*, 2007; Kwon *et al.*, 2008; Mariutti *et al.*, 2008; Souri *et al.*, 2008; Calliste *et al.*, 2010; Duan *et al.*, 2009; Akinboro *et al.*, 2011
1612.	*Myristica malabarica*	Myristicaceae	Fruits	Patro *et al.*, 2005
1613.	*Myrothamnus flabellifolia*	Myrothamnaceae		Kramer *et al.*, 2002
1614.	*Myrtus communis*	Myrtaceae	Fruits	Demo *et al.*, 1999; Romani *et al.*, 2004; Hayder *et al.*, 2004; Souri *et al.*, 2004; Davod *et al.*, 2006; Montoro *et al.*, 2006; Yadegarinia *et al.*, 2006; Wannes *et al.*, 2010; Mimica-Dukic *et al.*, 2010; Moussa *et al.*, 2011
1615.	*Nardostachys jatamanshi*	Valerianaceae	Rhizomes	Tripathi *et al.*, 1996; Lyle *et al.*, 2009a,b
1616.	*Nasturtium aquatica*	Brssicaceae	Herb	Lindsey *et al.*, 2006
1617.	*Nasturtium officinale*	Brassicaceae	Leaf, stalk	Maeda *et al.*, 2006; Wojcikowski *et al.*, 2007; Yazdanparast *et al.*, 2008; Ozen, 2009
1618.	*Nauclea latifolia*	Rubiaceae	Roots	Lagnika *et al.*, 2011
1619.	*Nelumbo nucifera*	Nelumbonaceae	Stamens, seeds	Jung *et al.*, 2003; Rai *et al.*, 2006; Huang *et al.*, 2010
1620.	*Neolitsea sericea* var. *aurita*	Magnoliaceae	Leaves	Lee *et al.*, 2006
1621.	*Nepeta cataria*	Lamiaceae		Adiguezel *et al.*, 2009
1622.	*Nepeta flavida*	Lamiaceae		Tepe *et al.*, 2007
1623.	*Nepeta foliosa*	Lamiaceae		Giamperi *et al.*, 2009
1624.	*Nepeta glomerulosa* subsp. *glomerulosa*	Lamiaceae		Souri *et al.*, 2004
1625.	*Nepeta ispahanica*	Lamiaceae	Aerial parts	Salehi *et al.*, 2007
1626.	*Nepeta meyeri*	Lamiaceae		Cigremis *et al.*, 2010
1627.	*Nepeta nuda* ssp. *nuda*	Lamiaceae	Leaves,infl-orescence	Gkinis *et al.*, 2010
1628.	*Nephelium lappaceum*	Sapindaceae		Palanisamy *et al.*, 2008; Thitilerdecha *et al.*, 2008, 2010
1629.	*Nephelium longan*	Sapindaceae	Leaf, stem	Ripa *et al.*, 2010
1630.	*Neptunia oleracea*	Mimosaceae		Chanwitheesuk *et al.*, 2005
1631.	*Nerium oleander*	Apocynaceae		Souri *et al.*, 2004

Contd...

Table 3.1–*Contd...*

Sl.No.	Plant Name	Family	Parts Used	References
1632.	*Newbouldia laevis*	Bignoniaceae		Akinmoladun *et al.*, 2010; Oloyede *et al.*, 2010
1633.	*Nicandra physalodes*	Solanaceae		David *et al.*, 2007
1634.	*Nigella sativa*	Ranunculaceae	Seeds	Burits and Bucar, 2000; Erkan *et al.*, 2008; Ramadan *et al.*, 2003; El-Ghorab, 2003; Kanter *et al.*, 2003, 2005; Suboh *et al.*, 2004; Souri *et al.*, 2008; Mehta *et al.*, 2009; O Huang *et al.*, 2010; Okeola *et al.*, 2010
1635.	*Nonea poulla*	Boraginaceae	Herb	Myagmar and Aniya, 2000; Namdeo *et al.*, 2010
1636.	*Nothapodytes nimmoniana*	Icacinaceae	Fruits, bark	Namdeo *et al.*, 2010; Pai *et al.*, 2010
1637.	*Nothopanax scutellarius*	Araliaceae		Andarwulan *et al.*, 2010
1638.	*Notopterygium incisum*	Apiaceae		Song *et al.*, 2010
1639.	*Nyctanthes arbor-tristis*	Oleaceae	Flowers	Nagavani *et al.*, 2010
1640.	*Nymphaea candida*	Nymphaeaceae	Flower	Zhao *et al.*, 2011
1641.	*Nymphaea caerulea*	Nymphaeaceae	Flower	Agnihotri *et al.*, 2008
1642.	*Nymphaea hybrid*	Nymphaeaceae	Flower	Ren *et al.*, 2010
1643.	*Nymphaea lotus*	Nymphaeaceae	Flowers	Saleem *et al.*, 2001
1644.	*Nymphaea nouchali*	Nymphaeaceae	Flower	Nagavani and Raghaava Rao, 2010
1645.	*Nymphaea stellata*	Nymphaeaceae	Flower	Rakesh *et al.*, 2010
1646.	*Ocimum americanum*	Lamiaceae		Chanwitheesuk *et al.*, 2005
1647.	*Ocimum basilicum*	Lamiaceae	Leaves, stems	Juliani and Simin, 2002; Javanmardi *et al.*, 2003; Souri *et al.*, 2004; Dasgupta *et al.*, 2004; Hinneburg *et al.*, 2005; Chanwitheesuk *et al.*, 2005; Abbas *et al.*, 2006; Mariassyova 2006; Wong *et al.*, 2006; Maeda *et al.*, 2006; Gülçin *et al.*, 2007; Wojcikowski *et al.*, 2007; Mariutti *et al.*, 2008; Zakaria *et al.*, 2008; Souri *et al.*, 2008; Meera *et al.*, 2009; Cioroi and Dumitriu, 2009; Wei *et al.*, 2010; Wei and Shibamoto, 2010; Narayanaswamy and Balakrishnan, 2011
1648.	*Ocimum citriodorum*	Lamiaceae	Herb	Juliani and Simon, 2002
1649.	*Ocimum gratissimum*	Lamiaceae	Herb, leaf	Oboh and Akindahunsi, 2004; Chanwitheesuk *et al.*, 2005; Odukoya *et al.*, 2005; Afolabi *et al.*, 2007; Akinmoladun *et al.*, 2010; Adefegha and Oboh, 2011; Salawu *et al.*, 2011

Contd...

Table 3.1–*Contd...*

Sl.No.	*Plant Name*	*Family*	*Parts Used*	*References*
1650.	*Ocimum micranthum*	Lamiaceae	Herb	Sacchetti *et al.*, 2004
1651.	*Ocimum minimum*	Lamiaceae		Telci *et al.*, 2009
1652.	*Ocimum sanctum*	Lamiaceae	Leaves	Devi and Ganasoundari, 1999; Kelm *et al.*, 2000; Umadevi, 2001; Juliani and Simon, 2002; Geetha *et al.*, 2003; Chandra *et al.*, 2003; Sethi *et al.*, 2004; Chanwitheesuk *et al.*, 2005; Juntachote and Berghofer, 2005; Arya *et al.*, 2006; Samson *et al.*, 2007; Kath and Gupta, 2007; Lukmanul Hakkim *et al.*, 2007; Leelapornpisid *et al.*, 2007; Mishra *et al.*, 2007; Halder *et al.*, 2009; Veeru *et al.*, 2009; Suanarunsawat *et al.*, 2010
1653.	*Oenanthe javanica*	Apiaceae.	Leaves	Zaibunnisa *et al.*, 2008; Huda-Faujan *et al.*, 2007, 2009
1654.	*Oenanthe stolonifera*	Apiaceae	Leavees	Chanwitheesuk *et al.*, 2005
1655.	*Oenothera biennis*	Onagraceae	Fruits, seeds	Knorr *et al.*, 2004; Budinevi *et al.*, 1995; Borchardt *et al.*, 2008
1656.	*Oenothera depressa*	Onagraceae		Budinevi *et al.*, 1995
1657.	*Oenothera lamarckiana*	Onagraceae	Herb	Knorr *et al.*, 2004
1658.	*Oenothera multicaulis*	Onagraceae		Lock *et al.*, 2005
1659.	*Oldenlandia diffusa*	Rubiaceae		Li *et al.*, 2008
1660.	*Olea europaea*	Oleaceae	Leaves	Owen *et al.*, 2000; McDonald *et al.*, 2001; Briante *et al.*, 2002; Somova *et al.*, 2003; Amro *et al.*, 2003; Atanassova *et al.*, 2005; Altarejos *et al.*, 2005; Efmorfopulou and Rodis, 2005; Aldini *et al.*, 2006; Marwah *et al.*, 2007; Mishra *et al.*, 2007; Wojcikowski *et al.*, 2007; Bouziz *et al.*, 2010; Haloui *et al.*, 2010
1661.	*Ononis spinosa*	Fabaceae	Herb	Souri *et al.*, 2004; Valyova *et al.*, 2009
1662.	*Onopordum acanthium*	Asteraceae	Herb	Kiselova *et al.*, 2006
1663.	*Onosma argenteum*	Boraginaceae	Roots	Ozgen *et al.*, 2004
1664.	*Ophiorrhiza discolor*	Rubiaceae		Ahmad *et al.*, 2010
1665.	*Ophiorrhiza major*	Rubiaceae		Ahmad *et al.*, 2010
1666.	*Ophryosporus heptanthus*	Asteraceae		Rosas-Romero and Saavedra, 2005; Barrero *et al.*, 2006
1667.	*Opuntia dillenii*	Cactaceae	Fruit	Chang Su-Feng *et al.*, 2008
1668.	*Opuntis ficus-indica*	Cactaceae	Fruit	Butera *et al.*, 2002; Bisson *et al.*, 2009

Contd...

Table 3.1–*Contd...*

Sl.No.	*Plant Name*	*Family*	*Parts Used*	*References*
1669.	*Opuntia polyacantha*	Cactaceae.	Fruits	Acuna *et al.*, 2002
1670.	*Opuntia vulgaris*	Cactaceae		Pal and Mitra, 2010
1671.	*Orbignya speciosa*	Arecaceae	Endocarp, mesocarp	Silva *et al.*, 2005
1672.	*Orchis mascula*	Orchidaceae		Kalim *et al.*, 2010
1673.	*Oriciopsis glaberrima*	Rutaceae	Stem bark	Wansi *et al.*, 2006
1674.	*Origanum acutidens*	Lamiaceae	Herb	Sokmen *et al.*, 2004; Alviano and Alviano, 2009; Goze *et al.*, 2010
1675.	*Origanum dictamnus*	Lamiaceae	Aerial parts	Couladis *et al.*, 2003
1676.	*Origanum ehrenbergii*	Lamiaceae	Herb	Loizzo *et al.*, 2009
1677.	*Origanum floribundum*	Lamiaceae	Essential oil	Hazzit *et a!.*, 2011
1678.	*Origanum glandulosum*	Lamiaceae	Essential oil	Hazzit *et al.*, 2011
1679.	*Origanum heracleoticum*	Lamiaceae	Leaves	Pieroni *et al.*, 2002
1680.	*Origanum majorana*	Lamiaceae	Aerial parts	Jun *et al.*, 2001; Vagi *et al.*, 2005; Nejla and Moncef, 2006; Mangathayaru *et al.*, 2007; Mossa and Nawwar, 2011
1681.	*Origanum minutiflorum*	Lamiaceae	Aerial pars	Kosar *et al.*, 2003; Dorman *et al.*, 2004
1682.	*Origanum onites*	Lamiaceae	Essential oil	Lagouri *et al.*, 1993; Kosar *et al.*, 2003; Dorman *et al.*, 2004; Kunduraci *et al.*, 2010; Ozkan *et al.*, 2010
1683.	*Origanum syriacum*	Lamiaceae	Essential oil	Alma *et al.*, 2003; Kosar *et al.*, 2003; Dorman *et al.*, 2004; Tepe *et al.*, 2004; Loizzo *et al.*, 2009
1684.	*Origanum vulgare*	Lamiaceae	Leaves	Laghouri *et al.*, 1993; Damien *et al.*, 1995; Kosar *et al.*, 2003; Dorman and Deans, 2004; Sahin *et al.*, 2004; Ivanova *et al.*, 2005; Mishra *et al.*, 2007; Mariutti, *et al.*, 2008; Cioroi and Dumitriu, 2009; Viuda-Martos *et al.*, 2010; Motamed and Naghibi, 2010; Mechergui *et al.*, 2010; Carlsen *et al.*, 2010; Spiridon *et al.*, 2011; Khanum *et al.*, 2011
1685.	*Ornithogalum alpigenum*	Liliaceae	Bulb	Makasci *et al.*, 2010
1686.	*Ornithogalum sintenisii*	Liliaceae	Bulb, aerial parts	Ebrahimzadeh *et al.*, 2010
1687.	*Ornithogalum umbellatum*	Liliaceae	Stem	Özen, 2010
1688.	*Orobanche caerulescens*	Orobanchaceae	Whole plant	Lin *et al.*, 2004
1689.	*Orostachys japonicus*	Crassulaceae	Bulb, leaf	Lee *et al.*, 2009

Contd...

Table 3.1–*Contd...*

Sl.No.	Plant Name	Family	Parts Used	References
1690.	*Oroxylum indicum*	Bignoniaceae		Song *et al.*, 2010
1691.	*Orthosiphon aristatus*	Lamiaceae	Herb	Hsu *et al.*, 2010
1692.	*Orthosiphon grandiflorus*	Lamiaceae		Chanwitheeesuk *et al.*, 2005
1693.	*Orthosiphon stamineus*	Lamiaceae	Stems, leaves	Akowuah *et al.*, 2003; Zakaria *et al.*, 2008; Ho *et al.*, 2010
1694.	*Oryza sativa*	Poaceae	Fruit	Moon *et al.*, 2002; Han *et al.*, 2004; Bajpai *et al.*, 2005; Rao and Muralikrishna, 2006; Chatchawan *et al.*, 2008; Sreeramulu *et al.*, 2009; Qiu *et al.*, 2010
1695.	*Osbeckia aspera*	Melastomataceae		Thabrew *et al.*, 1998
1696.	*Osmanthus fragrans*	Oleaceae	Pulp	Wu *et al.*, 2009; Wang *et al.*, 2010
1697.	*Osyris lanceolata*	Santalaceae	Root bark	Yeboah and Majinda, 2009
1698.	*Otostegia limbata*	Lamiaceae	Leaves	Ahmed *et al.*, 2010
1699.	*Otostegia michauxii*	Lamiaceae	Aerial parts	Firuzi *et al.*, 2010
1700.	*Otostegia persica*	Lamiaceae		Sherififar *et al.*, 2003; Yassa *et al.*, 2005
1701.	*Oxalis corniculata*	Oxalidaceae	Whole plant	Kathiriya *et al.*, 2010; Senthilkumar and Rajkapoor, 2010; Alam *et al.*, 2011
1702.	*Oxalis tuberosa*	Oxalidaceae	Tuber	Penarrieeta *et al.*, 2005; Campos *et al.*, 2006; Salluca *et al.*, 2008
1703.	*Oxystelma esculentum*	Asclepiadaceae		Ashokkumar *et al.*, 2008, 2009
1704.	*Oxytropis falcata*	Fabaceae		Jiang *et al.*, 2008
1705.	*Paederia foetida*	Rubiaceae	Herb	Osman *et al.*, 2009
1706.	*Paeonia emodi*	Paeoniaceae	Fruits	Riaz *et al.*, 2004
1707.	*Paeonia lactiflora*	Paeoniaceae		Cai *et al.*, 2004; Li *et al.*, 2008
1708.	*Paeonia suffruticosa*	Paeoniaceae		Kim *et al.*, 1997; Lee *et al.*, 2003; Cai *et al.*, 2004; Li *et al.*, 2008
1709.	*Paepalanthus brome-lioides*	Eriocaulaceae		Stasi *et al.*, 2004; Devienne *et al.*, 2009
1710.	*Pajanelia longifolia*	Bignoniaceae	Bark	Choudhury, 2010
1711.	*Palicourea acetosoides*	Rubiaceae		Mosquera *et al.*, 2009
1712.	*Palicourea angustifolia*	Rubiaceae		Mosquera *et al.*, 2009
1713.	*Palisota hirsuta*	Commelinaceae	Roots, leaves	Mamyrbekova-Bekoet *al.*, 2008
1714.	*Paliurus spina-christi*	Rhamnaceae		Souri *et al.*, 2004; Kirca and Arslan, 2008

Contd...

Table 3.1–*Contd...*

Sl.No.	Plant Name	Family	Parts Used	References
1715.	*Panax ginseng*	Araliaceae	Root	Masteikova *et al.*, 2007; Wojcikowski *et al.*, 2007; Kang *et al.*, 2007; Naval *et al.*, 2007; Mahour and Saxena, 2010
1716.	*Panax notoginseng*	Araliaceae	Root	Ng *et al.*, 2004; Wu and Wang, 2008; Gan *et al.*, 2010
1717.	*Panax quinquefolius*	Araliaceae	Root	Ng *et al.*, 2004; Kitts *et al.*, 2010
1718.	*Pandanus amaryllifolius*	Pandanaceae	Leaves	Nor *et al.*, 2008
1719.	*Pandanus odoratissimus*	Pandanaceae	Root	Sasikumar *et al.*, 2009; Panigrahi *et al.*, 2011
1720.	*Pandanus sanderi*	Pandanaceae	Leaves	Rahman *et al.*, 2010
1721.	*Papaver bracteatum*	Papaveraceae		Souri *et al.*, 2004
1722.	*Papaver somniferum*	Papaveraceae	Seed	Wojcikowski *et al.*, 2007; Souri *et al.*, 2008
1723.	*Parabarium huaitingii*	Apocynaceae	Plant	Tang *et al.*, 2009
1724.	*Parinari curatelifolia*	Chrysobalanaceae		Nhukarume *et al.*, 2010
1725.	*Parinari macrophylla*	Chrysobalanaceae		Cook *et al.*, 1998
1726.	*Parkia globosa*	Mimosaceae		Lamien-Meda *et al.*, 2008
1727.	*Parkia speciosa*	Mimosaceae		Wong *et al.*, 2006; Panpipat *et al.*, 2010
1728.	*Parthenocissus tricuspidata*	Vitaceae	Stem	Kim *et al.*, 2005
1729.	*Passiflora alata*	Passifloraceae	Fruit, leaves	Doyama *et al.*, 2005; Rudnicki *et al.*, 2007a,b
1730.	*Passiflora edulis*	Passifloraceae	Peel, leaves	Chen, C-H. *et al.*, 2009; Rudnicki *et al.*, 2007a; Narayanaswamy and Balakrishnan, 2011
1731.	*Passiflora liguralis*	Passifloraceae	Fruits	Tommonaro *et al.*, 2007
1732.	*Pastinara sativa*	Apiaceae		Budinevi *et al.*, 1995
1733.	*Patrinia scabiosaefolia*	Valerianaceae	Aerial parts	Cai *et al.*, 2004
1734.	*Patrinia villosa*	Valerianaceae		Li *et al.*, 2008
1735.	*Paullinia cupana*	Sapindaceae	Seeds	Basile *et al.*, 2005; Majhenic *et al.*, 2007
1736.	*Paullinia pinnata*	Sapindaceae	Leaves	Jimoh *et al.*, 2007
1737.	*Paulownia tomentosa*	Scrophulariaceae	Leaves	Smejkal *et al.*, 2007
1738.	*Pedilanthus tithymaloides*	Euphorbiaceae		Abreu *et al.*, 2006, 2008
1739.	*Peganum harmala*	Zygophyllaceae		Souri *et al.*, 2004
1740.	*Pelargonium endlicherianum*	Geraniaceae		Tepe *et al.*, 2006

Contd...

Table 3.1–*Contd...*

Sl.No.	Plant Name	Family	Parts Used	References
1741.	*Pelargonium graveolens*	Geraniaceae	Seeds	Sun *et al.*, 2005
1742.	*Pelargonium odoratissimum*	Geraniaceae		Moussa *et al.*, 2011
1743.	*Pelargonium reniforme*	Geraniaceae		Latte and Kolodziej, 2004
1744.	*Pelargonium zonale*	Geraniaceae	Leaves	Demo *et al.*, 1999
1745.	*Peltophorum africanum*	Mimosaceae		Bizimenyera *et al.*, 2006
1746.	*Peltophorum ferrugineum*	Caesalpiniaceae		Chanda *et al.*, 2011
1747.	*Pennisetum americanum* (Syn.: *Pennisetum typhoideum, Pennisetum glaucum*)	Poaceae	Fruit	Bajpai *et al.*, 2005; Sreeramulu *et al.*, 2009
1748.	*Penstemon pallidus*	Scrophulariaceae	Seeds	Borchardt *et al.*, 2008
1749.	*Pentacalia americana*	Asteraceae		Mosquera *et al.*, 2007
1750.	*Pera glabrata*	Euphorbiaceae		Santos *et al.*, 2009
1751.	*Pereskia grandifolia*	Cactaceae	Leaves	Adib *et al.*, 2010; Sim *et al.*, 2010
1752.	*Pergularia daemia*	Aclepiadaceae		Mirunalini and Karthishwaran, 2010
1753.	*Perilla frutescens*	Lamiaceae	Herb,seed	Masahiro *et al.*, 1996; Nakamura *et al.*, 1998; Jung *et al.*, 2001; Osakabe *et al.*, 2002; Wojcikowski *et al.*, 2007; Zekonis *et al.*, 2008; Meng *et al.*, 2009; Lin *et al.*, 2010; Song *et al.*, 2010
1754.	*Perilla ocymoides*	Lamiaceae	Leaf	Maeda *et al.*, 2006
1755.	*Perilla pankinensis*	Lamiaceae		Gülçin *et al.*, 2005
1756.	*Periploca laevigata*	Asclepiadaceae	Root bark	Hajji *et al.*, 2010
1757.	*Persea americana*	Lauraceae	Fruit, Leaves	Chen, C-H. *et al.*, 2009; Wang *et al.*, 2010
1758.	*Persea gamblei*	Lauraceae	Essential oil	Joshi *et al.*, 2010
1759.	*Persicaria tenella*	Polygonaceae		Abbas *et al.*, 2006
1760.	*Petasites formosanus*	Asteraceae	Leaves	Lin *et al.*, 2004
1761.	*Petrorhagia velutina*	Caryophyllaceae	Aerial parts	D'Abrosca *et al.*, 2010
1762.	*Petroselenium crispum*	Apiaceae	Leaf	Fejes *et al.*, 2000; Hinneburg *et al.*, 2005; Maeda *et al.*, 2006; Zhang *et al.*, 2006; Ozsoy-Sacan *et al.*, 2006; Wojcikowski *et al.*, 2007; Mariutti *et al.*, 2008
1763.	*Peucedanum japonicum*	Apiceae	Leaf	Maeda *et al.*, 2006; Thuong *et al.*, 2010
1764.	*Peucedanum longifolium*	Apiaceae	-	Tepe *et al.*, 2011
1765.	*Peucedanum palimbioides*	Apiaceae	-	Tepe *et al.*, 2011

Contd...

Table 3.1–*Contd...*

Sl.No.	Plant Name	Family	Parts Used	References
1766.	*Peucedanum praeruptorum*	Apiaceae		Song *et al.*, 2010
1767.	*Peumus boldus*	Monimiaceae	Leaves, bark	Speisky *et al.*, 1991; Krigstein and Cederbaum, 1995; Bannach *et al.*, 1996; Jimenez *et al.*, 2000; Schmeda-Hirschmann *et al.*, 2003; del Valle *et al.*, 2004
1768.	*Pfaffia glomerata*	Amaranthaceae	Roots	Daniel *et al.*, 2005
1769.	*Phalaria macrocarpa*	Thymeleaceae		Endrini, 2010
1770.	*Phaseolus aconitifolius*	Fabaceae	Seeds	Bajpai *et al.*, 2005
1771.	*Phaseolus aureus*	Fabaceae	Seeds	Lin *et al.*, 2001; Sreeramulu *et al.*, 2009
1772.	*Phaseolus calcaratus*	Fabaceae	Seeds	Lin *et al.*, 2001
1773.	*Phaseolus mungo*	Fabaceae	Seeds	Bajpai *et al.*, 2005; Sreeramulu *et al.*, 2009
1774.	*Phaseolus radiatus*	Fabaceae	Seeds	Lin *et al.*, 2001; Bajpai *et al.*, 2005;
1775.	*Phaseolus trilobus*	Fabaceae	Seeds	Fursule and Patil, 2010
1776.	*Phaseolus vulgaris*	Fabaceae	Seeds	Cardador-Martinez *et al.*, 2002; Jiratanam and Liu, 2004; Bajpai *et al.*, 2005; Nuria *et al.*, 2007; Granito *et al.*, 2008; Rocha *et al.*, 2007; Sreeramulu *et al.*, 2009; Xu and Chang, 2008, 2009; Oboh *et al.*, 2009
1777.	*Phaseolus vulgaris* var. *humulus*	Fabaceae	Pod	Maeda *et al.*, 2006
1778.	*Phaulopsis fascisepala*	Acanthaceae		Adesagun *et al.*, 2007
1779.	*Phellodendron amurense*	Rutaceae	Bark	Wu *et al.*, 2003; Cai *et al.*, 2004; Li *et al.*, 2008
1780.	*Phellodendron japonicum*	Rutaceae	Leaves	Chiu *et al.*, 2005
1781.	*Phillyrea latifolia*	Oleaceae	Leaves	Demo *et al.*, 1999
1782.	*Philodendron bipinnatifidum*	Araceae		Menezes *et al.*, 2004
1783.	*Philodendron solimoesense*	Araceae		Lock *et al.*, 2005
1784.	*Phlogocanthus tufiflorus*	Acanthaceae		Ramtake *et al.*, 2007
1785.	*Phlomis anisodonta*	Lamiaceae		Souri *et al.*, 2004
1786.	*Phlomis bruguieri*	Lamiaceae	Aerial parts	Firuzi *et al.*, 2010
1787.	*Phlomis elliptica*	Lamiaceae	Aerial parts	Firuzi *et al.*, 2010
1788.	*Phlomis lanata*	Lamiaceae	Aerial parts	Couladis *et al.*, 2003
1789.	*Phlomis olivieri*	Lamiaceae	Aerial parts	Firuzi *et al.*, 2010
1790.	*Phlomis persica*	Lamiaceae	Aerial parts	Firuzi *et al.*, 2010

Contd...

Table 3.1–*Contd...*

Sl.No.	Plant Name	Family	Parts Used	References
1791.	*Phodococcum vitis-idaea*	Vacciniaceae	Herb	Myagmar and Aniya, 2000
1792.	*Phoenix dactylifera*	Arecaceae	Fruits	Tariq and Al-Yahya, 1983; Vayalil 2002; Biglari *et al.*, 2008, 2009
1793.	*Phoradendron falcifrons*	Viscaceae		Menezes *et al.*, 2004
1794.	*Phoradendron piperoides*	Viscaceae		Vasconcellos *et al.*, 2009
1795.	*Phragmites australis*	Poaceae	Leaf, stalk	Kim *et al.*, 1997; Kahkonen *et al.*, 1999; Li *et al.*, 2008
1796.	*Phyllanthus acidus*	Euphorbiaceae	Fruit	Habib *et al.*, 2011; Usha Nagalakshmi, 2011
1797.	*Phyllanthus amarus*	Euphorbiaceae		Kumar and Kuttan, 2005; Kumaran and Karunakaran, 2007; Naaz *et al.*, 2007; Londhe *et al.*, 2008; Patnibul *et al.*, 2008
1798.	*Phyllanthus debilis*	Euphorbiaceae		Kumaran and Karunakaran, 2007
1799.	*Phyllanthus emblica* (=*Emblica officinalis*)	Euphorbiaceae	Fruits, leaves	Jose and Kuttan, 1995; Ghosal *et al.*, 1996; Vani *et al.*, 1997; Arunabh *et al.*, 1999; Bhattyacharya *et al.*, 1999, 2000; Scartezzini and Speroni, 2000; Bandhyopadhyay *et al.*, 2000; Khopde *et al.*, 2001; Sai Ram *et al.*, 2003; Jayajothi *et al.*, 2004; Naik *et al.*, 2005; Bajpai *et al.*, 2005; Scartezzini *et al.*, 2006; Om Prakash *et al.*, 2007; Pozharitskaya *et al.*, 2007a; Liu *et al.*, 2008, 2009; Shivaprasad *et al.*, 2008; Shukla *et al.*, 2009; Poltanov *et al.*, 2009; Jain and Agrawal, 2009; Povichit *et al.*, 2010; Niwano *et al.*, 2011
1800.	*Phyllanthus maderaspatensis*	Euphorbiaceae	Whole plant	Asha *et al.*, 2004
1801.	*Phyllanthus niruri*	Euphorbiaceae	Fruits	Mazumder *et al.*, 2005; Harish and Shivanandappa, 2006; Chatterjee *et al.*, 2006; Kumaran and Karunakaran, 2007; Mosquera *et al.*, 2009
1802.	*Phyllanthus polyphyllus*	Euphorbiaceae		BR *et al.*, 2008
1803.	*Phyllanthus reticulatus*	Euphorbiaceae	Herb	Ram *et al.*, 2008a,b; Maruthappana and Shreeb, 2010
1804.	*Phyllanthus sellowianus*	Euphorbiaceae	Herb	Hnatgyszyn *et al.*, 1996
1805.	*Phyllanthus urinaria*	Euphorbiaceae		Kumaran and Karunakaran, 2007; Fang *et al.*, 2008; Chen, C-H. *et al.*, 2009

Contd...

Table 3.1–*Contd...*

Sl.No.	Plant Name	Family	Parts Used	References
1806.	*Phyllanthus virgatus*	Euphorbiaceae		Kumaran and Karunakaran, 2007
1807.	*Phyllostachys bambusoides*	Poaceeae		Jun *et al.*, 2004
1808.	*Phyllostachys edulis*	Poaceae	Stem	Kweon *et al.*, 2001
1809.	*Phyllostachys nigra*	Poaceae	Leaves	Jung *et al.*, 2007
1810.	*Phyllostachys pubescens*	Poaceae	Culms	Jun *et al.*, 2004
1811.	*Physalis alkakengi*	Solanaceae		Song *et al.*, 2010
1812.	*Physalis angulata*	Solanaceae	Flower	Choi and Hwang, 2005
1813.	*Physalis peruviana*	Solanaceae	Leaves	Wu *et al.*, 2005
1814.	*Physospermum verticillatum*	Apiaceae		Tundis *et al.*, 2009
1815.	*Phytolacca americana*	Phytolaccaceae	Fruit	Souri *et al.*, 2004; Hosseinimehr *et al.*, 2007
1816.	*Phytolacca rivinoides*	Phytolaccaceae		De las Heras *et al.*, 1998
1817.	*Picrasma crenata*	Simaroubaceae	Roots	Menezes *et al.*, 2004
1818.	*Picrorrhiza kurroa*	Scrophulariaceae	Rhizomes, roots	Anandan and Devaki, 1999; Govindarajan *et al.*, 2003; Rajkumar *et al.*, 2011
1819.	*Picrorrhiza scrophulariiflora*	Scrophulariaceae	Rhizome	Cai *et al.*, 2004, 2006; Wang *et al.*, 2004; Li *et al.*, 2008
1820.	*Pilea melastemoides*	Urticaceae		Andarwulan *et al.*, 2010
1821.	*Pilea microphylla*	Urticaceae	Herb	Prabhakar *et al.*, 2007
1822.	*Piliostigma reticulatum*	Leguminosae		Aderogba *et al.*, 2004, 2005; Zerbo *et al.*, 2010
1823.	*Piliostigma thonningii*	Leguminosae		Aderogbaa *et al.*, 2004; Ajiboye, 2011
1824.	*Pimenta dioica*	Myrtaceae		Kikuzaki *et al.*, 1999; Ramos *et al.*, 2003; Mariutti, *et al.*, 2008; Miyajima *et al.*, 2008; Padmakumari *et al.*, 2011
1825.	*Pimpinella anisum*	Apiaceae	Fruit	Kunzemann and Herrmann, 1977; Gülçin *et al.*, 2003; Souri *et al.*, 2008; Hinneburg *et al.*, 2005; Nickavar and Abolhasani, 2009
1826.	*Pimpinella aurea*	Apiaceae	Aerial parts	Delazar *et al.*, 2006
1827.	*Pimpinella saxifraga*	Apiaceae	Budinevi	*et al.*, 1995
1828.	*Pinellia ternatea*	Araceae		Song *et al.*, 2010
1829.	*Piper betel*	Piperaceae	Leaves	Patel, 1979; Choudhary and Kale, 2002; Dasgupta and De, 2004; Wong *et al.*, 2006; Rathhe *et al.*, 2006; Manigauha *et al.*, 2009; Islam *et al.*, 2010; Chakraborty and Shah, 2011

Contd...

Table 3.1–*Contd...*

Sl.No.	*Plant Name*	*Family*	*Parts Used*	*References*
1830.	*Piper crassinervium*	Piperaceae	Fruits	Sacchetti *et al.*, 2005
1831.	*Piper cubeba*	Piperaceae	Fruits	Choi and Hwang, 2005; Khalaf *et al.*, 2008
1832.	*Piper fulvescens*	Piperaceae		Velazquez *et al.*, 2003
1833.	*Piper guineense*	Piperaceae	Seed	Odukoya *et al.*, 2005
1834.	*Piper kadsura*	Piperaceae		Gan *et al.*, 2010
1835.	*Piper leucophyllum*	Piperaceae		Ruiz-Terán *et al.*, 2008
1836.	*Piper longum*	Piperaceae	Fruit	Krishnamoorthy *et al.*, 2003; Bajpai *et al.*, 2005; Veeru *et al.*, 2009
1837.	*Piper nigrum*	Piperaceae	Fruits	Selvendian *et al.*, 2004; Vijayakumar *et al.*, 2004; Kaleem *et al.*, 2005; Gülçin, 2005; Suhaj *et al.*, 2006: Chatterjee *et al.*, 2007; Singh *et al.*, 2007, 2008; Khalaf *et al.*, 2008; Mariutti *et al.*, 2008
1838.	*Piper retrofractum*	Piperaceae	Leaf	Chanwitheesuk *et al.*, 2005; Maeda *et al.*, 2006; Sacchetti *et al.*, 2005
1839.	*Piper sarmentosum*	Piperaceae		Chanwitheesuk *et al.*, 2005; Hussain *et al.*, 2009, 2010; Panpipat *et al.*, 2010
1840.	*Piper umbellatum*	Piperaceae		Agbor *et al.*, 2005
1841.	*Pipturus arborescens*	Urticaceae		Peteros and Uy, 2010
1842.	*Pisonia alba*	Nyctaginaceae	Leaf	Subhasree *et al.*, 2009
1843.	*Pistacia atlantica*	Anacardiaceae		Souri *et al.*, 2004; Yousfi *et al.*, 2009; Gourine *et al.*, 2010a,b
1844.	*Pistacia lentiscus*	Anacardiaceae	Aerial parts	Barra *et al.*, 2007
1845.	*Pistasia terebinthus*	Anacardiaceae	Leaves	Demo *et al.*, 1999; Topcu *et al.*, 2007; Serteser *et al.*, 2009
1846.	*Pistacia weinmannifolia*	Anacardiaceae	Leaves	Zhao *et al.*, 2005
1847.	*Pisum sativum*	Fabaceae	Seeds	Kahkonen *et al.*, 1999; Torszynska *et al.*, 2002; Torszynska and Ciska, 2002; Han *et al.*, 2008; Agnieszka and Bozena, 2002; Nilsson *et al.*, 2004; Bajpai *et al.*, 2005; Sreeramulu *et al.*, 2009; Pownall *et al.*, 2010
1848.	*Pithecellobium dulce*	Mimosaceae	Fruit	Megala and Geetha, 2010.
1849.	*Pittosporum mannii*	Pittosporaceae		Momeni *et al.*, 2010
1850.	*Pittosporum moluccanum*	Pittosporaceae	Leaf	Lee *et al.*, 2006
1851.	*Pittosporum tobira*	Pittosporaceae	Leaves	Chen, C-H. *et al.*, 2009
1852.	*Plantago* sp.	Plantaginaceae		Galvez *et al.*, 2005a

Contd...

Table 3.1–*Contd...*

Sl.No.	*Plant Name*	*Family*	*Parts Used*	*References*
1853.	*Plantago afra*	Plantaginaceae		Galvez *et al.*, 2005a; Khalil *et al.*, 2007
1854.	*Plantago argentea*	Plantaginaceae		Beara *et al.*, 2009
1855.	*Plantago asiatica*	Plantaginaceae	Seed	Toda *et al.*, 1985a; Cai *et al.*, 2004; Gan *et al.*, 2010
1856.	*Plantago bellardii*	Plantaginaceae	Aerial parts	Galvez *et al.*, 2005b
1857.	*Plantago coronopus*	Plantaginaceae		Galvez *et al.*, 2005a
1858.	*Plantago holosteum*	Plantaginaceae		Beara *et al.*, 2009
1859.	*Plantago lagopus*	Plantaginaceae		Galvez *et al.*, 2005a
1860.	*Plantago lanceolata*	Plantaginaceae	Flowers	Galvez *et al.*, 2005a; Kardosova *et al.*, 2006
1861.	*Plantago major*	Plantaginaceae	Seed	Pourmorad *et al.*, 2006; Souri *et al.*, 2008; Beara *et al.*, 2009; Gan *et al.*, 2010
1862.	*Plantago maritima*	Plantaginaceae		Beara *et al.*, 2009
1863.	*Plantago media*	Plantaginaceae		Beara *et al.*, 2009
1864.	*Plantago ovata*	Plantaginaceae	Seed	Souri *et al.*, 2008
1865.	*Plantago serraria*	Plantaginaceae		Galvez *et al.*, 2005a
1866.	*Platostoma africanum*	Lamiaceae		Aladedunye *et al.*, 2008
1867.	*Platycodon grandiflorum*	Campanulaceae		Lee *et al.*, 2004; Jeong *et al.*, 2010; Song *et al.*, 2010
1868.	*Plazia daphnoides*	Asteraceae		Rosas-Romero and Saavedra, 2005
1869.	*Plectranthus amboinicus*	Lamiaceae	Herb	Kumaran and Karunakaran, 2005; Palani *et al.*, 2010; Anasuya and Gomathi, 2010; Khanum *et al.*, 2011
1870.	*Plectranthus barbatus*	Lamiaceae	Leaves	Maioli *et al.*, 2010
1871.	*Plectranthus hadiensis*	Lamiaceae		Mothana *et al.*, 2010
1872.	*Pluchea arabica*	Asteraceae	Whole plant	Marwah *et al.*, 2007
1873.	*Pluchea carolinensis*	Asteraceae	Leaves	Fernandez and Torres, 2006
1874.	*Pluchea indica*	Asteraceae		Patnibul *et al.*, 2008; Andarwulan *et al.*, 2010
1875.	*Plumbago zeylanica*	Plumbaginaceae	Root	Seo *et al.*, 2001; Tilak *et al.*, 2004; Natarajan *et al.*, 2006; Zahin *et al.*, 2009; Rajarajan *et al.*, 2010
1876.	*Plumeria alba*	Apocynaceae		Leelapornpisid *et al.*, 2007
1877.	*Plumeria rubra*	Apocynaceae		Ruiz-Terán *et al.*, 2008
1878.	*Podophyllum hexandrum*	Berberidaceae	Bark	Arora *et al.*, 2004, 2005; Chawla *et al.*, 2005; Raman *et al.*, 2005; Sagar *et al.*, 2006

Contd...

Table 3.1–*Contd...*

Sl.No.	*Plant Name*	*Family*	*Parts Used*	*References*
1879.	*Poincianella pyramidalis*	Fabaceae		De Melo *et al.*, 2010
1880.	*Poliomintha longifolia*	Lamiaceae		Ozkan, 2009
1881.	*Polycarpaea corymbosa*	Caryophyllaceae	Herb	Singh *et al.*, 2009
1882.	*Polygala alpestris*	Polygalaceae	Herb	Cervellati *et al.*, 2004
1883.	*Polygala chinensis*	Polygalaceae	Herb	Gurav *et al.*, 2000
1884.	*Polygala hongkogensis*	Polygalaceae	Herb	Wu *et al.*, 2008
1885.	*Polygala senega*	Polygalaceae		Amarowicz *et al.*, 2004
1886.	*Polygala tenuifolia*	Polygalaceae	Root	Cai *et al.*, 2004; Liu *et al.*, 2010
1887.	*Polygonatum multiflorum*	Liliaceae	Aerial parts	Chen *et al.*, 1999; Sengul *et al.*, 2009
1888.	*Polygonum amphibium*	Polygonaceae	Leaf	Özen, 2010
1889.	*Polygonum aviculare*	Polygonaceae	Whole plant	Cai *et al.*, 2004; Hsu, 2006; Katalinic *et al.*, 2006; Coruh *et al.*, 2007; Gan *et al.*, 2010
1890.	*Polygonum bistorta*	Polygonaceae	Roots	Demiray *et al.*, 2009
1891.	*Polygonum cuspidatum*	Polygonaceae	Rhizome	Cai *et al.*, 2004, 2006; Pan *et al.*, 2007; Gan *et al.*, 2010
1892.	*Polygonum hydropiper*	Polygonaceae	Herb	Peng *et al.*, 2003; Wong *et al.*, 2006
1893.	*Polygonum hyrcanicum*	Polygonaceae		Hosseinimehr *et al.*, 2007
1894.	*Polygonum minus*	Polygonaceae	Leaves	Huda-Faujan *et al.*, 2007, 2009; Zaibunnisa *et al.*, 2008
1895.	*Polygonum multiflorum*	Polygonaceae	Leaves, Root	Ip *et al.*, 1997; Chen *et al.*, 1997, 1999; Ryu *et al.*, 2002; Cai *et al.*, 2004, 2006; Wong *et al.*, 2006; Lv *et al.*, 2007; Luo *et al.*, 2011
1896.	*Polygonum orientale*	Polygonaceae	Fruit	Cai *et al.*, 2004; Gan *et al.*, 2010
1897.	*Polygonum paleaceum*	Polygonaceae	Rhizomes	Wang *et al.*, 2005
1898.	*Polyscias fruticosa*	Araliaceae		Chanwitheesuk *et al.*, 2005
1899.	*Polyscias pinnata*	Araliaceae		Andarwulan *et al.*, 2010
1900.	*Pongamia pinnata*	Fabaceae	Leaves	Punitha and Manoharan, 2006; Tenpe *et al.*, 2008; Ghosh *et al.*, 2009; Porwal, 2010; Gupta and Sharma, 2011
1901.	*Populus davidiana*	Salicaceae	Stem bark	Zhang *et al.*, 2006
1902.	*Populus nigra*	Salicaceae		Nadjet *et al.*, 2008; Mainar *et al.*, 2008; Dudonne *et al.*, 2009, 2011
1903.	*Populus tremula*	Salicaceae	Leaf, bark	Myer *et al.*, 1995; Kahkonen *et al.*, 1999
1904.	*Populus tremuloides*	Salicaceae	Bark	Diouf *et al.*, 2009
1905.	*Poulus ussuriensis*	Salicaceae	Bark	Si *et al.*, 2011

Contd...

Table 3.1–*Contd...*

Sl.No.	Plant Name	Family	Parts Used	References
1906.	*Portulaca grandiflora*	Portulacaceae	Herb	Singh *et al.*, 2009
1907.	*Portulaca oleracea*	Portulacaceae	Seed	Arruda *et al.*, 2004; Cai *et al.*, 2004; Abbas *et al.*, 2006; Maeda *et al.*, 2006; Li *et al.*, 2008; Souri *et al.*, 2008
1908.	*Potentilla alba*	Rosaceae	Herb	Oszmianski *et al.*, 2007; Tomczyk and Latte, 2009; Dorman *et al.*, 2011
1909.	*Potentilla anserina*	Rosaceae	Herb	Kahkonen *et al.*, 1999
1910.	*Potentilla argentea*	Rosaceae	Herb	Tomczyk and Latte, 2009
1911.	*Potentilla arguta*	Rosaceae	Seeds	Borchardt *et al.*, 2008
1912.	*Potentilla atrosanguinea*	Rosaceae	Aerial parts	Kalia *et al.*, 2008
1913.	*Potentilla chinensis*	Rosaceae		Wei *et al.*, 2010
1914.	*Potentilla erecta*	Rosaceae	Herb	Tomczyk and Latte, 2009
1915.	*Potentilla freyniana*	Rosaceae		Chen *et al.*, 2005
1916.	*Potentilla fruticosa*	Rosaceae	Flowers	Miliauskas *et al.*, 2004
1917.	*Potentilla recta*	Rosaceae	Herb	Tomczyk and Latte, 2009
1918.	*Potentilla reptans*	Rosaceae	Herb	Avci *et al.*, 2006; Tomczyk and Latte, 2009
1919.	*Potentilla fruticosa*	Rosaceae	Aerial parts	Miliauskas *et al.*, 2004
1920.	*Pouteria cambodiana*	Sapotaceae	Stem bark	Manosroi *et al.*, 2005
1921.	*Prangos ferulacea*	Apiaceae		Coruh *et al.*, 2007; Mavi *et al.*, 2004
1922.	*Prangos uloptera*	Apiaceae	Roots	Razavi *et al.*, 2009
1923.	*Primula elatior*	Primulaceae	Rhizome, root	Katalinic *et al.*, 2006
1924.	*Primula veris*	Primulaceae	Rhizome, root	Katalinic *et al.*, 2006
1925.	*Prosopis juliflora*	Mimosaceae	Leaves	Silva *et al.*, 2007; Sirmah *et al.*, 2009; Sathiya and Muthuchelian, 2010
1926.	*Prosopis stephaniana*	Mimosaceae		Souri *et al.*, 2004
1927.	*Prunella vulgaris*	Lamiaceae	Inflore-scence	Skottova *et al.*, 2001, 2004; Cai *et al.*, 2004; Shyur *et al.*, 2005; Lugasi *et al.*, 2006; Mariassyova 2006; Sarosi and Bernath, 2008; Li *et al.*, 2008; Mrudula *et al.*, 2010
1928.	*Prunus americana*	Rosaceae	Fruits	Acuna *et al.*, 2002
1929.	*Prunus armeniaca*	Rosaceae	Fruit	Tosun and Ustun, 2003; Durmez and Alpaslan, 2005; Leccese *et al.*, 2007; Madrau *et al.*, 2009; Yiit *et al.*, 2009; Song *et al.*, 2010

Contd...

Table 3.1–*Contd...*

Sl.No.	*Plant Name*	*Family*	*Parts Used*	*References*
1930.	*Prunus avium*	Rosaceae	Fruit	Goncalves *et al.*, 2004; Murtaza *et al.*, 2007; Usenik *et al.*, 2008; Karlidag *et al.*, 2009; Serra *et al.*, 2010
1931.	*Prunus campanulata*	Rosaceae	Leaves	Chen, C-H. *et al.*, 2009
1932.	*Prunus cerasus*	Rosaceae	Fruits	Tosun and Ustun, 2003; Kirakosyan *et al.*, 2010
1933.	*Prunus davidiana*	Rosaceae	Stem	Jung *et al.*, 2003
1934.	*Prunus divaricata*	Rosaceae	Fruit	Serteser *et al.*, 2009; Motamed and Naghibi, 2010
1935.	*Prunus domestica*	Rosaceae	Fruit	Wang *et al.*, 1996; Donovan *et al.*, 1998, 2003; Chun *et al.*, 2003; Kayano *et al.*, 2002, 2004; Lombardi-Boccia *et al.*, 2004; Kikuzaki *et al.*, 2004; Rop *et al.*, 2009
1936.	*Prunus dulcis*	Rosaceae	Seed skins	Monagas *et al.*, 2007
1937.	*Prunus insititia*	Rosaceae		Kahkonen *et al.*, 1999
1938.	*Prunus mahaleb*	Rosaceae	Seed cake	Mariod *et al.*, 2010
1939.	*Prunus malus*	Rosaceae	Fruits	Oabrrosa *et al.*, 2007
1940.	*Prunus mume*	Rosaceae	Flowers	Kim *et al.*, 1997; Shi *et al.*, 2009
1941.	*Prunus persica*	Rosaceae		Tosun and Ustun, 2003; Kardosova and Machova, 2006; Gan *et al.*, 2010
1942.	*Prunus serrulata*	Rosaceae	Leaves	Jung *et al.*, 2002
1943.	*Prunus spinosa*	Rosaceae	Fruits	Ionska-Rys *et al.*, 2009; Fraternale *et al.*, 2009
1944.	*Prunus virginiana*	Rosaceae	Flesh	Acuna *et al.*, 2002
1945.	*Psammogeton canescens*	Apiaceae		Gholivand *et al.*, 2010
1946.	*Pseudarthria viscida*	Fabaceae		Mathew and Sasikumar, 2007
1947.	*Pseudobombax ellipticum*	Bombacaceae		Ruiz-Terán *et al.*, 2008
1948.	*Pseudostellaria heterophylla*	Caryophyllaceae		Ng *et al.*, 2004
1949.	*Psidium guajava*	Myrtaceae	Fruits	Huang *et al.*, 2004a; Sacchetti *et al.*, 2005; Alfredo 2007; Tachakittirungrod *et al.*, 2007; Chen *et al.*, 2007; Ayoola *et al.*, 2008; Alothman *et al.*, 2009; Akinmoladun *et al.*, 2010; Isabelle *et al.*, 2010.
1950.	*Psophocarpus tetragonalobus*	Fabaceae		Yogalatha *et al.*, 2005

Contd...

Table 3.1–*Contd...*

Sl.No.	*Plant Name*	*Family*	*Parts Used*	*References*
1951.	*Psoralea corylifolia*	Fabaceae	Seeds	Haraguchi *et al.*, 2000, 2002; Cai *et al.*, 2004; Tang *et al.*, 2004; Park *et al.*, 2005; Shinde *et al.*, 2010
1952.	*Psoralea esculenta*	Fabaceae	Seeds	Borchardt *et al.*, 2008
1953.	*Psychotria brachyceros*	Rubiaceae		Nascimento *et al.*, 2007
1954.	*Psychotria griffithii*	Rubiaceae		Ahmad *et al.*, 2010
1955.	*Psychotria lasiocarpa*	Rubiaceae		Ahmad *et al.*, 2010
1956.	*Psychotria ophirensis*	Rubiaceae		Ahmad *et al.*, 2010
1957.	*Psychotria rostrata*	Rubiaceae		Saha *et al.*, 2004
1958.	*Pterocarpus erinaceus*	Fabaceae	Aerial parts	Karou *et al.*, 2005; Lagnika *et al.*, 2011
1959.	*Pterocarpus marsupium*	Fabaceae	Heartwood, leaves, bark	Bajpai *et al.*, 2005; Kasar Rahul *et al.*, 2006; Yeole *et al.*, 2008; Radhika *et al.*, 2010
1960.	*Pterocarpus mildbraedii*	Fabaceae	Leaf	Odukoya *et al.*, 2007;
1961.	*Pterocarpus santalinoides*	Fabaceae	Leaf	Odukoya *et al.*, 2007;
1962.	*Pterocarpus santalinus*	Fabaceeae	Heartwood	Narayan *et al.*, 2005; Halim and Mishra, 2011; Kumar, 2011
1963.	*Pterocarya fraxinifolia*	Juglandaceae		Souri *et al.*, 2004; Ebrahimzadeh *et al.*, 2009
1964.	*Pterocaulon cordobense*	Asteraceae		Borneo *et al.*, 2008
1965.	*Pterogyne nitens*	Fabaceae		Santos *et al.*, 2009
1966.	*Pteropyrum aucheri*	Polygonaceae		Souri *et al.*, 2004
1967.	*Pterospermum semisagittatum*	Sterculiaceae	Stem	Kshirsagar and Upadhyay, 2009
1968.	*Ptychopetalum olacoides*	Olacaceae	Roots	Siqueira *et al.*, 2007
1969.	*Pueraria lobata*	Fabaceae	Root	Jiang *et al.*, 2005; Cai *et al.*, 2006; Wu *et al.*, 2007; Cui Hengxiang *et al.*, 2008; Song *et al.*, 2010
1970.	*Pueraria mirifica*	Fabaceeae		Chanwitheesuk *et al.*, 2005; Phansawan and Poungbangpho, 2007
1971.	*Pueraria thomsonii*	Fabaceae	Root	Jiang *et al.*, 2005
1972.	*Pulicaria crispa*	Asteraceae	Whole plant	Marwah *et al.*, 2007
1973.	*Pulmonaria officinalis*	Boraginaceae		Ivanova *et al.*, 2005
1974.	*Pulsatilla chinensis*	Ranunculaceae	Root	Cai *et al.*, 2004; Li *et al.*, 2008
1975.	*Punica granatum*	Punicaceae	Seeds/ fruits Fruit peel Flowers	Singh *et al.*, 2002; Cai *et al.*, 2004; Wang *et al.*, 2004; Ricci *et al.*, 2006: Stangeland *et al.*, 2009, Al-Mustafa and Al-Thunibat, 2008; Bagri *et al.*, 2009; Hasan *et al.*, 2009; Ismail *et al.*, 2009; Sundararaju *et al.*, 2010; Kokila *et al.*, 2010; Niwano *et al.*, 2011; Moussa *et al.*, 2011

Contd...

Table 3.1–*Contd...*

Sl.No.	Plant Name	Family	Parts Used	References
1976.	*Putoria calabrica*	Rubiaceae	Roots	Tasdemir *et al.*, 2004
1977.	*Pyracantha koidzumi*	Rosaceae		Hou *et al.*, 2003
1978.	*Pyrola calliantha*	Pyrolaceae		Gan *et al.*, 2010
1979.	*Pyrus communis*	Rosaceae	Fruit	Wang *et al.*, 1996; Cai *et al.*, 2004
1980.	*Pyrus elaegnifolia*	Rosaceae	Fruit	Serteser *et al.*, 2009
1981.	*Pyrus taiwanensis*	Rosaceae		Hou *et al.*, 2003
1982.	*Quercus alba*	Fagaceae		McCune and Johns, 2002
1983.	*Quercus aliena*	Fagaceae		Jin *et al.*, 2005
1984.	*Quercus calliprinos*	Fagaceae	Leaves, fruit	Al-Mustafa and Al-Thunibat, 2008
1985.	*Quercus eduardii*	Fagaceae	Leaves	Rivas-Arreola *et al.*, 2010
1986.	*Quercus infectoria*	Fagaceae	Galls	Kaur *et al.*, 2008
1987.	*Quercus macrocarpa*	Fagaceae	Leaves	He *et al.*, 2011
1988.	*Quercus petraea*	Fagaceae	Cortex, leaf	Almeida *et al.*, 2008
1989.	*Quercus pubescens*	Fagaceae	Leaves	Demo *et al.*, 1999
1990.	*Quercus resinosa*	Fagaceae	Leaves	Rivas-Arreola *et al.*, 2010
1991.	*Quercus robur*	Fagaceae	Cortex, leaf	Almeida *et al.*, 2008
1992.	*Quercus salicina*	Fagaceae	Herb	Jung *et al.*, 2006
1993.	*Quercus sideroxyla*	Fagaceae	Leaves	Rivas-Arreola *et al.*, 2010
1994.	*Quisqualis indica*	Combretaceae		Leelapornpisid *et al.*, 2007
1995.	*Rabdosia japonica*	Lamiaceae	Leaves	Masuoka *et al.*, 2006
1996.	*Randia dumetorum*	Rubiaceaae	Fruit	Dharmishta *et al.*, 2009, 2010
1997.	*Randia echinocarpa*	Rubiaceae	Fruit	Santos-Cervantes *et al.*, 2007
1998.	*Randia hebecarpa*	Rubiaceae	Leaves	Nazari *et al.*, 2006
1999.	*Ranunculus marginatus*	Ranunculaceae		Kaya *et al.*, 2010
2000.	*Ranunculus repens*	Ranunculaceae	Herb	Kahkonen *et al.*, 1999
2001.	*Ranunculus sprunerianus*	Ranunculaceae		Kaya *et al.*, 2010
2002.	*Raphanus sativus*	Brassicaceae	Root, leaf	Kim *et al.*, 1997; Souri *et al.*, 2004; Lugasi *et al.*, 2005; Barillari *et al.*, 2006; Maeda *et al.*, 2006; Raghu *et al.*, 2010
2003.	*Rauvolfia caffra*	Apocynaceae	Roots	Erasto *et al.*, 2011
2004.	*Rauvolfia sellowii*	Apocynaceae		Menezes *et al.*, 2004
2005.	*Rauvolfia vomitoria*	Apocynaceae	Roots	Erasto *et al.*, 2011
2006.	*Rehmannia glutinosa*	Scrophulariaceae	Root	Cai *et al.*, 2004; Li *et al.*, 2008
2007.	*Remusatia vivipara*	Araceae	Tuber	Marwah *et al.*, 2007

Contd...

Table 3.1–*Contd...*

Sl.No.	*Plant Name*	*Family*	*Parts Used*	*References*
2008.	*Rennellia elliptica*	Rubiaceae		Ahmad *et al.*, 2010
2009.	*Retama raetum*	Fabaceae	Leaves, seeds	Conforti *et al.*, 2004; Edziri *et al.*, 2010
2010.	*Rhamnus cornifolia* subsp. *cornifolia*	Rhamnaceae		Souri *et al.*, 2004
2011.	*Rhamnus lycioides*	Rhamnaceae	Leaves	Demo *et al.*, 1999
2012.	*Rhapidophora pertusa*	Araceae	Stem	Sasikumar and Doss, 2006; Linnert *et al.*, 2010
2013.	*Rhaponticum carthamoides*	Asteraceae	Aerial parts	Miliauskas *et al.*, 2004, 2005
2014.	*Rhaponticum uniflorum*	Asteraceae		Li *et al.*, 2008
2015.	*Rhazya stricta*	Apocynaceae	Leaves	Iqbal *et al.*, 2006
2016.	*Rheum emodi*	Polygonaceae	Roots	Kren *et al.*, 2003; Rajkumar *et al.*, 2010a
2017.	*Rheum officinale*	Polygonaceae	Root	Cai *et al.*, 2004a,b, 2006; Nam and Kang, 2004
2018.	*Rheum palmatum*	Polygonaceae		Kim *et al.*, 1997; Wojcikowski *et al.*, 2007
2019.	*Rheum ribes*	Polygonaceae	Roots, stem	Ozturk *et al.*, 2007
2020.	*Rhinacanthus nasutus*	Acanthaceae	Leaves	Visweswara Rao and Dhanajaya Naidu, 2010
2021.	*Rhizophora conjugata*	Rhizophoraceae	Leaves, flowers	Vadlapudi and Naidu, 2009
2022.	*Rhizophora mangle*	Rhizophoraceae	Bark	Sanchez *et al.*, 2006; Berenguer *et al.*, 2006
2023.	*Rhizophora mucronata*	Rhizophoraceae	Leaves, flowers	Vadlapudi and Naidu, 2009; Suganthy *et al.*, 2009
2024.	*Rhodiola imbricata*	Crassulaceae	Root	Hagat *et al.*, 2004.
2025.	*Rhodiola rosea*	Crassulaceae	Roots	Kim *et al.*, 2008; Chen, C-H. *et al.*, 2009
2026.	*Rhodiola sachlinensis*	Crassulaceae		Zhang *et al.*, 2008
2027.	*Rhodiola sacra*	Crassulaceae		Ohsugi *et al.*, 1999; Wong *et al.*, 2006
2028.	*Rhodococcum vitis-idaea*	Ericaceae		Myagmar *et al.*, 2004
2029.	*Rhododendron arboretum*	Ericaceae	Leaves	Prakash *et al.*, 2007d
2030.	*Rhododendron baileyi*	Ericaceae	Leaves	Prakash *et al.*, 2007d
2031.	*Rhododendron barbatum*	Ericaceae	Leaves	Prakash *et al.*, 2007d
2032.	*Rhododendron camelliflorum*	Ericaceae	Leaves	Prakash *et al.*, 2007d

Contd...

Table 3.1–*Contd...*

Sl.No.	Plant Name	Family	Parts Used	References
2033.	*Rhododendron campanulatum*	Ericaceae	Leaves	Prakash *et al.*, 2007d
2034.	*Rhododendron ciliatum*	Ericaceae	Leaves	Prakash *et al.*, 2007d
2035.	*Rhododendron cinnabaricum*	Ericaceae	Leaves	Prakash *et al.*, 2007d
2036.	*Rhododendron dalhousiae*	Ericaceae	Leaves	Prakash *et al.*, 2007d
2037.	*Rhododendron decipiens*	Ericaceae	Leaves	Prakash *et al.*, 2007d
2038.	*Rhododendron falconeri*	Ericaceae	Leaves	Prakash *et al.*, 2007d
2039.	*Rhododendron grande*	Ericaceae	Leaves	Prakash *et al.*, 2007d
2040.	*Rhododendron griffithianum*	Ericaceae	Leaves	Prakash *et al.*, 2007d
2041.	*Rhododendron lepidotum*	Ericaceae	Leaves	Prakash *et al.*, 2007d
2042.	*Rhododendron maddenii*	Ericaceae	Leaves	Prakash *et al.*, 2007d
2043.	*Rhododendron mucronulatum*	Ericaceae	Stem	Lee *et al.*, 2005
2044.	*Rhododendron niveum*	Ericaceae	Leaves	Prakash *et al.*, 2007d
2045.	*Rhododendron pendulum*	Ericaceae	Leaves	Prakash *et al.*, 2007d
2046.	*Rhododendron ponticum*	Ericaceae	Leaves	Tasdemir *et al.*, 2004
2047.	*Rhododendron sallignum*	Ericaceae	Leaves	Prakash *et al.*, 2007d
2048.	*Rhododendron simsii*	Ericaceae		Takahashi *et al.*, 2001
2049.	*Rhododendron smirnovii*	Ericaceae	Leaves	Tasdemir *et al.*, 2004
2050.	*Rhododendron thomsonii*	Ericaceae	Leaves	Prakash *et al.*, 2007d
2051.	*Rhododendron ungernii*	Ericaceae	Leaves	Tasdemir *et al.*, 2004
2052.	*Rhododendron vaccinioides*	Ericaceae	Leaves	Prakash *et al.*, 2007d
2053.	*Rhododendron virgatum*	Ericaceae	Leaves	Prakash *et al.*, 2007d
2054.	*Rhododendron yedoense*	Ericaceae	Flower	Jung *et al.*, 2007
2055.	*Rhoicissus digitata*	Vitaceae		Fernandes *et al.*, 2004
2056.	*Rhoicissus rhomboidea*	Vitaceae	Leaf, stem, root	Opoku *et al.*, 2002
2057.	*Rhoicissus tomentosa*	Vitaceae		Fernandes *et al.*, 2004
2058.	*Rhoicissus tridentata*	Vitaceae	Leaf, stem, root, tuber	Opoku *et al.*, 2002; Fernandes *et al.*, 2004; Naidoo *et al.*, 2006
2059.	*Rhopalocnemis phalloides*	Balanophoraceae	Herb	She *et al.*, 2010
2060.	*Rhus chinensis*	Anacardiaceae	Gall	Cai *et al.*, 2004, 2006
2061.	*Rhus coriaria*	Anacardiaceae	Leaves, fruits	Bozan *et al.*, 2003; Ozcan, 2003a,b; Souri *et al.*, 2004; Candan and Sokmen, 2004; Koflar *et al.*, 2007; Serteser *et al.*, 2009

Contd...

Table 3.1–*Contd...*

Sl.No.	Plant Name	Family	Parts Used	References
2062.	*Rhus hirta*	Anacardiaceae		McCune, 1999
2063.	*Rhus verniciflua*	Anacardiaceae		Kim *et al.*, 1997; Lim *et al.*, 2001; Choi *et al.*, 2003; Cai *et al.*, 2006; Jung *et al.*, 2006
2064.	*Ribes grossularia*	Grossulariaceae	Fruit	Kahkonen *et al.*, 1999
2065.	*Ribes x nidigrolaria.*	Grossulariaceae	Fruit	Moyer *et al.*, 2002
2066.	*Ribes nigrum*	Grossulariaceae	Fruit	Constantino *et al.*, 1992; Kahkonen *et al.*, 1999; Moyer *et al.*, 2000, 2002; Landbo and Meyer, 2001; Kapasakalidis *et al.*, 2006; Currie *et al.*, 2006; Borges *et al.*, 2010
2067.	*Ribes rubrum*	Grossulariaceae	Fruit	Constantino *et al.*, 1992; Kahkonen *et al.*, 1999; Borges *et al.*, 2010
2068.	*Ribes uva-crispa*	Grossulariaceae	Fruit	Moyer *et al.*, 2002
2069.	*Ribes valdivianum*	Grossulariaceae	Fruit	Moyer *et al.*, 2000, 2002
2070.	*Ricinodendron heudelotii*	Euphorbiaceae		Momeni *et al.*, 2010
2071.	*Ricinus communis*	Euphorbiaceae		Shahwar *et al.*, 2010
2072.	*Ridolfia segetum*	Apiaceae	Root	Jabrane *et al.*, 2010
2073.	*Roemeria refracta*	Papaveraceae		Souri *et al.*, 2004
2074.	*Rorippa nasturtium-aquaticum*	Brassicaceae	Leaves	Lako *et al.*, 2006
2075.	*Rosa agrestis*	Rosaceae	Leaf	Bitis *et al.*, 2010
2076.	*Rosa canina*	Rosaceae	Seeds	Halvorsen *et al.*, 2002; Kumarswamy *et al.*, 2003; Souri *et al.*, 2004; Ghazghazi *et al.*, 2010
2077.	*Rosa chinensis*	Rosaceae	Flowers	Cai *et al.*, 2004, 2005; Gan *et al.*, 2010
2078.	*Rosa damascena*	Rosaceae	flower	Achutan *et al.*, 2003; Özkan *et al.*, 2004; Kalim *et al.*, 2010
2079.	*Rosa dumalis*	Rosaceae	Fruits	Gao *et al.*, 2000
2080.	*Rosa hybrida*	Rosaceae	Flower	Choi and Hwang, 2005
2081.	*Rosa nutkana*	Rosaceae	Fruit	Yi *et al.*, 2007
2082.	*Rosa pimpinellifolia*	Rosaceae		Mavi *et al.*, 2004
2083.	*Rosa pisocarpa*	Rosaceae	Fruit	Yi *et al.*, 2007
2084.	*Rosa rubiginosa*	Rosaceae		Moure *et al.*, 2001
2085.	*Rosa rugosa*	Rosaceae	Flowers	Cho *et al.*, 2003a,b; Ng *et al.*, 2005; Park *et al.*, 2009
2086.	*Rosa woodsii*	Rosaceae	Fruit	Yi *et al.*, 2007

Contd...

Table 3.1–*Contd...*

Sl.No.	Plant Name	Family	Parts Used	References
2087.	*Rosmarius officinalis*	Lamiaceae	Leaves/oil	Haraguchi *et al.*, 1995; Sara Lopez *et al.*, 1998; Demo *et al.*, 1999; Ibanez *et al.*, 1999, 2003; Patricia *et al.*, 2003; El-Ghorab, 2003; Rababah *et al.*, 2004; Etter, 2005; Sancheti and Goyal, 2006; Carrillo and Tena, 2006; Lugasi *et al.*, 2006; Schmourlo *et al.*, 2007; Biljana *et al.*, 2007; Evelyne *et al.*, 2007; Chen *et al.*, 2007; Gachkar *et al.*, 2007; Yesil-Celiktas *et al.*, 2007; Mishra *et al.*, 2007; Mariutti *et al.*, 2008; Bakirel *et al.*, 2008; Chang *et al.*, 2008; Erkan *et al.*, 2008; Gutierrez *et al.*, 2009; Viuda-Martos *et al.*, 2010; Yang *et al.*, 2010; Spiridon *et al.*, 2011
2088.	*Rotula aquatica*	Boraginaceae		Patil *et al.*, 2003
2089.	*Roylea cinerea*	Lamiaceae	Bark	Mathur *et al.*, 2010
2090.	*Rubia cordifolia*	Rubiaceae	Root	Pandey *et al.*, 1994; Tripathi *et al.*, 1997; Tripathi and Sharma, 1998; Joharapurkar *et al.*, 2003; Cai *et al.*, 2004a,b, 2006; Rawal *et al.*, 2004; Basu and Hazra, 2005, 2006; Kaur *et al.*, 2008; Gan *et al.*, 2010
2091.	*Rubia peregrina*	Rubiaceae	Roots rhizomes	Ozgen *et al.*, 2004
2092.	*Rubus caucasicus*	Rosaceae	Fruit	Deighton *et al.*, 2000
2093.	*Rubus chamamaemorus*	Rosaceae	Fruit, leaf	Kahkonen *et al.*, 1999
2094.	*Rubus chingii*	Rosaceae	Fruit	Cai *et al.*, 2004
2095.	*Rubus coreanus*	Rosaceae	Fruit	Deighton *et al.*, 2000; Kim *et al.*, 2010
2096.	*Rubus cyri*	Rosaceae	Fruit	Moyer *et al.*, 2002; Mishra *et al.*, 2007
2097.	*Rubus diversa*	Rosaceae		Ivanova *et al.*, 2005
2098.	*Rubus ellipticus*	Rosaceae	Fruit	Karuppusamy *et al.*, 2011
2099.	*Rubus fruticosus*	Rosaceae	Flowers, fruits	Katalinic *et al.*, 2006
2100.	*Rubus georgicus*	Rubiaceae	Fruit	Moyer *et al.*, 2002
2101.	*Rubus hunanensis*	Rosaceae	Fruit	Deighton *et al.*, 2000
2102.	*Rubus hybrid*	Rosaceae	Fruit	Moyer *et al.*, 2000
2103.	*Rubus idaeus*	Rosaceae	Fruit, flowers	Kahkonen *et al.*, 1999; Deiton *et al.*, 2000; Wang and Lin, 2000; Türemis *et al.*, 2003; Katalinic *et al.*, 2006; Özgen and Scheerens 2006; Venskutonis *et al.*, 2007; Borges *et al.*, 2010; Cekic and Ozgen, 2010; Rouanet *et al.*, 2010; Zhang *et al.*, 2010

Contd...

Table 3.1–*Contd...*

Sl.No.	*Plant Name*	*Family*	*Parts Used*	*References*
2104.	*Rubus innominatus*	Rosaceae	Fruit	Deighton *et al.*, 2000; Moyer *et al.*, 2002
2105.	*Rubus insularis*	Rosaceae	Fruit	Moyer *et al.*, 2002
2106.	*Rubus lambertianus*	Rosaceae	Fruit	Deighton *et al.*, 2000
2107.	*Rubus leucodermis*	Rosaceae	Fruit	Moyer *et al.*, 2000
2108.	*Rubus niveus*	Rosaceae	Fruit	Moyer *et al.*, 2000, 2002; Deighton *et al.*, 2000
2109.	*Rubus occidentalis*	Rosaceae	Fruit	Moyer *et al.*, 2000; Wang and Lin, 2000; Türemis *et al.*, 2003; Özgen *et al.*, 2008
2110.	*Rubus parvifolius*	Rosaceae	Fruit	Deighton *et al.*, 2000; Shyur *et al.*, 2005; Gao *et al.*, 2011
2111.	*Rubus sanctus*	Rosaceae		Motamed and Naghibi, 2010
2112.	*Rubus sumatranus*	Rosaceae	Fruit	Deighton *et al.*, 2000
2113.	*Rubus tsangorum*	Rosaceae	Fruit	Deighton *et al.*, 2000
2114.	*Rubus ulmifolius*	Rosaceae	Fruit	Deighton *et al.*, 2000
2115.	*Rubus ursinus*	Rosaceae	Fruit	Moyer *et al.*, 2000, 2002; Wang and Lin, 2000; Deighton *et al.*, 2000
2116.	*Rubus valdivianum*	Rosaceae	Fruit	Moyer *et al.*, 2000
2117.	*Rudbeckia fulgida*	Rosaceae	Leaves	Kardosova and Machova, 2006
2118.	*Ruellia tuberosa*	Acanthaceae	Aerial	Chen, F.A. *et al.*, 2006; Chen, C-H. *et al.*, 2009
2119.	*Rumex acetosa*	Polygonaceae	Flowers	Kahkonen *et al.*, 1999; Mantle *et al.*, 2000
2120.	*Rumex acetosella*	Polygonaceae	Leaf	Özen, 2010
2121.	*Rumex bucephalophorus*	Polygonaceae	Roots	Kerem *et al.*, 2003
2122.	*Rumex crispus*	Polygonaceae	Leaves, seeds	Yildirim *et al.*, 2001; Borchardt *et al.*, 2008
2123.	*Rumex hymenosepalus*	Polygonaceae		Van der Jagt *et al.*, 2002
2124.	*Rumex induratus*	Polygonaceae		Guerra *et al.*, 2008
2125.	*Rumex japonicus*	Polygonaceae	Aerial parts	Elzaawely *et al.*, 2005
2126.	*Rumex tuberosus*	Polygonaceae	Herb	Motamed and Naghibi, 2010
2127.	*Rumex verticillatus*	Polygonaceae	Seeds	Borchardt *et al.*, 2008
2128.	*Rupechtia fusca*	Polygonaceae		Ruiz-Terán *et al.*, 2008
2129.	*Ruscus aculeatus*	Ruscaceae	Rhizome	Seaver and Smith, 2004; Wojcikowski *et al.*, 2007
2130.	*Ruta chalepensis*	Rutaceae	Essential oil	Ahmed *et al.*, 2011
2131.	*Ruta graveolens*	Rutaceae	Leaves	Wong *et al.*, 2006; Ratheesh *et al.*, 2009; Pandey *et al.*, 2011

Contd...

Table 3.1–*Contd...*

Sl.No.	Plant Name	Family	Parts Used	References
2132.	*Ruta montana*	Rutaceae	Aerial parts	Kambouche *et al.*, 2008
2133.	*Saccharum officinarum*	Poaceae	Stem Juice	Kadam *et al.*, 2008; Ghosh *et al.*, 2008; Duarte-Almeida *et al.*, 2007; Ahmad *et al.*, 2011
2134.	*Saccharum spontaneum*	Poaceae	Flower	Gopinathan *et al.*, 2004; Ripa *et al.*, 2009
2135.	*Sacoglottis gabonensis*	Houmeriaceae	Stem bark	Maduka and OKoye, 2002
2136.	*Sageretia theezans*	Rhamnaceae	Leaves	Chung *et al.*, 2004
2137.	*Salacia fruticosa*	Celastraceae		Srinivasan *et al.*, 2009
2138.	*Salacia oblonga*	Celastraceae		Sabu and Kuttan, 2003
2139.	*Salacia pallescens*	Celastraceae	Leaf	Sofidya *et al.*, 2009
2140.	*Salicornia bracteata*	Chenopodiaceae	Aerial parts	Vadlapudi and Naidu, 2009
2141.	*Salicornia herbacea*	Chenopodiaceae		Ha *et al.*, 2006; Jang *et al.*, 2007
2142.	*Salix aegyptiaca*	Salicaceae		Sonboli *et al.*, 2010
2143.	*Salix alba*	Salicaceae	Bark, leaf	Kahkonen *et al.*, 1999; Pendry *et al.*, 2005; Katalinic *et al.*, 2006
2144.	*Salix caprea*	Salicaceae	Bark, leaf	Kahkonen *et al.*, 1999; Alam *et al.*, 2006
2145.	*Salix warburgii*	Salicaceae	Leaves	Lee *et al.*, 2006
2146.	*Salvadora persica*	Salvadoraceae	Stem	Souri *et al.*, 2008; Vadlapudi and Naidu, 2009
2147.	*Salvia aethiopia*	Lamiaceae		Tepe *et al.*, 2006; Tosun *et al.*, 2009
2148.	*Salvia africana-caerulea*	Lamiaceae	Aerial part	Kamatou *et al.*2010
2149.	*Salvia africana-lutea*	Lamiaceae	Aerial part	Kamatou *et al.*2010
2150.	*Salvia albicaulis*	Lamiaceae	Aerial part	Kamatou *et al.*2010
2151.	*Salvia aramiensis*	Lamiaceae		Kelen and Tepe, 2007
2152.	*Salvia aucheri* var. *aucheri*	Lamiaceae		Kelen and Tepe, 2007
2153.	*Salvia aurita*	Lamiaceae	Aerial part	Kamatou *et al.*2010
2154.	*Salvia brachystachya*	Lamiaceae	Plant	Esmaeili. and Soboli 2010
2155.	*Salvia caespitosa*	Lamiaceae		Tepe *et al.*, 2006
2156.	*Salvia candelabrum*	Lamiaceae	Aerial part	Zupkó *et al.*, 2001; Janicsak *et al.*, 2003
2157.	*Salvia candidissima*	Lamiaceae	Herb	Tosun *et al.*, 2009
2158.	*Salvia chamelaeagnea*	Lamiaceae	Aerial part	Kamatou *et al.*2010
2159.	*Salvia cryptantha*	Lamiaceae	Herb	Tepe *et al.*, 2004
2160.	*Salvia disermas*	Lamiaceae	Aerial part	Kamatou *et al.*2010
2161.	*Salvia dolomitica*	Lamiaceae	Aerial part	Kamatou *et al.*2010

Contd...

Table 3.1–*Contd...*

Sl.No.	*Plant Name*	*Family*	*Parts Used*	*References*
2162.	*Salvia eremophila*	Lamiaceae	Herb	Ebrahimabadi *et al.*, 2010; Firuzi *et al.*, 2010
2163.	*Salvia euphratica* subsp. *euphratica*	Lamiaceae		Tepe *et al.*, 2006
2164.	*Salvia fruticosa*	Lamiaceae	Shoot	Demo *et al.*, 1999; Al-Mustafa and Al-Thunibat, 2008
2165.	*Salvia garipensis*	Lamiaceae	Aerial part	Kamatou *et al.*2010
2166.	*Salvia glutinosa*	Lamiaceae	Aerial parts	Zupkó *et al.*, 2001; Miliauskas *et al.*, 2004
2167.	*Salvia halophila*	Lamiaceae		Bozan *et al.*, 2002; Albayrak *et al.*, 2008
2168.	*Salvia hydrangea*	Lamiaceae	Aerial part	Souri *et al.*, 2008; Firuzi *et al.*, 2010
2169.	*Salvia hypargeia*	Lamiaceae		Tepe *et al.*, 2006
2170.	*Salvia hypoleuca*	Lamiaceae		Souri *et al.*, 2004; Nickavar *et al.*, 2007
2171.	*Salvia lanceolata*	Lamiaceae	Aerial part	Kamatou *et al.*, 2010
2172.	*Salvia leriifolia*	Lamiaceae	Herb	Loizzo *et al.*, 2009, 2010
2173.	*Salvia limbata*	Lamiaceae	Herb	Tosun *et al.*, 2009
2174.	*Salvia macrosiphon*	Lamiaceae	Aerial parts	Souri *et al.*, 2004; Firuzi *et al.*, 2010
2175.	*Salvia microstegia*	Lamiaceae	Herb	Tosun *et al.*, 2009
2176.	*Salvia miltiorrhiza*	Lamiaceae	Root, leaf	Gordon and Weng, 1992; Weng and Gordon, 1992; Cai *et al.*, 2004; Zhao *et al.*, 2006; Wu *et al.*, 2007; Wojcikowski *et al.*, 2007; Matkowski *et al.*, 2008; Gan *et al.*, 2010; Zhang *et al.*, 2010
2177.	*Salvia mirzayani*	Lamiaceae	Herb	Moein *et al.*, 2007
2178.	*Salvia muirii*	Lamiaceae	Aerial part	Kamatou *et al.*2010
2179.	*Salvia multicaulis*	Lamiaceae		Tepe *et al.*, 2004; Erdemoglu *et al.*, 2006; Firuzi *et al.*, 2010
2180.	*Salvia namaensis*	Lamiaceae	Aerial part	Kamatou *et al.*2010
2181.	*Salvia nemorosa*	Lamiaceae	Herb	Zupkó *et al.*, 2001; Tosun *et al.*, 2009
2182.	*Salvia officinalis*	Lamiaceae	Leaves	Filipek, 1994; Wang *et al.*1998; Hohman *et al.*, 1999; Daukas *et al.*, 2001; Lu and Foo, 2001; Miura *et al.*, 2002; Bandoniene *et al.*, 2002; Santos-Gomes *et al.*, 2003; Miliauskas *et al.*, 2004; Lima *et al.*, 2005; Biljana *et al.*, 2007; Nickavar *et al.*, 2007; Mariutti *et al.*, 2008; Cioroi and Dumitriu, 2009; Viuda-Martos *et al.*, 2010; Spiridon *et al.*, 2011

Contd...

Table 3.1–*Contd...*

Sl.No.	Plant Name	Family	Parts Used	References
2183.	*Salvia pachystachys*	Lamiaceae	Herb	Tosun *et al.*, 2009
2184.	*Salvia palestina*	Lamiaceae	Herb	Firuzi *et al.*, 2010
2185.	*Salvia pilifera*	Lamiaceae		Kelen and Tepe, 2007
2186.	*Salvia plebeia*	Lamiaceae	Herb	Weng and Wang, 2000; Gu and Weng, 2001; Ai-Li and Wang, 2005
2187.	*Salvia pomifera*	Lamiaceae	Aerial parts	Demo *et al.*, 1999; Couladis *et al.*, 2003
2188.	*Salvia pratensis*	Lamiaceae	Aerial parts	Miliauskas *et al.*, 2004
2189.	*Salvia przewalskii*	Lamiaceae	Roots, leaves	Matkowski *et al.*, 2008
2190.	*Salvia psidica*	Lamiaceae	Herb	Ozkan *et al.*, 2010
2191.	*Salvia radula*	Lamiaceae	Aerial part	Kamatou *et al.*, 2010
2192.	*Salvia reflexa*	Lamiaceae	Herb	Malencic *et al.*, 2000
2193.	*Salvia repens*	Lamiaceae	Aerial part	Kamatou *et al.*, 2010
2194.	*Salvia reuterana*	Lamiaceae	Aerial parts	Nickavar *et al.*, 2007; Firuzi *et al.*, 2010
2195.	*Salvia ringens*	Lamiaceae	Aerial parts	Zupkó *et al.*, 2001; Couladis *et al.*, 2003
2196.	*Salvia runcinata*	Lamiaceae	Aerial part	Kamatou *et al.*2010
2197.	*Salvia sahendica*	Lamiaceae	Aerial parts	Salehi *et al.*, 2007
2198.	*Salvia santolinifolia*	Lamiaceae	Herb	Firuzi *et al.*, 2010
2199.	*Salvia schlechteri*	Lamiaceae		Kamatou *et al.*, 2010
2200.	*Salvia sclarea*	Lamiaceae	Aerial parts	Tang and Yuan, 1997; Miliauskas *et al.*, 2004; Tepe *et al.*, 2006; Kumar and Singh, 2011
2201.	*Salvia staminea*	Lamiaceae		Tepe, 2007
2202.	*Salvia stenophylla*	Lamiaceae	Aerial part	Kamatou *et al.*2010
2203.	*Salvia syriaca*	Lamiaceae	Aerial parts	Firuzi *et al.*, 2010
2204.	*Salvia tomentosa*	Lamiaceae	Essential oil	Zupkó *et al.*, 2001; Tepe *et al.*, 2005
2205.	*Salvia triloba*	Lamiaceae	Herb	Protogeras *et al.*, 1998; Yildirim *et al.*, 2000
2206.	*Salvia verbenaca*	Lamiaceae	Aerial parts	Khalil *et al.*, 2007; Tepe, 2007
2207.	*Salvia verticillata* subsp. *verticillata*	Lamiaceae		Souri *et al.*, 2004; Tepe *et al.*, 2007; Nickavar *et al.*, 2007; Matkowski *et al.*, 2008; Tosun *et al.*, 2009
2208.	*Salvia verticillata* subsp. *amasiaca*	Lamiaceae		Tepe *et al.*, 2007
2209.	*Salvia virgata*	Lamiaceae	Aerial parts	Nickavar *et al.*, 2007; Kosar *et al.*, 2008; Tepe, 2008; Tosun *et al.*, 2009; Firuzi *et al.*, 2010

Contd...

Table 3.1–*Contd...*

Sl.No.	*Plant Name*	*Family*	*Parts Used*	*References*
2210.	*Salvia viridis*	Lamiaceae	Aerial parts	Erdemoglu *et al.*, 2006; Khalil *et al.*, 2007
2211.	*Sambucus caerulea*	Caprifoliaceae,	Fruit	Acuna *et al.*, 2002
2212.	*Sambucus canadensis*	Caprifoliaceae		Ozgen *et al.*, 2010
2213.	*Sambucus ebulus*	Caprifoliaceae	Flowers	Kiselova *et al.*, 2006; Hosseinimehr *et al.*, 2007
2214.	*Sambucus nigra*	Caprifoliaceae	Flowers	Youdin *et al.*, 2004; Katalinic *et al.*, 2006
2215.	*Sanguisorba officinalis*	Rosaceae	Root	Cai *et al.*, 2004, 2006; Gan *et al.*, 2010
2216.	*Sanicula europaea*	Apiaceae		Le Claire *et al.*, 2005
2217.	*Sanicula graveolens*	Apiaceae		Viturro *et al.*, 1999
2218.	*Santalum album*	Santalaceae		Scartezzini and Speroni, 2000; Mishra *et al.*, 2007
2219.	*Sapindus emarginatus*	Sapindaceae	Leaf	Srikanth and Muralidharan, 2010
2220.	*Sapindus mukorossi*	Sapindaceae		Chen *et al.*, 2010
2221.	*Sapium macrocarpum*	Euphorbiaceae		Ruiz-Terán *et al.*, 2008
2222.	*Saposhnikovia divaricata*	Apiaceae		Song *et al.*, 2010
2223.	*Saraca asoca*	Caesalpiniaceae	Stem	Kshirsagar and Upadhyay, 2009
2224.	*Saraca thaipingensis*	Caesalpiniaceae		Leelapornpisid *et al.*, 2007
2225.	*Sarcolobus globosus*	Asclepiadaceae	Plant	Wangensteen *et al.*, 2006
2226.	*Sarcopterium spinosum*	Rosaceae	Root	Al-Mustafa and Al-Thunibat, 2008
2227.	*Sarcostemma brevistigma*	Asclepiadaceae		Surveswaran *et al.*, 2010
2228.	*Sargentodoxa cuneata*	Sargentodoxaceae		Li *et al.*, 2008
2229.	*Sasa borealis*	Poaceae	Leaves	Park *et al.*, 2007
2230.	*Satureja cuneifolia*	Lamiaceae	Aerial parts	Kosar *et al.*, 2003, 2008; Dorman *et al.*, 2004; Eminagaoglu *et al.*, 2007;
2231.	*Satureja forbesii*	Lamiaceae	Essential oil	Ortet *et al.*, 2009
2232.	*Satureja hortensis*	Lamiaceae	Leaf, twig, roots	Kosar *et al.*, 2003; Souri *et al.*, 2004; Mariutti, *et al.*, 2008
2233.	*Satureja intricata*	Lamiaceae	Essential oil	Jordan *et al.*, 2010
2234.	*Satureja macrostoma*	Lamiaceae		Gutierrez and Navarro, 2010
2235.	*Satureja mutica*	Lamiaceae	Herb	Motamed and Naghibi, 2010
2236.	*Satureja obovata*	Lamiaceae	Essential oil	Jordan *et al.*, 2010
2237.	*Satureja pilosa*	Lamiaceae		Nikolova and Dzhurmanski, 2009
2238.	*Satureja spicigera*	Lamiaceae		Eminagaoglu *et al.*, 2007

Contd...

Table 3.1–*Contd...*

Sl.No.	*Plant Name*	*Family*	*Parts Used*	*References*
2239.	*Satureja thymbra*	Lamiaceae	Essential oils	Lagouri *et al.*, 2003
2240.	*Satureja* x *delpozoi*	Lamiaceae	Essential oil	Jordan *et al.*, 2010
2241.	*Saurauia oldhami*	Saurauiaceae		Shyur *et al.*, 2005
2242.	*Sauropus andrygonus*	Euphorbiaceae	Leaves	Wong *et al.*, 2006; Zaibunnisa *et al.*, 2008; Subhasree *et al.*, 2009; Andarwulan *et al.*, 2010; Narayanaswamy and Balakrishnan, 2011
2243.	*Saururus chinensis*	Saururaceae	Aerial parts	Chang *et al.*, 2002; Kang *et al.*, 2005; Ma *et al.*, 2009; Jeong *et al.*, 2010
2244.	*Saussarea costus*	Asteraceae		Pandey *et al.*, 2005
2245.	*Saussurea lappa*	Asteraceae		Achliya *et al.*, 2003.
2246.	*Saxifraga stolonifera*	Saxifragaceae		Aoyagi *et al.*, 1995; Sohn *et al.*, 2008
2247.	*Scariola orientalis*	Asteraceae		Souri *et al.*, 2004
2248.	*Schlefflera actinophylla*	Araliaceae		Moussa *et al.*, 2011
2249.	*Schefflera leucantha*	Araliaceae	Leaves	Kruawan and Kangsadalampai, 2006
2250.	*Schima wallichii*	Theaceae	Leaf, stem	Didi Jauhari *et al.*, 2000; Kshirsagar and Upadhyay, 2009
2251.	*Schinus molle*	Anacardiaceae	Fruit	Bendaoud *et al.*, 2010
2252.	*Schinus terebinthifolia*	Anacardiaceae		Velazquez *et al.*, 2003; Bendaoud *et al.*, 2010
2253.	*Schinus weinmannifolia*	Anacardiaceae		Velazquez *et al.*, 2003
2254.	*Schisandra chinensis*	Schisandraceae	Fruit	Kim *et al.*, 1997; Cai *et al.*, 2004; Ko *et al.*, 2006; An *et al.*, 2006; Ma *et al.*, 2009
2255.	*Schizonepeta ternnuifolia*	Lamiaceae		Song *et al.*, 2010
2256.	*Schrankia leptocarpa*	Mimosaceae	Entire plant	Lagnika *et al.*, 2011
2257.	*Schrebera swietenioides*	Oleaceae	Root	Manda *et al.*, 2009; Kumar *et al.*, 2011
2258.	*Sclerocarya birrea*	Anacardiaceae	Young stem	Lamien-Meda *et al.*, 2008; Moyo *et al.*, 2010
2259.	*Scolopia braunii*	Flacourtiaceae	Stem	Mosaddik *et al.*, 2004
2260.	*Scoparia dulcis*	Scrophulariaceae	Plant	Ratnasooriya *et al.*, 2005
2261.	*Scrophularia auriculata*	Scrophulariaceae		Schinella *et al.*, 2002
2262.	*Scrophularia ningpoensis*	Scrophulariaceae		Li *et al.*, 2008
2263.	*Scrophularia striata*	Scrophulariaceae		Azadmehr *et al.*, 2009

Contd...

Table 3.1–*Contd...*

Sl.No.	Plant Name	Family	Parts Used	References
2264.	*Scutellaria baicalenis*	Lamiaceae	Leaves	Gao *et al.*, 1999; Lim *et al.*, 1999; Schinella *et al.*, 2002; Cai *et al.*, 2004, 2006; Shang *et al.*, 2006; Li *et al.*, 2008
2265.	*Scutellaria barbata*	Lamiaceae	Whole plant	Cai *et al.*, 2004; Gan *et al.*, 2010
2266.	*Scutellaria lateriflora*	Lamiaceae	Aerial parts	Wojcikowski *et al.*, 2007
2267.	*Scutellaria multicaulis*	Lamiaceae		Souri *et al.*, 2004; Firuzi *et al.*, 2010
2268.	*Scutellaria pinnatifida*	Lamiaceae		Souri *et al.*, 2004
2269.	*Scutellaria rehderiana*	Lamiaceae		Su *et al.*, 2004
2270.	*Scutellaria rivularis*	Lamiaceae		Shieh *et al.*, 2000
2271.	*Secale cereale*	Poaceae	Fruit	Kahkonen *et al.*, 1999; Korycinska *et al.*, 2009
2272.	*Secamone afzelii*	Asclepiadaceae	Aerial parts	Mensah *et al.*, 2004; Houghton *et al.*, 2005; Lagnika *et al.*, 2011
2273.	*Sechium edule*	Cucurbitaceae	Leaf, fruit	Huang *et al.*, 2004a; Ordonez *et al.*, 2006
2274.	*Securidaca longepedunculata*	Polygalaceae		Akinmoladun *et al.*, 2010
2275.	*Securigera securidaca*	Fabaceae	Seeds	Azad *et al.*, 2009
2276.	*Securinega leucopyrus*	Euphorbiaceae		Vidyadhar *et al.*, 2010
2277.	*Securinega virosa*	Euphorbiaceae		Dickson *et al.*, 2006; Sonogo *et al.*, 2009
2278.	*Sedum sarmentosum*	Crassulaceae	Aerial parts	Cai *et al.*, 2004; C.Y.Kim *et al.*, 2006; H.J.Kim *et al.*, 2008
2279.	*Sedum sempervioides*	Crassulaceae		Mavi *et al.*, 2004
2280.	*Semecarpus anacardium*	Anacardiaceae	Nut	Premalatha *et al.*, 1997; Pal *et al.*, 2008; Verma and Vinayak, 2009
2281.	*Sempervivum tectorum*	Crassulaceae		Sentjurc *et al.*, 2003; Blazovics *et al.*,1993; Kery *et al.*, 2008
2282.	*Senecio argunensis*	Asteraceae		Zhou *et al.*, 2008
2283.	*Senecio cineraria*	Asteraceae		Souri *et al.*, 2004
2284.	*Senecio herzogui*	Asteraceae		Rosas-Romero and Saavedra, 2005
2285.	*Senna alata*	Fabaceae		Chomnawang *et al.*, 2007
2286.	*Serenoa repens*	Arecaceae	Fruit	Wojcikowski *et al.*, 2007
2287.	*Sericostoma pauciflorum*	Boraginaceae	Whole plant	Singh *et al.*, 2009
2288.	*Serratula coronata*	Asteraceae	Herb	Báthori *et al.*, 2004
2289.	*Serratula strangulata*	Asteraceae	Herb	Cai *et al.*, 2002
2290.	*Serratula tinctoria*	Asteraceae	Herb	Báthori *et al.*, 2004
2291.	*Serratula wolffii*	Asteraceae	Herb	Báthori *et al.*, 2004

Contd...

Table 3.1–*Contd...*

Sl.No.	Plant Name	Family	Parts Used	References
2292.	*Sesamum indium*	Pedaliaceae	Seeds	Hu *et al.*, 2004; Suja *et al.*, 2004; Xu *et al.*, 2005;
2293.	*Sesbania grandiflora*	Fabaceae		Wong *et al.*, 2006; Gowri and Vasantha, 2010; Ouattara *et al.*, 2011; Padmaja *et al.*, 2011
2294.	*Sesbania pachycarpa*	Fabaceae	Leaf	Cook *et al.*, 1998
2295.	*Seseli rigidum*	Apiaceae	Flower	Stojkovic *et al.*, 2009
2296.	*Sesuvium portulacastrum*	Aizoaceae	Aerial parts	Magwa *et al.*, 2006; Vadlapudi and Naidu, 2009
2297.	*Setaria italica*	Poaceae	Grain	Suma and Urooj, 2011
2298.	*Setaria viridis*	Poaceae		Kwon *et al.*, 2002
2299.	*Shorea tumbaggaia*	Dipterocarpaceae		Ragini *et al.*, 2011
2300.	*Sida acuta*	Malvaceae	Whole plant	Karou *et al.*, 2005; Pieme *et al.*, 2010; Kiessoun *et al.*, 2010
2301.	*Sida alba*	Malvaceae	Herb	Kiessoun *et al.*, 2010
2302.	*Sida cordifolia*	Malvaceae	Leaves	Auddy *et al.*, 2003; Dhalwal *et al.*, 2005; Kubavat and Asdaq, 2009; Kiessoun *et al.*, 2010
2303.	*Sida retusa*	Malvaceae	Roots	Lissy *et al.*, 2006
2304.	*Sida rhombifolia*	Malvaceae	Herb	Dhalwal *et al.*, 2007; Kiessoun *et al.*, 2010
2305.	*Sida urens*	Malvaceae	Herb	Kiessoun *et al.*, 2010
2306.	*Sideritis caesarea*	Lamiaceae	Herb	Sagdic *et al.*, 2008
2307.	*Sideritis condensata*	Lamiacee	Herb	Ozkan *et al.*, 2005
2308.	*Sideritis erythrantha*	Lamiaceae		Ozkan *et al.*, 2005; Kose *et al.*, 2010
2309.	*Sideritis euboea*	Lamiaceae		Tsaknis and Lalas, 2005
2310.	*Sideritis libanotica*	Lamiaceae		Tepe *et al.*, 2006
2311.	*Siderites montana*	Lamiaceae		Koleva *et al.*, 2003; Khalil *et al.*, 2007; Firuzi *et al.*, 2010
2312.	*Sideritis ozturkii*	Lamiaceae	Herb	Sagdic *et al.*, 2008
2313.	*Sideritis raeseri*	Lamiaceae	Aerial parts	Gabrieli *et al.*, 2005; Pljevljakusic *et al.*, 2011
2314.	*Sideritis scardica*	Lamiaceae		Koleva *et al.*, 2003; Nikolova and Dzhurmanski, 2009
2315.	*Sideritis syriaca* ssp. *syriaca*	Lamiaceae	Aerial parts	Koleva *et al.*, 2003; Armata *et al.*, 2008
2316.	*Sideroxylon capiri*	Sapotaceae		Ruiz-Terán *et al.*, 2008
2317.	*Sideroxylon obtusifolium*	Sapotaceae		Desmarchelier *et al.*, 1999
2318.	*Siegesbeckia orientalis*	Asteraceae		Gan *et al.*, 2010
2319.	*Silene coronaria*	Caryophyllaceae		Souri *et al.*, 2004

Contd...

Table 3.1–*Contd...*

Sl.No.	Plant Name	Family	Parts Used	References
2320.	*Silene guntensis*	Caryophyllaceae		Mamadalieva *et al.*, 2011
2321.	*Silybum marianum*	Asteraceae		Valenzula *et al.*, 1989; Letteron *et al.*, 1990; Mira *et al.*, 1994; Dehmlow *et al.*, 1996; Filipe *et al.*, 1997; Skottova *et al.*, 2004; Samu *et al.*, 2004; Pendry *et al.*, 2005; Sersen *et al.*, 2006; Zielinska-Przyjemska and Wiktorowicz, 2006; Khalil *et al.*, 2007; Nencini *et al.*, 2007; Wojcikowski *et al.*, 2007; Hadaruga and Hadaruga, 2009
2322.	*Sinapis arvensis*	Brassicaceae		Coruh *et al.*, 2007
2323.	*Sinomonium acutum*	Menispermaceae		Shaw *et al.*, 2003; Gan *et al.*, 2010
2324.	*Siraitia grosvenori*	Cucurbitaceae	Fruits	Qi *et al.*, 2008
2325.	*Sisymbrium thellungii*	Brassicaceae	Herb	Lindsey *et al.*, 2006
2326.	*Smallanthus sonchifolius*	Asteraceae	Leaves, roots	Yan *et al.*, 1999; Valentová *et al.*, 2005
2327.	*Smilax aspera*	Smilacaceae	Leaves	Demo *et al.*, 1999
2328.	*Smilax china*	Smilacaceae	Roots	Lee *et al.*, 2001; Song *et al.*, 2006
2329.	*Smilax excelsa*	Smilacaceae	Leaves	Souri *et al.*, 2004; Ozsoy *et al.*, 2008; Özen, 2010
2330.	*Smilax glabra*	Smilacaceae	Root	Cai *et al.*, 2004; Li *et al.*, 2008
2331.	*Smilax glyciphylla*	Smilacaceae	Leaves	Cox *et al.*, 2005
2332.	*Smilax zeylanica*	Smilaceae	Leaves	Murali *et al.*, 2010
2333.	*Solanum aculeastrum*	Solanaceae	Fruit	Kodru *et al.*, 2006
2334.	*Solanum americanum*	Solanaceae		Iwalewa *et al.*, 2005
2335.	*Solanum brevifolium*	Solanaceae		Mosquera *et al.*, 2007
2336.	*Solanum centrale*	Solanaceae	Fruit	Konczak *et al.*, 2010
2337.	*Solanum deflexiflorum*	Solanaceae		Mosquera *et al.*, 2007
2338.	*Solanum fastigiatum*	Solanaceae		Sabir and Rocha, 2008
2339.	*Solanum indicum*	Solanaceae		Narayanaswamy and Balakrishnan, 2011
2340.	*Solanum leucocarpum*	Solanaceae		Mosquera *et al.*, 2007
2341.	*Solanum macrocarpon*	Solanaceae	Fruit	Oboh and Akindahunsi, 2004; Odukoya *et al.*, 2007; Morrison and Twumasi, 2010
2342.	*Solanum melongena*	Solanaceae	Fruits	Sudheesh *et al.*, 1999; Noda *et al.*, 2000; Cai *et al.*, 2004; Huang *et al.*, 2004a; Maeda *et al.*, 2006; Odukoya *et al.*, 2007; Chanasut and Rattanapanone, 2008; Nisha *et al.*, 2009; Akanitapichat *et al.*, 2010; Gacche *et al.*, 2010

Contd...

Table 3.1–*Contd...*

Sl.No.	*Plant Name*	*Family*	*Parts Used*	*References*
2343.	*Solanum nigrum*	Solanaceae	Leaves	Sultana *et al.*, 1995; Kumar *et al.*, 2001; Cai *et al.*, 2004; Jainu and Shyamala Devi, 2004; Heo *et al.*, 2004; Abbas *et al.*, 2006; Al-Fantimi *et al.*, 2007; Mohammed *et al.*, 2007; Adebooye *et al.*, 2008; Patnibul *et al.*, 2008; Veeru *et al.*, 2009; Jeong *et al.*, 2010; Maharana *et al.*, 2010
2344.	*Solanum ochranthum*	Solanceae		Mosquera *et al.*, 2007
2345.	*Solanum ovalifolium*	Solanaceae		Mosquera *et al.*, 2007
2346.	*Solanum pseudocapsicum*	Solanaceae	Leaves	Badami *et al.*, 2005
2347.	*Solanum quitoense*	Solanaceae	Fruit	Acosta *et al.*, 2009
2348.	*Solanum torvum*	Solanaceae	Seeds	Sivapriya and Leela, 2007; Chanasut and Rattanapanone, 2008; Kusirisin *et al.*, 2009
2349.	*Solanum trilobatum*	Solanaceae		Sini and Devi, 2004
2350.	*Solanum tuberosum*	Solanaceae	Tubers, peel	Sai Khan *et al.*, 1995; Kahkonen *et al.*, 1999; Penarrieeta *et al.*, 2005; Lachman *et al.*, 2008; Pihlanto *et al.*, 2008
2351.	*Solidago canadensis*	Asteraceae		McCune 1999; Kahkonen *et al.*, 1999; Wojcikowski *et al.*, 2007
2352.	*Solidago virgaurea*	Asteraceae	Herb	Filipek, 1994; Meyer *et al.*, 1995; Kahkonen *et al.*, 1999; Wojcikowski *et al.*, 2007
2353.	*Sonchus arvensis*	Asteraceae		Kahkonen *et al.*, 1999; Jimoh *et al.*, 2011
2354.	*Sonchus asper*	Asteraceae		Jimoh *et al.*, 2011
2355.	*Sonchus oleraceus*	Asteraceae		Yin *et al.*, 2007; Jimoh *et al.*, 2011
2356.	*Sonneratia apetala*	Sonneratiaceae	Leaves, flowers	Vadlapudi and Naidu, 2009
2357.	*Sophora alopecuroides*	Fabaceae		Souri *et al.*, 2004
2358.	*Sophora chrysophylla*	Fabaceae	Root	Toda and Shirataki, 2005
2359.	*Sophora flavescens*	Fabaceae	Root	Cai *et al.*, 2004; Kim *et al.*, 2006; Li *et al.*, 2008
2360.	*Sophora japonica*	Fabaceae	Flower buds	Paniwnyk *et al.*, 2001; Cai *et al.*, 2004, 2006; Gan *et al.*, 2010
2361.	*Sophora mooracraftiana*	Fabaceae		Toda and Shirataki, 2004
2362.	*Sophora tonkinensis*	Fabaceae	Root	Cai *et al.*, 2004, 2006
2363.	*Sorbus americana*	Rosaceae	Leaves	Acuna *et al.*, 2002
2364.	*Sorbus aria*	Rosaceae	Fruit, inflorescence, leaves	Olszewska and Michel, 2009

Contd...

Table 3.1–*Contd...*

Sl.No.	Plant Name	Family	Parts Used	References
2365.	*Sorbus aucuparia*	Rosaceae	Fruit, inflo-rescence, leaves	Kahkonen *et al.*, 1999; Olszewska and Michel, 2009; Ionska-Rys *et al.*, 2009
2366.	*Sorbus domestica*	Rosaceae		Termentzi *et al.*, 2006
2367.	*Sorbus intermedia*	Rosaceae	Fruit, inflo-rescence, leaves	Olszewska and Michel, 2009
2368.	*Sorbus torminalis*	Rosaceae	Fruit	Serteser *et al.*, 2009
2369.	*Sorghum bicolor* (Syn.: *Sorghum vulgare*)	Poaceae	Fruit	Awika *et al.*, 2003; Dykes *et al.*, 2005; Dlamini *et al.*, 2007; Sreeramulu *et al.*, 2009
2370.	*Soymida febrifuga*	Meliaceae	Leaves	Srinivas Reddy *et al.*, 2008a, b
2371.	*Sparganium stoloniferum*	Sparganiaceae		Gan *et al.*, 2010
2372.	*Spartium junceum*	Fabaceae		Yesilada *et al.*, 2000
2373.	*Spathodea campanulata*	Bignoniaceae	Aerial parts	Houghton *et al.*, 2005; Nazif, 2007; Shanmukha *et al.*, 2010
2374.	*Spathodea tilotica*	Bignoniaceae		Moussa *et al.*, 2011
2375.	*Spatholobus suberectus*	Fabaceae	Stem	Cai *et al.*, 2004
2376.	*Spergularia rubra*	Caryophyllaceae	Herb	Vinholes *et al.*, 2011
2377.	*Spermacoce articularis*	Rubiaceae		Saha *et al.*, 2004
2378.	*Spermacoce exilis*	Rubiaceae		Saha *et al.*, 2004; Ahmad *et al.*, 2010
2379.	*Spermacoce hispida*	Rubiaceae	Seeds	Kaviarasan *et al.*, 2008
2380.	*Spermacoce latifolia*	Rubiaceae		Ahmad *et al.*, 2010
2381.	*Sphaeranthus indicus*	Asteraceae		Tiwari and Khosa, 2009; Vadlapudi and Naidu, 2010
2382.	*Sphallerocarpus gracilis*	Apiaceae	Herb	Myagmar and Aniya, 2000;
2383.	*Sphenocentrum jollyanum*	Menisperma-ceae	Leaf, root	Oke and Hamburger, 2002
2384.	*Sphenostylis stenocarpa*	Fabaceae		Oboh, 2006a, 2009
2385.	*Splianthes acmella*	Asteraceae		Shirwaikar *et al.*, 2006; Wongsawatkul *et al.*, 2008; Tanwer *et al.*, 2010
2386.	*Spinacea oleracea*	Chenopodiaceae	Leaf	Cai *et al.*, 2004; Joseph *et al.*, 1999; Maeda *et al.*, 2006; Gacche *et al.*, 2010; Al-Dosani, 2010; Evanjelene and Natarajan, 2011.
2387.	*Spinacia turkestanica*	Chenopodiaceae	Herb	Motamed and Naghibi, 2010
2388.	*Spiraea alba*	Rosaceae	Seeds	Borchardt *et al.*, 2008
2389.	*Spiraea tomentosa*	Rosaceae	Seeds	Borchardt *et al.*, 2008
2390.	*Spirodela polyrrhiza*	Lemnaceae	Herb	Song *et al.*, 2010
2391.	*Spondias mangifera*	Anacardiaceae	Root	Acharyya *et al.*, 2010

Contd...

Table 3.1–*Contd...*

Sl.No.	*Plant Name*	*Family*	*Parts Used*	*References*
2392.	*Spondias mombin*	Anacardiaceae		Akinmoladun *et al.*, 2010
2393.	*Spondias pinnata*	Anacardiaceae	Stem bark	Hazra *et al.*, 2008
2394.	*Stachys acerosa*	Lamiaceae	Aerial parts	Salehi *et al.*, 2007
2395.	*Stachys alpina*	Lamiaceae	Herb	Haznagy-Radnai *et al.*, 2006
2396.	*Stachys alpina* ssp. *dinarica*	Lamiaceae	Herb	Kukic *et al.*, 2006
2397.	*Stachys anisochila*	Lamiaceae	Herb	Kukic *et al.*, 2006
2398.	*Stachys annua*	Lamiaceae	Herb	Haznagy-Radnai *et al.*, 2006
2399.	*Stachys beckeana*	Lamiaceae	Herb	Kukic *et al.*, 2006
2400.	*Stachys byzantina*	Lamiaceae	Aerial parts	Erdemoglu *et al.*, 2006; Khanavi *et al.*, 2009
2401.	*Stachys fruticulosa*	Lamiaceae	Aerial parts	Khanavi *et al.*, 2009
2402.	*Stachys iberica*	Lamiaceae	Herb	Tepe *et al.*, 2011
2403.	*Stachys inflata*	Lamiaceae	Aerial parts	Khanavi *et al.*, 2009; Ebrahimabadi *et al.*, 2010
2404.	*Stachys lavandulifolia*	Lamiaceae		Souri *et al.*, 2004
2405.	*Stachys laxa*	Lamiaceae	Aerial parts	Khanavi *et al.*, 2009
2406.	*Stachys macrantha*	Lamiaceae	Herb	Haznagy-Radnai *et al.*, 2006
2407.	*Stachys officinalis*	Lamiaceae	Herb	Haznagy-Radnai *et al.*, 2006; Matkowski and Piotrowska, 2006; Wojcikowski *et al.*, 2007
2408.	*Stachys persica*	Lamiaceae	Aerial parts	Khanavi *et al.*, 2009
2409.	*Stachys plumosa*	Lamiaceae	Herb	Kukic *et al.*, 2006
2410.	*Stachys prumeri*	Lamiaceae		Couladis *et al.*, 2003
2411.	*Stachys recta*	Lamiaceae	Herb	Haznagy-Radnai *et al.*, 2006
2412.	*Stachys setifera*	Lamiaceae	Aerial parts	Khanavi *et al.*, 2009
2413.	*Stachys subaphylla*	Lamiaceae	Aerial parts	Khanavi *et al.*, 2009
2414.	*Stachys sylvatica*	Lamiaceae	Herb	Haznagy-Radnai *et al.*, 2006
2415.	*Stachys spruneri*	Lamiaceae	Aerial parts	Couladis *et al.*, 2003
2416.	*Stachys trinervis*	Lamiaceae	Aerial parts	Khanavi *et al.*, 2009
2417.	*Stachys turcomanica*	Lamiaceae	Aerial parts	Khanavi *et al.*, 2009
2418.	*Stachytarpheta angustifolia*	Verbenaceae	Leaves	Awah *et al.*, 2010
2419.	*Stachytarpheta cayennensis*	Verbenaceae		De Souza *et al.*, 2011
2420.	*Stachytarpheta jamaicensis*	Verbenaceae	Leaves	Alvarez *et al.*, 2004
2421.	*Stellaria chamaejasme*	Caryophyllaceae	Herb	Myagmar and Aniya, 2000;

Contd...

Table 3.1–*Contd...*

Sl.No.	Plant Name	Family	Parts Used	References
2422.	*Stellaria dichotoma*	Caryophyllaceae		Li *et al.*, 2008; Lim *et al.*, 2008
2423.	*Stellaria media*	Caryophyllaceae		Pieroni *et al.*, 2002
2424.	*Stemmadenia bella*	Apocynaceae		Ruiz-Terán *et al.*, 2008
2425.	*Stemona burkillii*	Stemonaceae	Root	Brem *et al.*, 2004;
2426.	*Stemona cochinchinensis*	Stemonaceae	Root	Brem *et al.*, 2004
2427.	*Stemona collinasae*	Stemonaceae	Root	Brem *et al.*, 2004
2428.	*Stemona curtisii*	Stemonaceae	Root	Brem *et al.*, 2004
2429.	*Stemona kerrii*	Stemonaceae	Root	Brem *et al.*, 2004
2430.	*Stemona pierrei*	Stemonaceae	Root	Brem *et al.*, 2004
2431.	*Stemona tuberosa*	Stemonaceae	Root	Brem *et al.*, 2004
2432.	*Stemona sessilifolia*	Stemonaceae	Root	Cai *et al.*, 2004; Song *et al.*, 2010
2433.	*Stephania erecta*	Menispeermaceae		Patnibul *et al.*, 2008
2434.	*Stephania hemandifolia*	Menispermaceae		Sharma *et al.*, 2010
2435.	*Stephania japonica*	Menispermaceae	Leaf	Rahman *et al.*, 2011
2436.	*Stephania rotunda*	Menispermaceae		Gülçin *et al.*, 2010
2437.	*Stephania venosa*	Menispermaceae		Leewanich *et al.*, 2011
2438.	*Sterculia diversifolia*	Sterculiaceae		Moussa *et al.*, 2011
2439.	*Sterculia lychnophora*	Sterculiaceae		Lam *et al.*, 2011
2440.	*Sterculia scapigera*	Sterculiaceae		Song *et al.*, 2010
2441.	*Sterculia striata*	Sterculiaceae	Stem bark	Costa *et al.*, 2010
2442.	*Sterculia tragacantha*	Sterculiaceae		Udegbunam *et al.*, 2011
2443.	*Sterculia villosa*	Sterculiaceae		Kshirsagar and Upadhyay, 2009
2444.	*Stereospermum colias*	Bignoniaceae	Leaf	Vijaya Bharathi *et al.*, 2010
2445.	*Stereospermum kunthianum*	Bignoniaceae		Compaore *et al.*, 2011
2446.	*Stereospermum personatum*	Bignoniaceae	Stem, bark	Sampath Kumar *et al.*, 2005
2447.	*Stereospermum suaveolens*	Bignoniaceae		Chandrasekhar *et al.*, 2010; Balasubramanian and Chatterjee, 2010
2448.	*Stevia rebaudiana*	Asteraceae	Leaves	Tadhani *et al.*, 2007; Phansawan and Poungbangpho, 2007; Ghanta *et al.*, 2007; Shukla *et al.*, 2009; Kim *et al.*, 2011; Ahmad *et al.*, 2011

Contd...

Table 3.1–*Contd...*

Sl.No.	Plant Name	Family	Parts Used	References
2449.	*Streblus asper*	Moraceae		Kanti *et al.*, 2009; Riwari, 2010; Narayanaswamy and Balakrishnan, 2011; Choudhury *et al.*, 2011
2450.	*Striga asiatica*	Scrophulariaceae	Leaves	Badami *et al.*, 2003
2451.	*Striga hermonthica*	Scrophulariaceae	Herb	Kiendrebeogo *et al.*, 2005
2452.	*Striga orobanchoides*	Scrophulariaceae	Herb	Raj and Shalini, 1999; Badami *et al.*, 2003
2453.	*Strobilanthus crispus*	Acanthaceae	Leaves	Ismail *et al.*, 2000; Abu Bakar *et al.*, 2004
2454.	*Strophanthus hispidus*	Apocynaceae	Stem, root	Ayoola *et al.*, 2008
2455.	*Struchium sparganophora*	Asteraceae	Leaves	Oboh and Akindahunsi, 2004; Odukoya *et al.*, 2007
2456.	*Strychnos henningsii*	Loganiaceae	Bark	Oyedemi *et al.*, 2010
2457.	*Strychnos minor*	Loganiaceae		Patnibul *et al.*, 2008
2458.	*Strychnos potatorum*	Loganiaceae	Seeds	Sanmugapriya and Venkataraman, 2006; Dhobal *et al.*, 2008
2459.	*Strychnos spinosa*	Loganiaceae		Nhukarume *et al.*, 2010
2460.	*Strychnos vanprukii*	Loganiaceae	Stem	Thongphasuk *et al.*, 2004
2461.	*Stryphnodendron obovatum*	Fabaceae	Stem bark	Sanches *et al.*, 2005
2462.	*Stryphnodendron polyphyllum*	Fabaceae	Stem bark	Lopes *et al.*, 2005
2463.	*Styrax formosana*	Loganiaceae		Hou *et al.*, 2003
2464.	*Styrax japonica*	Styracaceae	Stem bark	Min *et al.*, 2004
2465.	*Succisa pratensis*	Dipsacaceae		Kahkonen *et al.*, 1999
2466.	*Sutherlandia frutescens*	Fabaceae		Fernandes *et al.*, 2004; Katere and Eloff, 2005
2467.	*Swartzia langsdorffi*	Fabaceae		Santos *et al.*, 2009
2468.	*Swertia chirata*	Gentianaceae		Scartezzini and Speroni, 2000; Chen *et al.*, 2011
2469.	*Swertia corymbosa*	Gentianaceae	Herb	Selvameena *et al.*, 2010
2470.	*Swertia decussata*	Gentianaceae	Herb	Patro *et al.*, 2005
2471.	*Swertia japonica*	Gentianaceae	Herb	Ashida *et al.*, 1994
2472.	*Swertia speciosa*	Gentianaceae	Rhizomes	Rawat *et al.*, 2004; Rana and Rawat, 2005
2473.	*Swietenia humilis*	Meliaceae		Ruiz-Terán *et al.*, 2008
2474.	*Swietenia macrophylla*	Meliaceae	Bark, leaves	Falah *et al.*, 2008; Tan *et al.*, 2009

Contd...

Table 3.1–*Contd...*

Sl.No.	Plant Name	Family	Parts Used	References
2475.	*Swietenia mahagoni*	Meliaceae	Bark, seeds	Sahgal *et al.*, 2009; Panda *et al.*, 2010; Hajra *et al.*, 2011
2476.	*Symphonia globulifera*	Clusiaceae	Seed shells	Ngouela *et al.*, 2006
2477.	*Symphytum asperum*	Boraginaceae		Barthomeuf *et al.*, 2001; Barbakadze *et al.*, 2005; Merlani *et al.*, 2010
2478.	*Symphytum caucasicum*	Boraginaceae		Barbakadze *et al.*, 2005; Merlani *et al.*, 2010
2479.	*Symphytum officinale*	Boraginaceae	Roots	Barbakadze *et al.*, 2009
2480.	*Symplocos cochinchinensis*	Symplocaceae		Sunil and Ignacimuthu, 2011
2481.	*Symplocos racemosa*	Symplocaceae		Vijayabaskaran *et al.*, 2010
2482.	*Synedrella nodiflora*	Asteraceae		Wijaya *et al.*, 2011
2483.	*Syringa pubescens*	Oleaceae		Xu *et al.*, 2010
2484.	*Syringa vulgaris*	Oleaceae	Flowers, cortex	Kostyuk *et al.*, 2008
2485.	*Syzygium anisatum*	Myrtaceae		Konczak *et al.*, 2010
2486.	*Syzygium aqueum*	Myrtaceae		Osman *et al.*, 2009; Palanisamy *et al.*, 2011
2487.	*Syzygium aromaticum*	Myrtaceae	Buds	Shymala *et al.*, 2003; Odukoya *et al.*, 2005; Wojdylo *et al.*, 2007; Viuda-Martos *et al.*, 2010; Niwano *et al.*, 2011
2488.	*Syzygium cerasoides*	Myrtaceae	Leaf, stem	Kshirsagar and Upadhyay, 2009
2489.	*Syzygium cumini*	Myrtaceae	Fruits, seeds	D'Mello *et al.*, 2000; Muruganandan *et al.*, 2001; Banerjee *et al.*, 2005; Bajpai *et al.*, 2005; Krishnamoorthy *et al.*, 2006; Ruan *et al.*, 2008; Reynertson *et al.*, 2008; Hasan *et al.*, 2009; Stangeland *et al.*, 2009; Mathur *et al.*, 2010; Udapurkar *et al.*, 2009; Hasan *et al.*, 2009; Farswan *et al.*, 2009
2490.	*Syzygium currani*	Myrtaceae	Fruits	Reynertson *et al.*, 2008;
2491.	*Syzygium formosanum*	Myrtaceae	Leaves	Lee *et al.*, 2006; Thuong *et al.*, 2006
2492.	*Syzygium gratum*	Myrtaceae	Leaf	Kukongviriyapan *et al.*, 2007
2493.	*Syzygium jambos*	Myrtaceae	Fruits	Reynertson *et al.*, 2008;
2494.	*Syzygium javanicum*	Myrtaceae	Fruits	Reynertson *et al.*, 2008;
2495.	*Syzygium malaccense*	Myrtaceae	Fruit	Lako *et al.*, 2006; Reynertson *et al.*, 2008
2496.	*Syzygium samarangense*	Myrtaceae	Fruit	Reynertson *et al.*, 2008; Chen, C-H. *et al.*, 2009
2497.	*Tabebuia impetiginosa*	Bignoniaceae	Volatile oil	Park *et al.*, 2003; Yan *et al.*, 1999

Contd...

Table 3.1–*Contd...*

Sl.No.	*Plant Name*	*Family*	*Parts Used*	*References*
2498.	*Tabebuia rosea*	Bignoniaceae	Leaves	Ramalakshmi and Muthuchelian, 2010
2499.	*Tabernaemontana divaricata*	Apocynaceae	Leaves	Jain *et al.*, 2007
2500.	*Tacca leontopetaloides*	Taccaceae		Habila *et al.*, 2011
2501.	*Tachigalia paniculata*	Caesalpiniaceae	Leaves	Cioffi *et al.*, 2002
2502.	*Tagetes erecta*	Asteraceae	Flowers	Wang *et al.*, 2006; Li *et al.*, 2007
2503.	*Tagetes lucida*	Asteraceae	Leaves	Aquino *et al.*, 2002
2504.	*Tagetes maxima*	Asteraceae	Flowers	Parejo *et al.*, 2005
2505.	*Tagetes mendocina*	Asteraceae	Aerial parts	Schmeda-Hirschmann *et al.*, 2004
2506.	*Tagetes multiflora*	Asteraceae		Rosas-Romero and Saavedra, 2005
2507.	*Talinum portulacifolium*	Portulacaceae	Leaves	Babu *et al.*, 2009
2508.	*Talinum triangulare*	Portulacaceae	Leaf, root	Estiasih and Kurniawan, 2006; Odukoya *et al.*, 2007; Andarwulan *et al.*, 2010; Morrison and Twumasi, 2010; Adefegha and Oboh, 2011; Liang *et al.*, 2011
2509.	*Tamarindus indica*	Caesalpiniaceae	Seed, Fruits (skin)	Tsuda *et al.*, 1994; Chanwitheesuk *et al.*, 2005; Sudjaroen *et al.*, 2005 ; Siddharaju *et al.*, 2007; Lamien-Meda *et al.*, 2008; Povichit *et al.*, 2010
2510.	*Tamarix aphylla*	Tamaricaceae	Aerial parts	Vadlapudi and Naidu, 2009
2511.	*Tamarix aralensis*	Tamaricaceae		Souri *et al.*, 2004
2512.	*Tamarix hispida*	Tamaricaceae	Aerial parts	Sulanova *et al.*, 2004
2513.	*Tamarix nilotica*	Tamaricaceae	Herb	Abouzid *et al.*, 2008
2514.	*Tanacetum budjnurdense*	Asteraceae		Esmaeili *et al.*, 2010
2515.	*Tanacetum balsamita*	Asteraceae		Nikolova and Dzhurmanski, 2009
2516.	*Tanacetum chiliophyllum*	Asteraceae		Esmaeili *et al.*, 2010
2517.	*Tanacetum densum*	Asteraceae		Tepe and Sokmen, 2006
2518.	*Tanacetum hololeucum*	Asteraceae		Esmaeili *et al.*, 2010
2519.	*Tanacetum kotschyi*	Asteraceae		Esmaeili *et al.*, 2010
2520.	*Tanacetum macrophyllum*	Asteraceae		Javidnia *et al.*, 2010
2521.	*Tanacetum parthenium*	Asteraceae		Wu *et al.*, 2006; Lopez *et al.*, 2008; Koganov, 2009; Nikolova and Dzhurmanski, 2009
2522.	*Tanacetum polycephalum*	Asteraceae		Hamzeh Amiri. 2007
2523.	*Tanacetum sonboli*	Asteraceae		Esmaeili *et al.*, 2010
2524.	*Tanacetum tabrisianum*	Asteraceae		Esmaeili *et al.*, 2010

Contd...

Table 3.1–*Contd...*

Sl.No.	*Plant Name*	*Family*	*Parts Used*	*References*
2525.	*Tanacetum vulgare*	Asteraceae	Aerial parts	Kahkonen *et al.*, 1999; Xie *et al.*, 2007; Juan-Badaturuqe *et al.*, 2009
2526.	*Tapinanthus globiferus*	Loranthaceae	Leaves	Cook *et al.*, 1998
2527.	*Taraxacum mongolicum*	Asteraceae	Whole plant	Cai *et al.*, 2004; Li *et al.*, 2008
2528.	*Taraxacum officinale*	Asteraceae	Root, herb, flower, leaf	Pendry *et al.*, 2005; Hu and Kitts, 2005; Katalinic *et al.*, 2006; Wojcikowski *et al.*, 2007; Sengul *et al.*, 2009; Choi *et al.*, 2010
2529.	*Tarenna attenuata*	Rubiaceae	Whole plant	Yang *et al.*, 2006
2530.	*Tasmannia lanceolata*	Magnoliaceae		Konczak *et al.*, 2010
2531.	*Tecoma radicans*	Bignoniaceae	Aerial parts	Hashem, 2008
2532.	*Tecoma stans*	Bignoniaceae		Govindappa *et al.*, 2011
2533.	*Tecomaria capensis*	Bignoniaceae		Moussa *et al.*, 2011
2534.	*Tectona grandis*	Verbenaceae	Leaf, bark, wood	Krishna and Nari, 2010, 2011
2535.	*Telfairia occidentalis*	Cucurbitaceae	Leaf	Oboh and Akindahunsi, 2004; Nwanna and Oboh, 2007; Odukoya *et al.*, 2007; Adefegha and Oboh, 2011
2536.	*Telosma minor*	Asclepiadaceae		Chanwitheesuk *et al.*, 2005
2537.	*Tephrosia egregia*	Fabaceae		Arriaga *et al.*, 2009
2538.	*Tephrosia purpurea*	Fabaceae	Leaves	Soni *et al.*, 2003, 2006; Patel *et al.*, 2010; Gunjegaonkar *et al.*, 2010; Nile and Khobragade, 2011
2539.	*Tephrosia tinctoria*	Fabaceae		Rajaram and Suresh Kumar, 2011
2540.	*Terminalia arjuna*	Combretaceae	Fruits, leaves, bark	Munasinghe *et al.*, 2001; Mau *et al.*, 2003; Tilak *et al.*, 2003; Chander *et al.*, 2004; Bajpai *et al.*, 2005; Gauthaman *et al.*, 2005; Sultan *et al.*, 2007; Koti Reddy *et al.*, 2008; Sai Ramesh *et al.*, 2010; Povichit *et al.*, 2010; Biswas *et al.*, 2010
2541.	*Terminalia bellerica*	Combretaceae	Fruits, leaves, bark	Vani *et al.*, 1997; Saleem *et al.*, 2001; Bajpai *et al.*, 2005; Naik *et al.*, 2005; Shivaprasad *et al.*, 2008; Jain and Agrawal, 2009; Sabu and Kuttan, 2009; Sherin *et al.*, 2010; Povichit *et al.*, 2010
2542.	*Terminalia brachystemma*	Combretaceae	Leaves	Masoko and Eloff, 2007
2543.	*Terminalia catappa*	Combretaceae	leaves	Ko *et al.*, 2002; Mau *et al.*, 2003; Kinoshita *et al.*, 2007

Contd...

Table 3.1–*Contd...*

Sl.No.	*Plant Name*	*Family*	*Parts Used*	*References*
2544.	*Terminalia chebula*	Combretaceae	Friuts, leaves, bark	Vani *et al.*, 1997; Kim *et al.*, 1997; Saleem *et al.*, 2001; Naik *et al.*, 2003; Cai *et al.*, 2004, 2006; Jayajothi *et al.*, 2004; Naik *et al.*, 2004, 2005; Lee *et al.*, 2005; Bajpai *et al.*, 2005; Shivaprasad *et al.*, 2008; Jain and Agrawal, 2009
2545.	*Terminalia crenulata*	Combretaceae	Bark	Jain *et al.*, 2010
2546.	*Terminalia gazensis*	Combretaceae	Leaves	Masoko and Eloff, 2007
2547.	*Terminalia mollis*	Combretaceae	Leaves	Masoko and Eloff, 2007
2548.	*Terminalia muelleri*	Combretaceae	Fruits, leaves, bark	Bajpai *et al.*, 2005
2549.	*Terminalia pallida*	Combretaceae		Gupta *et al.*, 2005
2550.	*Terminalia prunoides*	Combretaceae	Leaves	Masoko and Eloff, 2007
2551.	*Terminalia sambesiaca*	Combretaceae	Leaves	Masoko and Eloff, 2007
2552.	*Terminalia sericea*	Combretaceae	Leaves	Masoko and Eloff, 2007
2553.	*Terminalia triflora*	Combretaceae		Desmarchelier *et al.*, 1997
2554.	*Terorchidium andinum*	Euphorbiaceae		Mosquera *et al.*, 2007
2555.	*Tetracera loureiri*	Dilleniaceae		Kukongviriyapan *et al.*, 2003
2556.	*Tetracera potatoria*	Dilleniaceae		Oluwole *et al.*, 2008
2557.	*Tetracera volubilis*	Dilleniaceae		Lock *et al.*, 2005
2558.	*Teucrium chamaedrys*	Lamiaceae	Leaf, root	Kadifkova-Panovska *et al.*, 2005; Katalinic *et al.*, 2006; Pacifico *et al.*, 2009; Gursoy and Tepe, 2009
2559.	*Teucrium marum*	Lamiaceae	Essential oil	Ricci *et al.*, 2005
2560.	*Teucrium montanum*	Lamiaceae	Plant	Kadifkova-Panovska *et al.*, 2005; Djilas *et al.*, 2006; Canadanovic-Brunet *et al.*, 2006
2561.	*Teucrium polium*	Lamiaceae	Shoot	Couladis *et al.*, 2003; Suboh *et al.*, 2004; Arlorio *et al.*, 2005; Kadifkova-Panovska *et al.*, 2005; Ljubuncic *et al.*, 2006; Al-Mustafa and Al-Thunibat, 2008; Esmaeili *et al.*, 2009; Sharififar *et al.*, 2009; Firuzi *et al.*, 2010
2562.	*Thalictrum minus*	Ranunculaceae	Herb	Karyagina *et al.*, 2011
2563.	*Thelesperma megapotamicum*	Asteraceae		Borneo *et al.*, 2008
2564.	*Theobroma cacao*	Sterculiaceae	Fruits, leaves	Admson *et al.*, 1999; Lee *et al.*, 2003; Osman *et al.*, 2004; Arlorio *et al.*, 2005; Calder *et al.*, 2009; Othman *et al.*, 2010

Contd...

Table 3.1–*Contd...*

Sl.No.	*Plant Name*	*Family*	*Parts Used*	*References*
2565.	*Theobroma grandiflorum*	Sterculiaceae	Seeds	Yang *et al.*, 2003
2566.	*Thespesia populnea*	Malvaceae	Barks	Ilavaran *et al.*, 2003
2567.	*Thespesia populneoides*	Malvaceae	Leaves, flowers	Vadlapudi and Naidu, 2009
2568.	*Thevetia nerifolia*	Apocynaceae	Stem, leaf	Ayoola *et al.*, 2008
2569.	*Thonningia sanguinea*	Balanophoraceae		Gyamfi *et al.*, 1999; Ohtani *et al.*, 2000
2570.	*Thunbergia laurifolia*	Acanthaceae	Leaves	Chn and Lim, 2006; Oonsivilai *et al.*, 2008; Palipoch *et al.*, 2010
2571.	*Thymbra spicata*	Lamiaceae	Aerial parts	Kosar *et al.*, 2003; Dorman *et al.*, 2004; Avci *et al.*, 2006
2572.	*Thymophylla pentachaeta*	Asteraceae		Borneo *et al.*, 2008
2573.	*Thymus boveii*	Lamiaceae	Herb	Tepe *et al.*, 2011
2574.	*Thymus bracteosus*	Lamiaceae	Herb	Brantner *et al.*, 2005
2575.	*Thymus caespititius*	Lamiaceae	Essential oil	Miguel *et al.*, 2004; Dandlen *et al.*, 2010
2576.	*Thymus camphoratus*	Lamiaceae	Herb	Miguel *et al.*, 2004; Dandlen *et al.*, 2010
2577.	*Thymus capitatus*	Lamiaceae	Shoot	Al-Mustafa and Al-Thunibat, 2008
2578.	*Thymus capitellatus*	Lamiaceae	Herb	Dandlen *et al.*, 2010
2579.	*Thymus caramanicus*	Lamiaceae	Essential oil	Safaei-Ghomi *et al.*, 2009
2580.	*Thymus carnosus*	Lamiaceae	Essential oil	Dandlen *et al.*, 2010
2581.	*Thymus daenensis*	Lamiaceae	Aerial parts	Firuzi *et al.*, 2010
2582.	*Thymus eigii*	Lamiaceae	Aerial parts	Tepe *et al.*, 2004
2583.	*Thymus guyonii*	Lamiaceae	Essential oil	Hazzit *et al.*, 2006
2584.	*Thmus hirtus*	Lamiaceae	Essential oil	Ahmed *et al.*, 2011
2585.	*Thymus hyemalis*	Lamiaceae	Herb	Tepe *et al.*, 2011
2586.	*Thymus longicaulis* subsp. *longicaulis* var. *longicaulis*	Lamiaceae	Essential oil	Sarikurkcu *et al.*, 2010
2587.	*Thymus marschallinus*	Lamiaceae	Essential oil	Budinev *et al.*, 1995; Jia *et al.*, 2010
2588.	*Thymus mastichina*	Lamiaceae	Essential oil	Miguel *et al.*, 2004
2589.	*Thymus migricus*	Lamiaceae		Souri *et al.*, 2004
2590.	*Thymus mumbyanus*	Lamiaceae	Essential oil	Hazzit *et al.*, 2006
2591.	*Thymus pallescens*	Lamiaceae	Essential oil	Hazzit *et al.*, 2006
2592.	*Thymus pectinatus*	Lamiaceae	Methanol extracts	Vardar Unlu *et al.*, 2003
2593.	*Thymus praecox* subsp. *caucasicus* var. *caucasicus*	Lamiaceae	Herb	Orhan *et al.*, 2009

Contd...

Table 3.1–*Contd...*

Sl.No.	*Plant Name*	*Family*	*Parts Used*	*References*
2594.	*Thymus proximus*	Lamiaceae	Essential oil	Jia *et al.*, 2010
2595.	*Thymus pubescens*	Lamiaceae		Souri *et al.*, 2004
2596.	*Thymus pulegopides*	Lamiaceae	Essential oil	Dandlen *et al.*, 2010
2597.	*Thymus satureioides*	Lamiaceae	Leaves	Ismaili *et al.*, 2004
2598.	*Thymus serpyllum*	Lamiaceae	Herb	Katalinic *et al.*, 2006
2599.	*Thymus sipyleus*	Lamiaceae	Herb	Tepe *et al.*, 2005; Özgen *et al.*, 2011
2600.	*Thymus spathulifolius*	Lamiaceae	Essential oil	Sokmen *et al.*, 2004
2601.	*Thymus vulgaris*	Lamiaceae	Leaves	Haraguchi *et al.*, 1996; Kahkonen *et al.*, 1999; Zheng *et al.*, 2001; Dapkevivius *et al.*, 2002; Miura *et al.*, 2002; Sacchetti *et al.*, 2005; Agbor *et al.*, 2005; Mishra *et al.*, 2007; Wei and Shibomoto, 2007; Chizzola *et al.*, 2008: Mariutti *et al.*, 2008; Viuda-Martos *et al.*, 2010; Grosso *et al.*, 2010; Spiridon *et al.*, 2011
2602.	*Thymus zygis*	Lamiaceae	Essential oil	Soares *et al.*, 1997; Dorman and Deans, 2004; Dandlen *et al.*, 2010
2603.	*Tibouchia grossa*	Melastomata-ceae		Mosquera *et al.*, 2009
2604.	*Tilia argentea*	Tiliaceae	Leaves	Yildirim *et al.*, 2000; Demiray *et al.*, 2009
2605.	*Tilia cordata*	Tiliaceae	Flowers	Katalinic *et al.*, 2006
2606.	*Tilia platyphyllos*	Tiliaceae	Flowers	Katalinic *et al.*, 2006
2607.	*Tilia tomentosa*	Tiliaceae	Flowers	Yildirim *et al.*, 2000
2608.	*Tinospora cordifolia*	Menisperma-ceae	Stem and leaves	Mathew and Kuttan 1997; Bhattacharya *et al.*, 1997; Prince and Menon 1999; Subramanian *et al.*, 2002; Stanley *et al.*, 2004; Singh *et al.*, 2006; Jadhao *et al.*, 2009; Hasan *et al.*, 2009; Sadiq *et al.*, 2010; Lavanya *et al.*, 2010
2609.	*Tinospora crispa*	Menisperma ceae		Cavin *et al.*,1998; Amom *et al.*, 2009
2610.	*Toddalia asiatica*	Rutaceae	Stem, roots	Madhavan *et al.*, 2010; Pakutharivu and Suriyavadhana, 2010; Sithisarn and Jarikasem, 2010; Bala-subramaniam *et al.*, 2012
2611.	*Toona sinensis*	Meliaceae	Leaves, shoots	Wang *et al.*, 2007; Chen *et al.*, 2007; Hseu *et al.*, 2008; Jiang *et al.*, 2009; Yang *et al.*, 2011
2612.	*Toona sureni*	Meliaceae	Leaves	Ekaprasada *et al.*, 2009
2613.	*Tordylium apulum*	Apiaceae		Pieroni *et al.*, 2002

Contd...

Table 3.1–*Contd...*

Sl.No.	*Plant Name*	*Family*	*Parts Used*	*References*
2614.	*Torenia concolor*	Scrophulariaceae		Chou *et al.*, 2009
2615.	*Torenia fournieri*	Scrophulariaceae	Flower	Shindo *et al.*, 2008
2616.	*Torreya grandis*	Scrophulariaceae	Seeds	Chen *et al.*, 2006; Shi *et al.*, 2009
2617.	*Torreya nucifera*	Scrophulariaceae	Leaves	Lee, W.S. *et al.*, 2006
2618.	*Tournefortia bicolor*	Boraginaceae		Da Silva *et al.*, 2010
2619.	*Trachelospermum jasminoides*	Apocynaceae		Gan *et al.*, 2010
2620.	*Trachyspermum ammi* (Syn.: *Carum copticum*)	Apiaceae	Fruit	Rajesh and Mehta Joseph, 1994; Souri *et al.*, 2008; Anilakumar *et al.*, 2009; Khanum *et al.*, 2011
2621.	*Trachyspermum copticum*	Apiaceae	Fruit	Souri *et al.*, 2008; Nickavar and Abolhasani, 2009
2622.	*Trachystemon orientalis*	Boraginaceae	Stem, petiole	Özen, 2010
2623.	*Tragia pungens*	Euphorbiaceae		Mothana *et al.*, 2010
2624.	*Tragopogon aureus*	Asteraceae		Coruh *et al.*, 2007
2625.	*Trapa bispinosa*	Trapaceae		Kim *et al.*, 1997;
2626.	*Trapa natans*	Trapaceae	Fruits	Malaviya *et al.*, 2010
2627.	*Trapa pseudoincisa*	Trapaceae	Herb	Song *et al.*, 2007
2628.	*Trapa taiwanensis*	Trapaceae	Pericarp	Chiang and Ciou, 2009
2629.	*Trewia nudiflora*	Euphorbiaceae	Leaves, fruits	Prakash *et al.*, 2007
2630.	*Trewia polycarpa*	Euphorbiaceae	Roots	Chamundeshwari *et al.*, 1992
2631.	*Trianthema decandra*	Aizoaceae	Whole plant	Singh *et al.*, 2009
2632.	*Trianthema portulacastrum*	Aizoaceae	Herb	Mandal *et al.*, 1996.
2633.	*Tribulus terrestris*	Zygophyllaceae	Aerial parts	Singh *et al.*, 2009; Sengul *et al.*, 2009
2634.	*Trichilia emetica*	Meliaceae	Root	Germano *et al.*, 2006
2635.	*Trichilia prieuriana*	Meliaceae		Kuglerova *et al.*, 2008
2636.	*Trichilia roka*	Meliaceae		Germano *et al.*, 2001
2637.	*Tricholepis glaberrima*	Asteraceae		Naphade *et al.*, 2009
2638.	*Trichopus zeylanicus*	Dioscoreaceae	Seeds	Tharakan *et al.*, 2005
2639.	*Trichosanthes cucumeriana*	Cucurbitaceae	Fruit	Adebooye, 2007
2640.	*Trichosanthes kirilowii*	Cucurbitaceae	Peel	Cai *et al.*, 2004; Li *et al.*, 2008; Song *et al.*, 2010
2641.	*Tridax procumbens*	Asteraceae	Leaves	Ravikumar *et al.*, 2005; Nwanjo, 2008; Habila *et al.*, 2010
2642.	*Trifolium arvense*	Fabaceae	Herb	Kahkonen *et al.*, 1999
2643.	*Trifolium hybridum*	Fabaceae		Kahkonen *et al.*, 1999

Contd...

Table 3.1–*Contd...*

Sl.No.	Plant Name	Family	Parts Used	References
2644.	*Trifolium pannonicum*	Fabaceae	Herb	Godevac *et al.*, 2007
2645.	*Trifolium pratense*	Fabaceae	Herb	Kahkonen *et al.*, 1999
2646.	*Trifolium repens*	Fabaceae	Herb	Kahkonen *et al.*, 1999
2647.	*Trigonella foenum-graecum*	Fabaceae	Seeds	Sabu and Kuttan, 2003; Suboh *et al.*, 2004; Rababah *et al.*, 2004; Dixit *et al.*, 2005, 2007; Choudhary, 2006; Kaviarasan *et al.*, 2007; Khalaf *et al.*, 2008; Souri *et al.*, 2008; Subhasree *et al.*, 2009; Meera *et al.*, 2009; Gacche *et al.*, 2010
2648.	*Tripleurospermum disciforme*	Asteraceae	Herb	Souri *et al.*, 2006
2649.	*Tripleurospermum inodorum*	Asteraceae		Kahkonen *et al.*, 1999
2650.	*Tripleurospermum oreades*	Asteraceae		Özgen *et al.*, 2004
2651.	*Tripterygium wilfordii*	Celastraceae	Root	Cai *et al.*, 2004; Gan *et al.*, 2010
2652.	*Triticum aestivum*	Poaceae	Grains, wheat grass	Kahkonen *et al.*, 1999; Garcia *et al.*, 2000; Martinez-Tome *et al.*, 2004; Bajpai *et al.*, 2005; Kulakarini *et al.*, 2006; Shyam *et al.*, 2007; Siebenhandl *et al.*, 2007; Huang *et al.*, 2009; Sreeramulu *et al.*, 2009; Korycinska *et al.*, 2009; Shukla *et al.*, 2009; Alvarez-Jubete *et al.*, 2010
2653.	*Triticum durum*	Poaceae	Grains	Onyeneho and Hettiarachchy, 1992
2654.	*Triumfetta pilosa*	Tiliaceae		Ramakrishna *et al.*, 2011
2655.	*Triumfetta rhomboidea*	Tiliaceae	Roots	Lissy *et al.*, 2006; Sivakumar *et al.*, 2008
2656.	*Trollius chinensis*	Ranunculaceae		Zhao *et al.*, 2009; Li, H. *et al.*, 2011
2657.	*Trollius ledeori*	Ranunculaceae	Flower	Yang *et al.*, 2008
2658.	*Tropaeolum majus*	Tropaeolaceae	Flowers	Garzon and Wrolstad, 2009
2659.	*Tropaeolum tuberosum*	Tropaeolaceae	Tuber	Chirinos *et al.*, 2008; Campos *et al.*, 2006; Salluca *et al.*, 2008
2660.	*Tulbaghia violacea*	Liliaceae		Naidoo *et al.*, 2008
2661.	*Turnera diffusa*	Turneraceae	Aerial parts	Wojcikowski *et al.*, 2007; Garza-Juarez *et al.*, 2011
2662.	*Turnera ulmifolia*	Turneraceae		Nascimento *et al.*, 2006
2663.	*Tussilago farfara*	Asteraceae	Flowers	Cho *et al.*, 2005; Katalinic *et al.*, 2006; Özen, 2010; Song *et al.*, 2010
2664.	*Tylophora indica*	Asclepiadaceae	Leaves	Gupta *et al.*, 2011
2665.	*Typha capensis*	Typhaceae	Leaves, rhizome	Henkel *et al.*, 2011

Contd...

Table 3.1–*Contd...*

Sl.No.	*Plant Name*	*Family*	*Parts Used*	*References*
2666.	*Typha latifolia*	Typhaceae		Kahkonen *et al.*, 1999
2667.	*Typhonium flagelliforme*	Araceae		Mohan *et al.*, 2008
2668.	*Ullucus tuberosus*	Basellaceae	Tuber	Penarrieeta *et al.*, 2005; Campos *et al.*, 2006; Salluca *et al.*, 2008
2669.	*Ulmus* sp.	Ulmaceae		Nadjet *et al.*, 2008
2670.	*Ulmus davidiana*	Ulmaceae		Guo and Wang, 2007; Ko *et al.*, 2007; Jung *et al.*, 2008, 2010
2671.	*Ulmus pumila*	Ulmaceae		Kim *et al.*, 2010
2672.	*Uncaria macrophylla*	Rubiaceae	Branch	Cai *et al.*, 2004
2673.	*Uncaria rhynchophylla*	Rubiaceae	Stems	Liu and Mori, 1992; Kim *et al.*, 2010
2674.	*Uncaria sinensis*	Rubiaceae	Hoos, stems	Mahakundakorn *et al.*, 2004
2675.	*Uncaria tomentosa*	Rubiaceae	Bark	Desmarchelier *et al.*, 1997; Goncalves *et al.*, 2005; Pilarski *et al.*, 2006; Wojcikowski *et al.*, 2007
2676.	*Uraria crinita*	Fabaceae		Yen *et al.*, 2001
2677.	*Urena lobata*	Malvaceae	Leaves	Lissy *et al.*, 2006; Adeloye *et al.*, 2007; Omonkhua and Onoagbe, 2008; Pieme *et al.*, 2010
2678.	*Urginea indica*	Liliaceae	Bulbs	Murthy *et al.*, 2011
2679.	*Urginea maritima*	Liliaceae		Mammadov *et al.*, 2010
2680.	*Urophyllum ferrugineum*	Rubiaceae		Ahmad *et al.*, 2010
2681.	*Urophyllum glabrum*	Rubiaceae		Ahmad *et al.*, 2010
2682.	*Urophyllum griffithianum*	Rubiaceae		Ahmad *et al.*, 2010
2683.	*Urtica dioica*	Urticaceae	Herb, root	Pieroni *et al.*, 2002; Trouillas *et al.*, 2003; Mavi *et al.*, 2004; Gulcin *et al.*, 2004; Kanter *et al.*, 2003, 2005; Avci *et al.*, 2006; Lindsey *et al.*, 2006; Pourmorad *et al.*, 2006; Wojcikowski *et al.*, 2007; Celik and Tuluce, 2007; Kamkar *et al.*, 2010
2684.	*Urtica parviflora*	Urticaceae	Herb	Pandey *et al.*, 2010
2685.	*Urtica pilulifera*	Urticaceae	Root, seed, flower, leaf	Mahmoud *et al.*, 2005; Ozen *et al.*, 2010
2686.	*Urtica urens*	Urticaceae	Herb	Abdel Wahab *et al.*, 2006
2687.	*Uvaria grandiflora*	Annonaceae		Aminimoghadamfarouj *et al.*, 2011
2688.	*Uvaria semecarpinifolia*	Annonaceae	Stem bark, leaves	Puvanendran *et al.*, 2011
2689.	*Uvaria sphenocarpa*	Annonaceae	Stem bark, leaves	Puvanendran *et al.*, 2011
2690.	*Vaccinium angustifolium*	Ericaceae	Fruit	Prior *et al.*, 1998; Moyer *et al.*, 2002

Contd...

Table 3.1–*Contd...*

Sl.No.	*Plant Name*	*Family*	*Parts Used*	*References*
2691.	*Vaccinium arctostaphylos*	Ericaceae	Fruit	Koca and Karadeniz, 2009
2692.	*Vaccinium ashei*	Ericaceae	Fruit	Prior *et al.*, 1998; Sellappan *et al.*, 2002; Moyer *et al.*, 2002
2693.	*Vaccinium constablaei X V.ashei*	Ericaceae	Fruit	Moyer *et al.*, 2002
2694.	*Vaccinium corymbosum*	Ericaceae	Fruit	Prior *et al.*, 1998; Cao *et al.*, 1999; Wang and Jiao, 2000; Ehlenfeldt and Prior, 2001; Moyer *et al.*, 2002; Sellappan *et al.*, 2002; Howard *et al.*, 2003; Madhujith and Shahidi, 2003; Zheng and Wang, 2003; Castrejon *et al.*, 2008; Giovanelli and Buratti, 2008; Koca and Karadeniz, 2009; Borges *et al.*, 2010
2695.	*Vaccinium macrocarpon*	Ericaceae	Herb	Villareal *et al.*, 2007
2696.	*Vaccinium membranaceum*	Ericaceae	Fruit	Lee *et al.*, 2004; Moyer *et al.*, 2002
2697.	*Vaccinium myrtilloides*	Ericaceae	Fruit	Moyer *et al.*, 2002
2698.	*Vaccinium myrtillus*	Ericaceae	Fruit, flowers, leaf	Salvayre *et al.*, 1982; Morazzoni and Malandrino, 1988; Prior *et al.*, 1998; Kahkonen *et al.*, 1999; Pietta *et al.*, 1998; Trouillas *et al.*, 2003; Giovanelli and Buratti, 2008; Rouanet *et al.*, 2010; Özen, 2010
2699.	*Vaccinium ovalifolium*	Ericaceae	Fruit	Moyer *et al.*, 2002
2700.	*Vaccinium ovatum*	Ericaceae	Fruit	Lee *et al.*, 2004; Moyer *et al.*, 2002
2701.	*Vaccinium oxycoccus*	Ericaceae	Fruit	Kahkonen *et al.*, 1999; Borges *et al.*, 2010
2702.	*Vaccinium parvifolium*	Ericaceae	Fruit	Acuna *et al.*, 2002; Moyer *et al.*, 2002
2703.	*Vaccinium stamineum*	Ericaceae	Fruit	Wang and Ballington, 2007
2704.	*Vaccinium uliginosum*	Ericaceae	Fruit	Kahkonen *et al.*, 1999; Masuoka *et al.*, 2007
2705.	*Vaccinium vitis-idaea*	Ericaceae	Fruit	Kahkonen *et al.*, 1999; Cai *et al.*, 2006
2706.	*Valeriana officinalis*	Valerianaceae	Root	Mantle *et al.*, 2000; Sudati *et al.*, 2009
2707.	*Valeriana wallichii*	Valerianaceae		Kalim *et al.*, 2010
2708.	*Valeriana prinophylla*	Valerianaceae	Root	Piccinelli *et al.*, 2004
2709.	*Vanda caerulea*	Orchidaceae		Simmler *et al.*, 2010
2710.	*Vangueria infausta*	Rubiaceae	Leaves	Abeer, 2011
2711.	*Vanilla planifolia*	Orchidaceae	Beans	Shyamala *et al.*, 2007
2712.	*Varthemia iphionoides*	Asteraceae	Shoot	Al-Dabbas *et al.*, 2006a,b, 2010; Al-Mustafa and Al-Thunibat, 2008

Contd...

Table 3.1–*Contd...*

Sl.No.	Plant Name	Family	Parts Used	References
2713.	*Ventilago maderaspatana*	Rhamnaceae		Basu and Hazra, 2006
2714.	*Vepris heterophylla*	Rutaceae		Momeni *et al.*, 2010
2715.	*Verbascum alceoides*	Scrophulariaceae		Souri *et al.*, 2004
2716.	*Verbascum bottae*	Scrophulariaceae		Mothana *et al.*, 2010
2717.	*Verbascum densiflorum*	Scrophulariaceae	Flowers	Mariassyova 2006; Katalinic *et al.*, 2006
2718.	*Verbascum macrurum*	Scrophulariaceae	Aerial parts	Aligiannis *et al.*, 2003
2719.	*Verbascum phlomides*	Scrophulariaceae	Flowers	Katalinic *et al.*, 2006; Spiridon *et al.*, 2011
2720.	*Verbascum thapsus*	Scrophulariaceae	Flowers	Katalinic *et al.*, 2006; Wojcikowski *et al.*, 2007; Narayanaswamy and Balakrishnan, 2011
2721.	*Verbascum weidemannianum*	Scrophulariaceae		Tepe *et al.*, 2006
2722.	*Verbena littoralis*	Verbenaceae	Aerial parts	Mensor *et al.*, 2001; Castro-Gamboa and Castro, 2004
2723.	*Verbena officinalis*	Verbenaceae	Herb	Deepak and Handa, 2000; Speroni *et al.*, 2007; Lopez *et al.*, 2008; Casanova *et al.*, 2008; Rehecho *et al.*, 2011
2724.	*Verbena rigida*	Verbenaceae	Aerial parts	El-Hela and Abdullah, 2010
2725.	*Verbena tenera*	Verbenaceae	Aerial parts	El-Hela and Abdullah, 2010
2726.	*Verbena venosa*	Verbenaceae	Aerial parts	El-Hela and Abdullah, 2010
2727.	*Verbesina enceloides*	Asteraceae	Whole plant	Jain *et al.*, 2008; Singh *et al.*, 2009
2728.	*Vernonia ambigua*	Asteraceae		Builders *et al.*, 2011
2729.	*Vernonia amygdalina*	Asteraceae	Leaves	Igile *et al.*, 1994; Oboh and Akindahunsi, 2004; Iwalewa *et al.*, 2005; Iwalokun *et al.*, 2006; Odukoya *et al.*, 2007; Ayoola *et al.*, 2008; Owolabi *et al.*, 2008; Salawu *et al.*, 2011; Farombi and Owoeye, 2011
2730.	*Vernonia anthelmintica*	Asteraceae		Sabu and Kuttan, 2003
2731.	*Vernonia canescens*	Asteraceae		Mosquera *et al.*, 2007
2732.	*Vernonia cinerea*	Asteraceae		Kumar and Kuttan, 2009
2733.	*Vernonia cognata*	Asteraeae		Mota *et al.*, 2011
2734.	*Veronica anagallis-aquatica*	Plantaginaceae	Aerial parts	Kwak *et al.*, 2009
2735.	*Veronica baranetzkii*	Plantaginaceae	Aerial parts	Kwak *et al.*, 2009
2736.	*Veronica chamaedrys*	Plantaginaceae	Aerial parts	Kwak *et al.*, 2009
2737.	*Veronica cuneifolia*	Plantaginaceae	Aerial parts	Kwak *et al.*, 2009

Contd...

Table 3.1–*Contd...*

Sl.No.	*Plant Name*	*Family*	*Parts Used*	*References*
2738.	*Veronica cymbalaria*	Plantaginaceae	Aerial parts	Kwak *et al.*, 2009
2739.	*Veronica fuhsii*	Plantaginaceae	Aerial parts	Kwak *et al.*, 2009
2740.	*Veronica hederifolia*	Plantaginaceae	Aerial parts	Kwak *et al.*, 2009
2741.	*Veronica longifolia*	Plantaginaceae		Jensen *et al.*, 2010
2742.	*Veronica officinalis*	Plantaginaceae	Herb	Katalinic *et al.*, 2006; Casanova *et al.*, 2008
2743.	*Veronica orientalis*	Plantaginaceae	Herb	Harput *et al.*, 2011; Kwak *et al.*, 2011
2744.	*Veronica peduncularis*	Plantaginaceae	Aerial parts	Kwak *et al.*, 2009
2745.	*Veronica peregrina*	Plantaginaceae	Aerial parts	Kwak *et al.*, 2009
2746.	*Veronica persica*	Plantaginaceae	Aerial parts	Kwak *et al.*, 2009
2747.	*Veronica polita*	Plantaginaceae	Aerial parts	Kwak *et al.*, 2009
2748.	*Veronica serpyllifolia*	Plantaginaceae	Aerial parts	Kwak *et al.*, 2009
2749.	*Vetiveria zizanioides*	Poaceae	Root	Kim *et al.*, 2005; Luqman *et al.*, 2009
2750.	*Viburnum dilatatum*	Caprifoliaceae		Iwai *et al.*, 2001, 2003; Kim *et al.*, 2003
2751.	*Viburnum ebulus*	Caprifoliaceae	Flowers	Kiselova *et al.*, 2006
2752.	*Viburnum lantana*	Caprifoliaceae	Fruit, stem	Altun *et al.*, 2008; Serteser *et al.*, 2009
2753.	*Viburnum opulus*	Caprifoliaceae	Stem	Andreeva *et al.*, 2004; Sagdic *et al.*, 200; Altun *et al.*, 2008; Rop *et al.*, 2010
2754.	*Viburnum propinquum*	Caprifoliaceae	Stems, leaves	Wang *et al.*, 2009
2755.	*Vicia canescens*	Fabaceae	–	Nabavi, 2009
2756.	*Vicia sativa* (=*Vicia faba*)	Fabaceae	Seeds	Gamal-Eldeen *et al.*, 2004; Bajpai *et al.*, 2005; Amarowicz *et al.*, 2008
2757.	*Vigna cylindrica*	Fabaceae	Seeds	Bajpai *et al.*, 2005
2758.	*Vigna mungo*	Fabaceae	Seeds	Sreeramulu *et al.*, 2009
2759.	*Vigna radiata*	Fabaceae	Seeds	Lin *et al.*, 2001; Sreeramulu *et al.*, 2009
2760.	*Vigana subterraneana*	Fabaceae		Oboh *et al.*, 2009
2761.	*Vigna umbellata*	Fabaceae	Seeds	Lin *et al.*, 2001; Bajpai *et al.*, 2005; Gan *et al.*, 2010
2762.	*Vigna unguiculata*	Fabaceae	Seeds	Oboh, 2006a
2763.	*Viola serpens*	Violaceae	Herb	Kumar *et al.*, 2011
2764.	*Viola tricolor*	Violaceae	Herb	Katalinic *et al.*, 2006
2765.	*Viola yedoensis*	Violaceae	Whole plant	Cai *et al.*, 2004; Li *et al.*, 2008
2766.	*Virola sebifera*	Myristichaceae		Renzende *et al.*, 2005

Contd...

Table 3.1–*Contd...*

Sl.No.	*Plant Name*	*Family*	*Parts Used*	*References*
2767.	*Viscum album*	Loranthaceae	Aerial parts, leaves, fruits, stems	Orhan *et al.*, 2005; Avci *et al.*, 2006; Wojcikowski *et al.*, 2007; Janakat and Al-Thnaibat, 2008; Oluwaseum and Oboh, 2008; Sengul *et al.*, 2009; Vicas *et al.*, 2009; Alpsoy *et al.*, 2010
2768.	*Viscum album* ssp. *album*	Loranthaceae	Leaves	Onay-Ucar *et al.*, 2006; Avci *et al.*, 2006
2769.	*Viscum coloratum*	Loranthaceae	Stem	Cai *et al.*, 2004
2770.	*Vitellaria paradoxa*	Sapotaceae	Fruit	Lamien-Meda *et al.*, 2008
2771.	*Vitex agnus-castus*	Verbenaceae		Sarikurkcu *et al.*, 2009
2772.	*Vitex altissima*	Verbenaceae	Leaves	Sridhar *et al.*, 2004
2773.	*Vitex doniana*	Verbenaceae		Muanda *et al.*, 2011
2774.	*Vitex mollis*	Verbenaceae		Ruiz-Terán *et al.*, 2008
2775.	*Vitex negundo*	Verbenaceae		One *et al.*, 2004; Kulkarni *et al.*, 2008; Alam *et al.*, 2009; Guha *et al.*, 2010a; Zargar *et al.*, 2011
2776.	*Vitex obovata*	Verbenaceae		Nyligira *et al.*, 2008
2777.	*Vitex peduncularis*	Verbenaceae		Hewamanna *et al.*, 2004; Mathew *et al.*, 2004
2778.	*Vitex polygama*	Verbenaceae		Mensor *et al.*, 2001
2779.	*Vitex pooara*	Verbenaceae		Nyligira *et al.*, 2008
2780.	*Vitex rehmannii*	Verbenaceae		Nyligira *et al.*, 2008
2781.	*Vitex rotundifolia*	Verbenaceae		Song *et al.*, 2010
2782.	*Vitex trifolia*	Verbenaceae	Fruit	Cai *et al.*, 2004; Moussa *et al.*, 2011
2783.	*Vitis thunbergii*	Vitaceae	Roots	Huang *et al.*, 2005; Shyur *et al.*, 2005
2784.	*Vitis vinifera*	Vitaceae	Fruits, Seeds, Seed extract, leaves	Pietta *et al.*, 1990; Da Silva *et al.*, 1991; Maffei-Facino *et al.*, 1994; Wang *et al.*, 1996; Bagchi *et al.*, 1998; Blois, 2001; Burns, 2001; Jay Prakash *et al.*, 2001; Lee and Talcott, 2004; Yilmaz and Toledo, 2004; Rababah *et al.*, 2004; Balu and Panneselvan, 2004; Sendogdu *et al.*, 2006; Busserolles *et al.*, 2006; Chiou *et al.*, 2007; Orhan *et al.*, 2007; Kedage *et al.*, 2007; Kelen and Tepe, 2007; Mishra *et al.*, 2007; Parker *et al.*, 2007; Saada *et al.*, 2008; de Campos *et al.*, 2008; Chedea *et al.*, 2010; Katalinic *et al.*, 2010; Breska *et al.*, 2010; Mitic *et al.*, 2010
2785.	*Voacanga africana*	Liliaceae	Leaf	Oke and Hamburger, 2002; Ayoola *et al.*, 2008

Contd...

Table 3.1–*Contd...*

Sl.No.	Plant Name	Family	Parts Used	References
2786.	*Vochysia discolor*	Vochysiaceae	Flower	Khalil *et al.*, 2006
2787.	*Vochysia tucanorum*	Vochysiaceae		Gomes *et al.*, 2009
2788.	*Waldsteinia geoides*	Rosaceae	Roots	Oszmianski *et al.*, 2007
2789.	*Walsura robusta*	Meliaceae		Voravuthikunchai *et al.*, 2010
2790.	*Walsura yunnanensis*	Meliaceae		Luo *et al.*, 2006
2791.	*Warburgia ugandensis*	Canellaceae	Bark	Kuglerova *et al.*, 2008
2792.	*Washingtonia filifera*	Arecaceae	Leaves	El-Sayed *et al.*, 2006; Chen, C-H. *et al.*, 2009
2793.	*Wattakaka volubilis*	Asclepiadaceae	Leaf	Divya *et al.*, 2009
2794.	*Wedelia chinensis* (Syn.: *Wedelia calendulacea*)	Asteraceae	Herb	Kataki *et al.*, 2012; Senthilkumar *et al.*, 2008
2795.	*Wedelia trilobata*	Asteraceae		Govindappa *et al.*, 2011
2796.	*Wendlandia formosana*	Rubiaceae	Leaves	Raju *et al.*, 2004
2797.	*Wendlandia wallichii*	Rubiaceae	Twig, stem	Kshirsagar and Upadhyay, 2009
2798.	*Wilbrandea ebracteata*	Rubiaceae		Menezes *et al.*, 2004
2799.	*Withania qaratica*	Solanaceae	Whole plant	Marwah *et al.*, 2007
2800.	*Withania somnifera*	Solanaceae	Roots, Fruits	Battacharaya *et al.*, 1997, 2001; Dhuley, 1998; Scartezzini and Speroni, 2000; Parihar and Hemnani, 2003; Gajera *et al.*, 2005; Mishra *et al.*, 2007; Zafir and Banu, 2007; Wojcikowski *et al.*, 2007; Prakash *et al.*, 2007; Jain and Agrawal, 2009; Jatwa and Kar, 2009; Mahour and Saxena, 2010
2801.	*Woodfordia fruticosa*	Lythraceae	Flower	Kumaraswamy and Satish, 2008; Kumaraswamy *et al.*, 2010
2802.	*Wrightia tinctoria*	Apocynaceae		Kumar *et al.*, 2011.
2803.	*Xanthium indicum*	Asteraceae	Leaves	Akter *et al.*, 2009; Hasan *et al.*, 2009
2804.	*Xanthium sibiricum*	Asteraceae	Fruit	Cai *et al.*, 2004; Song *et al.*, 2010
2805.	*Xanthosoma sagittifolium*	Araceae	Leaves	Arruda *et al.*, 2004; Schmourlo *et al.*, 2007; Morrison and Twumasi, 2010
2806.	*Xanthosoma violaceum*	Araceae	Leaves	Picerno *et al.*, 2003
2807.	*Xerophyta viscosa*	Velloziaceae		Mowla *et al.*, 2002
2808.	*Ximenia americana*	Olacaceae	Fruit	Lamien-Meda *et al.*, 2008
2809.	*Xylocarpus granatum*	Malvaceae	Leves, flowers	Vadlapudi and Naidu, 2009
2810.	*Xylocarpus moluccensis*	Meliaceae	Leaves	Adib *et al.*, 2010

Contd...

Table 3.1–*Contd...*

Sl.No.	Plant Name	Family	Parts Used	References
2811.	*Xylopia aethiopica*	Annonaceae	Fruit	Odukoya *et al.*, 2005
2812.	*Xylopia championii*	Annonaceae	Stem bark	Puvanendran *et al.*, 2007, 2008
2813.	*Xylopia nigricans*	Annonaceae	Stem bark, seeds, leaves	Puvanendran *et al.*, 2007
2814.	*Yucca schidigera*	Agavaceae	Bark	Piacente *et al.*, 2004
2815.	*Zanthoxylum alatum*	Rutaceae		Babu *et al.*, 2002; Dhobal *et al.*, 2008
2816.	*Zanthoxylum americanum*	Rutaceae	Seeds	Borchardt *et al.*, 2008
2817.	*Zanthoxylum armatum*	Rutaceae		Ranawat *et al.*, 2010
2818.	*Zanthoxylum chalybeum*	Rutaceae		Kuglerova *et al.*, 2008
2819.	*Zanthoxylum nitidum*	Rutaceae	Root, stem	Cai *et al.*, 2004; Shyur *et al.*, 2005; Bhattacharya *et al.*, 2010; Gan *et al.*, 2010
2820.	*Zanthoxylum piperitum*	Rutaceae		Kim *et al.*, 1997; Chon *et al.*, 2009
2821.	*Zataria multiflora*	Lamiaceae	Leaf, twig	Souri *et al.*, 2008; Saei-Dehkordi *et al.*, 2010
2822.	*Zea mays*	Poaceae	Corn silk, Grains	Maksimovic and Kovacevic, 2003; Bajpai *et al.*, 2005; El-Ghorab *et al.*, 2007; Narman and Omaye, 2007; Wojcikowski *et al.*, 2007; Sreeramulu *et al.*, 2009; Mohsen and Ammar, 2009
2823.	*Zelkova serrata*	Ulmaceae		Kim *et al.*, 2004
2824.	*Zingiber cassumnar*	Zingiberaceae	Rhizome	Chanwitheesuk *et al.*, 2005; Leelapornpisid *et al.*, 2007
2825.	*Zingiber officinale*	Zingiberaceae	Rhizome	Lee and Ahn, 1985; Zhou and Xu, 1992; Cao *et al.*, 1993; Kikuzaki and Nakatani, 1993; Kikuzaki *et al.*, 1994; Sujatha and Srinivas, 1995; Guo *et al.*, 1997; Pearson *et al.*, 1997; Jeyakumar *et al.*, 1999; Ahmed *et al.*, 2000; Dang *et al.*, 2001; Jagetia *et al.*, 2003; Patricia *et al.*, 2003; Cai *et al.*, 2004; Murcia *et al.*, 2004; Rababah *et al.*, 2004; Wu *et al.*, 2004; Bhandari *et al.*, 2005; Chanwitheesuk *et al.*, 2005; Hinneburg *et al.*, 2005; Singh *et al.*, 2005; Odukoya *et al.*, 2005; Sacchetti *et al.*, 2005; Kruawan and Kangsadalampai, 2006; Lako *et al.*, 2006; Ajith *et al.*, 2007; Kim *et al.*, 2007; Leelapornpisid *et al.*, 2007; Mishra *et al.*, 2007; Singh *et al.*, 2008; Stoilova *et al.*, 2007; Mariutti

Contd...

Table 3.1–*Contd...*

Sl.No.	Plant Name	Family	Parts Used	References
				et al., 2008; Khalaf *et al.*, 2008; Ahmed *et al.*, 2008; Ansari and Bhandari, 2009; Jain and Agrawal, 2009; El-Ghorab *et al.*, 2010; Suresh *et al.*, 2010; Oboh *et al.*, 2012
2826.	*Zingiber ottensii*	Zingiberaceae		Leelapornpisid *et al.*, 2007
2827.	*Zingiber zerumbet*	Zingiberaceae	Rhizome	Chanwitheesuk *et al.*, 2005; Lako *et al.*, 2006
2828.	*Zizania aquatica*	Poaceae	Grain	Wu *et al.*, 1994; Asamarai *et al.*, 1996; Qiu *et al.*, 2010
2829.	*Zizania latifolia*	Poaceae	Young stalk	Maeda *et al.*, 2006
2830.	*Ziziphora clinopodioides* subsp. *rigida*	Lamiaceae	Aerial parts	Souri *et al.*, 2004; Salehi *et al.*, 2005; Gursoy *et al.*, 2009
2831.	*Ziziphus amolle*	Rhamnaceae		Ruiz-Terán *et al.*, 2008
2832.	*Ziziphus hajarensis*	Rhamnaceae	Whole plant	Marwah *et al.*, 2007
2833.	*Ziziphus joazeiro*	Rhamnaceae	Inner bark	Alviano *et al.*, 2008
2834.	*Ziziphus jujuba*	Rhamnaceae	Fruit	Kamilioglu *et al.*, 2009
2835.	*Ziziphus mauritiana*	Rhamnaceae	Fruit, flesh	Lamien-Meda *et al.*, 2008; Bhuiyan *et al.*, 2009; Pakutharivu and Suriyavadhana, 2010
2836.	*Ziziphus rugosa*	Rhamnaceae	Fruit	Karuppusamy *et al.*, 2011
2837.	*Zornia gibbosa*	Fabaceae	Herb	Laxane *et al.*, 2008.
2838.	*Zygophyllum coccineum*	Zygophyllaceae		Abouzid *et al.*, 2008
2839.	*Zygophyllum potalini*	Zygopyllaceae	Herb	Bavarma and Ambaga, 2010
		Algae		
2840.	*Aphanizomenon flos-aquae*	Nostocaceae	Thallus	Kuriakose and Kurup, 2010.
2841.	*Aulosira fertilissima*	Cyanophyceae	Thallus	Gini and Kurup, 2008
2842.	*Bifurcaria bifurcata*	Fucaceae	Thallus	Zubia *et al.*, 2009
2843.	*Callophyllis japonica*	Kallymeniaceae	Thallus	Kang *et al.*, 2005
2844.	*Caulerpa racemosa*	Caulerpaceae	Thallus	Lako *et al.*, 2006
2845.	*Caulerpa racemosa* var. *macrophysa*	Caulerpaceae	Thallus	Yangthong *et al.*, 2009
2846.	*Chlamydomonas nivalis*	Chlamydo-monadaceae	Thallus	Li *et al.*, 2006
2847.	*Chlorella stigmatophora*	Chlorellaceae	Thallus	Guzman *et al.*, 2001
2848.	*Chlorella vulgaris*	Chlorellaceae	Thallus	Wu *et al.*, 2005; Hu *et al.*, 2007: Vijayavel *et al.*, 2007; Rodriguez-Garcia and Guil-Guerrero, 2007

Contd...

Table 3.1–*Contd...*

Sl.No.	*Plant Name*	*Family*	*Parts Used*	*References*
2849.	*Cystoseira tamariscifolia*	Fucaceae	Thallus	Zubia *et al.*, 2009
2850.	*Desmarestia viridis*	Desmarestiaceae	Thallus	Xiaojun *et al.*, 1996
2851.	*Dunaliella salina*	Dunaliellaceae	Thallus	Murthy *et al.*, 2005
2852.	*Ecklonia cava*	Lessoniaceae	Thallus	Athukorala *et al.*, 2006
2853.	*Ecklonia stolonifera*	Lessoniaceae	Thallus	Lee *et al.*, 1996; Iwai, 2008
2854.	*Fucus ceranoides*	Fucaceae	Thallus	Zubia *et al.*, 2009
2855.	*Fucus vesiculosus*	Fucaceae	Thallus	Zaragozá *et al.*, 2008
2856.	*Gelidium amansii*	Gelidiaceae	Thallus	Xiaojun *et al.*, 1996
2857.	*Gloiosiphonia capillaris*	Ceramiaceae	Thallus	Xiaojun *et al.*, 1996
2858.	*Gracilaria* sp.	Gracilariaceae	Thallus	Lako *et al.*, 2006
2859.	*Gracilaria tenuispicata* var. *tenuispicata*	Gracilariaceae	Thallus	Yangthong *et al.*, 2009
2860.	*Gracilaria verrucosa*	Gracilariaceae	Thallus	Heo *et al.*, 2006
2861.	*Halidrys siliquosa*	Fucaceae	Thallus	Zubia *et al.*, 2009
2862.	*Haslea ostrearia*	Naviculaceae	Thallus	Pouvreau *et al.*, 2008
2863.	*Ishige okamurae*	Ishigeaceae	Thallus	Heo *et al.*, 2008; Zou *et al.*, 2008
2864.	*Laminaria japonica*	Laminariaceae	Thallus	Zhao *et al.*, 2004; Wang *et al.*, 2008; Song *et al.*, 2010
2865.	*Neorhodomela aculeata*	Rhodomelaceae	Thallus	Lim *et al.*, 2006
2866.	*Nostoc ellipsosporum*	Nostocaceae	Thallus	Gini *et al.*, 2006
2867.	*Nostoc sphaeroides*	Nostocaceae	Thallus	Gini *et al.*, 2006
2868.	*Palmaria palmata*	Palmariaceeae	Thallus	Yuan *et al.*, 2005
2869.	*Phaeodactylum tricornutum*	Phaeodactylaceae	Thallus	Guzman *et al.*, 2001; Rodriguez-Garcia and Guil-Guerrero 2007
2870.	*Phormidium tenue*	Phormidiaceeae	Thallus	Nagasathya and Thajuddin, 2008
2871.	*Polysiphonia urceolata*	Polysiphoniaceae	Thallus	Xiaojun *et al.*, 1996; Duan *et al.*, 2006; Li *et al.*, 2007
2872.	*Porphyra haitanensis*	Bangiaceae	Thallus	Zhang *et al.*, 2003; Zhao *et al.*, 2006
2873.	*Porphyridium cruentum*	Porphyridaceae	Thallus	Rodriguez-Garcia and Guil-Guerrero, 2007
2874.	*Rhodomela confervoides*	Rhodomelaceae	Thallus	Wang *et al.*, 2009
2875.	*Rhodomela teres*	Rhodomelaceae	Thallus	Xiaojun *et al.*, 1996
2876.	*Sargassum filipendula*	Sargassaceeae	Thallus	Costa *et al.*, 2011
2877.	*Sargaassum fulvellum*	Sargassaceae	Thallus	Kim *et al.*, 2007
2878.	*Sargassum fusiforme*	Sargassaceae		Zhou *et al.*, 2008
2879.	*Sargassum kjellmanium*	Sargassaceae	Thallus	Xiaojun *et al.*, 1996
2880.	*Sargassum micranthum*	Sargassaceae	Thalus	Iwashima *et al.*, 2005; Mari *et al.*, 2005

Contd...

Table 3.1–*Contd...*

Sl.No.	Plant Name	Family	Parts Used	References
2881.	*Sargassum myriocystum*	Sargassaceae	Thallus	Badrinathan *et al.*, 2011
2882.	*Sargassum siliquastrum*	Sargassaceae	Thallus	Cho *et al.*2008
2883.	*Sargassum thunbergii*	Sargassaceae	Thallus	Park *et al.*, 2005
2884.	*Sargassum sp.*	Sargassaceae	Thallus	Varadaraja Bhat and Madhyastha, 2000; Yangthong *et al.*, 2009
2885.	*Spirulina maxima*	Oscillatoriales	Thallus	Mirada *et al.*, 1998; Zhi-Gang *et al.*, 1997
2886.	*Spirulina platensis*	Oscillatoriales	Thallus	Manoj *et al.*, 1992; Pinero Estrada *et al.*, 2001; Herreo *et al.*, 2004; Mendiola *et al.*, 2005: Wu *et al.*, 2005; Wojcikowski *et al.*, 2007
2887.	*Symphyocladia latiuscula*	Rhodomelaceae	Thallus	Duan *et al.*, 2007
2888.	*Synechococcus* sp.	Synechococcaceae	Thallus	Li *et al.*, 2006
2889.	*Turbinaria conoides*	Sargassaceae	Thallus	Chattopadhyay *et al.*, 2010
2890.	*Ulva lactuca*	Chlorophyceae	Thallus	Abd El-Baky *et al.*, 2004, 2008; Yangthong *et al.*, 2009
2891.	*Ulva pertusa*	Chlorophyceae	Thallus	Qi *et al.*, 2005a,b, 2006, 2010
2892.	*Undaria pinnatifida*	Alariaceae	Thallus	Je *et al.*, 2009
		Bacteria		
2893.	*Botryococcus braunii*	Extract		Rao *et al.*, 2006
2894.	*Haematococcus pluvialis*	Cells		Cerón *et al.*, 2007; Kamath *et al.*, 2008.
		Bryophytes		
2895.	*Ceratodon purpureus*	Ditrichaceae	Gametophyte	Chobot *et al.*, 2006
2896.	*Dicranum polysetum*	Dicranaceae	Gametophyte	Chobot *et al.*, 2006
2897.	*Dicranum scoparium*	Dicranaceae	Gametophyte	Chobot *et al.*, 2006
2898.	*Leucobryum glaucum*	Lucobryaceae	Gametophyte	Chobot *et al.*, 2006
2899.	*Mnium marginatum*	Mniaceae	Gametophyte	Chobot *et al.*, 2006
2900.	*Plagiochasma appendiculatum*	Aytoniaceae	Gametophyte	Singh *et al.*, 2006
2901.	*Lunularia cruciata*	Lunulariaceae	Thallus	Lelpo *et al.*, 1998

Contd...

Table 3.1–*Contd...*

Sl.No.	*Plant Name*	*Family*	*Parts Used*	*References*
		Lichens		
2902.	*Cetraria islandica*	Lecanorineae	Lichen	Gulcin *et al.*, 2002
2903.	*Cladonia foliacea*	Cladoniaceae	Lichen	Aslan *et al.*, 2006
2904.	*Dermatocarpon miniatum*	Verrucariaceae	Lichen	Aslan *et al.*, 2006
2905.	*Evernia divaricata*	Parmeliaceae	Lichen	Aslan *et al.*, 2006
2906.	*Evernia prunastri*	Parmeliaceae	Lichen	Aslan *et al.*, 2006
2907.	*Lethariella cashmeriana*	Usneaceae	Lichen	Kinoshita *et al.*, 2010
2908.	*Lethariella sernanderi*	Usneaceae	Lichen	Kinoshita *et al.*, 2010
2909.	*Lethariella sinensis*	Usneaceae	Lichen	Kinoshita *et al.*, 2010
2910.	*Neofuscelle pulla*	Parmeliaceae	Lichen	Aslan *et al.*, 2006
2911.	*Parmelia saxatilis*	Parmeliaceae	Lichen	Gulluce *et al.*, 2006
2912.	*Peltigera rufescens*	Peltigeraceae	Lichen	Tanas *et al.*, 2010
2913.	*Platismatia glauca*	Parmeliaceae	Lichen	Gulluce *et al.*, 2006
2914.	*Ramalina lacera*	Usneaceae	Lichen	Weisman *et al.*, 2005
2915.	*Ramalina pollinaria,*	Usneaceae	Lichen	Gulluce *et al.*, 2006
2916.	*Ramalina polymorpha*	Usneaceae	Lichen	Gulluce *et al.*, 2006
2917.	*Rimelia reticulata*	Parmeliaceae	Lichen	Meera *et al.*, 2008
2918.	*Umbilicaria nylanderiana*	Umbilicariaceae	Lichen	Gulluce *et al.*, 2006
2919.	*Usnea articulata*	Usneaceae	Lichen	Lohézic-Le Dévéhat *et al.*, 2007
2920.	*Usnea ghattensis*	Usneaceae	Lichen	Behera *et al.*, 2005; Verma *et al.*, 2008
2921.	*Usnea longissima*	Usneaceae	Lichen	Halici *et al.*, 2005; Bayir *et al.*, 2006; Odabasoglu *et al.*, 2006
2922.	*Usnea undulata*	Usneaceae	Lichen	Susithra *et al.*, 2011
		Fungus		
2923.	*Agaricus arvensis*	Agaricaceae	Mushroom	Barros *et al.*, 2007
2924.	*Agaricus bisporus*	Agaricaceae	Mushroom	Anguiana *et al.*, 2007
2925.	*Agaricus blazei*	Agaricaceae	Mushroom	Oliveira *et al.*, 2007: Soares *et al.*, 2009: Izawa and Inoue, 2004
2926.	*Agrocybe aegerita*	Strophariaceae	Edible mushroom	Zhang *et al.*, 2003
2927.	*Amanita cesaria*	Amanitaceae	Mushroom	Anguiano *et al.*, 2007
2928.	*Antrodia camphorata*	Polyporaceae		Hsiao *et al.*, 2003
2929.	*Antrodia cinnamomia*	Polyporaceae	Carpophore	Chen, C-H. *et al.*, 2009
2930.	*Antrodia salmonea*	Polyporaceae	Basidiomata	Shen *et al.*, 2006
2931.	*Auricularia auricula*	Auriculariaceae	Thallus	Acharya *et al.*, 2004
2932.	*Boletopsis grisea*	Boletaceae	Mushroom	Liu *et al.*, 2004

Contd...

Table 3.1–*Contd...*

Sl.No.	*Plant Name*	*Family*	*Parts Used*	*References*
2933.	*Boletus edulis*	Boletaceae		Anguiamo *et al.*, 2007; Sarikurkcu *et al.*, 2008
2934.	*Clitocybe geotropa*	Tricholomataceae	Mushroom	Sarikurkcu *et al.*, 2010
2935.	*Cordyceps sinensis*	Clavicipitaceae	Mycelia	Li *et al.*, 2001; Wojcikowski *et al.*, 2007
2936.	*Dermatocarpon miniatum*	Verrucariaceae		Aslan *et al.*, 2006
2937.	*Eurotium herbariorum*	Trichochomaceae	Filaments	Miyake *et al.*, 2009
2938.	*Fistulina hepatica*	Schizophyllaceae	Mushroom	Ribeiro *et al.*, 2007
2939.	*Flammulina velutipes*	Physalacriaceae	Mushroom	Bao *et al.*, 2008, 2010
2940.	*Ganoderma applanatum*	Ganodermataceae		Acharya *et al.*, 2005
2941.	*Ganoderma boninense*	Ganodermataceae		Mitchell *et al.*, 2009
2942.	*Ganoderma formosanum*	Ganodermataceae		Lin *et al.*, 1995
2943.	*Ganoderma lucidium*	Ganodermataceae	Mushroom	Lin *et al.*, 1995; Sun *et al.*,2004; Wong *et al.*, 2004; Wojcikowski *et al.*, 2007; Joseph *et al.*, 2008; Ng 20.
2944.	*Ganoderma neo-japonicum*	Ganodermataceae		Lin *et al.*, 1995
2945.	*Ganoderma tsugae*	Ganodermataceae		Yen and Wu, 1999
2946.	*Grifola frondosa*	Meripilaceae	Edible mushroom	Zhang *et al.*, 2002; Mau *et al.*, 2004
2947.	*Hydnellum geogenirum*	Thelephoraceae	Inedible mushroom	Hashimoto *et al.*, 2006
2948.	*Hydnellum suaveolens*	Thelephoraceae	Indedible mushroom	Hashimoto *et al.*, 2006
2949.	*Hypsizigus marmorensis*		Edible mushroom	Yoshino *et al.*, 2008
2950.	*Inonotus obliquus*	Hymenochaetaceae	Mushroom	Nakajima *et al.*, 2007; Ju *et al.*, 2010
2951.	*Inonotus xeranticus*	Hymenochaetaceae	Mushroom	Lee *et al.*, 2006a,b; Lee and Yun, 2006, 2007
2952.	*Lactarius deterrimus*	Russulaceae		Puttaraju *et al.*, 2006; Sarikurkcu *et al.*, 2008
2953.	*Laetiporus sulphureus*	Polyporaceae		Turkoglu *et al.*, 2007
2954.	*Lasiosphaera fenzlii*	Lycoperdaceae		Li *et al.*, 2008

Contd...

Table 3.1–*Contd...*

Sl.No.	Plant Name	Family	Parts Used	References
2955.	*Lentinula edodes*	Marasmiaceae	Mushroom	Anguiano *et al.*, 2007
2956.	*Lepista nuda*	Tricholomataceae	Mushroom	Mercan *et al.*, 2006
2957.	*Leucoagaricus pudicus*	Agaricaceae	Mushroom	Sarikurkcu *et al.*, 2010
2958.	*Leucopaxillus giganteus*	Tricholomataceae	Mushroom	Barros *et al.*, 2007
2959.	*Morchella angusticeps*	Morchellaceae	Mushroom	Gursoy *et al.*, 2009
2960.	*Morchella conica*	Morchellaceae		Turkoglu *et al.*, 2006
2961.	*Morchella esculenta*	Morchellaceae	Mushroom	Anguiana *et al.*, 2007; Gursoy *et al.*, 2009
2962.	*Morchella rotunda*	Morchellaceae		Gursoy *et al.*, 2009
2963.	*Morchella esculenta*	Morchellaceae	Mushroom	Mau *et al.*, 2004; Elmastas *et al.*, 2006; Anguiano *et al.*, 2007
2964.	*Morchella vulgaris*	Morchellaceae		Elmastas *et al.*, 2006
2965.	*Phellinus linteus*	Hymenochaetaceae		Song *et al.*, 2003
2966.	*Phellinus merrillii*	Hymenochaetaceae	Thallus	Chang *et al.*, 2007
2967.	*Phellinus rimosus*	Hymenochaetaceae	Thallus	Ajith and Janardhanan, 2001, 2002; Ajith *et al.*, 2006; Rony *et al.*, 2008
2968.	*Phellinus sp.*	Hymenochaetaceae	Thallus	Hwang *et al.*, 2006
2969.	*Pleurotus florida*	Pleurotaceae	Mushroom	Jose *et al.*, 2002; Selvi and Chinnaswamy, 2007
2970.	*Pleurotus ostreatus*	Pleurotaceae		Jayakumar *et al.*, 2006, 2007, 2008
2971.	*Pleurotus* sp.	Pleurotaceae	Thallus	Anguiana *et al.*, 2007
2972.	*Polyporus umbellatus*	Polyporaceae		Gan *et al.*, 2010
2973.	*Polyporus valvatus*	Polyporaceae	Thallus	Orhan and Ustun, 2010.
2974.	*Poria cocos*	Polyporaceae	Thallus	Schinella *et al.*, 2002; Wu *et al.*, 2005; Gan *et al.*, 2010
2975.	*Ramaria flava*	Gomphaceae		Gursoy *et al.*, 2010
2976.	*Rhizopogon roseolus*	Rhizopogonaceae		Gursoy *et al.*, 2010
2977.	*Russula delica*	Russulaceae	Mushroom	Turkoglu *et al.*, 2007; Yaltirak *et al.*, 2009; Gursoy *et al.*, 2010
2978.	*Saccharomyces cerevisiae*	Saccharomycetaceae		Jaehrig *et al.*, 2007
2979.	*Sarcodon imbricatus*	Bankeraceae	Mushroom	Barros *et al.*, 2007
2980.	*Suillus collitinus*	Suillaceae		Sarikurkcu *et al.*, 2008
2981.	*Termitomyces albuminosus*	Lyophyllaceae	Mycelium	Mau *et al.*, 2004
2982.	*Thelephora aurantiotincta*	Thelephoraceae	Mushroom	Liu *et al.*, 2004

Contd...

Table 3.1–*Contd...*

Sl.No.	Plant Name	Family	Parts Used	References
2983.	*Thelephora ganbajun.*	Thelephoraceae	Mushroom	Yang *et al.*, 2004; Liu *et al.*, 2004
2984.	*Thelephora vialis*	Thelephoraceae	Mushroom	Xie *et al.*, 2005
2985.	*Xerocomus chrysenteron*	Boletaceae		Sarikurkcu *et al.*, 2008
2986.	*Xylaria nigripes*		Mycelia	Ko *et al.*, 2009
	Pteridophytes			
2988.	*Abacopteris penangiana*	Thelypteridaceae		Zhongxiang *et al.*, 2007
2989.	*Actiniopteris semiflabellata*	Actiniopteridaceae	Aerial parts	Mothana *et al.*, 2010
2990.	*Adiantum capillus-veneris*	Adiantaceae		Pourmorad *et al.*, 2006; Pourmorad *et al.*, 2006; Kumar, 2009
2991.	*Adiantum trapeziforme*	Adiantaceae	Fronds	Limaye *et al.*, 2010
2992.	*Adiantum pedatum*	Adiantaceae		Chandrappa *et al.*, 2011
2993.	*Allantodia himalayensis*	Woodsiaceae		Ding *et al.*, 2008
2994.	*Alsophila costularis*	Cyatheaceae		Ding *et al.*, 2008
2995.	*Angiopteris evecta*	Angiopteridaceae	Rhizome	Hutadilok-Towatana *et al.*, 2006; Povichit *et al.*, 2010
2996.	*Arachiniodes exilis*	Dryopteridaceae		Daonian *et al.*, 2010
2997.	*Arthromeris wallichiana*	Polypodiaceae		Ding *et al.*, 2008
2998.	*Aspidium cicutarium*	Aspidiaceae		Ghoghari *et al.*, 2006
2999.	*Asplenium nudus*	Aspleniaceae	Young leaf	Maeda *et al.*, 2006
3000.	*Athyrium esculentum*	Athyriaceae	Leaves	Lako *et al.*, 2006
3001.	*Blechnum orientale*	Blechnaceae	Leaves	Lai *et al.*, 2009
3002.	*Brainea insignis*	Blechnaceae		Ding *et al.*, 2008
3003.	*Camptosorus sibiricus*	Aspleniaceae		Li *et al.*, 2008
3004.	*Cheilanthes glauca*	Sinopteridaceae		Pastene *et al.*, 2003
3005.	*Cibotium barometz*	Cyatheaceae	Leaf	Lai *et al.*, 2009
3006.	*Cyathea phalerata*	Cyatheaceae		Hort *et al.*, 2008
3007.	*Cyclogramma neoauriculata*	Thelypteridaceae		Ding *et al.*, 2008
3008.	*Cyrtomium hookerianum*	Dryopteridaceae		Ding *et al.*, 2008
3009.	*Davallia divaricata*	Davalliaceae	Rhizome	Chang *et al.*, 2007
3010.	*Davallia mariesii*	Davalliaceae		Chang *et al.*, 2007
3011.	*Davallia solida*	Davalliaceae	Rhizome	Chang *et al.*, 2007; Y.-H. Chen *et al.*, 2008
3012.	*Dicranopteris linearis*	Gleicheniaceae	Leaf	Lai *et al.*, 2009
3013.	*Diplazium esculentum*	Polypodiaceae		Wong *et al.*, 2006

Contd...

Table 3.1–*Contd...*

Sl.No.	Plant Name	Family	Parts Used	References
3014.	*Drynaria fortunei*	Polypodiaceae	Root	Cai *et al.*, 2004; Chang *et al.*, 2007
3015.	*Drynaria quercifolia*	Polypodiaceae		Beknal *et al.*, 2010
3016.	*Dryopsis heterolaena*	Dryopteridaceae		Ding *et al.*, 2008
3017.	*Dryopsis mariformis*	Dryopteridaceae		Ding *et al.*, 2008
3018.	*Dryopteris conjugata*	Dryopteridaceae		Ding *et al.*, 2008
3019.	*Dryopteris crassirhizoma*	Dryopteridaceae	Rhizome	Lee *et al.*, 2003
3020.	*Dryopteris neorosthornii*	Dryopteridaceae		Ding *et al.*, 2008
3021.	*Equisetum arvense*	Equisetaceae	Stem	Myagmar and Aniya, 2000; Trouillas *et al.*, 2003; Dos Santos *et al.*, 2005; Oh *et al.*, 2004; Stajner *et al.*, 2006; Mimica-Dukic *et al.*, 2008; Cetojevic-Simon *et al.*, 2010
3022.	*Equisetum hiema*	Equisetaceae		Song *et al.*, 2010
3023.	*Equisetum maximum*	Equisetaceae		Pourmorad *et al.*, 2006
3024.	*Equisetum ramosissimum*	Equisetaceae	Stem	Stajner *et al.*, 2006
3025.	*Equisetum telemateia*	Equisetaceae	Stem	Correia *et al.*, 2005; Lopez *et al.*, 2008; Stajner *et al.*, 2006
3026.	*Equisetum* spp	Equisetaceae		Amarowicz *et al.*, 2004
3027.	*Helminthostachys zeylanica*	Ophioglossaceae	Rhizomes	Huang *et al.*, 2003; Chen *et al.*, 2003
3028.	*Humata griffithiana*	Davalliaceae	Rhizome	Chang *et al.*, 2007
3029.	*Humata platylepis*	Davalliaceae		Ding *et al.*, 2008
3030.	*Lindsaea odorata*	Lindsaeaceae		Ding *et al.*, 2008
3031.	*Lygodium flexuosum*	Lygodiaceae	Herb	Jeetendra and Manish, 2011
3032.	*Lygodium japonicum*	Lygodiaceae		Li *et al.*, 2006
3033.	*Matteuccia orientalis*	Onocleaceae		Cai *et al.*, 2004; Ding *et al.*, 2008
3034.	*Matteuccia struthiopteris*	Onocleaceae	Rhizome	Kimura *et al.*, 2004
3035.	*Nothoperanema hendersonii*	Dryopteridaceae		Ding *et al.*, 2008
3036.	*Osmunda cinnamomia*	Osmundaceae		Ding *et al.*, 2008
3037.	*Osmunda claytoniana*	Osmundaceae		Ding *et al.*, 2008
3038.	*Parathelypteris nipponica*	Thelypteridaceae		Fu *et al.*, 2010
3039.	*Phlegmariurus henryi*	Lycopodiaceae		Ding *et al.*, 2008
3040.	*Phymatosorus cuspidatus*	Polypodiaceae		Ding *et al.*, 2008
3041.	*Plagiogyria maxima*	Plagiogyriaceae		Ding *et al.*, 2008
3042.	*Plagiogyria distinctissima*	Plagiogyriaceae		Ding *et al.*, 2008

Contd...

Table 3.1–*Contd...*

Sl.No.	*Plant Name*	*Family*	*Parts Used*	*References*
3043.	*Polypodiodes amoena*	Polypodiaceae		Ding *et al.*, 2008
3044.	*Polypodiodes mengtzeense*	Polypodiaceae		Ding *et al.*, 2008
3045.	*Polypodium leucotomos*	Polypodiaceae	Aerial parts	Gonzalez *et al.*, 1994; Gombau *et al.*, 2005; Garcia *et al.*, 2006; Tedesco *et al.*, 1997
3046.	*Polypodium vulgare*	Polypodiaceae	Rhizome	Souri *et al.*, 2008
3047.	*Polystichum pycnopterum*	Dryopteriaceae		Ding *et al.*, 2008
3048.	*Polystichum semifertile*	Dryopteridaceae		Ding *et al.*, 2008
3049.	*Pseudodrynaria coronans*	Polypodiaceae	Rhizome	Chang *et al.*, 2007
3050.	*Pteridium aquilinum*	Dennstaedtiaceae	Herb	Xu *et al.*, 2009
3051.	*Pteris multifida*	Pteridaceae		Cai *et al.*, 2004; Wang *et al.*, 2007
3052.	*Pteris yunnanensis*	Pteridaceae		Ding *et al.*, 2008
3053.	*Pyrrosia lingua*	Polypodiaceae	Leaf	Cai *et al.*, 2004; Ding *et al.*, 2008
3054.	*Pyrrosia petiolosa*	Polypodiaceae		Hsu, 2008
3055.	*Pyrrosia sheari*	Polypodiaceae		Ding *et al.*, 2008; Gan *et al.*, 2010
3056.	*Pyrrosia subfuracea*	Polypodiaceae		Ding *et al.*, 2008
3057.	*Selaginella delicatula*	Selaginellaceae	Aerial parts	Gayathri *et al.*, 2005
3058.	*Selaginella involvens*	Selaginellaceae	Aerial parts	Gayathri *et al.*, 2005
3059.	*Selaginella tamariscina*	Selaginellaceae		Gan *et al.*, 2010
3060.	*Sphaeropteris brunoniana*	Cyatheaceae		Ding *et al.*, 2008
3061.	*Woodwardia japonica*	Blechnaceae		Ding *et al.*, 2008
3062.	*Woodwardia unigemmata*	Blechnaceae		Ding *et al.*, 2008
		Gymnosperms		
3063.	*Abies alba*	Pinaceae	Essential oil	Yang *et al.*, 2009
3064.	*Abies pindrow*	Pinaceae	Leaves	Kalim *et al.*, 2010; Gupta *et al.*, 2011
3065.	*Araucaria angustifolia*	Araucariaceae	Needles	Yamaguchi *et al.*, 2005; Seccon *et al.*, 2010
3066.	*Araucaria excelsa*	Araucariacae	Needles	Michael *et al.*, 2010
3067.	*Biota orientalis*	Cupressaceae		Gan *et al.*, 2010
3068.	*Calocedrus formosana*	Cupressaceae	Heartwood, bark, leaf	Wang *et al.*, 2004
3069.	*Calocedrus macrolepis*	Cupressaceae	Leaves	Chen, C-H. *et al.*, 2009
3070.	*Cedrus deodara*	Pinaceae		Tiwari *et al.*, 2001
3071.	*Chamaecyparis formosensis*	Cupressaceae	Leaves	Chen, C-H. *et al.*, 2009
3072.	*Chamaecyparis obtusa* var. *formosana*	Cupressaceae	Leaves	Marimuthu *et al.*, 2008; Chen, C-H. *et al.*, 2009
3073.	*Cupressus arizonica*	Cupressaceae	Essential oil	Emami *et al.*, 2010

Contd...

Table 3.1–*Contd...*

Sl.No.	*Plant Name*	*Family*	*Parts Used*	*References*
3074.	*Cupressus macrocarpa*	Cupressaceae	Leaves	El Ghorab *et al.*, 2007
3075.	*Cupressus semipervirens*	Cupressaceae	Leaves, fruits	Emami *et al.*, 2007
3076.	*Ephedra americana*	Ephedraceae	Herb	Rivera and Parja, 2001
3077.	*Ephedra gerardiana*	Ephedraceae	Herb	Kumar and Singh, 2011
3078.	*Ephedra nebrodensis*	Ephedraaceae	Herb	Ballero *et al.*, 2010
3079.	*Ephedra pachyclada*	Ephedraceae	Herb	Parsaeimehr *et al.*, 2010
3080.	*Ephedra procera*	Ephedraceae	Herb	Parsaeimehr *et al.*, 2010
3081.	*Ephedra sinica*	Ephedraceae	Herb	Song *et al.*, 2010
3082.	*Ephedra strobilacea*	Ephedraceae	Herb	Parsaeimehr *et al.*, 2010
3083.	*Ephedra viridis*	Ephedraceae	Whole plant	Pullela *et al.*, 2005
3084.	*Ginkgo biloba*	Ginkgoaceae	Leaves	Marcocci *et al.*, 1994a,b; Kose and dogan, 1995; Wojcicki and Samochowiec, 1996; Butnaru *et al.*, 1997; Kose *et al.*, 1997; Pietta *et al.*, 1998; Hibatallah *et al.*, 1999; Yoshikawa *et al.*, 1999; Daiamond *et al.*, 2000; Lugasi *et al.*, 1999; Mantle *et al.*, 2003: Bridi *et al.*, 2001: Rababah *et al.*, 2004: Sarikcioglu *et al.*, 2004; Ellnain-Wojtaszek *et al.*, 2003; Mishra *et al.*, 2007; Naik and Panda, 2007; Naik *et al.*, 2006; Masteikova *et al.*, 2007; Tunali-Akbay *et al.*, 2007; Feng *et al.*, 2009; Song *et al.*, 2010
3085.	*Gnetum bucholzianum*	Gnetaceae	Leaf	Odukoya *et al.*, 2007
3086.	*Gnetum gnemon*	Gnetaceae		Iliya *et al.*, 2003; Wong *et al.*, 2006; Santoso *et al.*, 2010; Siswoyo *et al.*, 2011; Wazir *et al.*, 2011
3087.	*Gnetum macrostachyum*	Gnetaceae		Saisin *et al.*, 2009
3088.	*Gnetum montanum*	Gnetaceae	Stem	Thuong *et al.*, 2006
3089.	*Juniperus chinensis*	Cupressaceae		Lim *et al.*, 2002
3090.	*Juniperus communis*	Cupressaceae	Leaves	Hinneburg *et al.*, 2005; Elmastas *et al.*, 2006; Emami *et al.*, 2007; Wojcikowski *et al.*, 2007; Al-Mustafa and Al-Thunibat, 2009
3091.	*Juniperus excelsa*	Cupressaceae	Leaves, cones	Emami *et al.*, 2007; Öztürk *et al.*, 2011
3092.	*Juniperus foetidissima*	Cupressaceae	Leaves, cones	Emami *et al.*, 2007; Emami *et al.*, 2011
3093.	*Juniperus oblonga*	Cupressaceae	Leaves, cones	Emami *et al.*, 2007
3094.	*Juniperus oxycedrus*	Cupressaceae	Fruit	Öztürk *et al.*, 2011
3095.	*Juniperus phoenicea*	Cupressaceae	Leaves, berries	Ennajar *et al.*, 2009, 2010; Öztürk *et al.*, 2011
3096.	*Juniperus procera*	Cupressaceae		Burits *et al.*, 2001

Contd...

Table 3.1–*Contd...*

Sl.No.	*Plant Name*	*Family*	*Parts Used*	*References*
3097.	*Juniperus sabina*	Cupressaceae	Leaves, cones	Emami *et al.*, 2007; Öztürk *et al.*, 2011
3098.	*Larix gmelinii*	Pinaceae	Wood	Kolhir *et al.*, 1996
3099.	*Larix sibirica*	Pinaceae	Bark	Khairullina *et al.*, 2006
3100.	*Metasequoia glyptostroboides*	Taxodiaceaee	Leaves	Bajpai *et al.*, 2009a,b
3101.	*Picea abies*	Pinaceae	Needle	Kahkonen *et al.*, 1999; Lee *et al.*, 2006; Co *et al.*, 2011
3102.	*Pinus* sp.	Pinaceae	Pollen	Lee *et al.*, 2008
3103.	*Pinus abies*	Pinaceae		Saarium *et al.*, 2000
3104.	*Pinus armandii*	Pinaceae	Essential oil	Yang *et al.*, 2010
3105.	*Pinus densiflora*	Pinaceae		Jung *et al.*, 2003; Ku *et al.*, 2007
3106.	*Pinus halepens*	Pinaceae	Pollen	Al-Mustafa and Al-Thunibat, 2008
3107.	*Pinus koraiensis*	Pinaceae		Su *et al.*, 2009
3108.	*Pinus maritima*	Pinaceae		Packer *et al.*, 1999; Kam and Park, 2004; Busserolles *et al.*, 2006
3109.	*Pinus massoniana*	Pinaceae	Bark	Cui *et al.*, 2005; Yu *et al.*, 2008; Pan *et al.*, 2009
3110.	*Pinus morrisonicola*	Pinaceae	Needle, bark, cone	Hsu *et al.*, 2005
3111.	*Pinus mungo*	Pinaceae	Essential oil	Grassman *et al.*, 2003
3112.	*Pinus nigra*	Pinaceae		Lee *et al.*, 2009: Gulcin *et al.*, 2003
3113.	*Pinus pinaster*	Pinaceae	Bark	Jerez *et al.*, 2007
3114.	*Pinus radiata*	Pinaceae	Bark	Jerez *et al.*, 2007; Ku *et al.*, 2007
3115.	*Pinus rigida*	Pinaceae	Bark	Ku *et al.*, 2007
3116.	*Pinus roxburghii*	Pinaceae	Needles	Puri *et al.*, 2010
3117.	*Pinus sylvestris*	Pinaceae		Kahkonen *et al.*, 1999; Saleem *et al.*, 2003
3118.	*Pinus tabulaeformis*	Pinaceae		Gan *et al.*, 2010
3119.	*Platycladus orientalis*	Cupressaceae	Leaves, cones	Emami *et al.*, 2007
3120.	*Podocarpus elongatus*	Podocarpaceae		Abdillahi *et al.*, 2011
3121.	*Podocarpus falcatus*	Podocarpaceae		Abdillahi *et al.*, 2011
3122.	*Podocarpus henkelii*	Podocarpaceae		Abdillahi *et al.*, 2011
3123.	*Podocarpus latifolius*	Podocarpaceae		Abdillahi *et al.*, 2011
3124.	*Podocarpus nagi*	Podocarpaceae		Haraguchi *et al.*, 1996, 1997
3125.	*Sabina przewalski*	Cupressaceae		Ip *et al.*, 2004
3126.	*Taxus baccata*	Taxaceae	Fruit, leaves	Emami *et al.*, 2007; Serteser *et al.*, 2009; Kucukboyaci and Sener, 2010
3127.	*Taxus cuspidata*	Taxaceae	Wood, roots	Veselova *et al.*, 2007
3128.	*Taxus yunnanensis.*	Taxaceae	Wood	Banskota *et al.*, 2003
3129.	*Taiwania cryptomerioides*	Taxodiaceae	Bark	Kuo *et al.*, 2002
3130.	*Thuja occidentalis*	Cupressaceae	Aerial parts	Deb *et al.*, 2009; Dubey and Batra, 2009a,b; Biswas *et al.*, 2011

Aquilegia vulgaris L. Family: Ranunculaceae

Two extracts (ethyl acetate and ethanol) and isocytisoside obtained from *Aquilegia vulgaris* were tested by Murias *et al.* (2005) for their antioxidant and free radical scavenging activity *in vitro*. Inhibition both non-enzymatic and enzymatic microsomal lipid peroxidation was observed, the extracts being more active than isocytisoside. They said that hepatoprotective properties of this plant may be ascribed to the antioxidant activity of its constituents.

Rats pretreated with acetaminophen (N-acetyl-p-aminophenol, APAP) (600 ml/kg b.w.p.o.) were administered with ethanol and ethyl acetate extracts as well as with isocytisoside (100mg/kg b.w., p.o.) obtained from *Aquilegia vulgaris* herb. The substances tested decreased enzymatic, non-enzymatic and uninduced microsomal lipid peroxidation (LPO) in liver of rats treated with APAP by 18-48 per cent. Activity of the antioxidant enzymes in the liver inhibited by APAP was increased in the majority of groups after administration of the substances tested: catalase (CAT) by 55 per cent, glutathione peroxidase (GPx) by 50 per cent, glutathione reductase (GR) by 35 per cent and glutathione S-transferase by 60 per cent (Jodynis-Liebert *et al.*, 2005).

Arachis hypogaea L. Family: Fabaceae (Papilionaceae)

English Name: Peanut, Ground nut.

The antioxidant activity of ethanolic extracts of peanut seed testa (EEPST) and its antioxidative component, ethyl protocatechuate (EP) was examined by Yen *et al.* (2005). It was found that EEPST and EP showed a dose dependent activity on the inhibition of liposome peroxidation. EEPST and EP in the range of 50-500µg/ml were effective in protecting protein against oxidative damage. EEPST and EP at 100 mg/l showed 92.6 per cent and 84.6 per cent scavenging effect, respectively, on DPPH radical, indicating that they act as a primary antioxidant. In addition, EEPST and EP, at a dose of 200 mg/l, showed 70.6 per cent and 67.7 per cent scavenging effect respectively, on the hydroxyl radical. EEPST also exhibited a metal-binding ability, while EP did not. The inhibitory effect of EEPST on linoleic peroxidation correlated with their polyphenolic content. Lee *et al.* (2006) investigated the effect of far infrared radiation and heat treatment on the antioxidant activity of water extracts from peanut hulls. Duncan *et al.* (2006) investigated the phytochemical content and antioxidant capacity of water-soluble isolates from peanuts.

Free radical and reactive oxygen species scavenging activities of water soluble extract from defatted peanut skins (EPS) was 0.097g GAE/g, and total flavonoid content of EPS was 0.065 g catechin equivalent/g of skin tested. EPS at 500µg/ml exhibited 97 per cent of DPPH radical scavenging activity and 98 per cent of inhibition of erythrocyte hemolysis and showed 98.6 per cent, 89.1 per cent and 85.3 per cent scavenging activities on O_2, H_2O_2 and OH, respectively and also effectively exerted 76.6 per cent of Fe^{2+} chelating activity (Wang *et al.*, 2007).

Aralia continentalis Kitagawa Family: Araliaceae

Antioxidant activity of the young leaves of *Aralia continentalis* was determined by measuring lipid peroxide produced when a mouse liver homogenate was exposed to the air at 37°C, using TBA and by evaluation of the radical scavenging activity on

DPPH radical (Kim *et al.*, 1998). Chromatographic separation of active fraction led to the isolation of six flavonoids, among which quercetin hyperoside and kaempferol showed strong antioxidant activity, while 6"-O-acetyl, astragalin and trifolin were inactive. Adenosine oleanolic acid 28-O-glucosyl ester and salsoloside C methyl ester isolated from the somewhat active BuOH fraction did not exhibit antioxidant activity.

Aralia cordata Thunb. Family: Arliaceae

Microwave-assisted extraction techniques of flavonoids from *Aralia cordata* were studied by Liu *et al.* (2010) by the single factor analysis and orthogonal test, then the property antioxidant activities of flavonoids extracts were evaluated by two antioxidant assays, for 1,1-diphenyl-2-picrylhydrazyl (DPPH) radicals scavenging and reducing power. The results showed that the optimum conditions were as follows: twice extracting *Aralia cordata* with 75 times volume (V/W) of 30 per cent ethanol at 140W of microwave for 60 min at 80 degrees C. Under these conditions, the extraction yield was 1.35 per cent. The flavonoids extracts could significantly scavenge DPPH and superoxide anion free radicals in dose-dependent manner, meanwhile a linear correlation between its concentration and reducing power was also observed.

Aralia elata Seem. Family: Araliaceae

Chung and Jung (2003) reported that ethanol fraction of *Aralia elata* enhances antioxidant activity and lowers serum lipids in rats when administered with benzo(a)pyrene.

Chung *et al.* (2005) investigated the aldose reductase inhibitory activity, antioxidant activity and anticataract capacity of *Aralia elata* water extract using various experimental systems. The resulting IC_{50} value as an antioxidant was 25.1 µg/ml.

Hu and Wang (2009) investigated the *in vitro* antioxidant potential of dichloromethane, ethyl acetate (EtOAc), n-butanol (n-BuOH) and water fraction of 70 per cent methanolic (MeOH) extract of *Aralia elata* seeds using different antioxidant tests, such as total phenolic content, 1,1-diphenyl-2-pricylhydrazyl (DPPH) radical-scavenging assay, hydroxyl-radical scavenging assay, reducing power assay, metal-chelating activity and lipid peroxidant assay. Among these fractions, EtOAc fraction exhibited higher antioxidant potency than other fractions in DPPH free radical-scavenging, hydroxyl radical-scavenging activity, reducing power assay and lipid peroxidant assay. It was concluded that the *A. elata* seeds fractions obtained from 70 per cent MeOH extract are highly suitable for medical application due to their efficient antioxidant properties.

Aralia taibaiensis Z.Z. Wang et H.C. Zheng Family: Araliaceae

Xi *et al.* (2008, 2010) has demonstrated that the antidiabetic activity of the extract of root bark of *Aralia taibaiensis* (EAT) was correlated with its combined antioxidant and antiglycation properties. To confirm further the constituents responsible, 12 triterpenoid saponins were isolated from EAT and examined for their antioxidant and antiglycation activities. The antioxidant activities of the pure compounds and EAT were evaluated by studying the inhibition of lipid peroxidation in rat liver microsomes induced by ascorbate/Fe(2+), cumine hydroperoxide (CHP) or CCl(4)/

reduced form of nicotinamide-adenine dinucleotide phosphate (NADPH). The antioxidant capacities were also evaluated by studying the scavenging of DPPH free radical. EAT outperformed other compounds in all the assays. The compounds with best antioxidant (TA7, TA24 and TA35) and antiglycation (TA21, TA9 and TA24) activities in different assays were screened out. The results suggest that the antioxidant and antiglycation properties of EAT could be explained, at least in part, by the synergistic effect of pure compounds isolated from it.

***Arbutus andrachne* L. Family: Ericaceae**

English Name: The Greek Strawberry tree.

Kirca and Arslan (2008) investigated the antioxidant capacity and total phenol contents of *Arbutus andrachne*. The antioxidant capacity of methanolic extracts prepared from various parts of plants was evaluated by both trolox equivalent antioxidant capacity (TEAC) and DPPH assays, while the total phenolics were determined using the Folin-Ciocalteu method. *Arbutus andrachne* showed higher antioxidant activities in both TEAC and DPPH method.

Serce *et al.* (2010) determined the chemical composition, antioxidant activities and total phenolic content of *Arbutus andrachne* fruits from Turkey. Among the *Arbuts andrachne* accessions tested, the total phenolics ranged from 2422 to 4102 μg Gallic acid/g fresh weight with an average of 3343 μg gallic acid equivalent g/fw. Antioxidant activities, determined by TEAC and FRAP (Ferric Reducing Ability of Plasma), reached up to 36.1 and 29.5 μ mol Trolox equivalent (TE) g/fresh weight, respectively. It can thus be concluded that *A. andrachne* is a good source of phenols and antioxidants (Serce *et al.*, 2010).

***Arbutus unedo* L. Family: Ericaceae**

English Name: Strawberry tree

The ethanol and methanol extracts of *Arbutus unedo* leaves were screened for antioxidant activity by Pabuccuoglu *et al.* (2003). The antioxidant activity was determined by an improved assay based on the decolorization of the radical monocation of ABTS+. The ethanol and methanol extracts of leaves displayed potent antioxidant activity. Phytochemical analysis showed that the extracts contained flavonol glycosides and tannins but not phenol glycosides and vitamin E.

The fruits of the strawberry tree (*Arbutus unedo*) are consumed mainly as processed product, but may be a good source of antioxidants if consumed as fresh fruit. Pallauf *et al.* (2008) carried out investigation to identify and quantify the antioxidant components present in strawberry tree fruits, including flavonoids, vitamin C and E and carotenoids. The fruits are a very good source of antioxidants precisely because they have a high flavonoid content (32.37mg/100g edible portion) and, within this group of antioxidant compounds, proanthocyanidins are the most abundant, representing more than 80 per cent of the total flavonoid contents.

Oliveira *et al.* (2009) assessed the antioxidant and free radical scavenging potential of water, ethanol, methanol and diethyl ether extracts of *Arbutus unedo* leaves. Antioxidant activity was evaluated using three different methods: reducing power of iron(III)/ferricyanide complex assay, scavenging effect on DPPH radicals and scavenging effect on superoxide radicals by using PMS-NADH-nitroblue tetrazolium system. Ethanol extracts of leaves were the highest in reducing power (IC_{50} = 232.7µg/ml) and DPPH scavenging effect (IC_{50} = 63.2 µg/ml) followed by water extracts (with IC_{50} = 287.7 and 73.7 µg/ml); whereas diethyl ether extracts were the lowest. In the scavenging on superoxide radical assay, methanol extracts obtained the best results (IC_{50} = 6.9 µg/ml). For all the methods tested the antioxidant activity was concentration dependent. In accordance with the antioxidant activity, highest total phenols content were found in ethanol, followed by water, methanol and diethyl ether extract.

Arcangelisia flava **(L.) Merr. Family: Menispermaceeae**

Keawpradub *et al.* (2005) evaluated the antioxidant activities of *Arcangelisia flava* employing DPPH radical scavenging assay. Methanol extract exhibited moderate antioxidant activity with EC_{50} values of 25-55µg/ml.

Archidendron jiringe **I.C.Nielsen Family: Mimosaceae**

The phenolic compounds of five soutern Thai indigenous vegetables including djenkol tree (*Archidendron jiringe*) fruit were extracted using different solvents. The extracts were analyzed for total phenolic content using the Folin-Ciocalteu procedure, free radical scavenging capacity by using ABTS and DPPH methods, and reducing capacity by using FRAP assay. The acetone extract possessed the highest total phenolic content followed by methanolic and distilled water extracts, respectively. The acetone extract showed higher free radical scavenging capacity and reducing capacity than those of their methanolic and water extracts counterparts respectively (Panpipat *et al.*, 2010).

Archontophoenix alaxandrae **(F.Muell.) H. Wendl. et Drude Family: Arecaceae (Palmae)**

Liu *et al.* (2008) evaluated the antioxidant activity of water, 50 per cent ethanol and 70 per cent acetone extracts of flowers of 15 plants of Hainan. Water extracts of *Archontophoenix alexandrae* had better effects on eliminating OH radical.

Arctium lappa **L. Family: Asteraceae (Compositae)**

English Name: Burdock.

Kardosova and Machova (2006) investigated the ability of polysaccharides, isolated from leaves, to inhibit peroxidation of soyabean lecithin liposomes by OH radicals. All polysaccharides exhibited antioxidant activity.

Cai *et al.* (2004, 2006) evaluated the antioxidant activity and phenolic compounds in traditional Chinese medicinal plants associated with anticancer. The improved ABTS.+ method was used to systematically assess the total antioxidant capacity (Trolox equivalent antioxidant capacity, TEAC) of the medicinal extracts. The TEAC values and total phenolic content for methanolic extract of fruit of *Arctium lappa* were 513.3 µmol Trolox equivalent/100g dry weight (DW), and 3.74 g of gallic acid equivalent/100g DW respectively. A positive significant linear relationship between antioxidant activity and total phenolic content showed that phenolic compounds were the dominant antioxidant components in the tested herbs. Major types of phenolic compounds in *Arctium lappa* include phenolic acids (chlorogenic acid), lignans (arctigenin, arctiin).

Antioxidant capacities of 56 selected Chinese medicinal plants, including *Arctium lappa* were evaluated by Song *et al.* (2010) using the TEAC and FRAP assays, and their total phenolic content was measured by the Folin-Ciocaleteu method. The TEAC value was 74.66μmol Trolox/g, FRAP value was 223.68μmol Fe^{2+}/g and phenolic content was 16.94mg GAE/g.

Arctostaphylos uva-ursi (L.) Spreng. Family: Ericaceae

English Name: Bearberry.

Ethanolic extracts of leaves of bearberry were prepared and evaluated by Amarowicz *et al.* (2004) for their free-radical scavenging capacity and their antioxidant activity, by a number of chemical assays. Assays employed included β-carotene linoleic acid (lineolate) model system, reducing power, scavenging effect on the DPPH free radical and capacity to scavenge hydroxyl free radical (HO), by use of electron paramagnetic resonance (EPR) spectroscopy. The bearberry-leaf extract consistently exhibited the highest antioxidant activity based on the tests performed, and seems to be a promising source of natural antioxidants. The polyphenolic constituents appear to be responsible, at least in part, for the extract's radical-scavenging capacity.

Wojcikowski *et al.* (2007) evaluated the antioxidant activity of aerial parts of *Arctostaphylos uva-ursi*. Oxygen radical absorbance capacity of the methanol extract and aqueous extracts were 184.36 and 517.95μmol Trolox equivalent per g of dried starting material, respectively.

Ardisia colorata Wall. ex Roxb. Family: Myrsinaceae

Novel alkylphenols, ardisiphenols A-C (1-3) and a novel bergenin derivative, demethoxybergenin (10) were isolated from the fruits of *Ardisia colorata*, together with known alkylresorcinols (4-6), embelin (7), myricetin (8), quercetin (9), bergenin (11), norbergenin (12), kaempferol (13), quercetin-3-O-beta-D-glucopyranoside (14) and gallic acid (15). Ardisiphenols showed moderate scavenging activities toward DPPH radicals and showed cytotoxicity against the murine breast cancer cell line, FM3A (Sumino *et al.*, 2001, 2002).

Ardisia compressa Kunth Family: Myrsinaceae

The *in vitro* antioxidant activity of crude methanolic(Me) and Hexane (He) extracts of stem bark of *Ardisia compressa* (AC) were investigated by Ramirez-Mares *et al.* (2010) using DPPH method. Ascorbic acid was substantially more powerful than extracts from the steem bark of AC. The antiradical efficiency of ascorbic acid was 249-fold that for Hexane extract and 8.3 fold for methanolic extract.

Ardisia gigantifolia Stapf Family: Myrsinaceae

Yang *et al.* (2008) investigated the chemical constituents of *Ardisia gigantifolia* and their antioxidant activity by DPPH method. Gallic acid and epicatechin isolated from *Ardisia gigantifolia* showed moderate antioxidant activities.

Ardisia japonica (Horrst.) Bl. Family: Myrsinaceae

Antioxidant capacities of 56 selected Chinese medicinal plants, including *Ardisia japonica* were evaluated by Song *et al.* (2010) using the TEAC and FRAP assays, and their total phenolic content was measured by the Folin-Ciocaleteu method. The TEAC value was 164.19μmol Trolox/g, FRAP value was 170.2μmol Fe^{2+}/g and phenolic content was 13.58mg GAE/g. The strong correlation between TEAC value and FRAP value suggested that the antioxidants in these plants possess free radical scavenging activity and reducing power, and the high positive correlation between antioxidant capacities and total phenolic content implied that phenolic compounds are a major contributor to the antioxidant activity of these plants.

Areca catechu L. Family: Arecaceae (Palmae)

Methanol aqueous extract of *Areca catechu* was screened by Kim *et al.* (1997) for anti-oxidative activity using Fenton's reagent/ethyl lineolate system. The extract showed antioxidant activity.

Cai *et al.* (2004) evaluated the antioxidant activity and phenolic compounds in traditional Chinese medicinal plants associated with anticancer. The improved ABTS.+ method was used to systematically assess the total antioxidant capacity (Trolox equivalent antioxidant capacity, TEAC) of the medicinal extracts. The TEAC values and total phenolic content for methanolic extract of seed of *Areca catechu* were 2674 μmol Trolox equivalent/100g dry weight (DW), and 11.83 g of gallic acid equivalent/100g DW respectively. A positive significant linear relationship between antioxidant activity and total phenolic content showed that phenolic compounds were the dominant antioxidant components in the tested herbs. Major types of phenolic compounds in *Areca catechu* include condensed tannins (leucoanthocyanidins), flavonols (catechin).

Areca catechu L. var. dulcissima Family: Arecaceae (Palmae)

Methanol extract of *Areca catechu* var. *dulcissima* showed relatively high levels of DPPH radical scavenging activity (IC_{50} <6.0 μg/ml) (Lee *et al.*, 2003). The activities of superoxide dismutase, catalase and glutathione peroxidase were dose-dependently enhanced in V79-4 cells treated with the plant extract. The extract showed higher antioxidant activity than resveratrol in all experiments.

Argania spinosa (L.) Skeels Family: Sapotaceae

English Name: Argan

El Babili *et al.* (2009) evaluated the potential antioxidant and antimalarial activity and also activity against breast cancer cells (MCF 7) of Argan fruit extracts using *in vitro* models. Polyphenols (89.4 – 218.5 GAE mg/g dry), tannins (39.3-214.0 eq catechin mg/g dry), flavonoids (3.4 – 11.1 eq quercetin mg/g dry) and anthocyanins (0.74 – 10.92 eq cyanidin were quantified. A good (ethyl acetate and decoction) and moderate (petroleum ether) antioxidant activity were obtained for DPPH (IC_{50} = 32.3 – 600.8μg/ml) and ABTS (IC_{50} = 11.9 -988.8μg/ml) assays. The results indicated a good correlation between anthocyanins quantity and the potential antioxidant and also antimalarial activity.

Argemone mexicana L. Family: Papaveraceae

Lagnika *et al.* (2011) evaluated the antioxidant activity of *Argemone mexicana* using DPPH radical scavenging assay. The extract showed good radical scavenging activity with IC_{50} value of 1.73μg/ml while the standard L-Ascorbic acid had IC_{50} value of 1.1μg/ml.

***Argostemma cf. viscidum* Family: Rubiaceae**

Ahmad *et al.* (2010) evaluated the antioxidant potential of 22 species of medicinal plants from Malaysian Rubiaceae including *Argostemma cf. viscidum*. FTC, TBA, TPC and the DPPH assays were employed. The tested extracts showed strong antioxidant potential with percent inhibition of 91.7 per cent in the FTC, and 78.1 per cent in TBA assays. The TPC of the extract was 18.54mg GAE/g. A good correlation was observed between total phenolic content and radical-scavenging activities.

***Argostemma yappi* King Family: Rubiaceae**

Ahmad *et al.* (2010) evaluated the antioxidant potential of 22 species of medicinal plants from Malaysian Rubiaceae including *Argostemma yappi*. FTC, TBA, TPC and the DPPH assays were employed. The tested extracts showed strong antioxidant potential with per cent inhibition of 96.3 per cent in the FTC, and 91.9 per cent in TBA assays. The TPC of the extract was 11.39mg GAE/g PE. A good correlation was observed between total phenolic content and radical-scavenging activities.

***Argyreia cymosa* R. Sweet Family: Convolvulaceae**

Badami *et al.* (2008) studied the *in vitro* antioxidant activity of bark extracts of *Argyreia cymosa*. The petroleum ether extract has shown antioxidant activity in ABTS, nitric oxide, hydroxyl radical and lipid peroxidation methods. The ethyl acetate extract has shown antioxidant activity in DPPH, H_2O_2 and hydroxyl radical scavenging methods.

***Argyreia nervosa* (Burm.f.) Boj. Family: Convolvulaceae**

Aqueous extract of *Argyreia nervosa* was studied by Shreedhara *et al.* (2009) for its *in vitro* radical scavenging activity by different methods *viz.* DPPH, ABTS, lipid peroxidation, iron chelating activity, superoxide scavenging, total antioxidant capacity, nitric oxide scavenging and non-enzymatic glycosylation of Haemoglobin assay. Its antioxidant activity was estimated by IC_{50} values and the values were 88.86μg/ml for DPPH radical scavenging. Total antioxidant capacity was found to be 22μg/l. In haemoglobin glycosylation the per cent scavenging was found to be 43.77 per cent and 64.42 per cent for concentration 0.5 mg/ml and 1.0 mg/ml respectively. In all the methods, the extract showed its ability to scavenge free radicals in a concentration dependent manner.

***Argyreia speciosa* (Burm.f.) Boj. Family: Convolvulaceae**

Ethanol extract of *Argyreia speciosa* exhibited significant scavenging effect of DPPH, hydroxyl radical and LPO (Habbu *et al.*, 2010).

Arisaema consanguineum Schott. Family: Araceae

Antioxidant capacities of 56 selected Chinese medicinal plants, including *Arisaema consanguineum* were evaluated by Song *et al.* (2010) using the TEAC and FRAP assays, and their total phenolic content was measured by the Folin-Ciocalteu method. The TEAC value was very low (0.78μmol Trolox/g), FRAP value was 1.05μmol Fe^{2+}/g and phenolic content was 0.24mg GAE/g.

Aristolochia bracteolata Lam. Family.: Aristolochiaceae

English Name: Worm killer, Indian Birthwort

The ethanol extract of shade-dried leaves was studied for its effect on wound healing in rats, using incision, excision and dead-space wound models. The plant showed a definite, positive effect on wound healing, with a significant increase in the level of two powerful antioxidant enzymes, superoxide dismutase and catalase, in the granuloma tissue. Preliminary phytochemical screening revealed the presence of alkaloids, saponins, glycosides, steroids, tannins, phenolic compounds and flavonoid glycosides (Shirwaikar *et al.*, 2003).

Thirugnanasampandan *et al.* (2008) investigated the antioxidant activity of five Aristolochiaceae species. *Aristolochia bracteolata* showed good antioxidant, DPPH radical scavenging and reducing power.

Aristolochia brasiliensis Mart. et Zucc. Family: Aristolochiaceae

Thirugnanasampandan *et al.* (2008) investigated the antioxidant activity of five Aristolochiaceae species. Among them *Aristolochia brasiliensis* showed lowest antioxidant, DPPH radical scavenging and reducing power.Total phenolic compounds were also lowest.

Aristolochia cymbifera Mart. Family: Aristolochiaceae

Alviano *et al.* (2008) evaluated the antioxidant activity of alcoholic extract of rhizomes of *Aristolochia cymbifera* by DPPH photometric assay. The extract showed good antioxidant activity with EC_{50} value of 124.5μg/ml.

Aristolochia giberti Hook. Family: Aristolochiaceae

The antioxidant property of *Aristolochia giberti* was studied using free radical-generating systems (Velazquez *et al.*, 2003). The methanol extract protected against enzymatic and non-enzymatic lipid peroxidation in microsomal membranes of rat.

Aristolochia indica L. Family: Aristolochiaceae

Thirugnanasampandan *et al.* (2008) investigated the antioxidant activity of five Aristolochiaceae species. *Aristolochia indica* showed good antioxidant, DPPH radical scavenging and reducing power.

Aristolochia mollissima Hance Family: Aristolochiaceae

Cai *et al.* (2004) evaluated the antioxidant activity and phenolic compounds in traditional Chinese medicinal plants associated with anticancer. The improved ABTS.+method was used to systematically assess the total antioxidant capacity (Trolox equivalent antioxidant capacity, TEAC) of the medicinal extracts. The TEAC values and total phenolic content for methanolic extract of *Aristolochia mollissima* were 88.7 µmol Trolox equivalent/100g dry weight (DW), and 0.42 g of gallic acid equivalent/100g DW respectively. A positive significant linear relationship between antioxidant activity and total phenolic content showed that phenolic compounds were the dominant antioxidant components in the tested herbs. Major types of phenolic compounds in *Aristolochia mollissima* include phenolic acids (vanillic acid).

Aristolochia tagala Cham. Family: Aristolochiaceae

Thirugnanasampandan *et al.* (2008) investigated the antioxidant activity of five Aristolochiaceae species. *Aristolochia tagala* showed highest antioxidant, DPPH radical scavenging and reducing power. Amount of total phenolic compounds were highest in *Aristolochia tagala*.

Armoracia rusticana Gaertn. *et al.*, Family: Brassicaceae (Cruciferae)

English Name: Horseradish.

The antioxidative activity of phenolic extracts from fruits of horseradish was examined by autoxidation of methyl linoleate by Kahkonen *et al.* (1999). The total phenolics were very very low (0.9mg/g GAE). The antioxidant activity was practically nil.

Arnica chamissonis Less. var. *foliosa* Family: Asteraceae (Compositae)

Water extract of seeds of plantation-grown *Arnica chamissonis* var. *foliosa* (AC) contain higher levels of phenolic acids and flavonoids (Gawlikdziki *et al.*, 2009). Fractions of phenolics were determined in seeds, comprising chlorogenic, caffeic acid, quercetin and kaempferol. All samples showed high antioxidant activity, however their levels depend on the extraction procedure and used material. Studied samples showed positive correlations between inhibition of lipid peroxidation ability and total flavonoids and phenolic acids content.

Arnica montana L. Family: Asteraceae (Compositae)

Water extract of seeds of plantation-grown *Arnica montana* contain higher levels of phenolic acids and flavonoids (Gawlikdziki *et al.*, 2009). Fractions of phenolics were determined in seeds, comprising chlorogenic, caffeic acid, quercetin and kaempferol. All samples showed high antioxidant activity, however their levels depend on the extraction procedure and used material. Studied samples showed positive correlations between inhibition of lipid peroxidation ability and total flavonoids and phenolic acids content.

Aronia melanocarpa Elliot Family: Rosaceae

English Name: Aronia, blackchokeberry.

The antioxidative activity of phenolic extracts from fruits of *Aronia melanocarpa* was examined by autoxidation of methyl linoleate by Kahkonen *et al.* (1999). Remarkable high antioxidant activity and high total phenolic content (GAE >20mg/g) were found in aronia.

The red pigment fraction of the black chokeberry contained three main components, one of which was identified as cyanidin 3-O-β-glucoside. The black chokeberry red pigment fraction scavenged less than 44 per cent DPPH radicals at a concentration of 25 μg/ml compared to the control solution (Matsumoto *et al.*, 2004)

Kedzierska *et al.* (2009) studied the effects of polyphenol rich extract of *Aronia melanocarpa* (final concentration of 50μg/ml, 5 min, 37°C) on superoxide anion radicals and glutathione (GSH) in platelets from patients with breast cancer and in healthy group *in vitro*. The study demonstrated the protective role of the extract in patients with breast cancer *in vitro*.

Arrabidaea samydoides (Cham.) Sandwith. Family: Bignoniaceae

Three new C-glucosylxanthones were isolated from the stems of *Arrabidaea samydoides* by Pauletti *et al.* (2003). Xanthones showed moderate free radical scavenging activity against DPPH as well as antioxidant activity evidenced by redox properties measured on EICD-HPLC.

Arracacia xanthorhiza Bauer Family: Apiaceae (Umbelliferae)

Penarrieeta *et al.* (2005) measured the Total Antioxidant Capacity (TAC) in some Andean foods. Eight Andean foods, including *Arracacia xanthorhiza* were analyzed by two methods ABTS and FRAP to assess TAC. TAC value of *Arracacia xanthorhiza* was good in both water-soluble fraction and water-insoluble fraction.

Artemisia absinthium L. Family: Asteaceae (Compositae)

English Name: Wormwood.

Antioxidant and DPPH radical scavenging activities, reducing powers and the amount of total phenolic compounds of some medicinal Asteraceae species were studied by Özgen *et al.* (2004). Water extract of *Artemisia absinthium* showed good antioxidant activity with 65.7 per cent inhibition at 500μg/ml concentration. The DPPH radical scavenging activity was 86.5 per cent at 500μg/ml concentration and reducing power was 22.9 μg/ml Ascorbic acid equivalent. Total phenolic amount was 24.1μg/ml Gallic acid equivalent.

Canadanic-Brunet *et al.* (2005) screened the antioxidant activity of *Artemisia absinthium* by measuring DPPH free radical and reactive hydroxyl radical during Fenton reaction trapped by DMPO. Results demonstrated that the antiradical and antioxidative activity depend on the type and concentration of applied extracts and increased in the order ethyl acetate>methanol>n-butanol>chloroform>petroleum ether> remaining water extracts. The investigation showed that the antiradical activity increased with increasing concentration of all extracts. The high contents of total phenolic compounds (25.6 mg/g) and total flavonoids (13.06 mg/g) indicated that these compounds contribute to the antiradical and antioxidative activity.

Artemisia abyssinica Schultz.-Bip. Family: Asteraceae (Compositae)

The essential oil of *Artemisis abyssinica* was examined for its potential radical scavenging activities (Burits *et al.*, 2001) in DPPH assay. Oil has shown positive result.

Artemisia afra Jacq. Family: Asteraceae (Compositae)

The essential oil of *Artemisia afra* was examined for its potential radical scavenging activities (Burits *et al.*, 2001) in DPPH assay. The oil of *Artemisia afra* was effective hydroxyl radical scavenging agent when assessed in the deoxyribose degradation assay. In the *in vitro* assay for non-enzymatic lipid peroxidation in liposomes, the oil of *A. afra* also displayed antioxidant potential. The compounds that contribute to the radical scavenging activities were identified and then assessed for their effects in various test systems.

Artemisia annua L. Family: Asteraceae (Compositae)

Juteau *et al.* (2002) evaluated the antioxidant activity of the essential oil by chemilumescence method. This oil has shown an antioxidant activity equivalent to 18 per cent of the reference compound (α-tocopherol).

Cai *et al.* (2004) evaluated the antioxidant activity and phenolic compounds in traditional Chinese medicinal plants associated with anticancer. The improved ABTS.+ method was used to systematically assess the total antioxidant capacity (Trolox equivalent antioxidant capacity, TEAC) of the medicinal extracts. The TEAC values and total phenolic content for methanolic extract of aerial parts of *Artemisia annua* were 582.3 μmol Trolox equivalent/100g dry weight (DW), and 3.51 g of gallic acid equivalent/100g DW respectively. A positive significant linear relationship between antioxidant activity and total phenolic content showed that phenolic compounds were the dominant antioxidant components in the tested herbs. Major types of phenolic compounds in *Artemisia annua* include flavones (luteolin, luteolin-7-glucoside), flavonols (kaempferol, quercetin, rutin), coumarins (coumarin, 6,7-dimethoxy-coumarin).

Artemisia annua has a high content of flavonoid compounds which are responsible for its high antioxidant activity. *A. annua* leaves (Zheng and Wang, 2001) and crude extracts have been reported to be a good source of antioxidants (Cai *et al.*, 2004), being among the four medicinal plants with highest ORAC (oxygen radical absorbance capacity) level among other medicines (Zheng and Wang, 2001). This high antixodiant capacity is probably due to the high content (Bilia *et al.*, 2006) and diversity of its leaf flavonoids, including the newly reported C-glycosyl flavonoids as a possible component of the antioxidant and antiviral activity (Han *et al.*, 2008). Flavanoids have been reported to be responsible for more of the antioxidant activity of leafy vegetables and herbs than vitamin E, C or glutathione (Cao *et al.*, 1996).

Artemisia anomala S.Moore Family: Asteraceae (Compositae)

Gan *et al.* (2010) evaluated the antioxidant activities of 40 medicinal plants associated with prevention and treatment of cardiovascular and cerebrovascular diseases using FRAP and TEAC assays and their total phenolic contents were measured by the Folin-Ciocalteu methods. The antioxidant activity of *Artemisia anomala* was 10.94μmol Fe(II)/g in FRAP assay, 32.77μmol Trolox/g and Total phenolic content was 3.15 mg GAE/g. A high correlation between antioxidant capacity and total phenolic content indicated that phenolic compounds could be the main contributor of the antioxidant activity of these plants.

Artemisia apiacea Hance Family: Asteraceae (Compositae)

Li *et al.* (2008) evaluated the antioxidant properties of 45 medicinal plants including *Artemisia apiacea* using FRAP and TEAC assays and the total phenolic contents of these plants were measured. *Artemisia apiacea* showed 48.72 µmol Fe(II)/g FRAP value, 30.83µmol Trolox/g of TEAC value and 8.57 mg GAE/g of Phenolic content. A high correlation was observed between antioxidant capacities and their phenolic contents.

Artemesia arborescens L. Asteraceae (Compositae)

Dessi *et al.* (2001) carried out studies on the effect of methanol extract and essential oil on lipid peroxidation. Essential oil and methanol extract showed antioxidant activity.

Wojcikowski *et al.* (2007) evaluated the antioxidant activity of aerial parts of *Artemisia arborscens.* Oxygen radical absorbance capacity of the ethyl extract, methanol extract and aqueous extracts were 43.68, 237.20 and 267.17µmol Trolox equivalent per g of dried starting material, respectively.

Artemisia argyi Levl. et Vant. Family: Asteraceae (Compositae)

Cai *et al.* (2004, 2006) evaluated the antioxidant activity and phenolic compounds in traditional Chinese medicinal plants associated with anticancer. The improved ABTS.+ method was used to systematically assess the total antioxidant capacity (Trolox equivalent antioxidant capacity, TEAC) of the medicinal extracts. The TEAC values and total phenolic content for methanolic extract of leaves of *Artemisia argyii* were 328.5 µmol Trolox equivalent/100g dry weight (DW), and 1.94 g of gallic acid equivalent/100g DW respectively. A positive significant linear relationship between antioxidant activity and total phenolic content showed that phenolic compounds were the dominant antioxidant components in the tested herbs. Major types of phenolic compounds in *Artemisia argyii* include flavones (apigenin, eupatilin, chrysoeriol).

Gan *et al.* (2010) evaluated the antioxidant activities of 40 medicinal plants associated with prevention and treatment of cardiovascular and cerebrovascular diseases using FRAP and TEAC assays and their total phenolic contents were measured by the Folin-Ciocalteu methods. The antioxidant activity of *Artemisia argyi* was 241.19µmol Fe(II)/g in FRAP assay, 127.73µmol Trolox/g and Total phenolic content was 12.87 mg GAE/g. A high correlation between antioxidant capacity and total phenolic content indicated that phenolic compounds could be the main contributor of the antioxidant activity of these plants.

Artemisia austriaca Jacq. Family: Asteraceae (Compositae)

Antioxidant and DPPH radical scavenging activities, reducing power and the amount of total phenolic compounds of some medicinal Asteraceae species were studied by Özgen *et al.* (2004). Methanolic extract of *Artemisia austriaca* showed good antioxidant activity with 81.6 per cent inhibition at 500µg/ml concentration. The DPPH radical scavenging activity was 87.8 per cent at 500µg/ml concentration and reducing power was 22.1 µg/ml Ascorbic acid equivalent. Total phenolic amount was 18.2µg/ml Gallic acid equivalent.

Artemisia campestris L. Family: Asteraceae (Compositae)

English Name: Field Mugwort; Japanese name: Ryukyuyomogi.

Aniya *et al.* (2000) examined the antioxidant action of *Artemisia campestris in vitro* and *in vivo*. A water extract of *A. campestris* showed a strong scavenging action of DPPH, hydroxyl and superoxide anion radicals. When the extract was given intraperitoneally to mice prior to carbon tetrachloride (CCl_4) treatment, CCl_4-induced liver toxicity, as seen by an elevation of serum aspartase aminotransferase and alanine aminotransferase activities, was significantly reduced. Depression of the elevation of serum enzyme levels after CCl_4–treatment was also observed by oral administration of the extract. In that case, CCl_4-derived lipid peroxidation in the liver was decreased by the extract treatment. These results suggest that the extract of *A.campestris* scavenges radicals formed by CCl_4 treatment resulting in protection against CCl_4–induced liver toxicity.

Maeda *et al.* (2006) examined the phenolic contents of leaf of *Artemisia campestris* and their radical scavenging activities by using DPPH assay. High contents of phenolic compounds and strong DPPH scavenging activities were observed in extracts from this species. They also showed radical scavenging activities for the superoxide radical and tert-butylhydroperoxyl radical (t-BuOO·).

Serteser *et al.* (2009) studied the antioxidant activity of 50 per cent aqueous methanol extracts of *Artemisia campestris* including free radical scavenging, hydrogen peroxide scavenging and metal (Fe^{2+}) chelating activities.The DPPH radical scavenging effects of leaves extract was 0.405. Fe^{2+} chelating activity (per cent) of leaves extract was 25.42 and Hydrogen peroxide inhibition activity of leaves extracts was 28.32.

Artemisia capillaris Thunb. Family: Asteraceae (Compositae)

Cai *et al.* (2004) evaluated the antioxidant activity and phenolic compounds in traditional Chinese medicinal plants associated with anticancer. The improved ABTS.+ method was used to systematically assess the total antioxidant capacity (Trolox equivalent antioxidant capacity, TEAC) of the medicinal extracts. The TEAC values and total phenolic content for methanolic extract of seedling of *Artemisia capillaris* were 395.2 μmol Trolox equivalent/100g dry weight (DW), and 2.63 g of gallic acid equivalent/100g DW respectively. A positive significant linear relationship between antioxidant activity and total phenolic content showed that phenolic compounds were the dominant antioxidant components in the tested herbs. Major types of phenolic compounds in *Artemisia capillaris* include coumarins (6,7-dimethoxy coumarin), phenolic acids (chlorogenic acid, caffeic acid).

Gan *et al.* (2010) evaluated the antioxidant activity of *Artemisia capillaris* using the FRAP and TEAC assays, and its total phenolic content was measured by the Folin-Ciocalteu method. For FRAP assay the value was 158.87 μmol Fe(II)/g dry weight. For TEAC assay, the value was 106.55μmol Trolox/g dry weight. Total phenolic content was 8.38 mg gallic acid equivalent/g dry weight of plant material.

Artemisia dracunculus L. Family: Asteraceae (Compositae)

English Name: Tarragon.

Souri *et al.* (2004) investigated the antioxidant activity of methanol extracts of garlic, used as vegetable in Iranian diet against linoleic acid peroxidation. The extracts showed good antioxidant activity. The antioxidant activity of the aerial part was lower than dl-α-tocopherol but higher than quercetin.

Mariutti *et al.* (2008) evaluated the antioxidant activity of garlic employing ABTS and DPPH radical assay. DPPH EC_{50} value was 62.4g/kg while DPPH-TEAC value was 3.34mM/g. ABTS-TEAC value was 5.6mM/g.

Artemisia gorgonum Webb. Family: Asteraceae (Compositae)

The essential oil of *Artemisia gorgonum* was tested for its antioxidant properties and was found to exhibit free-radical scavenging on DPPH, prevention of lipid peroxidation-*in vitro* by TBARS assay (Ortet *et al.*, 2010).

Artemisia herba-alba Asso. Family: Asteraceae (Compositae)

Suboh *et al.* (2004) investigated the antioxidant activity effects of various concentrations of methanol extracts of *Artemisia herba-alba* and six other medicinal plants. They reported that the extract did not protect erythrocytes against lipid peroxidation.

Al-Mustafa and Al-Thunibat (2008) evaluated the antioxidant activity of Jordanian medicinal plant *Artemisia herba-alba* shoot. The level of antioxidant activity was determined by DPPH and ABTS assay in relation to total phenolic contents of the medicinally used part. They concluded that *Artemisia herba-alba* shoot possesses moderate antioxidant capacity (DPPH-TEAC range 20-80mg/g).

Mighri *et al.* (2010) studied the antioxidant activity of the essential oils of *Artemisia herba-alba* cultivated in Southern Tunisia. Diverse methods were used. All the oils exhibited weak antioxidant abilities for preventing linoleic acid oxidation. They attributed such results to the absence of the non-phenolic compounds.

Artemisia judaica L. Family: Asteraceae (Compositae)

Al-Mustafa and Al-Thunibat (2008) evaluated the antioxidant activity of Jordanian medicinal plant *Artemisia judaica* stem and leaves. The level of antioxidant activity was determined by DPPH and ABTS assay in relation to total phenolic contents of the medicinally used part. They concluded that *Artemisia judaica* shoot possess low antioxidant capacity (DPPH-TEAC range 20-80mg/g).

Artemisia maritima L. Family: Asteraceae (Compositae)

A study was carried out by Budinevi *et al.* (1995) on the antioxidative activity of *Artemisia maritima*. Using the ethanolic extracts of plant material good results were obtained.

Artemisia montana (Nakai) Pamp. Family: Asteraceae (Compositae)

The antioxidant activity of *Artemisia montana* was determined by Kim *et al.* (2000) by measuring the radical scavenging effect on DPPH radical and inhibitory activity against free radical generation of hepatocytes. The methanol extract of *A. montana* showed strong radical scavenging activity at a concentration of 10.1µg/ml. Esculetin and luteolin 7-O-rutinoside (scolymoside) were isolated as the active principles from the EtOAc and interphase fractions, respectively. The antioxidant activity of these compounds were comparable to that of Ascorbic acid.

Artemisia princeps Pamp. Family: Asteraceae (Compositae)

English Name: Mugwort; Japanese name: Nishiyomogi.

Maeda *et al.* (2006) examined the phenolic contents of leaf of *Artemisia princeps* and their radical scavenging activities by using DPPH assay. High contents of phenolic compounds (786.3 mg-gallic acid equivalent/100g) and strong DPPH scavenging activities were observed in extracts from this species. They also showed radical scavenging activities for the superoxide radical and tert-butylhydroperoxyl radical (t-BuOO·).

Artemisia santonicum L. Family: Asteraceae (Compositae)

Serteser *et al.* (2009) studied the antioxidant activity of 50 per cent aqueous methanol extracts of *Artemisia santonicum* including free radical scavenging, hydrogen peroxide scavenging and metal (Fe^{2+}) chelating activities.The DPPH radical scavenging effects of leaves extract was 0.400. Fe^{2+} chelating activity (per cent) of leaves extracts was 26.02 and Hydrogen peroxide inhibition activity of leaves extracts was 26.17.

Artemisia scoparia Waldst et Kit. Family: Asteraceae (Compositae)

The essential oils from young and mature leaves of *Artemisia scoparia* displayed a high degree of hydroxyl radical scavenging activity (Singh *et al.*, 2010). However, the scavenging activity of principal constituents, β-myrcene and *p*-cymene, was very less compared to that of oils. The same authors also evaluated the antioxidant activity of the samples by the DPPH method and also reported less activity of the main constituents of the oils than the essentials oils, nevertheless the mature leaf oils were more effective as antioxidants than young leaf oils.

Artemesia sieversiana Family: Asteraceae (Compositae)

Mangathayaru *et al.* (2007) reported antioxidant activity of aerial parts of *Artemisia sieversiana.*

Artemisia vulgaris L. Family: Asteraceae (Compositae)

Temrej and El-Tantawy (2008) evaluated the *in vitro* and *in vivo* antioxidant activities of aqueous extracts of *Artemisia vulgaris.* The plant extract was tested for DPPH radical scavenging, nitric oxide radical scavenging, reducing power assays, total phenol, flavonoid and flavonol content. The extract exhibited scavenging potential with IC_{50} value of 11.4µg/ml for DPPH, the value were found close to the standard rutin (10 µg/ml). On the other hand *Artemisia vulgaris* extract exhibited nitric oxide scavenging activity with IC_{50} value of 125mg/ml. The reducing power of the extract dependends on the amount of extract. The content of the phenolic compounds (mg/g) in aqueous extract was found 19 mg GAE/g. The flavonoid and phenolic contents were found to be 7.96 and 3.4 mg/g respectively in rutin equivalent. The treatment of rats with aqueous extract of *Artemisia vulgaris* resulted in a significant increase in blood glutathione level, SOD activity and serum ascorbic acid level as compared to their corresponding controls.

Artocarpus anisophyllus Miq. Family: Moraceae

Rahmani *et al.* (2009) evaluated the antioxidant activity of bark extract of *Artocarpus anisophyllus* by means of DPPH colorimetric method. The extract showed highest activity ith IC_{50} values ranging from 76-342μg/ml.

Artocarpus heterophyllus Lam. Family: Moraceae

English Name: Jackfruit.

Jagtap *et al.* (2010) evaluated the antioxidant activity of jackfruit pulp. The methanolic extract exhibited a high scavenging activity against DPPH (IC_{50} = 0.4-0.7 mg/ml), FRAP (0.851 optical density value, O.P.) and DMPD (IC_{50} 3.43 mg/ml). The antioxidant activities of Jackfruit pulp extracts were correlated with the total phenolic and flavonoid content.

Artocarpus lakoocha Roxb. Family: Moraceae

Hydromethanol extracts of 15 Bangladeshi medicinal plants, including *Artocarpus lakoocha*, were evaluated by Hasan *et al.* (2009) for antioxidant potential using DPPH radical scavenging assay. *Artocarpus lakoocha* exhibited radical scavenging activity with an IC_{50} value of 54.57 μg/ml (leaves) and 39.93 μg/ml (pericarp) compared to the IC_{50} value of 5.15μg/ml as shown by the reference antioxidant ascorbic acid, in a dose dependent fashion.

Povichit *et al.* (2010) investigated phenolic content and free radical scavenging effect of heartwood of *Artocarpus lakoocha* by DPPH and ABTS assays, anti-lipid peroxidation activity by TBARS and for antiglycation activity. The results revealed that the total phenolic content showed good correlation with free radical scavenging by ABTS and anti-lipid peroxidation by TBARS, but showed no correlation with antiglycation. The extract of *Artocarpus lakoocha* significant antioxidant effect and also showed a promising antiglycation effect. The IC_{50} were 0.16mg/ml for the DPPH method, TEAC value of 1.72 mg/Trolox/mg sample for the ABTS method, IC_{50} of 1.22 mg/ml for the TBARS method and IC_{50} of 3.20 for the antiglycation method.

The antioxidant activity of *Artocarpus lakoocha* heartwood extract was investigated by Singhatong *et al.* (2010) from ethanol extraction by ABTS decolorization, DPPH radical and H_2O_2 scavenging assay. Polyphenolic, total phenolic, flavonoids and tannins were measured. Anti-oxidative stress was studied by AAPH-oxidized blood glutathione (GSH), and NDA was evaluated. Results showed that antioxidant activities were 128, 55.86 and 463.49 μmol Trolox/g extracted from ABTS, DPPH and H_2O_2 scavenging methods. One gram of extract contained total phenol (325.63 mgGE), flavonoids (521.98 mg GE) and tannins (124.03 g TE), including rutin and resorcinol. In the blood system, a low concentration of extract inhibited MDA progression and improved GSH, which was in contrast to a high concentration with its toxicity effect.

Artocarpus nobilis Thw. Family: Moraceae

Geranyl chalcone derivatives and stilbene derivatives isolated from the n-butanol extract of the leaves of *Artocarpus nobilis* showed high radical scavenging activity towards DPPH radical in TLC bio-autography method (Jayasinghe *et al.*, 2004a,b).

Artocarpus obtusus F.M.Jarrett Family: Moraceae

Rahmani *et al.* (2009) evaluated the antioxidant activity of bark extract of *Artocarpus obtusus* by means of DPPH colorimetric method. The extract showed highest activity with IC_{50} values ranging from 76-342μg/ml.

Arum dioscoridis Sm. Family: Araceae

Janakkat and Al-Thanibat (2008) evaluated the antilipoperoxidative activity of the aqueous, boiled and non boiled and methanolic extracts of leaves of *Arum dioscoridis* against lipid peroxidation (LPO) induced by a rat liver homogenate model. Nonboiled aqueous extracts exhibited the highest LPO inhibition per cent followed by methanolic and boiled aqueous extracts. Moreover, LPO inhibition per cent was dose dependent, where 1.65×10^{-2} mg/ml caused the highest inhibition per cent in all aqueous extracts.

Arum palaestinum Boiss. Family: Araceae

Al-Mustafa and Al-Thunibat (2008) evaluated the antioxidant activity of Jordanian medicinal plant *Arum palaestinum* leaves. The level of antioxidant activity was determined by DPPH and ABTS assay in relation to total phenolic contents of the medicinally used part. They concluded that *Arum palaestinum* leaves possess moderate antioxidant capacity (DPPH-TEAC range 20-80mg/g).

Arundo donax L. Family: Poaceae (Graminae)

Menezes *et al.* (2004) investigated the antioxidant activity of hydroalcoholic extracts of *Arundo donax* by DPPH assay and phosphomolybdenum method. The relative activity in phosphomolybdenum complex was 0.98 while scavenging activity against DPPH in IC_{50} (μg/ml) was 399.10.

Aspalathus linearis (N.L.Burm.) R.Dahlgr. Family: Fabaceae (Leguminosae)

English Name: Rooibos.

Unfermented rooibos tea is known to contain higher levels of total polyphenols and flavonoids than its fermented counterpart, making it the obvious choice for the preparation of flavonoid enriched fractions. Joubert *et al.* (2005) evaluated aqueous extracts and crude polyphenolic fractions of unfermented and fermented rooibos for anti- and/or pro-oxidant activities, using a linoleic acid-Tween-buffer emulsion for lipid peroxidation and the deoxyribose degradation assay, based on a Fenton reaction model system containing $FeCl_3$-EDTA and H_2O_2 for the generation of hydroxyl radicals. Except for the ethyl acetate fraction, with the highest total polyphenol content and offering the least protection presumably due to pro-oxidant activity, the inhibition of lipid peroxidation by the samples correlated moderately with their TP content in a linear relationship. Fermentation (*i.e.*, oxidation) of rooibos decreased the pro-oxidant activity of aqueous extracts, which was contributed to a decrease in their dihydrochalcone content. The *in vitro* pro-oxidant activity displayed by flavonoid-enriched fractions of rooibos demonstrates that one must be aware of the potential adverse biological properties of potent antioxidant extracts utilized as dietary supplements (Joubert *et al.*, 2005).

Rooibos is a popular tisane in its native South Africa and having a world wide market. The herbal tea is used traditionally for medicinal purposes and is rich in popyphenols and a rare source of the dietary dihydrochalcones, aspalathin and nothogagin. McKay and Blumberg (2006) investigated in model studies which indicate that the herbal tea possesses potent antioxidant, immune-modulating and chemopreventive actions. The herbal tea showed potent antioxidant and antimutagenic activities. No adverse effect of rooibos consumption as tisane has been reported.

In a screening of South African indigenous food plants for antioxidant activity by testing for inhibition of lipid peroxidation the teas of *Aspalathus linearis* had high activity (Lindsey *et al.*, 2006).

Asparagus racemosus Willd. Family: Liliaceae

Kamat *et al.* (2000) have examined the antioxidant effects of crude extract as well as purified polysaccharide fraction of *Asparagus racemosus* against membrane damage induced in rat mitochondria and liver by free radicals generated during γ-irradiation. They found that these materials had potent antioxidant properties *in vitro*.

Chloroform extracts of the *Asparagus racemosus* yielded a new antioxidant activity compound named racemofuran along with two known compounds asparagamine and racemosol (Wiboonpun *et al.*, 2004). Racemofuran revealed antioxidant property against DPPH with IC_{50} value of 130 µM.

Patnibul *et al.* (2008) screened 18 vegetables for antioxidant activity using silica gel Thin Layer chromatography followed by spraying with DPPH. *Asparagus racemosus* exhibited strong antioxidant activity.

Jain and Agrawal (2009) reported antioxidant activity of the methanolic extract of *Asparagus racemosus* with IC_{50} value of 9.28 µg/ml.

Asperula stylosa (Trin.) Boiss. Family: Rubiaceae

Souri *et al.* (2004) evaluated the antioxidative activity of *Asperula stylosa* by linoleic acid peroxidation test using 1,3-diethyl-2-thiobarbituric acid as the reagent. Methanolic extract showed 82.65 per cent inhibition at 40µg concentration with IC_{50} value of 4.51µg.

Aspilia quinquenervis Blake. Family: Asteraceae (Compositae)

Mosquera *et al.* (2009) evaluated the antioxidant activity of forty six methanol plant extracts using DPPH radical scavenging assay. *Aspilia quinquenervis* showed 26.1 per cent antioxidant activity.

Aster scaber Thunb. Family: Asteraceae (Compositae)

Heo *et al.* (2009) investigated the total phenolic content, antioxidant activity and cytotoxicity of methanol extracts from aerial parts of 11 Korean medicinal salad plants. Methanol extract of aerial parts showed highest phenolic content (17.1mg/100g) and DPPH radical scavenging activity at 50 microg/ml was 83.3 per cent. In conclusion, the studied salad plants have a high phenolics content and high antioxidant activity. These plants dose-dependently increased DPPH free radical scavenging activity. The total phenolics level was highly correlated with free radical scavenging activity.

Aster tataricus L.f. Family: Asteraceae (Compositae)

Cai *et al.* (2004) evaluated the antioxidant activity and phenolic compounds in traditional Chinese medicinal plants associated with anticancer. The improved ABTS.+ method was used to systematically assess the total antioxidant capacity (Trolox equivalent antioxidant capacity, TEAC) of the medicinal extracts. The TEAC values and total phenolic content for methanolic extract of root of *Aster tataricus* were 81.8 μmol Trolox equivalent/100g dry weight (DW), and 0.60 g of gallic acid equivalent/100g DW respectively. A positive significant linear relationship between antioxidant activity and total phenolic content showed that phenolic compounds were the dominant antioxidant components in the tested herbs. Major types of phenolic compounds in *Aster tataricus* include flavonols (quercetin, quercetrin).

Antioxidant capacities of 56 selected Chinese medicinal plants, including *Aster tataricus* were evaluated by Song *et al.* (2010) using the TEAC and FRAP assays, and their total phenolic content was measured by the Folin-Ciocalteu method. The TEAC value was 47.38μmol Trolox/g, FRAP value was 14.77μmol Fe^{2+}/g and phenolic content was 5.56mg GAE/g.

Astianthus humilis Baill. Family: Bignoniaceae

Ruiz-Terán *et al.* (2008) evaluated the antioxidant activity of aerial parts of *Astianthus humilis* using DPPH and β-carotene-linoleic acid bleaching assay. The extract displayed antioxidant activity comparable to the commercial antioxidants BHA and TBHQ.

Astragalus complanatus R.Br. Family: Fabaceae (Leguminosae)

Cai *et al.* (2004, 2006) evaluated the antioxidant activity and phenolic compounds in traditional Chinese medicinal plants associated with anticancer. The improved ABTS.+ method was used to systematically assess the total antioxidant capacity (Trolox equivalent antioxidant capacity, TEAC) of the medicinal extracts. The TEAC values and total phenolic content for methanolic extract of seed of *Astragalus complanatus* were 119.5 μmol Trolox equivalent/100g dry weight (DW), and 0.98 g of gallic acid equivalent/100g DW respectively. A positive significant linear relationship between antioxidant activity and total phenolic content showed that phenolic compounds were the dominant antioxidant components in the tested herbs. Major types of phenolic compounds in *Astragalus complanatus* include flavonoids (astragalin, myricetin, kaempferol, glycosides).

Astragalus hamosus L. Family: Fabaceae (Leguminosae)

English Name: Milk vetch.

Souri *et al.* (2008) investigated the antioxidant activity and radical scavenging activity of methanolic extract of fruit. The antioxidant activity of this plant extract against linoleic acid peroxidation expressed as IC_{50} was 7.65ng/ml and radical scavenging activity against DPPH expressed as IC_{50} was 32.36 µg/ml.

Astragalus membranaceus (Fisch.) Bunge. Family: Fabaceae (Leguminosae)

Wojcikowski *et al.* (2007) evaluated the antioxidant activity of aerial parts of *Astragalus membranaceus.* Oxygen radical absorbance capacity of the ethyl extract, methanol extract and aqueous extracts were 27.47, 16.04 and 134.21µmol Trolox equivalent per g of dried starting material, respectively.

Huang *et al.* (2009) carried out studies on principal components and antioxidant activity of different Radix Astragali samples using HPLC/electrospray ionization multiple stage tandem mass spectrometry.The correlations of the isoflavonoids and astragalosides components and the antioxidant activities of Radix Astragali were summarized.

Chen *et al.* (2009) studied the antioxidant activity of roots extract towards xanthine oxidase, lipoxygenase and tyrosinase activities. The inhibitory effect of acetone extract on xanthine oxidase activity was about 40 per cent, tyrosine activity was more than 50 per cent and lipoxygenase activity was more than 60 per cent.

Astragalus mongholicus Bunge. Family: Fabaceae (Leguminosae)

English Name: Milk vetch root.

Cai *et al.* (2004, 2006) evaluated the antioxidant activity and phenolic compounds in traditional Chinese medicinal plants associated with anticancer. The improved ABTS.+ method was used to systematically assess the total antioxidant capacity (Trolox equivalent antioxidant capacity, TEAC) of the medicinal extracts. The TEAC values and total phenolic content for methanolic extract of root of *Astragalus mongholicus* were 120.8 µmol Trolox equivalent/100g dry weight (DW), and 0.77 g of gallic acid equivalent/100g DW respectively. A positive significant linear relationship between antioxidant activity and total phenolic content showed that phenolic compounds were the dominant antioxidant components in the tested herbs. Major types of phenolic compounds in *Astragalus mongholicus* include flavonoids (astragalin, calycosin, formononetin).

Astrodaucus orientalis (L.) Drude Family: Apiaceae (Umbelliferae)

Souri *et al.* (2004) evaluated the antioxidative activity of *Astrodaucus orientalis* by linoleic acid peroxidation test using 1,3-diethyl-2-thiobarbituric acid as the reagent. Methanolic extract showed 83.11 per cent inhibition at 40µg concentration with IC_{50} value of 2.86µg.

Astronium urundeuva Engl. Family: Anacardiaceae

The *in vitro* antioxidant and free radical scavenging properties of bark extract of *Astronium urundeuva* used as anti-inflammatory agent in the Brazilian state of Bahis, was studied using different bioassays (Desmarchelier *et al.*, 1999). The total reactive antioxidant potential (TRAP) of the aqueous and methanolic extracts was determined by monitoring the intensity of luminal-enhanced chemiluminescence (CL), using 2,2′-azo-bis(2-amidinopropane) as a peroxyl radical source.The extract studied was active in this method. Lipid peroxidation was assessed by means of the production of thiobarbituric acid reactive substances (TBARS) and hydroperoxide-initiated CL in rat liver homogenates. The extract tested was effective in reducing the production of TBARS.

Athamanta turbith (L.) Brot. Family: Apiaceae (Umbelliferae)

The mature fruits of *Athamanta turbith* ssp. *hungarica* and *Athamanta turbith* ssp. *haynaldii* are rich in essential oils, myristicin being the main component of both oils (Tomi *et al.*, 2010). The capacity for scavenging hydroxyl radicals was evaluated and the authors found that both oils had similar ability. When the activity was evaluated by the DPPH method, the authors reported that *Athamanta turbith* ssp. *hungarica* was the most effective, albeit with less ability than the references used (quercetin and ascorbic acid)

Athrixia phylicoides DC. Family: Asteraceae (Compositae)

English Name: Bush tea.

Extracts of *Athrixia phylicoides* and *Athrixia elata* were tested for antioxidant activity by McGaw *et al.* (2007). Antioxidant activity comparable to that found in rooibos was established.

Atractylodes lancea (Thunb.) DC. Family: Asteraceae (Compositae)

Cai *et al.* (2004) evaluated the antioxidant activity and phenolic compounds in traditional Chinese medicinal plants associated with anticancer. The improved ABTS.+ method was used to systematically assess the total antioxidant capacity (Trolox equivalent antioxidant capacity, TEAC) of the medicinal extracts. The TEAC values and total phenolic content for methanolic extract of floral bud of *Tripterygium wilfordii* were 1858.3 μmol Trolox equivalent/100g dry weight (DW), and 6.58 g of gallic acid equivalent/100g DW respectively. A positive significant linear relationship between antioxidant activity and total phenolic content showed that phenolic compounds were the dominant antioxidant components in the tested herbs.

Gan *et al.* (2010) evaluated the antioxidant activity of *Atractylodes lancea* using the FRAP and TEAC assays, and its total phenolic content was measured by the Folin-Ciocalteu method. For FRAP assay the value was 28.15 μmol Fe(II)/g dry weight. For TEAC assay, the value was 17.61μmol Trolox/g dry weight. Total phenolic content was 2.41 mg gallic acid equivalent/g dry weight of plant material.

Avena sativa L. Family: Poaceae (Graminae)

English Name: Oats.

The antioxidative activity of phenolic extracts from oats was examined by autoxidation of methyl linoleate by Kahkonen *et al.* (1999). Oats extracts did not show remarkable antioxidant activity. Oat bran exerted a moderate effect. In oat flour, *p*-hydroxybenzoic, protocatechuic, vanillic, trans-*p*-coumaric, (*p*-hydroxyphenyl)-acetic, syringic, trans-sinapic, caffeic and ferulic acids were found, with ferulic acid being the most abundant (White and Xing, 1997). Ferulic acid showed moderate, caffeic acid showed strong and *p*-coumaric acid showed negligible antioxidant activity in a study by Terao *et al.* (1993), where the activity was measured by autooxidation of bulk MeLo. Oats contain a series of anionic substituted cinnamic acid conjugates, avenanthramides, which are also potential antioxidants (Dimberg *et al.*, 1993). The processing seems to be important when evaluating the antioxidant activity of cereals as brans were more active than other products from the same cereal species. This is obviously due to localization of the phenolics in the grain: the outer layers of the grain (husk, pericarp, testa, and aleurone cells) contain the greatest concentrations of total phenolic, whereas their concentrations are considerably lower in the endosperm layers.

***Averrhoa bilimbi* L. Family: Averrhoaceae**

English Name: Bilimbi.

Abbas *et al.* (2006) evaluated the antioxidant activity of *Averrhoaceae* using DPPH assay. Methanolic extract showed DPPH radical scavenging activity.

***Avicennia alba* Bl. Family: Avicenniaceae**

Vadlapudi and Naidu (2009) evaluated the antioxidant potential of *Avicennia alba*, a mangrove plant. The SOD was 0.96U/mg, Catalase was 0.98U/mg, Ascorbic acid was 160mg/100g, DPPH radical inhibition was 30.2 per cent and FRAP units were 625.

***Avicennia marina* (Forsk.) Vierh. Family: Avicenniaceae**

Vadlapudi and Naidu (2009) evaluated the antioxidant potential of *Avicennia marina*, a mangrove plant. The SOD was 1.44U/mg, Catalase was 0.43U/mg, Ascorbic acid was 150mg/100g, DPPH radical inhibition was 56 per cent and FRAP units were 1772.

***Avicennia officinalis* L. Family: Avicenniaceae**

Vadlapudi and Naidu (2009) evaluated the antioxidant potential of *Avicennia officinalis*, a mangrove plant. The SOD was 1.24U/mg, Catalase was 1.81U/mg, Ascorbic acid was 140mg/100g, DPPH radical inhibition was 42.2 per cent and FRAP units were 960.

***Azadirachta indica* A.Juss. Family: Meliaceae**

English Name: Neem

The water-soluble constituents of alcoholic extract of neem leaves reduced the level of lipid peroxide(s) (Chattopadhyay, 1998). An antioxidant principle was isolated from neem seed using high pressure liquid chromatography with hydrophobic reverse-phase column (Rao *et al.*, 1998). It was found to be a potent inhibitor of plant lipoxygenases. However, such data on mammalian systems are not yet available.

A significant decrease in malondialdehyde (MDA) levels and a significant increase in antioxidant activity were observed by Chandra *et al.* (2003) when *Azadirachta indica* was administered to diabetic rats.

Total phenol, antioxidant and free radical scavenging activities of bark of *Azadirachta indica* were studied by Prakash *et al.* (2007). Bark was found to have high TPC (89.8 mg/g) and high antioxidant activity (84.8 per cent).

***Azima tetracantha* Lam. Family: Salvadoraceae**

The chloroform and ethanolic extracts of *Azima tetracantha* leaves were studied to evaluate the hepatoprotective and antioxidant activities in CCl_4-induced hepatotoxic rats (Ekbote *et al.*, 2010). Ethanolic extract exhibited significant effect showing increased levels of enzymatic and non-enzymatic parameters, *viz.*, catalase, GSH, total thiols and decreased level of malondialdehyde (MDA).

Baccaurea ramiflora Lour. Family: Euphorbiaceae

Two new compounds, 6′-vanilloylisotachioside (1) and 6′vanilloyltachioside (2), together with nine known compounds, were isolated by Yang *et al.* (2007) from the leaves of *Baccaurea ramiflora*. Compounds 1-10 were tested for antioxidant activities by using MTT and DPPH assays. Seven compounds, 1,2 and 4-8, revealed potent antioxidant activities against H_2O_2-induced impairment in PC12 cells, and exhibited significant DPPH radical-scavenging activities with IC_{50} values of 86.9, 142.9, 15.2, 37.6, 35.9, 30.2 and 79.8μM.

Hydromethanol extracts of 15 Bangladeshi medicinal plants, including *Baccaurea ramiflora*, were evaluated by Hasan *et al.* (2009) for antioxidant potential using DPPH radical scavenging assay. *Baccaurea ramiflora* fruit pericarp exhibited radical scavenging activity with an IC_{50} value of 31.38 μg/ml compared to the IC_{50} value of 5.15μg/ml as shown by the reference antioxidant ascorbic acid, in a dose dependent fashion.

Three new and four known compounds were isolated by Yang *et al.* (2010) from the stems of *Baccaurea ramiflora*. The new compounds were identified as 4-O-(6-O-vanilloyl)-β-D-glucopyranosyl tachioside D, 6′-O-vanilloylpicraquassioside D and 6-O-vanilloylicariside B5. Compounds 1 exhibited significant DPPH radical scavenging activity, with IC_{50} value of 36.9 μm, while compound 4 revealed weak antioxidant ctivity against H_2O_2-induced impairment in PC12 cells (Yang *et al.*, 2010).

Baccharis cordifolia DC. Family: Asteraceae (Compositae)

Mongelli *et al.* (1997) reported *in vitro* antioxidant and cytotoxic activity of extracts of *Baccharis cordifolia*.

The antioxidant and prooxidant activities of lyophilized extracts of 19 Argentine plants were studied using the hydroperoxide-initiated chemilumiscence assay in liver homogenates. Six extracts were found to have and effective antioxidative activity in the range of 10-100μg dry weight/ml and were further evaluated using TBARS assay. *Baccharis cordifolia* showed the higher antioxidant activity with both assay systems (Desmarchelier *et al.*, 1997).

Baccharis crispa Spreng. Family: Asteraceae (Compositae)

The aqueous extract of aerial parts of *Baccharia crispa* displayed significant radical scavenging activity in a DPPH/TLC assay. The fractions were monitored by the DPPH/TLC assay and a series of radical-scavenging quinic acid derivatives could be identified.

The antioxidant and prooxidant activities of lyophilized extracts of 19 Argentine plants were studied using the hydroperoxide-initiated chemilumiscence assay in liver homogenates. Six extracts were found to have effective antioxidative activity in the range of 10-100μg dry weight/ml and were further evaluated using TBARS assay.

Baccharis crispa showed the higher antioxidant activity with both assay systems (Desmarchelier *et al.*, 1997).

Baccharis grisebacchi Hieron Family: Asteraceae (Compositae)

Crude extract of *Baccharis grisebachi* displayed strong free radical activity which is attributed to their flavonoid constituents (Tapia *et al.*, 2004).

Baccharis salicifolia (Ruiz. et Pavon) Pers. Family: Asteraceae (Compositae)

The aerial parts of 17 Bolivian plants including *Baccharis grisebacchi* and *Baccharis salicifolia* were screened by Rosas-Romero and Saavedra (2005) to determine antioxidant activity. A methanol extract of each plant was prepared and portioned sequentially with hexane, chloroform, and ethyl acetate, leaving an aqueous solution. All extracts and their 5 fractions for a total of 102 samples, were evaluated using two techniques: an adaptation of the β-carotene bleaching technique using an emulsion of linoleic acid in water as the oxidizable substrate, and the DPPH free radical trapping technique. The results of the β-carotene bleaching technique were more discriminating and better related to the rancidity process under normal conditions; with this assay, 11 species provided at least one fraction with highly promising antioxidant activity. All species gave good results under the DPPH technique, and in most cases they performed better than BHA, which was used as a reference antioxidant (Rosas-Romero and Saavedra, 2005).

Baccharis sessiliflora Vahl Family: Asteraceae (Compositae)

Borneo *et al.* (2008) evaluated the antioxidant activity of ethanolic extract of *Baccharis sessiliflora*. The FRAP value was 629.5μmol of Fe(II)/g and DPPH radical scavenging activity was EC_{50} = 529.5.

Baccharis trimera (Less.) DC. Family: Asteraceae (Compositae)

The aqueous extract of aerial parts of *Baccharis trimera* displayed significant radical scavenging activity in a DPPH/TLC assay. The fractions were monitored by the DPPH/TLC assay and a series of radical-scavenging quinic acid derivatives could be identified (Simoes-Pires *et al.*, 2005).

Baccharis trinervis Pers. Family: Asteraceae (Compositae)

Ethanolic extracts from 15 plant species used in traditional medicine in Ecuador were evaluated by De Las Heras *et al.* (1998) for anti-inflammatory and antioxidant activities. *Baccharis trinervis* extract was active as antioxidant.

Baccharis usterii Heering Family: Asteraceae (Compositae)

The aqueous extract of aerial parts of *Baccharis usterii* displayed significant radical scavenging activity in a DPPH/TLC assay. Six quinic acids derivatives were isolated from *B. usterii* by MPLC. The fractions were monitored by the DPPH/TLC assay and a series of radical-scavenging quinic acid derivatives could be identified (Simoes-Pires *et al.*, 2005).

Backhousia citriodora F. Muell. Family: Myrtaceae

English Name: Lemon myrtle.

The antioxidant capacities and phenolic composition in six native, commercially grown, Australian herbs and spices were investigated by Konczak *et al.* (2010). Lemon myrtle contained phenolics 31.4 mg gallic acid equivalents (GAE)/g dry weight. Lemon myrtle exhibited good oxygen radical absorbance capacity (ORAC assay).

Bacopa monnieri (L.) Pennel Family: Scrophulariaceae

Free radical scavenging capacity of a methanol extract of *Bacopa monnieri* and the effect on DNA cleavage induced by H_2O_2 UB-photolysis were investigated by Russo *et al.* (2003). It showed a dose-dependent free-radical scavenging capacity and a protective effect on DNA clevage. The antioxidant capacity of *Bacopa monnieri* may explain, at least in part, the reported antistress, immunomodulatory, cognition-facilitating anti-inflammatory and antiaging effects produced by it in experimental animals and in clinical situations and may be useful in the treatment of human pathologies in which free radical production plays a key role.

Antioxidative property and tumor inhibitive property of *Bacopa monnieri* (20mg/kg body wt, sc) was examined in 3-methylcholanthrene induced fibrosarcoma rats by Rohini *et al.* (2004). Antioxidant enzymes, levels of glutathione (GSH) and the rate of lipid peroxidation (LPO) in the liver and kidney tissues were assessed. They concluded that extract promoted the antioxidant status, reduced the rate of lipid peroxidation and the markers of tumor progression in the fibrosarcoma bearing rats.

Balanites aegyptiaca (L.) Del. Family: Balanitaceae

Using Trolox assay, Cook *et al.* (1998) estimated the antioxidant activity of 17 wild edible plants of Niger Republic used for food and traditional medicine. They observed that *Balanites aegyptiaca* fruits/seeds possessed strong antioxidant activity.

Balanophora laxiflora Hemsl. Family: Balanophoraceae

The DPPH assay on the acetone extract of the fresh female plant of *Balanophora laxiflora* displayed considerable radical-scavenging activity (SC_{50} = 16.4 µg/ml). Further purification of the extract led to the isolation of two new phenolic glycosides, balaxiflorins A and B, together with 17 known phenolic compounds. The free radical scavenging activity of the isolated compounds was examined by DPPH assay (She *et al.*, 2009).

Balanophora polyandra Griff. Family: Balanophoraceae

The 80 per cent acetone extract of *Balanophora polyandra* was found to exhibit high radical-scavenging activity (SC_{50} = 14.48 mug/ml) towards DPPH radicals (Wang *et al.*, 2006). Further chemical investigation led to the isolation of two new hydrolysable tannins, balapolyphorins A (1) and B (2), together with 20 known phenolic compounds. The radical-scavenging properties of all isolated compounds were determined by DPPH assay.

Baliospermum montanum Muell.-Arg. Family: Euphorbiaceae

Free radical scavenging potential of the aqueous extract of roots of *Baliospermum montanum* was evaluated by Desai *et al.* (2008) by using DPPH assay and NO scavenging assay. In the DPPH method the plant extract showed an IC_{50} of 26.22µg/

ml while standard ascorbic acid showed an IC_{50} of 19.90μg/ml. In the NO scavenging method plant extract showed IC_{50} of 36.66μg/ml while standard ascorbic acid showed IC_{50} of 5.62μg/ml. The aqueous extract showed remarkable and concentration dependent free radical-scavenging activity.

Ballota acetabulosa (L.) Benth. Family: Lamiaceae (Labiatae)

Couladis *et al.* (2003) screened Greek aromatic plants from the Lamiaceae family for the antioxidant activity. Of the 21 plants they tested they found ethanol extracts prepared from *Ballota acetabulosa*, as well as 9 other plants exhibited the same antioxidant activity as α-tocopherol in their ability to inhibit bleomycin-Fe(II) complex-induced arichidonic acid superoxidation to MDA.

Ballota aucheri Boiss. Family: Lamiaceae (Labiatae)

Souri *et al.* (2004) evaluated the antioxidative activity of *Ballota aucheri* by linoleic acid peroxidation test using 1,3-diethyl-2-thiobarbituric acid as the reagent. Methanolic extract showed 90.54 per cent inhibition at 40μg concentration with IC_{50} value of 5.09μg.

Firuzi *et al.* (2010) evaluated the antioxidant activity and total phenolic content of 24 Lamiaceae species growing in Iran. *Ballota aucheri* showed low antioxidant activity with 14.1μM/g DW FRAP value and IC_{50} = 783.6μg/ml of DPPH radical scavenging activity. It also showed low phenolic content of 7.8 mg CE/g.

Ballota nigra L. Family: Lamiaceae (Labiatae)

The ability of *Ballota nigra* infusion to act as a scavenger of DPPH radical, reactive oxygen species (superoxide radical, hydroxyl radical, hypochlorous acid) and nitric oxide was investigated by Vrchovska *et al.* (2007). The tested infusion mainly exhibited a potent scavenging effect on DPPH, nitric oxide and superoxide radicals. In hydroxyl radical assay a potent pro-oxidant activity was noticed. No effect was found against hypochlorous acid. Seven phenolic compounds (chlorogenic acid, caffeic acid, and caffeoylmalic acids, ballotetroside, forsythoside B, verbascoside and allysonoside) and eight organic acids (oxalic, aconitic, citric, ascorbic, malic, quinic, shikimic and fumaric acids) were identified and quantified. Forsythoside B and quinic acid were revealed to be the main compounds in this infusion.

Matkowski *et al.* (2008) evaluated the antioxidant activity of aerial parts of *Ballota nigra* using DPPH, OH radical scavenging assays, reducing power assay and inhibition of deoxyribose degradation. EC_{50} value of DPPH radical scavenging was 29.06μg/ml for MeOH extract, 49.91μg/ml for Dichloromethane extract, 12.55μg/ml for BuOH extract and 10.32μg/ml for ethyl acetate extract.

Ballota pseudodictamnus (L.) Benth. Family: Lamiaceae (Labiatae)

Couladis *et al.* (2003) screened Greek aromatic plants from the Lamiaceae family for the antioxidant activity. Of the 21 plants they tested they found ethanol extracts prepared from *Ballota pseudodictamnus*, as well as 9 other plants exhibited the same antioxidant activity as α-tocopherol in their ability to inhibit bleomycin-Fe(II) complex-induced arichidonic acid superoxidation to MDA.

Ballota rotundifolia L. Family: Lamiaceae (Labiatae)

Gursoy and Tepe (2009) examined the antioxidant activities and the amount of total phenolics of the methanol extracts of *Ballota rotundifolia*. The subfractions were screened for their antioxidant activities by two complementary tests, namely DPPH free radical-scavenging and β-carotene/linoleic acid assays. Non polar extracts of the plant species remained inactive in both test systems. On the other hand polar extracts showed remarkable antioxidant activities.

Balsamita major Desf. Family: Asteraceae (Compositae)

English Name: Costmary.

The natural compounds found in the acetone and water extracts exhibit radical scavenging activity (RSA) and antioxidant capacity (ORAC) (Ena *et al.*, 2009). Significant RSA and ORAC activities are attributed to phenolic compounds, as flavonoids, mainly quercetin esters.

Bambusa breviflora Munro Family: Poaceae (Graminae)

Antioxidant capacities of 56 selected Chinese medicinal plants, including *Bambusa breviflora* were evaluated by Song *et al.* (2010) using the TEAC and FRAP assays, and their total phenolic content was measured by the Folin-Ciocaleteu method. The TEAC value was 82.46μmol Trolox/g, FRAP value was 115.74μmol Fe^{2+}/g and phenolic content was 9.03mg GAE/g.

Bambusa vulgaris Schrad. ex Wendl. Family: Poaceae (Graminae)

The antioxidant activity of the methanolic extract of *Bambusa vulgaris* leaves was evaluated by Goyal *et al.* (2010) using DPPH free radical scavenging assay. The antioxidant activity of the investigated extract has a scavenging ability of hydroxyl peroxide radicals (421.74mg/ml) and DPPH radical scavenging activity (around 95 per cent). The high content of phenolic compounds (22.69mg GAE/g of dry extract) and total flavonoids (159.80 mg quercetin/g of dry extract) indicated that these compounds contribution to the antioxidative activity.

Baphia nitida Afzel. Family: Fabaceae (Leguminosae)

English Name: Cam-wood

Kouakou-Siransy *et al.* (2010) evaluated the antioxidant properties through a scavenger effect on reactive oxygen species (ROS), such as hydrogen peroxide and hypochlorous acid. The plant showed dose-dependent antioxidant activity. Rates of total phenolic, flavonoid and proanthocyanidin contents were evaluated. The highest rates were to be found in the most active extracts, indicating that antioxidant activity could be influenced by these phytochemical groups.

Baphicanthus cusia (Nees) Bremek. Family: Acanthaceae

Li *et al.* (2008) evaluated the antioxidant properties of 45 medicinal plants including using FRAP and TEAC assays and the total phenolic contents of these plants were measured. *Baphicanthus cusia* showed very low antioxidant activity with 1.23 μmol Fe(II)/g FRAP value, 0.97μmol Trolox/g of TEAC value and 1.15 mg GAE/g of Phenolic content.

Barringtonia acutangula L. Family: Lecythidaceae

Babre *et al.* (2010) evaluated the antioxidant activity of hydroalcoholic extract of *Barringtonia acutangula* roots on streptozotocin-induced diabetic rats by the method non enzymatic antioxidant glutathione (GSH), catalae, SOD and lipid peroxidation (malondialdehyde) activities. All the *in vitro* models showed dose dependent antioxidant activity.

Barringtonia racemosa Roxb. Family: Lecythidaceae

Anti-lipid peroxidation effect of *Barringtonia racemosa* was studied *in vitro* by Shikha *et al.* (2010) following the modified method of Yoshiuki *et al.* (1981) and Masao *et al.* (1993) *Barringtonia racemosa* prevented the rise in lipid peroxides (MDA production) showing its significant anti-lipid peroxidant effect.

Basella alba L. Family: Basellaceae

Oboh and Akindahunsi (2004) investigated the effect of preservation of green leafy vegetables of Nigeria on the antioxidant phytoconstituents (Vitamin C and Total phenol) and activity (reducing property and free radical scavenging ability) including *Basella alba*. The result of the study revealed that sun-drying of green leafy vegetables cause a significant decrease in the vitamin C content (16.67-64.68 per cent loss). Conversely it leads to a significant increase in the total phenol content (6.45 - 223.08 per cent grain) reducing property (16.00-362.50 per cent grain) and free radical scavenging ability (126.00-5757.00 per cent gain) of green leafy vegetables.

Adefegha and Oboh (2011) reported that cooking decreases the vitamin C contents in the leafy vegetable *Basella alba,* while it increased the phenolic content and antioxidant activities.

Bauhinia forficata Link Family: Caesalpiniaceae

In vitro antioxidant properties or action against reactive oxygen species of kaempferol-3,7-O-(α)-dirhamnoside from *Bauhinia forficata* leaves was evaluated by de Sousa *et al.* (2004). The compound showed high reactivity with DPPH, IC_{50} of 84.0 ± 7.8 μM inhibited myeloperoxidase activity. Damasceno *et al.* (2004) also reported antioxidant activity of *Bauhinia forficata*.

Bauhinia microstachya (Raddi) Macbr. Family: Caesalpiniaceae

Aqueous (AE) and hydro-ethanolic extracts (HE) of *Bauhinia microstachya* leaves were investigated for their phenolic content and phytochemical profile (Silva *et al.*, 2007). The antioxidant activities and free radical scavenging potential of isolates were determined in different *in vitro* systems (TAP, TEAC, TBARS, nitric oxide, superoxide and hydroxyl radical). HE presented a 27.4 per cent higher content of phenolics when compared to AE and a distinct phytochemical profile was observed. Both extracts exhibited antioxidant activities and their capacity and efficiency varied according to the radical-generating system. In general, HE was more effective than AE in avoiding ROS-generating damage and in scavenging the various radicals formed. Neverthless, when results were normalized to total phenolic content, a different profile of antioxidant activities and free radical scavenging potential was observed, particularly against oxidative lipid damage and superoxide radical (Silva *et al.*, 2007).

Menezes *et al.* (2004) investigated the antioxidant activity of *Bauhinia microstachya*, against DPPH free radicals and by the phosphomolybdenum method. The extract was found to be the most potent in both models.

Bauhinia monandra Kurz. Family: Caesalpiniaceae

Argolo *et al.* (2004) evaluated the radical scavenging activity of leaf extracts from *Bauhinia monandra*. The chloroform and ethyl acetate extracts were the most appropriate as sources of antioxidant compounds as shown by their inhibition concentration and inhibition percentage (IP) values. The chloroform and ethyl acetate extracts exhibited an IC_{50} of approximately 2 mg/g DPPH and IP values in the range of 60-65 per cent.

Aderogba *et al.* (2006) investigated the antioxidant constituents in the ethyl acetate soluble leaves extract of *Bauhinia monandra*. Bioassay directed fractionation led to isolateion of two active compounds identified as: quercetin-3-O-rutinoside (1) and quercetin (2). DPPH method was followed for the antioxidant activity assay. Compound 2 had higher antioxidant activity while compound 1 had lower activity than L-ascorbic acid which was used as a standard.

Bauhinia purpurea L. Family: Caesalpiniaceae

Chanwitheesuk *et al.* (2005) investigated the antioxidant activity of *Bauhinia purpurea* using a β-carotene bleaching method. The contents of plant chemicals such as vitamin C, Vitamin E, carotenoids, tannin and total phenolics, were also determined. Methanolic extract showed antioxidant activity. *Bauhinia purpurea* has antioxidant index of 3.73, Vitamin C 13.9 mg per cent, Vitamin E 0.0103mg per cent, Total carotenes 2.55 mg per cent, Total xanthophylls 3.72mg per cent, Tannins 24.3mg per cent and Total phenolics of 59.5mg per cent.

Zakaria (2007) evaluated the free radical scavenging properties of *Bauhinia purpurea* using DPPH and superoxide anion radical scavenging assays. The extract was found to show reemarkaable antioxidant activity in both assays. The DPPH radical scavenging activity was 94.80 per cent and superoxide scavenging activity was 83.70 per cent. Phytochemicals screening of this plant demonstrated the presence of flavonols, triterpenes, saponins and steroids.

An investigation was carried out by Jatwa and Kar (2009) to reveal the possible ameliorative role of *Bauhinia purpurea* plant extract on an antidiabetic drug-induced hypothyroidism in Type 2 diabetic animals. The results revealed that the plant extract showed antioxidant activity.

Bauhinia racemosa Lam. Family: Caesalpiniaceae

Kumar *et al.* (2005) evaluated the antioxidant activities of the methanol extract of *Bauhinia racemosa* (MEBR) stem bark in various systems. DPPH radical, superoxide anion radical, nitric oxide radical and hydroxyl radical scavenging assays were carried out to evaluate the antioxidant potential of the extract. The antioxidant activity of the methanol extract increased in a concentration-dependent manner. About 50, 100, 250 and 500 μg MEBR inhibited the peroxidation of linoleic acid emulsion by 62.43, 67.21, 71.04 and 76.83 per cent, respectively. Similarly, the effect of MEBR on reducing power increased in a concentration-dependent manner. In DPPH radical scavenging assays the IC_{50} value of the extract was 152.29 µg/ml. MEBR inhibited the nitric oxide radicals generated from sodium nitropursside with and IC_{50} of 78.34 µg/ml, as opposed to 20.4µg/ml for curcumin. More over MEBR scavenged the superoxide generated by the PMS/NADH-NBT system. MEBR also inhibited the hydroxyl radical generated by Fenton's reaction, with an IC_{50} value of more than 1000µg/ml, as compared to 5µg/ml for catechin. The amounts of total phenolic compounds were also determined and 64.7 µg pyrocatechol phenol equivalents were detected in MEBR (1 mg).

***Bauhinia tarapotensis* Benth. Family: Caesalpiniaceae**

Braca *et al.* (2001) evaluated the antioxidant activity of the compounds isolated from the leaves of *Bauhinia tarapotensis* using DPPH and TEAC methods and the coupled oxidation of β-carotene and linoleic acid bioassay. Cyclohexane and caffeoyl ester derivative were active in the coupled oxidation of β-carotene and linoleic acid bioassay while (-)-isolariciresinol 3-α-O- β-D-glucopyranoside (3), (+)-hydroxypinoresinol 1-O-β-D-glucopyranoside (4) and isoacetoside (5) showed good activities in the DPPH and TEAC tests.

***Bauhinia variegata* L. Family: Caesalpiniaceae**

Chemoprevention of cytotoxic effect of *Bauhinia variegata* against N-nitrosodiethylamine induced liver tumors and human cancer cell lines was investigated by Rajkapoor *et al.* (2006). The ethanol extract produced an increase in enzymatic antioxidant (superoxide dismutase and catalase) levels and total proteins when compared to those in liver tumor bearing rats.

Rajani and Ashok (2009) evaluated the antioxidant activity of the ethanolic and aqueous extracts of *Bauhinia variegata* stem bark and root by various methods namely total reducing power, scavenging of various free radicals such as DPPH, superoxide, nitric oxide and hydrogen peroxide. Significant antioxidant activity was observed in all the methods for reducing power and for scavenging DPPH, super oxide, nitric oxide and hydrogen peroxide radicals.

The chloroform and methanolic extracts of 124 Egyptian plant species belonging to 56 families were investigated by Moussa *et al.* (2011) and compared their antioxidant activity by DPPH scavenging assay. Among the 124 plant species tested 18 exhibited extremely high antiradical activity (>80 per cent inhibition). The IC_{50} value of DPPH radical scavenging of *Bauhinia variegata* was 19.36µg/ml, while total phenolic and flavonoid contents were 109.76 mg Tannic acid equivalent/g extract and 226.0 mg Rutin equivalent/g extract, respectively. Correlation coefficient between DPPH radical scavenging activity and the total phenolic and flavonoid contents suggest that phenolics and flavonoids in the extracts were partly responsible for the antiradical activities.

***Beaumontia grandiflora* Wall. Family: Apocynaceae**

Investigation of the chemical constituents of the aerial parts of *Beaumontia grandiflora* in pet. ether extract led to the isolation and identification of twelve fatty acids in which linoleic acid (29.60 per cent) and palmitic acid (19.12 per cent) were the main acids. The unsaponifiable fraction was found to contain a series of hydrocarbons in addition to β-sitosterol, stigmasterol, cholesterol, campasterol and α-amyrine. The flavonoidal constituents were isolated from the ethyl acetate fraction of the alcoholic extract. Four flavonoids were identified as : quercetin (1), quercetin-3-O-galactoside (2), diosmetin-7-O-rhamnoside (3) and kaempferol-3-O-neohesperidoside (4). The crude extracts (total alcoholic, ethyl acetate and butanol extracts) and the isolated compounds were screened for their antioxidant activity using DPPH radical scavenging method. Ethyl acetate extract and compound 1 showed high antioxidant activity compared to Trolox, 91, 89 per cent, respectively.

Becium dhofarense Sebald Family: Lamiaceae (Labiatae)

Marwah *et al.* (2007) investigated the antioxidant activity of wound-healing herbs using *in vitro* DPPH and phosphomolybdenum assay methods. Out of the 19 plants screened, the aqueous ethanol extract of *Becium dhofarense* showed the best DPPH scavenging activities with per cent IP value 93.1 per cent, after 15 min of incubation at an effective test concentration of 50μg/ml. The Total Antioxidant Capacity as gallic acid equivalent of 510mg/g of ethanol extract was obtained in the phosphomolybdenum assay. Total phenolic content was 137 mg GAE/g of ethanol extract.

Beilschmedia manni (Meisn.) Benth. et Hook.f. Family: Lauraceae

Odukoya *et al.* (2007) evaluated the antioxidant activity and total phenolic contents of Nigerian green leafy vegetables including *Beilschmedia manni*. Potential free radical scavenging activity of these vegetables was confirmed by spraying spots of the extracts with DPPH (yellow color on purple background). Antioxidant activity was assayed in linoleic acid model system. Total polyphenols as Tannic Acid Equivalent (TAE) and Ascorbic Acid. There was low correlation between AA/TAE, AA/ASC and TAE/ASC.

Belamcanda chinensis (L.) DC. Family: Iridaceae

Tectorigenin and tectoridin isolated by Jung *et al.* (2004) from the rhizomes of *B. chinensis* was found to possess not only antioxidative, but also the hepatoprotective activities. Both compounds showed strong increase in the antioxidant enzymes such as hepatic cytosolic superoxide dismutase (SOD) and glutathione peroxidase (GSH-px) activities in CCl_4-intoxicated rats.

Li *et al.* (2008) evaluated the antioxidant properties of 45 medicinal plants including using FRAP and TEAC assays and the total phenolic contents of these plants were measured. *Belamcanda chinensis* showed 71.39 μmol Fe(II)/g FRAP value, 76.83μmol Trolox/g of TEAC value and 20.14 mg GAE/g of Phenolic content.

Bellis perennis L. Family: Asteraceae (Compositae)

Siatka and Kasparova (2010) investigated the total phenolics and flavonoid contents as well as antioxidant activity of *Bellis perennis* flowers by DPPH radical scavenging assay. The contents of flavonoids varied from 0.31 to 0.44 mg quercetin equivalent/100g dw (method 1) and from 1.37 to 2.20 mg pigenin-7-glucoside equivalent/100mg dry weight (method 2). Total phenolics ranged from 2.81 to 3.57

mg/GAE/100mg dw.The antioxidant activity expressed as IC_{50} values varied from 66.03 to 89.27μg/ml; it is about 50, 30, 20 and 10 times lower as compared with quercetin, ascorbic acid, Trolox and BHT, respectively and about five times higher in comparison with apigenin-7-glucoside. There is a significant correlation between antioxidant activity and total phenolics. No correlation between total flavonoid content and antioxidant activity was observed.

Benincasa hispida (Thunb.) Cogn. Family: Cucurbitaceae

English Name: Wax gourd; Japanese: Tougan.

Huang *et al.* (2004a) evaluated the antioxidant activities of various fruits and vegetables produced in Taiwan. Wax gourd seeds showed the highest antioxidant activity. At the level of 1 g fresh sample, low-density lipoprotein peroxidation was inhibited by at least 90 per cent by wax gourd meat, core and seed. The total phenolic content was significantly correlated with antioxidant activities measured by TBA and Iodometric assays. Huang *et al.* (2004b) investigated the abilities of antioxidant and inhibition of angiotension-converting enzyme (ACE) activity of wax gourd pulp, core, seed and peel prepared by different extraction methods. The fresh weights required to reach 50 per cent inhibition of linoleic acid oxidation were higher in fresh extracts, compared to other extraction methods. The higher antioxidant capacity of the seed may result from the higher total phenolics contents and superoxide dismutase activity. Maeda *et al.* (2006) examined the phenolic contents of fruit of *Benincasa hispida* and its radical scavenging activities for DPPH radical and found them to be very low. The polyphenol content was 5.8 mg-gallic acid equivalent/100 g) and DPPH scavenging activity was 96.2 μmol-Trolox eq./100 g).

Gan *et al.* (2010) evaluated the antioxidant activity of *Benincasa hispida* using the FRAP and TEAC assays, and its total phenolic content was measured by the Folin-Ciocalteu method. For FRAP assay the value was 57.28 μmol Fe(II)/g dry weight. For TEAC assay, the value was 40.81μmol Trolox/g dry weight. Total phenolic content was 4.21 mg gallic acid equivalent/g dry weight of plant material.

A study was carried out by Raghu *et al.* (2010) to determine the antioxidant activity of aqueous extracts of ten selected common vegetables. For *Benincasa hispida* the IC_{50} value for DPPH scavenging was 3.79mg/ml while standard was 2.45μg/ml. Reducing power of the extract was moderate.

Berberis aristata DC. Family: Berberidaceae

Singh and Kakkar (2009) evaluated the antioxidant effect of 50 per cent aqueous ethanolic extract of *Berberis aristata* root. The administration of the extract increased CAT, SOD, GPx, GR activity significantly and reduced lipid peroxidation (41.6 per cent) and protein carbonylation (30.15 per cent).

Berberis crataegina DC. Family: Berberidaceae

Souri *et al.* (2004) evaluated the antioxidative activity of *Berberis crataegina* by linoleic acid peroxidation test using 1,3-diethyl-2-thiobarbituric acid as the reagent. Methanolic extract showed 90.29 per cent inhibition at 40μg concentration with IC_{50} value of 0.81μg which is approximately in the range of α-tocopherol (IC_{50} = 0.60μg).

Serteser *et al.* (2009) studied the antioxidant activity of 50 per cent aqueous methanol extracts of *Berberis crataegina* including free radical scavenging, hydrogen peroxide scavenging and metal (Fe^{2+}) chelating activities.The DPPH radical scavenging effects of fruit extract was 0.743. Fe^{2+} chelating activity (per cent) of leaves extracts was 28.36 while fruit extracts showed 44.76 per cent inhibition. Hydrogen peroxide inhibition activity of leaves extracts was 32.76 per cent and fruit extracts was 59.43 per cent.

Berberis integerrima Bunge. Family: Berberidaceae

Serteser *et al.* (2009) studied the antioxidant activity of 50 per cent aqueous methanol extracts of *Berberis integerrima* including free radical scavenging, hydrogen peroxide scavenging and metal (Fe^{2+}) chelating activities.The DPPH radical scavenging effects of leaves extract was 0.413 while fruit extract was 0.708. Fe^{2+} chelating activity (per cent) of leaves extracts was 27.12 while fruit extracts showed 43.65 per cent inhibition. Hydrogen peroxide inhibition activity of leaves extracts was 30.46 per cent and fruit extracts was 57.34 per cent.

Berberis vulgaris L. Family: Berberidaceae

English Name: Barberry.

Motalleb *et al.* (2005) evaluated the phenolic content and total antioxidant activity of barberry fruit extracts. The extract was prepared with distilled water and ethanol and methanol (80 per cent), respectively. The carotene bleaching and DPPH assay were used to determine antioxidant properties of barberry fruit by measuring the decrease in absorbance at 470 and 517 nm. In distilled water, *Berberis vulgaris* showed 82.52 per cent free radical scavenging activity. In ethanol, *Berberis vulgaris* fruit extract showed 73.62 per cent free radical scavenging activity, BHT 67.50 and vitamin C 88.56, respectively. *Berberis vulgaris* fruit extract exhibited the higher free radical scavenging activity in distilled water with 82.52 per cent and $EC_{50} = 0.64$ mg/ml. The methanol 80 per cent extract showed 60.15 per cent carotene bleaching assay and vitamin C 91.17 per cent. Fruit extract in 80 per cent methanol had the highest phenolic content followed by extract in water.

Serteser *et al.* (2009) studied the antioxidant activity of 50 per cent aqueous methanol extracts of *Berberis vulgaris* including free radical scavenging, hydrogen peroxide scavenging and metal (Fe^{2+}) chelating activities.The DPPH radical scavenging effects of leaves extract was 0.421 while fruit extract was 0.687. Fe^{2+} chelating activity (per cent) of leaves extracts was 28.02 while fruit extracts showed 42.98 per cent inhibition. Hydrogen peroxide inhibition activity of leaves extracts was 35.54 per cent and fruit extracts was 55.67 per cent.

Bergenia ciliata (Haw.) Sternb. Family: Saxifragaceae

Rajkumar *et al.* (2010d) evaluated antioxidant activities of methanolic and aqueous extracts of *Bergenia ciliata* rhizome. Free radical (DPPH and OH) scavenging potential of the extracts revealed that both extracts to be active radical scavengers. Reducing ($Fe^{+3} - Fe^{+2}$) power and lipid peroxidation inhibition efficiency (TBARS assay) of both extracts were also evaluated and both extracts showed promising activity in preventing lipid peroxidation and might prevent oxidative damages to biomolecules.

Bergenia crassifolia (L.) Fristsch. Saxifragaceae

Ethanol extracts from *Bergenia crassifolia* green leaves exhibit antioxidant properties (Shilova *et al.*, 2006). The most significant activity has been observed for 70 per cent ethanol extract, which is a typical acceptor of peroxy radicals. Fractionation of the active extract yielded chloroform and ethyl acetate fractions, which also showed a pronounced antioxidant action. Phenolic compounds (hydroquinone, quercetin, dihydroquercetin, rutin, gallic acid, esculetin) also exhibited significant antioxidant activity comparable with that of ionol.

An aqueous ethanol extract of *Bergenia crassifolia* rhizomes strongly inhibited human pancreatic lipase activity and increased scavenging of DPPH free radicals *in vitro* (Ivanova *et al.*, 2010). Chromatographic separation of this extract led to isolation of the hydrolysable tannins (+)-catechin 3,5-di-O-gallate (1) and (+)-catechin 3-O-gallate. This compound strongly inhibited human pancreatic lipase (with an IC_{50} value of 0.42 µg/ml) and exhibited a remarkable free radical-scavenging ability (with an SC_{50} value of 1.04 µg/ml).

Bergenia ligulata (Wall.) Engl. Family: Saxifragacae

Bashir and Gilani (2009) reported that *Bergenia ligulata* rhizome exhibited antioxidant effect against DPPH free radical and lipid peroxidation in the *in vitro*.

Beta vulgaris L. var. conditiva. Family: Chenopoidaceae

English Name: Table beets.

Jiratanam and Liu (2004) reported antioxidant activity of Table beets.

Beta vulgaris L. var. esculenta. Family: Chenopodiaceae

English Name: Beet root.

The antioxidative activity of phenolic extracts from beetroot was examined by autoxidation of methyl linoleate by Kahkonen *et al.* (1999). Beetroot peel showed remarkable antioxidant activity. Beetroot peel extract inhibited hydrogen peroxide formation by 98 per cent. However the phenolics are very low in beetroot peel (4.3mg/g GAE).

Beta vulgaris L. var. altissima. Family: Chenopodiaceae

English Name: Sugar beet.

The antioxidative activity of phenolic extracts from sugarbeet was examined by autoxidation of methyl linoleate by Kahkonen *et al.* (1999). Sugarbeet peel showed remarkable antioxidant activity.

Beta vulgaris var. cicla Family: Chenopodiaceae

The antioxidative activity of the 80 per cent ethanol extract obtained from eleven commonly consumed wild edible plants was determined according to the phosphomolybdenum method, reducing power, metal chelating, superoxide anion and free radical scavenging activity (Özen, 2010). *Beta vulgaris* var. *cicla* had the highest antioxidant capacities. The results indicated that the antioxidant activity of vegetables seems to be due to the presence of polyphenols, flavonoid and anthocyanoside. Total phenolics in *Beta vulgaris* var. *cicla* were 41.0mg pyrocatechol equivalent/g dw, Flavonoids were 7.9 mg quercetin equivalent/g dw and anthocyanins 0.5 mg cyanidin glycoside equivalent/g dw.

Betula pendula Roth Family: Betulaceae

English Name: Silver birch.

The antioxidative activity of phenolic extracts from silver birch was examined by autoxidation of methyl linoleate by Kahkonen *et al.* (1999). Silver birch phloem was one of the strongly active tree materials with high phenolic content. The outer bark extract resulted in a very low amount of total phenolics (2mg/g GAE), but still it inhibited the oxidation of MeLo effectively at the 500 ppm concentration. On the contrary, the antioxidative effect was weak when tested at the concentration of 5000 ppm of the original sample, which indicates that the recovery of phenolic extract was low. Phenols of three different types were found in silver birch inner bark (phloem): arylbutanoid glycosides, benzoic acid derivatives, and catechins. Suberin and the triterpenes betulin (0-30 per cent of the outer bark), betulinic aldehyde, and acid, lupeol, erythdiol, oleanic acid, and several triterpene caffeates were identified in outer bark of silver birch (Pan, 1995). The latter compounds have previously been shown to have antioxidant (Takagi and Iida, 1980) and anti-inflammatory properties (Otsuka *et al.*, 1981).

Main phenolics in silver birch leaves have been identified by Ossipov *et al.* (1996). Total phenolic content in the leaves of silver birch leaves have been identified by Ossipov *et al.* (1996). Total phenolic content in the leaves analyzed by HPLC method was 27.1mg/g dry mass

Betula platyphylla Sucaczew. var. _japonica_ Family: Betulaceae

The antioxidant and anticancer properties of a medicinal plant, *Betula platyphylla* var. *japonica* were investigated by Ju *et al.* (2004). The total methanol extract of *B. platyphylla* var. *japonica* had a protective effect against hydrogen peroxide (H_2O_2) in the Chinese hamster lung fibroblast (V79-4) cell line and induced apoptotic cell death in human promyelocytic leukemia (HL-60) cells, a cancer cell line. *B. platyphylla* var. *japonica* extract significantly increased cell viability against H_2O_2. The extract also showed high DPPH radical scavenging activity IC_{50} 2.4 microg/ml) and lipid peroxidation inhibitory activity (IC_{50} below 4.0μg/ml). Further more *B.platyphylla* var. *japonica* extract reduced the number of V79-4 cells arrested in G2/M in response to H_2O_2 treatment and increased the activities of several cellular antioxidant enzymes, including superoxide dismutase, catalase and glutathione peroxidase. Treatment with *B.platyphylla* var. *japonica* extract induced cytotoxicity and apoptosis in HL-60 cells, as shown by nucleosomal DNA fragmentation, increases in the subdiploid cell population, and fluorescence microscopy. *B. platyphylla* var. *japonica* extract gradually increased the expression of proapoptotic Bax and led to the activation of caspase-3 and cleavage of PARP. These findings suggests that *B. platyphylla* var. *japonica* exhibits potential antioxidant and anticancer properties.

Betula pubescens Ehrh. Family: Betulaceae

English Name: White birch.

The antioxidative activity of phenolic extracts from beet root was examined by autoxidation of methyl linoleate by Kahkonen *et al.* (1999). The tree extract had high antioxidant activity of over 90 per cent. The total phenolics were 38.7mg/g GAE.

Betula utilis D.Don Family: Betulaceae

English Name: Himalayan Birch, Bhojpatra

Antioxidant activity of *Betula utilis* was evaluated by Kumaraswamy and Satish (2008) using DPPH, ABTS and lipoxygenase inhibition assay. In both DPPH and ABTS scavenging activities methanol and water extract showed good scavenging activities (8.4, 35.08µg/ml IC_{50} for DPPH and 83.18, 37.14µg/ml IC_{50} for ABTS assay but very mild activity against lipoxygenase inhibition activity (18.74 and 28.78 per cent inhibition at 1.0 mg/ml).

Bidens andicola Kunth. var. _decomposita_ Family: Asteraceae (Compositae)

Borneo *et al.* (2008) evaluated the antioxidant activity of ethanolic extract of *Bidens andicola* var. *decomposita*. The FRAP value was 457.7µmol of Fe(II)/g and DPPH radical scavenging activity was EC_{50} = 176.2.

Bidens tripartita L. Family: Asteraceae (Compositae)

Wolniak *et al.* (2007) studied the antioxidant activity of extracts of flowers of *Bidens tripartita* and isolated compounds using DPPH assay. The content of flavonoids in flowers is half that found in the herb, however, the extract from flowers showed that the antioxidant activity was almost two times higher there. Some extracts (n-BuOH fraction) showed long lasting radical scavenging activity.

Biebersteina multifida DC. Family: Berberidaceae

English Name: Bieberstein.

Souri *et al.* (2008) investigated the antioxidant activity and radical scavenging activity of methanolic extract of root. The antioxidant activity of this plant extract against linoleic acid peroxidation expressed as IC_{50} was 1.28ng/ml and radical scavenging activity against DPPH expressed as IC_{50} was 37.14 µg/ml. The anti-inflammatory and analgesic activities of this plant have been studied (Farsam *et al.*, 2000) and there may be a relationship between these two effects.

Biophytum reinwardtii DC. Family: Oxalidaceae

The antioxidant potential of 80 per cent aqueous methanol and methanol extracts was evaluated by Srinivasan *et al.* (2008) by *in vitro* DPPH free radical scavenging activity and the extracts exhibited poor scavenging activity (725.5 and 560µg/ml respectively). In this study, the comparatively lower free radical scavenging activity of methanol extracts was observed, even though the plant has a good source of non-enzymatic antioxidants such as phenols, flavonoids and tannins. Ethyl palmitate, lupeol, β-sitosterol and linoleic acid were isolated from the plant (Srinivasan *et al.*, 2008).

Biophytum sensitivum (L.) DC. Family: Oxalidaceae

An extract of *Biophytum sensitivum* was evaluated for its antioxidant potential *in vitro* and *in vivo*. *Biophytum sensitivum* was found to scavenge superoxide radicals generated by the photoreduction of riboflavin and hydroxyl radicals generated by the Fenton reaction and inhibited *in vitro* lipid peroxidation at concentrations of 50, 95, and 20 mg/ml (50 per cent inhibition [IC_{50}]), respectively. The drug also scavenged nitric oxide (IC_{50} = 100 mg/ml. The extract also induced the dose-dependant scavenging of nitric oxide in culture. Intraperitoneal administration of *Biophytum* extract inhibited superoxide generation in macrophages *in vivo*. The administration of *B. sensitivum* to mice significantly increased the catalase activity. The extract produced a significant increase in glutathione levels in blood and liver. The levels of glutathione-S transferase and glutathione reductase increased and those of glutathione peroxidase decreased after administering the *Biophytum* extract.

Bletilla striata (Thunb.) Reichb.f. Family: Orchidaceae

Gan *et al.* (2010) evaluated the antioxidant activities of 40 medicinal plants associated with prevention and treatment of cardiovascular and cerebrovascular diseases using FRAP and TEAC assays and their total phenolic contents were measured by the Folin-Ciocalteu methods. The antioxidant activity of *Bletilla striata* was 27.24μmol Fe(II)/g in FRAP assay, 38.47μmol Trolox/g and Total phenolic content was 2.75 mg GAE/g. A high correlation between antioxidant capacity and total phenolic content indicated that phenolic compounds could be the main contributor of the antioxidant activity of these plants.

Blumea balsamifera (L.) DC. Family: Asteraceae (Compositae)

Effects of pet-ether, chloroform, and methanol extracts and flavonoids of *Blumea balsamifera* leaves on nonenzymatically (phenazine methosulphate/NADH) generated superoxide radical were evaluated by Fazilatun *et al.* (2004). The methanol extract (93.91 per cent) exhibited higher radical scavenging activity than the chloroform extract (84.58 per cent). The pet-ether extract was inactive toward nonenzymatically generated superoxide radicals. The superoxide radical scavenging capacity of flavonoids (100 μ M) as determined was decreased in the order quercetin > luteolin > rhamnetin > tamarixetin > luteolin-7-methylether > dihydroquercetin-4′-methyl ether > dihydroquercetin-4′,7-dimethyl ether. It was concluded that flavonoids with a free hydroxyl group were more active than methylated compounds and the flavonoid content of the extracts contributed to their free radical scavenging activity (Fazilatun *et al.*, 2004).

Blumea napifolia DC. Family: Asteraceae (Compositae)

Patnibul *et al.* (2008) screened 18 vegetables for antioxidant activity using silica gel Thin Layer chromatography followed by spraying with DPPH. *Blumea napifolia* exhibited strong antioxidant activity.

Boehmeria nivea (L.) Gaud. Family: Urticaceae

Gan *et al.* (2010) evaluated the antioxidant activities of 40 medicinal plants associated with prevention and treatment of cardiovascular and cerebrovascular diseases using FRAP and TEAC assays and their total phenolic contents were measured by the Folin-Ciocalteu methods. The antioxidant activity of *Boehmeria nivea* was 139.10µmol Fe(II)/g in FRAP assay, 132.78µmol Trolox/g and Total phenolic content was 10.07 mg GAE/g. A high correlation between antioxidant capacity and total phenolic content indicated that phenolic compounds could be the main contributor of the antioxidant activity of these plants.

Boerhavia diffusa L. Family: Euphorbiaceae

Administration of *Boerhavia diffusa* leaf extract (200mg/kg) for 4 weeks resulted in a significant reduction in thiobarbutiric acid reactive substances and hydroperoxides, with an increase in reduced glutathione, superoxide dismutase, catalase etc. in liver and kidney of induced diabetic rats (Satheesh and Pari, 2004). These results suggest that *Boerhavia* has remarkable antidiabetic activity and can improve antioxidant status in diabetic rats.

In order to find their potential sources, the crude extracts of different parts of some selected food and medicinal plants, including *Boerhavia diffusa*, were studied by Bajpai *et al.* (2005) for total phenolic contents (TPC) and antioxidant activity (AOA). AOA was assayed by auto-oxidation of β-carotene and linoleic acid and expressed as the per cent of inhibition relative to control. TPC in roots of *Boerhavia diffusa* was 3.4 mg/g (GAE) while antioxidant activity was 36.2 per cent.

Kouakou-Siransy *et al.* (2010) evaluated the antioxidant properties of *Boerhavia diffusa* leaves through a scavenger effect on reactive oxygen species (ROS), such as hydrogen peroxide and hypochlorous acid. The plant showed dose-dependent antioxidant activity. Rates of total phenolic, flavonoid and proanthocyanidin contents were evaluated. The highest rates were to be found in the most active extracts, indicating that antioxidant activity could be influenced by these phytochemical groups.

Extracts of *Boerhavia diffusa* leaves were evaluated for antioxidant and hepatoprotective properties in the acetaminophen-induced liver damage model by Olaleye *et al.* (2010). Antioxidative evaluation of ethanolic extract gave total phenolic content, total flavonoid content, vitamin C content and vitamin E content and the levels of selenium and zinc as 6.6 mg/g TAE, 0.092 mg/g quercetin equivalent, 0.21 mg/g, 0.054 mg/g, 0.52 ppm and 9.28 ppm, respectively. The DPPH scavenging capacity and the reductive potential were 78.32 per cent and 0.65 mg/g ascorbic acid, respectively.

Boesenbergia pandurata (Roxb.) Schltr. Family: Zingiberaceae

English Name: Wild ginger.

Chanwitheesuk *et al.* (2005) investigated the antioxidant activity of rhizome of *Boesenbergia pandurata* using a β-carotene bleaching method. The contents of plant chemicals such as vitamin C, Vitamin E, carotenoids, tannin and total phenolics, were also determined. Methanolic extract showed antioxidant activity. *Boesenbergia pandurata* has antioxidant index of 0.91, Vitamin C 0.98 mg per cent, Vitamin E 0.0032mg per cent, Total carotenes 0.64 mg per cent, Total xanthophylls 0.52mg per cent, Tannins 4.48mg per cent and Total phenolics of 20.5mg per cent.

Bolusanthus speciosus (Bolus) Harms Family: Fabaceae (Leguminosae)

Three new flavonoids – 5,7-4′trihydroxy-6{1-hydroxy-2-methylbuten-2-yl} isoflavone (isogancaonin C), 7,2′-dihydroxy-4′-methoxyisoflav-3-ene (bolusanthin III), 6,6′-dihydroxy-4′-methoxy-2-arylbenzofuran (bolusanthin IV), in addition to eight known flavonoids; derrone, medicarpin, genistin, wighteone, lupiwighteone, gancaonin C, 7-hydroxy-4′-methoxyisoflavone and 7,3′-dihydroxy-4′-methoxyisoflavone were isolated by Erasto *et al.* (2004) from the root wood of *Bolusanthes speciosus*. The isolated compounds showed moderate to strong radical scavenging properties against DPPH radical with the highest activities shown by the 2-arylbenzofuran, the isoflav-3-ene and 7,3′-dihydroxy-4′-methoxyisoflavone in decreasing order.

Bombax costatum Pellegr. et Vuillet Family: Bombacaceae

Using Trolox assay, Cook *et al.* (1998) estimated the antioxidant activity of 17 wild edible plants of Niger Republic used for food and traditional medicine. They observed that *Bombax costatum* possessed strong antioxidant activity.

Bombax malabaricum DC. Family: Bombaceae

The chloroform and methanolic extracts of 124 Egyptian plant species belonging to 56 families were investigated by Moussa *et al.* (2011) and compared their antioxidant activity by DPPH scavenging assay. Among the 124 plant species tested 18 exhibited extremely high antiradical activity (>80 per cent inhibition). The IC_{50} value of DPPH radical scavenging of *Bombax malabaricum* was 18.80μg/ml, while total phenolic and flavonoid contents were 157.96 mg Tannic acid equivalent/g extract and 214.06 mg Rutin equivalent/g extract, respectively. Correlation coefficient between DPPH radical scavenging activity and the total phenolic and flavonoid contents suggest that phenolics and flavonoids in the extracts were partly responsible for the antiradical activities.

Borago officinalis L. Family: Boraginaceae

English Name: Borage.

Wettasinghe and Shahidi (1999) evaluated the antioxidant and free radical-scavenging properties of ethanolic extracts of defatted seeds of Boragee. Borage meal exerted a concentration-dependent antioxidant activity in a meat model system. At 2 per cent (w/w), it inhibited TBARS, hexanal and total volatile formation in meat by 26.5, 30.5 and 18.6 per cent, respectively. Hydrogen peroxide, hydroxyl radical and superoxide radical-scavenging properties of the extract were somewhat less than, but comparable to, those observed in trans-sinapic acid at similar concentrations of phenolics. At 200 ppm, a 100 per cent quenching of the hydroxyl radical and superoxide radical was evident. The extract scavenged 29-75 per cent of the hydrogen peroxide in assay media after 10 min of assay as compared to 3 per cent reduction in the control.

The rapid evaluation of antioxidant activity of crude borage (*Borago officinalis*) extract was determined by using DPPH free radical method (Bandoniene and Murkovic, 2002). The borage extract resulted in a rapid decrease of the absorbance and showed very high hydrogen-donating capacity towards the DPPH radical. A new HPLC-DPPH on-line method was applied for a screening of several radical scavenging components in their borage extract as well as for quantitative analysis. The HPLC analysis of borage extract revealed the presence of several radical scavenging components in the borage extract. The results obtained from the chromatograms suggest that some compounds present in the extract possess high radical quenching ability. The dominant antioxidative compound in the crude extract of borage leaves was identified as rosmarinic acid (Bandoniene and Murkovic, 2002).

Boscia senegalensis (Pers.) Lam. ex Poiret Family: Capparaceae

Using Trolox assay, Cook *et al.* (1998) estimated the antioxidant activity of 17 wild edible plants of Niger Republic used for food and traditional medicine. They observed that *Boscia senegalensis* possessed strong antioxidant activity.

Boswellia carteri Birdw. Family: Burseraceae

English Name: Frankinense

Yang *et al.* (2010) reported antioxidant activity of the essential oil of *Boswellia carteri.*

Boswellia dioscorides Thul. et Gibri Family: Burseraceae

Mothana *et al.* (2009) evaluated the antioxidant activity of 26 Yemeni medicinal plants using DPPH assay. Methanolic extracts of *Boswellia dioscorides* showed good antioxidant potential at low concentrations (more than 80 per cent at 50 μg/ml).

Boswellia serrata Triana et Planch Family: Burseraceae

Sharma *et al.* (2011) evaluated the antioxidant activity of aqueous extract of *Boswellia serrata.* The aqueous extract showed free radical scavenging activity in a concentration dependent manner. At the concentration of 500 mg/ml, 94.6 per cent diphenyl picryl hydrazyl, 80.9 per cent nitric oxide, 40.97 per cent hydroxide and 86.2 per cent super oxide radicals were scavenged by the aqueous extract. At the concentration of 500 mg/ml extract showed maximum reducing ability (converting Fe^{3+} to Fe^{2+}).

Assimopoulou *et al.* (2005) evaluated the antioxidant activity of natural resins and bioactive triterpenes in oil substrates. *Boswellia serrata* resins and the triterpenes, ursolic acid oleanolic acid, presented satisfactory antioxidant activity in sunflower oil. In lard *B. serrata* showed good antioxidant activity.

Boswellia socotrana Balf. f. Family: Burseraceae

Mothana *et al.* (2009) evaluated the antioxidant activity of 26 Yemeni medicinal plants using DPPH assay. Methanolic extracts of *Boswellia socotrana* showed good antioxidant potential at low concentrations (more than 80 per cent at 50g/ml).

Bougainvillea glabra Family: Nyctaginaceae

Antioxidant activity of *Bougainvillea glabra* and four other species have been investigated by Adib *et al.* (2010) by DPPH assay. *B. glabra* was found to have the moderate antioxidant activity. Tert-butyl-1-hydroxytoulene fraction of leaves showed free radical scavenging activity with an IC_{50} value of 91.5µg/ml, Hexane soluble fraction of methanolic extract was with IC_{50} value of 104µg/ml, Carbon tetrachloride soluble fraction of methanolic extract was with IC_{50} value of 97µg/ml and chloroform soluble fraction of methanolic extract had IC_{50} value of 323µg/ml.

Brassica alba (L.) Boiss. Family: Brassicaceae (Cruciferae)

Methanol aqueous extract of *Brassica alba* was screened by Kim *et al.* (1997) for anti-oxidative activity using Fenton's reagent/ethyl lineolate system. The extract showed antioxidant activity.

Antioxidant capacities of 56 selected Chinese medicinal plants, including *Brassica alba* were evaluated by Song *et al.* (2010) using the TEAC and FRAP assays, and their total phenolic content was measured by the Folin-Ciocaleteu method. The TEAC value was 53.51µmol Trolox/g, FRAP value was 64.87µmol Fe^{2+}/g and phenolic content was 3.34mg GAE/g.

Brassica campestris L. Family: Brassicaceae (Cruciferae)

English Name: Chinese mustard. Japanese: Komatsuna.

Maeda *et al.* (2006) examined the phenolic contents of mustard leaves and its radical scavenging activities for DPPH radical. The polyphenol content was 30.4 mg-gallic acid equivalent/100 g) and DPPH scavenging activity was 137.3 µmol-Trolox eq./100 g).

Brassica campestris L. var. _japonica_ Hara Family: Brassicaceae (Cruciferae)

A simple but novel compound, S-1-methoxy-1(3,5-dimethoxy-4-hydroxyphenylethane, was isolated by Nagatsu *et al.* (2004) as a moderately antioxidant compound from rape (*Brassica campestris* subsp. *japonica*) oil cake together with 5 known compounds. Three of these compounds, indolacetonitrile, 4-hydroxyindoleacetonitrile and 4-hydroxyphenylacetonitrile, showed strong antioxidative activity evaluated by the ferric thiocyanate method.

Brassica juncea Czern et Coss. Family: Brassicaceae (Cruciferae)

English Name: Mustard. Japanese: Karashina.

Khan *et al.* (1996, 1997) studied the antioxidant effects of *Brassica juncea* seeds fed with high fat diet. The study revealed that this species alter the peroxidation (TBA reactive substances) level to a beneficial extent. Concentration of malondialdehyde showed a significant decrease, while hydroperoxides and conjugated dienes were significantly increased in liver and heart of both the experimental groups. SOD and Catalase activity was found to be increased in liver and heart and kidney were lowered in rats administered this species. Glutathione reductase, glutathione peroxidase and glutathione S-transferase activity showed a sharp increase in the experimental groups compared with controls.

Antioxidant effects of isorhamnetin 3,7-di-O-β-D-glucopyranoside isolated from mustard leaf (*Brassica juncea*) in rats with streptozotocin-induced diabetes was studied by Yokozawa *et al.* (2002). On the basis of the oral and intraperitoneal results, it was hypothesized that isorhamnetin diglucoside was converted to its metabolite *in vivo*, and its conversion to its aglycone, isorhamnetin, by β-glucosidase was confimed; isorhamnetin acted as an antioxidant. Moreover, it was observed that isorhamnetin diglucoside had no effect on the DPPH radical, whereas isorhamnetin showed a potent antioxidant effect *in vitro*. Furthermore, lipid peroxidation in blood, liver, and kidney associated with diabetes mellitus declined after the administration of isorhamnetin. These results suggest that isorhamnetin diglucoside is metabolized *in vivo* by intestinal bacteria to isorhamnetin and that isorhamnetin plays an important role as an antioxidant.

From the leaves of *Brassica juncea* an radical scavenging isorhamnetin 7-O-glucoside on peroxynitite and DPPH was isolated and characterized. The compound showed the peroxynitrite and DPPH scavenging activities with IC_{50} values of 2.07 and 13.3μM, respectively (Choi *et al.*, 2002).

Maeda *et al.* (2006) examined the phenolic contents of leaves of mustard and its radical scavenging activities for DPPH radical. The polyphenol content was 35.9 mg-gallic acid equivalent/100 g and DPPH scavenging activity was 174.7 μmol-Trolox eq./100 g).

From the leaves of *Brassica juncea*, three kaempferol glycosides, kaempferol-3-O-(2-O-sipoyl)-β-D-glucopyranosyl-(1→2)-beta-D-glucopyranoside-7-O-beta-D-glucopyranoside (1), kaempferol-3-O-beta-D-glucopyranosyl-(1→2)-beta-D-glucopyranoside-7-O-beta-D-glucopyranosyl-(1→6)-beta-D-glucopyranoside (2) and kaempferol-3-O(2-O-sianpoyl)- β -D-glucopyranosyl-(1→2)-beta-D-glucopyranoside-7-O-beta-D-glucopyranosyl-(1→6)-β-D-glucopyranoside (3) were isolated. Antioxidant activities were determined by measuring the scavenging activities on DPPH radical and peroxynitrite (ONOO(-). Compounds 1 and 3 showed good antioxidant activities with respective IC_{50} values of 28.61 and 36.93 microM toward DPPH; respective IC_{50} values of 9.79 and 11.40 μM toward ONOO(-). However, compound 2 showed no DPPH scavenging activity and weak ONOO(-) scavenging activity with IC_{50} value of 32.00 microM (Jung *et al.*, 2009).

Brassica napus L. Family: Brassicaceae (Cruciferae)

English Name: Canola

Brettonnet *et al.* (2010) studied phenolic acids composition and antioxidant activity of canola extracts in cooked beef, chicken and pork. Crude polyphenol extracts from canola meal reduced the formation of 2-thiobarbituric acid-reactive substances (TBRAS) in pre-cooked beef, pork and chicken. Although the individual activities of salicylic, *p*-hydroxybenzoic and *p*-coumaric acids were marginal, all significant synergisms included them.

Brassica oleracea L. Convar. _Capitata_ var. _alba_ Family: Brassicaceae (Cruciferae)

English Name: White cabbage.

Souri *et al.* (2004) investigated the antioxidant activity of methanol extracts of white cabbage, used as vegetable in Iranian diet against linoleic acid peroxidation. The extracts showed good antioxidant activity.

Brassica oleracea L. Family: Brassicaceae

English Name: Sprouting Broccoli.

Souri *et al.* (2004) investigated the antioxidant activity of methanol extracts of sprouting broccoli, used as vegetable in Iranian diet against linoleic acid peroxidation. The extracts showed good antioxidant activity.

Pellegrini *et al.* (2010) evaluated the effect of common cooking practices (*i.e.*, boiling, microwaving, and basket and oven steaming on the polyphenol content (carotenoids, chlorophylls, glucosinolates, polyphenols, and ascorbic acid), total antioxidant capacity (TAC), and color changes of three generally consumed Brassica vegetables analyzed fresh and frozen. Among cooking procedures, boiling determined an increase of fresh broccoli carotenoids and fresh Brussels sprout polyphenols, whereas decrease of almost all other phytochemicals in fresh and frozen samples was observed. Steaming procedures determined a release of polyphenols in both fresh and frozen samples. Microwaving was the best cooking method for maintaining the color of both fresh and frozen vegetables and obtaining good retention of glucosinolates. During all cooking procedures, ascorbic acid was lost in great amount from all vegetables. The overall results of this study demonstrate that fresh Brassica vegetables retain phytochemicals and TAC better than frozen samples.

Brassica oleracea L. var. *botrytis* Family: Brassicaceae (Cruciferae)

Antioxidant potential of selected vegetables commonly used in diet in Asian subcontinent, including *Brassica oleracea* L. var. *botrytis* inflorescence has been evaluated by Gacche *et al.* (2010). The DPPH radical scavenging activity was 67.2 per cent, inhibition of lipid peroxidation was 18.1 per cent, Vitamin C content was 5.63 mg/100g and total phenols was 4.99 mg/g.

Brassica oleracea L. var. *botrytis* subvar. *cauliflora* A.P.DC. Family: Brassicaceae

English Name: Cauliflower.

Souri *et al.* (2004) investigated the antioxidant activity of methanol extracts of cauliflower, used as vegetable in Iranian diet against linoleic acid peroxidation. The extracts showed good antioxidant activity.

Koksal and Gulcin (2008) reported that the free radical scavenging activity of water and ethanolic extracts of cauliflower were found as 64.6 and 51.2 per cent, respectively at 30μg/ml concentration.

Brassica oleracea L. var. *capitata* L. Family: Brassicaceae (Cruciferae)

English Name: Cabbage. Japanese: Kyabetsu.

Maeda *et al.* (2006) examined the phenolic contents of Cabbage and its radical scavenging activities for DPPH radical. The polyphenol content was 27.9 mg-gallic acid equivalent/100 g and DPPH scavenging activity was 98.9 μmol-Trolox eq./100 g.

Antioxidant potential of selected vegetables commonly used in diet in Asian subcontinent, including *Brassica oleracea* L. var. *capitata* shoot apex has been evaluated by Gacche *et al.* (2010). The DPPH radical scavenging activity was 42.4 per cent, inhibition of lipid peroxidation was 58.4 per cent, Vitamin C content was 12.28 mg/100g and total phenols was 3.00 mg/g.

A study was carried out by Raghu *et al.* (2010) to determine the antioxidant activity of aqueous extracts of ten selected common vegetables. For *Brassica oleracea* var. *capitata* the IC_{50} value for DPPH scavenging was 0.95mg/ml while standard was 2.45µg/ml. Reducing power of the extract was moderate.

Brassica oleracea **L. var.** ***caulocarpa*** **L. Family: Brassicaceae (Cruciferae)**

English Name: Knol khol.

Andallu *et al.* (2003) assessed the antidiabetic and antioxidant role of knol khol, commonly consumed vegetable in NIDDM subjects. Knol khol therapy significantly improved the glycemic control and also significantly brought down lipid peroxidation in erythrocytes (33 per cent), plasma (21 per cent) and urine (16 per cent) and significantly elevated the activity of glutathione reductase (9 per cent) and glutathione-S-transferase (53 per cent) while decreasing the activity of catalase in erythrocyte in experimental group compared to the control group. Besides, knol khol improved remarkably by the levels of serum antioxdants, *i.e.*, ascorbic acid 140 per cent), uric acid (69 per cent), and ceruloplasmin (81 per cent) in experimental subjects. This data shows that knol khol is antidiabetic and antioxidant by virtue of polyphenols, tannins and ascorbic acid present in it.

Brassica oleracea **L. var.** ***gemmifera*** **A.P.DC. Family: Brassicaceae (Cruciferae)**

English Name: Brussels sprouts.

Souri *et al.* (2004) investigated the antioxidant activity of methanol extracts of Brussels sprouts, used as vegetable in Iranian diet against linoleic acid peroxidation. The extracts showed good antioxidant activity.

Pellegrini *et al.* (2010) evaluated the effect of common cooking practices (*i.e.*, boiling, microwaving, and basket and oven steaming) on the polyphenol content (carotenoids, chlorophylls, glucosinolates, polyphenols, and ascorbic acid), total antioxidant capacity (TAC), and color changes of three generally consumed Brassica vegetables analyzed fresh and frozen. Among cooking procedures, boiling determined an increase of fresh broccoli carotenoids and fresh Brussels sprout polyphenols, whereas decrease of almost all other phytochemicals in fresh and frozen samples was observed. Steaming procedures determined a release of polyphenols in both fresh and frozen samples. Microwaving was the best cooking method for maintaining the color of both fresh and frozen vegetables and obtaining good retention of glucosinolates. During all cooking procedures, ascorbic acid was lost in great amount from all vegetables. The overall results of this study demonstrate that fresh Brassica vegetables retain phytochemicals and TAC better than frozen samples.

Brassica oleracea L. var. *gongylodes* Family: Brassicaceae

A study was carried out by Raghu *et al.* (2010) to determine the antioxidant activity of aqueous extracts of ten selected common vegetables. For *Brassica oleracea* var. *gongylodes* the IC_{50} value for DPPH scavenging was 3.54mg/ml while standard was 2.45μg/ml. Reducing power of the extract was moderate.

Brassica oleracea L. var. *italica* Plenck. Family: Brassicaceae (Cruciferae)

English Name: Broccoli. Japanese name: Burokkori.

Cai *et al.* (2004) evaluated the antioxidant activity and phenolic compounds in traditional Chinese medicinal plants and vegetables associated with anticancer. The improved ABTS.+ method was used to systematically assess the total antioxidant capacity (TEAC) of the medicinal extracts. The TEAC values and total phenolic content for methanolic extract of fruit of broccoli were 101.4 μmol Trolox equivalent/100g dry weight (DW), and 0.63 g of gallic acid equivalent/100g DW respectively.

Maeda *et al.* (2006) examined the phenolic contents of broccoli and its radical scavenging activities for DPPH radical. The polyphenol content was 93.3 mg-gallic acid equivalent/100 g) and DPPH scavenging activity was 346.7 μmol-Trolox eq./ 100 g).

Brassica rapa L. var. *rapa* Family: Brassicaceae (Cruciferae)

English Name: Turnip.

The phenolic acids and organic acids of turnip edible parts (leaves and stems, flower buds and roots) were determined by Fernandes *et al.* (2007). A screening of the antioxidative potential was also performed by means of DPPH radical scavenging assay. Turnip flower buds exhibited the strongest antioxidant capacity.

Brickellia veronicaefolia A. Gray Family: Asteraceae (Compositae)

The antioxidant and antiradical activities of 5,7,3′-trihydroxy-3,6,4′-trimethoxyflavone or centaureidin isolated and characterized from *Brickellia veronicaefolia* were elucidated by Perez *et al.* (2004) by heat-induced oxidation in a β-carotene and linoleic acid system and by DPPH decolouration test. The centaureidin (32.1 per cent) exhibited antioxidative activity less than that of BHT (95.5 per cent) and α-tocopherol (94.6 per cent) in DPPH decoloration test was found.

Bridelia airy-shawii Li Family: Euphorbiaceae

Tatiya *et al.* (2008) investigated the antioxidant and anti-inflammatory activities of the petroleum ether, chloroform, methanol and aqueous extracts from stem bark of *Bridelia airy-shawii*. The antioxidant activity of all the extracts was determined by using Nitric oxide scavenging assay. The stem bark extracts showed significant free radical scavenging activity against nitric oxide induced release of free radical.

Broussonetia papyrifera Vent. Family: Moraceae

Fractionation of the EtOH extract fro the fruits of *Broussonetia papyrifera* led to the isolation of 15 phenolic compounds. Antioxidant activities of compounds 2-15 against H_2O_2-induced injury in SY5Y cells and DPPH radical scavenging activities were evaluated by Zhou *et al.* (2010).

***Brucea amarissima* (Lour.) Merr. Family: Simaroubaceae**

The crude methanol extracts of four Philippine medicinal plants namely *Brucea amarissima* bark; *Intsia bijuga, Laportea meyeniana* and *Pipturus arborescens* were screened for their antioxidant activity using DPPH free radical assay. EC_{50} values for DPPH radical scavenging activity ranged from 343 to 1846 µg/ml, with *Brucea amarissima* and *Intsia bijuga* having the lowest and highest values, respectively.

***Brucea javanica* (L.) Merr. Family: Simaroubaceae**

Cai *et al.* (2004) evaluated the antioxidant activity and phenolic compounds in traditional Chinese medicinal plants associated with anticancer. The improved ABTS.+ method was used to systematically assess the total antioxidant capacity (Trolox equivalent antioxidant capacity, TEAC) of the medicinal extracts. The TEAC values and total phenolic content for methanolic extract of fruit of *Brucea javanica* were 93.6 µmol Trolox equivalent/100g dry weight (DW), and 0.57 g of gallic acid equivalent/100g DW respectively. A positive significant linear relationship between antioxidant activity and total phenolic content showed that phenolic compounds were the dominant antioxidant components in the tested herbs. Major types of phenolic compounds in *Brucea javanica* include flavonoids (quercetin and luteolin glycosides), phenolic acids (vanillic acid).

***Bruguiera cylindrica* L. Family: Rhizophoraceae**

Vadlapudi and Naidu (2009) evaluated the antioxidant potential of *Bruguiera cylindrica*, a mangrove plant. The SOD was 1.27U/mg, Catalase was 0.81U/mg, Ascorbic acid was 320mg/100g, DPPH radical inhibition was 64 per cent and FRAP units were 576.

***Bruguiera gymnorrhiza* (L.) Lamk. Family: Rhizophoraceae**

Vadlapudi and Naidu (2009) evaluated the antioxidant potential of *Bruguiera gymnorrhiza*, a mangrove plant. The SOD was 0.87U/mg, Catalase was 0.60U/mg, Ascorbic acid was 200mg/100g, DPPH radical inhibition was 60 per cent and FRAP units were 400.

A new cyclic 4-hydroxy-dithiosulfonate, bruguiesulfurol (1) and two known 4-hydroxydithiolane 1-oxides, brugierol (2) and isobrugierl (3) were isolated from flowers of *Bruguiera gymnorrhiza* (Homhual *et al.*, 2006). With stably-transfected HepG2 cells, the three isolates activated antioxidant response element luciferase activation with (EC_{50}) values of 56.7, 3.7 and 1.8µM, respectively. Compounds 2 and 3 also inhibited phorbol ester-induced NF-kappaB (nuclear factor-kappaB) luciferase activity with IC_{50} values of 85.0 and 14.5 µM respectively. In addition compound 2 inhibited enzyme cyclooxygenase-2 activity with an IC_{50} value of 6.1 µM.

***Bryonia alba* L. Family: Cucurbitaceae**

Karagezian (1981) examined the action of *Bryonia alba* on lipid peroxidation in microsomes and on fatty acid composition of individual lipid fractions in the liver of rats with alloxan diabetes. Administration of the extract was found to produce an appreciably normalizing effect on the biochemical indices of liver function.

***Bryophyllum calycinum* Salisb. Family: Crassulaceae**

Jain *et al.* (2010) evaluated the antioxidant activities of alcoholic and aqueous extracts of *Bryophyllum calycinum* leaf. The DPPH and nitric oxide free radical scavenging methods were used to detect oxidative activity. The results of DPPH method showed 50 per cent inhibition rate at the 144.23μg/ml and 117.42μg/ml with aqueous and alcoholic extract, respectively. Nitric oxide scavenging inhibition showed 50 per cent inhibition rate at the 525.92μg/ml and 460.48μg/ml with aqueous and alcoholic extract respectively.

***Buddleja asiatica* Lour. Family: Buddlejaceae**

The methanol extract of the leaves of *Buddleja asiatica* showed antioxidant activity toward well known *in vitro* antioxidant tests such as total antioxidant capacity by the DPPH assay and hydrogen peroxide scavenging methods. Due to high scavenging activity of the n-butanol successive fraction toward DPPH and H_2O_2, this extract was subjected to chromatographic separation and isolation. Four non-phenolic compounds were isolated and identified. The four compounds were evaluated as antioxidant agents using the three antioxidant bioassay tests (El-Sayed *et al.*, 2008).

***Buddleja globosa* Lam. Family: Buddlejaceae**

Houghton *et al.* (2005) reported antioxidant activities of *Buddleja globosa*. Inhibition of superoxide anion, lipoperoxidation and DPPH bleaching effect was found in the methanol serial and global extracts.

Backhouse *et al.* (2008) reported inhibition of superoxide anion, lipoperoxidation and DPPH bleaching effect in the methanol serial and global extracts of *Buddleja globosa*. Verbascoside and luteolin-7-O-glucoside were the major components of the methanol extract, apigenin 7- O-glucoside was also detected.

***Buddleja officinalis* Maxim. Family: Buddlejaceae**

Luteolin and acteoside isolated from flowers of *Buddleia officinalis* exhibited the most potent antioxidative activity on the NBT superoxide scavenging assay (Piao *et al.*, 2003).

Li *et al.* (2008) evaluated the antioxidant properties of 45 medicinal plants including *Buddleja officinalis* using FRAP and TEAC assays and the total phenolic contents of these plants were measured. *Buddleja officinalis* showed 284.19 μmol Fe(II)/g FRAP value, 130.82μmol Trolox/g of TEAC value and 28.82 mg GAE/g of Phenolic content. A high correlation between antioxidant capacities and their phenolic contents indicated that phenolic compounds were a major contributor of antioxidant activity of these plants.

The antioxidant activities of microwave-assisted extract of *Buddleja officinalis* and Soxhlet extract at the optimum extraction conditions were evaluated and compared with synthetic antioxidant butylated hydroxytoulene by Pan *et al.* (2010) employing DPPH free radical assay, ABTS assay, total antioxidant activity and reducing power. Both the extracts had stronger antioxidant activities than BHT in all assays except reducing power. The total phenolic contents of Microwave-assisted extract and soxhlet extract reached 113.56mg/and 100.94 mg/g dry weight of extract, respectively, expressed as pyrocatechol equivalents, while the total flavonoid contents were 75.33 mg/g and 62.56 mg/g dry weight of extract, respectively, expressed as catechin equivalents. Following activity oriented separation, luteolin was isolated as an active principle, which exhibited excellent free radical scavenging activities with DPPH IC_{50} 3.09 µg/ml and ABTS IC_{50} 2.20µg/ml.

Houghton *et al.* (2005) reported antioxidant activities of *Buddleja officinalis.*

Bunchosia canesens **(Aiton) DC. Family: Malpighiaceae**

Ruiz-Terán *et al.* (2008) evaluated the antioxidant activity of aerial parts of *Bunchosia canesens* using DPPH and β-carotene-linoleic acid bleaching assay. The extract displayed antioxidant activity comparable to the commercial antioxidant BHA.

Bunium persicum **(Boiss.) B. Fedstch. Family: Apiaceae (Umbelliferae)**

English Name: Wild caraway

Souri *et al.* (2008) evaluated the antioxidant activity against linoleic acid peroxidation using 1,3-diethyl-2-thiobarbituric acid as reagent. Antioxidant activity (IC_{50}) against peroxidation of linoleic acid (2mg/ml) was 82.25 and phenolic content was 214.03 mg/100g dry weight. The results of this study showed that there is no significant correlation between antioxidant activity and phenolic content of the studied plant materials and phenolic content could not be a good indicator of antioxidant capacity.

Antioxidant activities of ethanol extracts from seven Apiaceae fruits have been studied by Nickavar and Abolhasani (2009) by the DPPH radical scavenging test. All the studied extracts, including *Bunium persicum,* showed antioxidant capability. For *Bunium persicum* IC_{50} value of the DPPH scavenging activity was 208.40µg/ml and total flavonoid content was 20.21µg/mg. A positive correlation was found between the antioxidant potency and flavonoid content of the fractions.

Bupleurum chinense **DC. Family: Apiaceae (Umbelliferae)**

Antioxidant capacities of 56 selected Chinese medicinal plants, including *Bupleurum chinense* were evaluated by Song *et al.* (2010) using the TEAC and FRAP assays, and their total phenolic content was measured by the Folin-Ciocaleteu method. The TEAC value was 19.93µmol Trolox/g, FRAP value was 32.05µmol Fe^{2+}/g and phenolic content was 3.41mg GAE/g.

Bupleurum falcatum **L. Family: Apiaceae (Umbelliferae)**

An anti-ulcer pectic polysaccharide with antioxidant activity has been isolated from *Bupleurum falcatum* by Matsumoto *et al.* (1993).

Wojcikowski *et al.* (2007) evaluated the antioxidant activity of aerial parts of *Bupleurum falcatum.* Oxygen radical absorbance capacity of the ethyl extract, methanol extract and aqueous extracts were 5.11, 47.40 and 69.63µmol Trolox equivalent per g of dried starting material, respectively.

Bupleurum marginatum Wall. ex DC. Family: Apiaceae (Umbelliferae)

Ashour *et al.* (2009) investigated the chemical composition and antioxidant activity of *Bupleurum marginatum*. The ability of the oil to reduce DPPH and to prevent degradation of deoxyribose were used to evaluate the antioxidant activity and the corresponding IC_{50} values were found to be 3.66 mg/ml and 17.4µg/ml, respectively.

Bupleurum scorzonerifolium Willd. Family: Apiaceae (Umbelliferae)

Cai *et al.* (2004) evaluated the antioxidant activity and phenolic compounds in traditional Chinese medicinal plants associated with anticancer. The improved ABTS.+ method was used to systematically assess the total antioxidant capacity (Trolox equivalent antioxidant capacity, TEAC) of the medicinal extracts. The TEAC values and total phenolic content for methanolic extract of root of *Bupleurum scrozonerifolium* were 75.3 µmol Trolox equivalent/100g dry weight (DW), and 0.47 g of gallic acid equivalent/100g DW respectively. A positive significant linear relationship between antioxidant activity and total phenolic content showed that phenolic compounds were the dominant antioxidant components in the tested herbs. Major types of phenolic compounds in *Bupleurum scorzonerifolium* include flavonols (quercetin, rutin, isoquercetin).

Burkea africana Hook. f. Family: Fabaceae (Leguminosae)

The stem bark of *Burkea africana*, a sub-saharan Africa medicinal plant has been shown to possess antioxidant and anti-radical scavenging activity (Mathisen *et al.*, 2002). The antioxidant component identified by the authors include fisetinidol-(4-α-8)-catechin-3-gallate and bis-fisetinidol-(4-α-6, 4-α-8)-catechin-3-gallate and to a lesser extent, monomeric flavor-3-ols (catechin, epicatechin and fisetinidol).

Bursera grandifolia Engl. Family: Burseraceae

Ruiz-Terán *et al.* (2008) evaluated the antioxidant activity of aerial parts of *Bursera grandifolia* using DPPH and β-carotene-linoleic acid bleaching assay. The extract displayed antioxidant activity comparable to the commercial antioxidant BHA.

Butea monosperma (Lam.) Kuntze Family: Fabaceae

Stigmasterol, isolated from the bark was evaluated for its thyroid hormone and glucose regulatory efficacy in mice by Panda *et al.* (2009). Its administration at 2.6mg/kg/d for 20 days reduced serum triidothyronine (T_3), thyroxin (T_4) and glucose concentrations as well as the activity of hepatic glucose-6-phosphatase (G-6-Pase) with a concomitant increase in insulin indicating its thyroid inhibiting and hypoglycemic properties. A decrease in the hepatic lipid peroxidation (LPO) and an increase in the activities of catalase (CAT), superoxide dismutase (SOD) and glutathione (GSH) suggested its antioxidative potential. The highest concentration tested (5.2mg/kg) evoked pro-oxidative activity. In conclusion the findings reveal the hitherto unknown thyroid inhibitory and insulin stimulatory nature of stigmasterol, isolated from *B. monosperma* and suggest its antidiabetic and antiperoxidase properties.

The possible protective effect of ethanolic extract of *Butea monosperma* leaves (BMEE) on diabetes and diabetes-induced oxidative stress was evaluated by Sharma and Garg (2009) in alloxan-induced diabetic male adult mice. The activities of antioxidant enzymes (CAT and GSH-Px) and GSH level in hepatic, pancreatic and renal tissues increased significantly in BMEE-treated mice, but the activity of SOD was not improved significantly. BMEE treatment also reduced the TBARS levels and lowered serum lipid profile significantly.

Hydromethanol extracts of 15 Bangladeshi medicinal plants, including *Butea monosperma*, were evaluated by Hasan *et al.* (2009) for antioxidant potential using DPPH radical scavenging assay. *Butea monosperma* leaves exhibited radical scavenging activity with an IC_{50} value of 25.96 µg/ml compared to the IC_{50} value of 5.15µg/ml as shown by the reference antioxidant ascorbic acid, in a dose dependent fashion.

Buxus hyrcana **Pojark Family: Buxaceae**

Souri *et al.* (2004) evaluated the antioxidative activity of *Buxux hyrcana* by linoleic acid peroxidation test using 1,3-diethyl-2-thiobarbituric acid as the reagent. Methanolic extract showed 86.94 per cent inhibition at 40µg concentration with IC_{50} value of 1.15µg which is approximately in the range of α-tocopherol (IC_{50} = 0.60µg).

Caesalpinia benthamiana **(Baill.) Herend et Zarucchi (=*Mezoneuron benthamianum* Baill.) Family: Caesalpiniaceae**

Benthamin 2 isolated from the root bark of *Caesalpinia benthamiana* showed antioxidant activity with IC_{50} values 42.7µM and 74.2µM for the DPPH and TBA assays, respectively (Dickson *et al.*, 2007). Deoxycaesaldekarin C, also isolated from the bark possessed both antibacterial and antioxidant activities. The relatively stronger antioxidant activity of benthamin 2 may be associated with the presence of an exocyclic methylene function (Dickson *et al.*, 2007).

Caesalpinia bonduc **(L.) Roxb. (=*Caesalpinia bonducella* Flem.) Family: Caesalpiniaceae**

English Name: Bonduc nut.

To assess the anticarcinogenic activity of a methanol extract of *Caesalpinia bonducella* leaves containing flavonoids and triterpenoids, the antioxidant defense system has been evaluated by Gupta *et al.* (2005). The effect of the extract on lipid peroxidation end-product MDA, enzymatic antioxidants were analyzed in the liver of control and experimental animals. After drug administration, there was a marked increase in lipid peroxidation levels. Gupta *et al.* (2005) concluded that the observed anticarcinogenic properties of the extract may also be explained by its strong antioxidant capacity and capability to induce an *in vitro* antioxidant system.

Shukla *et al.* (2009) assessed the *in vitro* potential of ethanolic extract of *Caesalpinia bonducella* seeds as a natural antioxidant. The DPPH activity of the extract (20, 40, 50, 100 and 200 µg/ml) was increased in a dose dependent manner, which was found in the range of 38.93-74.77 per cent as compared to ascorbic acid in DPPH radical scavenging assay were obtained to be 74.73 and 26.68µg/ml respectively. The ethanolic extract was also found to scavenge the superoxide generated by EDTA/NBT system. Measurement of total phenolic content of the ethanolic extract was achieved using Folin-Ciocalteau reagent containing 62.50 mg/g of phenolic content, which was found to be significantly higher when compared to reference standard gallic acid. The ethanolic extract also inhibited the hydroxyl radical, nitric oxide, superoxide anions, with IC_{50} values of 109.85, 102.65 and 89.84 µg/ml respectively. However, the IC_{50} values for the standard ascorbic acid were noted to be 70.79, 65.98 and 36.68 µg/ml respectively.

Caesalpinia digyna Rottl. Family: Caesalpiniaceae

Various extracts and flash chromatographic fractions of *Caesalpinia digyna* were evaluated by Sundaram and Mitra (2006) by *in vitro* and *in vivo* experimental models. Out of all the extracts prepared, the ethyl acetate soluble portion of methanolic extract offered maximum protection. Flash chromatographic fractionation of CDM-I yielded 16 fractions. All the fractions showed very good activity when assessed for *in vitro* lipid peroxidation, out of which the fractions 3-16 showed the presence of major phenolics by TLC (Sundaram and Mitra, 2006).

The antioxidant properties of three successive extracts of *Caesalpinia digyna* root (CDM) and isolated compound bergenin, were tested using standard *in vitro* and *in vivo* models by Srinivasan *et al.* (2007). The successive methanol extract of root exhibited strong scavenging effect on DPPH free radical, ABTS radical cation, hydrogen peroxide, nitric oxide, hydroxyl radical and inhibition of lipid peroxidation. The free radical scavenging effect of CDM was comparable with that of reference antioxidants. The CDM has the highest content of phenolic compounds and strong free radical scavenging effect.

Caesalpinia mimosoides Lam. Family: Caesalpiniaceae

Chanwitheesuk *et al.* (2005) investigated the antioxidant activity of *Caesalpinia mimosoides* using a β-carotene bleaching method. The contents of plant chemicals such as vitamin C, Vitamin E, carotenoids, tannin and total phenolics, were also determined. Methanolic extract showed antioxidant activity. *Caesalpinia mimosoides* has antioxidant index of 5.43, Vitamin C 20.1 mg per cent, Vitamin E 0.0105mg per cent, Total carotenes 1.92 mg per cent, Total xanthophylls 5.28mg per cent, Tannins 484mg per cent and Total phenolics of 1924mg per cent.

Caesalpinia pulcherrima (L.) Sw. Family: Caesalpiniaceae

Liu *et al.* (2008) evaluated the antioxidant activity of water, 50 per cent ethanol and 70 per cent acetone extracts of flowers of 15 plants of Hainan. All the extracts of *Caesalpinia pulcherrima* showed good antioxidant activity in eliminating DPPH, O^{2-} and OH radicals.

Hydromethanol extracts of 15 Bangladeshi medicinal plants, including *Caesalpinia pulcherrima*, were evaluated by Hasan *et al.* (2009) for antioxidant potential using DPPH radical scavenging assay. *Caesalpinia pulcherrima* leaves exhibited radical scavenging activity with an IC_{50} value of 16.0 µg/ml compared to the IC_{50} value of 5.15µg/ml as shown by the reference antioxidant ascorbic acid, in a dose dependent fashion.

Caesalpinia pyramidalis Tul. Family: Caesalpiniaceae

Alviano *et al.* (2008) evaluated the antioxidant activity of leaves of *Caesalpinia pyramidalis* by DPPH photometric assay. The extract showed good antioxidant potential with EC_{50} value of 15.2µg/ml.

Caesalpinia sappan L. Family: Caesalpiniaceae

Popyphenolic compounds isolated from heartwood showed antioxidant activity (Badami *et al.*, 2003). Two antioxidant compounds were isolated by Safitri *et al.* (2003) from *Caesalpinia sappan*. These were identified as 1′,4′-dihydro-spiro[benzofuron-3(2H),3′-(3H-2)benzopyran]-1′,6′,6′,7′-tetrol and 3[[4,5-dihydroxy-2(hydroxymethyl) phenyl]-methyl]-2,3-dihydro-3,6-benzofurandiol. These two compounds inhibited xanthine oxidase activity and scavenged superoxide anion and hydroxyl radicals. These two compounds possessed similar radical scavenging activities as ascorbic acid and they were more effective than other well-known antioxidants such as alpha-tocopherol, beta-carotene and BHT.

Cai *et al.* (2004, 2006) evaluated the antioxidant activity and phenolic compounds in traditional Chinese medicinal plants associated with anticancer. The improved ABTS.+ method was used to systematically assess the total antioxidant capacity (Trolox equivalent antioxidant capacity, TEAC) of the medicinal extracts. The TEAC values and total phenolic content for methanolic extract of heartwood of *Caesalpinia sappan* were 1301 µmol Trolox equivalent/100g dry weight (DW), and 7.92 g of gallic acid equivalent/100g DW respectively. A positive significant linear relationship between antioxidant activity and total phenolic content showed that phenolic compounds were the dominant antioxidant components in the tested herbs. Major types of phenolic compounds in *Caesalpinia sappan* include chalcones, flavonols (quercetin), tannins.

Wetwitayaklung *et al.* (2005) evaluated the antioxidant activity of *Caesalpinia sappan* heartwood in various ages. Its *in vitro* antioxidation in DPPH) nitric oxide and oil oxidation and *in vivo* antioxidation have been evaluated. These results showed that the sappan crude extracts have high antioxidant property. The antioxidant activity was obtained by trolox equivalent antioxidant capacity (TEAC) assay. The CH_3OH:CH_2Cl_2 (1:3) was the best ratio for extraction solution and gave extracts with high amount of brazilin and antioxidant property. The older heartwood had higher quantities of antioxidant activity.

Sarumathy *et al.* (2010) evaluated the protective effect of *Caesalpinia sappan* on acetaminophen induced nephrotoxicity and oxidative stress in male albino rats. Ethanolic extract significantly increased the activities of renal superoxide dismutase, catalase and glutathione peroxidase and decreased Malondialdehyde content of APAP treated rats.

Gan *et al.* (2010) evaluated the antioxidant activities of 40 medicinal plants associated with prevention and treatment of cardiovascular and cerebrovascular diseases using FRAP and TEAC assays and their total phenolic contents were measured by the Folin-Ciocalteu methods. The antioxidant activity of *Caesalpinia sappan* was 313.50µmol Fe(II)/g in FRAP assay, 417.48µmol Trolox/g and Total phenolic content was 40.974 mg GAE/g. A high correlation between antioxidant capacity and total phenolic content indicated that phenolic compounds could be the main contributor of the antioxidant activity of these plants.

Cajanus cajan (L.) Millsp. (=*Cajanus indicus* Spreng.)

Family: Fabaceae (Papilionaceae)

English Name: Red gram, pigeon pea.

The antioxidant properties of some commonly consumed pigeon pea (*Cajanus cajan*) were assessed with regard to its vitamin C, total phenol and phytate content, as well as antioxidant activity as typified by its reducing power and free radical scavenging ability (Oboh, 2006a). Vitamin C content of the pigeon peas was 0.9 mg/100g. The phenol content of the pigeon peas was 0.4 (white) and 1.2 mg/g (brown). The phytate content of pigeon pea was 2.0-2.4 per cent analyzed were generally high while the reducing power and free radical scavenging ability was 54.5 per cent (pigeon pea brown).

Sreeramulu *et al.* (2009) evaluated the antioxidant activity of red gram dhal (pulse) by DPPH, scavenging assay, FRAP assay and reducing power. Phenolic content of the dhal was 229.33 mg/100g. DPPH radical scavenging activity expressed as trolox equivalent, FRAP (μmol/g) and reducing power (mg/g) were 0.61, 102 and 1.63, respectively in red gram dhal.

The antioxidant activity, reducing power, nitric oxide scavenging activity, superoxide radical scavenging activity, anti lipid peroxidation activity and total phenolics of the ethanolic extract of leaves of *Cajanus cajan* were studied by Ghosh *et al.* (2009) *in vitro* as parameters to investigate its antioxidant potential. The reducing power of the extract was found to be concentration dependent. The absorbance increased with increasing amount of the sample. The antioxidant activity, oxide scavenging activity, superoxide radical scavenging activity and anti-lipid peroxidation activity were also concentration dependent with IC_{50} value being 324.59 μg/ml, 46.42μg/ml, 35.24 μg/ml and 338.2 μg/ml respectively. The activities were found to be comparable with the reference drugs. Amount of total phenolics was found to be 39.8 μg of pyrocatechol equivalent per mg of extract.

The effect of fermentation on the polyphenol distribution and antioxidant activity of pigeon pea was investigated by Oboh *et al.* (2009). The results of the study revealed that fermentation caused a significant decrease in the bound phenol content of the legume. Free soluble phenol from both the fermented and unfermented legumes had a significantly higher reducing power.

Ghosh and Sil (2007) investigated the protective effect of a protein isolated from the herb *Cajanus cajan (=Cajanus indicus)*, against acetaminophen-induced hepatic and renal toxicity. Application of acetaminophen to hepatocytes increased reactive oxygen species production and reduced the total antioxidant activity of the treated hepatocytes. It also reduced the levels of antioxidant enzymes. Treatment with the protein significantly reversed these changes to almost normal. Results suggest that the protein protects hepatic and renal tissues against oxidative damages and could be used as an effective protector against acetaminophen induced hepato-nephrotoxicity.

Cakile maritima Scope Family: Brassicaceae (Cruciferae)

The study of the flavonoidal constituents isolated from the chloroform, ethyl acetate and butanol fractions of aqueous alcoholic extract of *Cakile maritima* revealed the isolation of quercetin, quercetin-3-O-rhamnoside, kaempferol-7-O-glucoside, luteolin, luteolin-7-O-glucoside and caffeic acid. The radical scavenging effects of the tested extracts and isolated compounds on DPPH free radical were observed, in which the flavonoidal compounds showed a strong antioxidant activity (Shams *et al.*, 2010).

Calamintha glandulosa (Req.) Benth. Family: Lamiaceae (Labiatae)

Couladis *et al.* (2003) screened Greek aromatic plants from the Lamiaceae family for the antioxidant activity. Of the 21 plants they tested they found ethanol extracts prepared from *Calamintha glandulosa*, as well as 9 other plants exhibited the same antioxidant activity as α-tocopherol in their ability to inhibit bleomycin-Fe(II) complex-induced arichidonic acid superoxidation to MDA.

Calceolaria chelidonioides Kunth. Family: Scrophulariaceae

The aerial parts of *Calceolaria chelidonioides* showed antioxidant activity using *in vitro* DPPH model (Falcao *et al.*, 2006).

Calea angosturana Hieron Family: Asteraceae (Compositae)

Mosquera *et al.* (2007) studied the antioxidant activity of the extracts using DPPH free-radical scavenging method. The percentage of the antioxidant activity of the n-Hexane fraction was 26.984, Dichloromethane fraction was 38.39 and methanol fraction was 50.94.

Calendula officinalis L. Family: Asteraceae (Compositae)

English Name: Pot marigold.

Cordova *et al.* (2002) studied the protective properties of butonolic extract of the marigold against lipid peroxidation of rat liver microsomes and action of free radical scavenger. Miliuskas *et al.* (2004) studied the radical scavenging activity using DPPH and ABTS assays. Mariassyova (2006) studied antioxidant activity of marigold extracts in rapeseed and sunflower oils. Antioxidant concentrates from marigold were added to sunflower and rapeseed oil, respectively, at concentrations ranging from 2 mg/kg to 200 mg/kg oil. The Rancimat assay was used for the optimization of antioxidant doses of individual plant concentrates. The most effective antioxidant in both sunflower and rapeseed oil was the marigold concentrate (optimal dose was 10 mg/kg and 30 mg/kg respectively).

Khalil *et al.* (2007) studied the growth, phenolic compounds and antioxidant activity of marigold grown under organic farming condition. Antoxdant activity of ethanol extract (inhibition per cent) was 63.11 at 200μl extract.

Braga *et al.* (2009) while studying the antioxidant activity of *Calendula officinalis* polypropylene glycol extract reported inhibitory effects on chemiluminiscence of human neutrophil bursts. The *C. officinalis* extract exerted its anti-ROS and anti-RNS activity in a concentration-dependent manner, with significant effects being observed at even very low concentrations.

Calliandra haematocephala Family: Mimosaceae

Liu *et al.* (2008) evaluated the antioxidant activity of water, 50 per cent ethanol and 70 per cent acetone extracts of flowers of 15 plants of Hainan. All the extracts of *Calliandra haematocephala* had better effects on eliminating DPPH, O_2 and OH radical.

Calligonum leucocladum Bunge Family: Polygonaceae

Stilbene derivative (E)-resveratrol 3-(6"-galloyl) O-beta-D glucopyranoside isolated from *Calligonum leucocladum* showed antioxidant activity (Okasaka *et al.*, 2004).

Calluna vulgaris (L.) Hull. Family: Ericaceae

English Name: Heather.

Antioxidant capacity of the chloroform, ethyl acetate, n-butanol and water fractions of the aerial parts of *Calluna vulgaris* was assessed by Deliorman-orhan *et al.* (2009). Antioxidant capacity of the plant was screened by assays of DPPH, superoxide anion and hydrogen peroxide scavenging, metal-chelating activity and reducing power. Butylated hydroxyanisole and ethylene diamine tetraacetic acid were used as reference in the assay of metal-chelating activity. Total phenolic contents of the fractions were determined by the Folin-Ciocalteu method. Liquid chromatography/diode array detection/mass spectrometry was used for phytochemical identification of the fractions. Kaempferol-3-O-β-D-galactoside was found to be the major constituent in the ethyl acetate fraction (37.1±0.9 per cent) followed by the n-butanol fraction (4.6±0.1 per cent). High occurrence of antioxidant capacity, with the exception of metal-chelating activity, was observed in the ethyl acetate and chloroform fractions as well as in kaempferol-3-O-β-D-galactoside of *C. vulgaris*. The major flavonoid, kaempferol-3-O-β-D-galactoside of this species shown high antioxidant capacity in various assays.

The antioxidative activity of phenolic extracts from heather was examined by autoxidation of methyl linoleate by Kahkonen *et al.* (1999). Remarkable high antioxidant activity and high total phenolic content were found in heather. Kaempferol, quercetin, myricetin, herbacetin, and isocutellarein and their glycosides were identified in heather. In addition dihydroflavonols (callunin and taxifolin-3-glucoside) and cyanidin 3-glucoside have been identified in heather flowers (Allais *et al.*, 1991, 1995).

Calotropis gigantea R.Br. Family: Asclepiadaceae

Rathod *et al.* (2009) reported the effect of chloroform extracts of *Calotropis gigantea* leaf and flower on free radical scavenging activity, and lipid profile in streptozotozin-induced diabetic rats. The lipid peroxidation, superoxide dismutase, and catalase were measured in liver homogenate and serum glutamic pyruvic transaminase, serum glutamic oxaloacetic transaminase, alkaline phosphatase, lipid profile were measured in blood serum. Administration of single dose of streptozotozin (55 mg/kg, i.p.) caused significant increases in lipid peroxidation, serum glutamic pyruvic transaminase, serum glutamic oxaloacetic transaminase, alkaline phosphatase, cholesterol and triglyceride levels, while superoxide dismutase and catalase levels were significantly decreased. Further, administration of chloroform extracts of *Calotropis gigantea* leaf and flower to streptozotocin-induced diabetes rats at a dose of 10, 20 and 50 mg/kg orally for 27 d lead to a significant decrease in lipid peroxidation, serum glutamic pyruvic transaminase, serum glutamic oxaloacetic transaminase, alkaline phosphatase, cholesterol and triglyceride levels. Consequently, superoxide dismutase and catalase levels were significantly increased. Glibenclamide was used as a positive control (10 mg/kg). It was observed that the effect of chloroform extracts of *Calotropis gigantea* on alkaline phosphatase, cholesterol, superoxide dismutase, serum glutamic pyruvic transaminase, serum glutamic oxaloacetic transaminase, levels were comparable to that of those produced by the positive control.

Calotropis procera **(Ait.) R.Br. Family: Asclepiadaceae**

Padhy *et al.* (2007) reported that aqueous suspension of dried latex of *Calotropis procera* (DL) affords protection against carbon tetrachloride induced hepatotoxicity in rats. The DL treatment normalized various biochemical parameters of oxidative stress.

Joshi *et al.* (2010) analyzed total phenols, flavonoids and antioxidant potential of the root and leaf extracts and latex of field grown as well as tissue cultured *Calotropis procera* plants. Total phenols and flavonoids were determined by Folin Ciocalteu reagent and aluminium chloride method, respectively. Free radical scavenging activity was determined by DPPH. Total phenols and flavonoids content were high in latex of field grown plants and lowest in the extracts of *in vitro* roots and *in vivo* leaves. The highest antioxidant capacity was exhibited by extracts of lyophilized latex (IC_{50} = 0.060 mg/ml) and the lowest (IC_{50} = 0.27 mg/ml) was in root extracts of field grown plants.

Calpurnia aurea **(Ait.) Benth. Family: Fabaceae (Leguminosae)**

Adedapo *et al.* (2008) evaluated the antibacterial, antioxidant and phenolic contents of methanol extracts of the leaves and stems of *Calpurnia aurea* using *in vitro* standard methods. The antioxidant activities of the stem extract were determined by ABTS, DPPH, and FRAP methods. The results from this study showed that the antioxidant activities of the stem extract of *Calpurnia aurea* as determined by the total phenol, flavonoids, and FRAP methods were higher than that of the leaves. On the other hand the leaf extract of the plant has higher level of total flavonols and proanthocyanidins. The leaf extract also has higher radical scavenging activity as shown in DPPH and ABTS assay.

Calycotome villosa **L. Family: Fabaceae (Leguminosae)**

Dessi *et al.* (2001) carried out studies on the effect of methanol extract and essential oil on lipid peroxidation. Essential oil and methanol extract showed antioxidant activity.

Camellia sinensis **(L.) O.Kuntze (Syn.:** ***Thea sinensis*** **L.) Family: Theaceae**

English Name: Tea, Green tea, black tea.

Tanizawa *et al.* (1984) have isolated antioxidant compounds from green tea leaves.

The flavonoids found in green tea have been found to be powerful antioxidants *in vitro* (Xu *et al.*, 1992; Wang *et al.*, 1989).

Green tea is used by the cosmetic industry for a variety of reasons. The activity of green tea as an antioxidant both within the product itself (as a preservative) and on the skin surface helps reduce sun damage to the skin with perceived benefits such as anti-aging and reduce wrinkles. Skin care products with green tea extract were also believed to increase the effectiveness of sunscreens and prevent skin irritation (http://www.kovic.com/Green Tea2.html).

Benzie and Szeto (1999) compared the *in vitro* antioxidant power of different types of tea using ferric reducing/antioxidant power (FRAP) assay. Results showed that different teas had widely different *in vitro* antioxidant power and that the antioxidant capacity was strongly correlated with the total phenolics content of the tea. Expressed as µmol/g of antioxidant power/g dried tea leaves, values ranged as 132-654µmol/g for black (fermented) teas, 233-532 µmol/g for Oolong (semifermented) teas, and 272-1144 µmol/g for green (nonfermented) teas. One cup of tea of usual strength (1-2 per cent), therefore, can provide the same potential for improving antioxidant status as around 150mg of pure ascorbic acid (vitamin C).

Yildirim *et al.* (2000) reported that although the reducing power of a substance may be an indicator of its potential antioxidant activity, there may not always be a linear correlation between these two activities. They reported that young shoot extract of *Camellia sinensis* was the most effective antioxidant. Flower extract also showed antioxidant activity.

Cai *et al.* (2004, 2006) evaluated the antioxidant activity and phenolic compounds in traditional Chinese medicinal plants associated with anticancer. The improved ABTS.+ method was used to systematically assess the total antioxidant capacity (Trolox equivalent antioxidant capacity, TEAC) of the medicinal extracts. The TEAC values and total phenolic content for methanolic extract of young leaves of *Camellia sinensis* were 5268.6 µmol Trolox equivalent/100g dry weight (DW), and 17.4 g of gallic acid equivalent/100g DW respectively. A positive significant linear relationship between antioxidant activity and total phenolic content showed that phenolic compounds were the dominant antioxidant components in the tested herbs. Major types of phenolic compounds in *Camellia sinensis* include flavanols (catechin, epicatechin), flavonols (kaempferol and quercetin glycosides), condensed tannins.

Katsube *et al.* (2004) reported high levels of antioxidant activity in *Camellia sinensis*.

The antioxidant potential of high-grown black tea was measured by Abeywickrama *et al.* (2005) both *in vitro* (radical scavenging activity using DPPH and TBARS assays) and *in vivo* (using DPPH assays in rat serum, following oral administration for three months). The study also estimated total polyphenols, caffeine, theaflavin, thearubigin and total and individual levels of major catechins. The results showed that black tea possesses mild but dose-dependent antioxidant activity *in vitro*. The *in vivo* activity was both dose- and time-dependent. The antioxidant activity of serum was elevated only as the tea was administered to the rat.

Extracts of *Camellia sinensis* leaf showed remarkable free radical scavenging activity with IC_{50} value of 8.5µg/ml and significant inhibitory effects on lipid peroxidation with 96.2 per cent inhibition at the concentration of 50ng/ml (Thoung *et al.*, 2006).

Green tea extract was tested by Morteza-Semnani *et al.* (2007) for antioxidative activity in comparison with antioxidants (sodium metabisulfate and BHT) at 0.1, 0.5, 1.0 and 2.0 per cent w/w hydroquinone cream. The green tea extract at 1.0 and 2.0 per cent can be used as a double action (both water and oil soluble) antioxidant having 77 and 81 per cent (at 25°C) and 58 and 62 per cent (at 45°C) hydroquinone remaining, respectively, after 3 months.

Wojcikowski *et al.* (2007) evaluated the antioxidant activity of aerial parts of *Camellia sinensis*. Oxygen radical absorbance capacity of the ethyl extract, methanol extract and aqueous extracts were 3.96, 577.49 and 45.68μmol Trolox equivalent per g of dried starting material, respectively.

Khalaf *et al.* (2008) also reported antioxidant activity of methanolic crude extracts of green tea and black tea by DPPH assay. Antioxidant activities of different factions of polysaccharide conjugates from green tea were studied by Chen Haixia *et al.* (2008).

Chen *et al.* (2009) studied the antioxidant activity of leaves extract towards xanthine oxidase, lipoxygenase and tyrosinase activities. The inhibitory effect of acetone extract on xanthine oxidase activity was more than 50 per cent, tyrosine activity was more than 70 per cent and lipoxygenase activity was more than 60 per cent.

Tea extract, decreases cellular tissue damage thus prevent ageing. It also supports immune system. The green tea extract Epigallocatechin-3-gallate (EGCG) is thought to have anticancerous properties and reduced liver cancer. EGCG and polyphenols are the main antioxidative constituents of green tea (Mishra *et al.*, 2007).

Raihan *et al.* (2009) investigated the effect of aqueous extracts of black and green teas in arsenic-induced toxicity in rabbits. Arsenic causes oxidative stress in the body. Its administration (3mg/kg/day) for 14 days in rabbits resulted in a significant reduction of whole blood glutathione (GSH), and elevation of thiobarbituric acid reactive substances (TBARS) and the index nitrite/nitrate (NOx) levels. These are the markers of oxidative stress. Both black tea (BT) and green tea (GT), when administered to the arsenic-treated rabbits for 14 days, caused a significant elevation of depleted GSH level to 53.12 per cent and 57.47 per cent respectively. On the contrary in the placebo group the level was 26.59 per cent. The BT and GT reduced the elevated TBARS level to 43.27 per cent and 62.28 per cent respectively, whereas the corresponding level in the placebo groups was 21.24 per cent. The NOx levels were also reduced to 63.6 per cent, 67.67 per cent and 58.94 per cent in BT, GT and placebo groups, respectively. When arsenic and black tea were given concurrently to another group the results were even more pronounced. The polyphenol components of black and green tea were 27.69 per cent and 29.71 per cent of the dry weight of the total extracts, respectively. These results indicated that arsenic-induced toxicities in rabbits were significantly lowered by the black and green tea polyphenols. The greater activity of green tea than that of black tea correlates with the slightly higher content of polyphenols in green tea.

A study was designed by Celik and Tuluce (2007) to investigate the protective effect of *Camellia sinensis*, beverage used against chemical carcinogen trichloroacetic acid (TCA)-exposure in rats. The prevention potential of the plant infusion was evaluated by measuring the level of antioxidant defense systems, reduced glutathione (GSH), glutathione reductase (GR), superoxide dismutase (SOD), glutathione-S-transferase (GST) and catalase (CAT); and lipid peroxidation (malondialdehyde, MDA) content in rats. The study revealed that the constituents present in *Camellia sinensis* impart protection against carcinogenic chemical induced oxidative injury that may result in the development of cancer.

The antioxidant effects of Myanmar tea were determined for 3-minute and 30-minutes infusion (Htay *et al.*, 2007). The latter infusion is Myanmar's way of drinking tea and has 3 times higher antioxidant activity than 3-minute infusion. Antioxidant activity of green tea is higher than black tea and pickled tea in both 3 min and 30 min infusions.

The effect of green and black tea on early atherosclerosis in hamsters was investigated by Rouanet *et al.* (2010). They received an atherogenic diet at the same time a tea at a daily dose corresponding to the consumption of 275 ml by a 70 kg human. After 12 weeks teas inhibited aortic lipid deposition by 79-96 per cent and triggered reduced activity of hepatic antioxidant enzymes, not accompanied by a lowered plasma cholesterol. These findings suggest that moderate consumption of green and black teas can help prevent the development of early atherosclerosis.

Divya *et al.* (2009) reported anti iron toxicity and antioxidant effect of *Camellia sinensis*. Povichit *et al.* (2010) investigated phenolic content and free radical scavenging effect of leaf of *Camellia sinensis* by DPPH and ABTS assays, anti-lipid peroxidation activity by TBARS and for antiglycation activity. The results revealed that the total phenolic content showed good correlation with free radical scavenging by ABTS and anti-lipid peroxidation by TBARS, but showed no correlation with antiglycation. The extract of *Camellia sinensis* demonstrated a significant antioxidant effect and also showed a promising antiglycation effect. The IC_{50} were 0.10mg/ml for the DPPH method, TEAC value of 2.83 mg/Trolox/mg sample for the ABTS method, IC_{50} of 1.11 mg/ml for the TBARS method and IC_{50} of 0.04 for the antiglycation method.

Campsis grandiflora K.Schum. Family: Bignoniaceae

Exposure of human dermal fibroblasts to 50 per cent EtOH extract of *Campsis grandiflora* flower at 10 and 100µg/ml showed significant protective effect against hydrogen peroxide (300µM) (Cui *et al.*, 2006). The extract not only protected cell survival from H_2O_2-induced leakage of lactate dehydrogenase enzyme release and DNA ragmentation significantly. The extract showed scavenging activities of radicals and reactive oxygen species with IC_{50} values of 20µg/ml against DPPH radical and 52µg/ml against superoxide radicals in the xanthine/xanthine oxidase system, respectively (Cui *et al.*, 2006).

Gan *et al.* (2010) evaluated the antioxidant activities of 40 medicinal plants associated with prevention and treatment of cardiovascular and cerebrovascular diseases using FRAP and TEAC assays and their total phenolic contents were measured by the Folin-Ciocalteu methods. The antioxidant activity of *Campsis grandiflora* was 96.68μmol Fe(II)/g in FRAP assay, 101.59μmol Trolox/g and Total phenolic content was 8.93 mg GAE/g. A high correlation between antioxidant capacity and total phenolic content indicated that phenolic compounds could be the main contributor of the antioxidant activity of these plants.

Campsis radicans (L.) Seem Family: Bignoniaceae

Cai *et al.* (2004) evaluated the antioxidant activity and phenolic compounds in traditional Chinese medicinal plants associated with anticancer. The improved ABTS.+method was used to systematically assess the total antioxidant capacity (Trolox equivalent antioxidant capacity, TEAC) of the medicinal extracts. The TEAC values and total phenolic content for methanolic extract of *Campsis radicans* were 446.2 μmol Trolox equivalent/100g dry weight (DW), and 2.17 g of gallic acid equivalent/100g DW respectively. A positive significant linear relationship between antioxidant activity and total phenolic content showed that phenolic compounds were the dominant antioxidant components in the tested herbs. Major types of phenolic compounds in *Campsis radicans* include flavonoids (quercetin, apigenin, cyanidin-3-rutoside) and phenolic acids (ferulic acid).

Camptotheca acuminata Decne. Family: Nyssaceae

Chen *et al.* (2009) studied the antioxidant activity of leaves extract towards xanthine oxidase, lipoxygenase and tyrosinase activities. The inhibitory effect of acetone extract on xanthine oxidase activity was about 20 per cent, tyrosine activity was less than 20 per cent and lipoxygenase activity was about 50 per cent.

Cananga odorata (Lam.) Hook. f. et Thoms. Family: Family: Annonaceae

Leelapornpisid *et al.* (2007) reported antioxidant activity of the absolutes (obtained by solvent extraction) *Cananga odorata*.

Cannabis sativa L. Family: Cannabinaceae

Methanol aqueous extract of *Cannabis sativa* was screened by Kim *et al.* (1997) for anti-oxidative activity using Fenton's reagent/ethyl lineolate system. The extract showed antioxidant activity.

Cansjera rheedii J.Gmelin Family: Opiliaceae

The hepatoprotective activity of ethanol (95 per cent) extract of (250 mg kgG^{-1}) *Cansjera rheedii* whole plant was evaluated against paracetamol induced hepatotoxicity by evaluating biochemical parameters such as Serum Glutamate Pyruvate Transaminase (SGPT), Serum Glutamate Oxaloacetate Transaminase (SGOT), Alkaline Phosphatase (ALP), Total bilirubin, Total Protein and Gamma Glutamate Transpeptidase (GGTP). A 10 per cent of liver homogenate was used for estimation of enzyme such as Superoxide Dismutase (SOD), Glutathione S-Transferase (GST), Lipid Peroxidase (LPO) and Glutathione Peroxidase (GPx) for antioxidant study. Treatment of rats with ethanol extract significantly altered serum marker enzymes and antioxidants level near to normal against paracetamol intoxicated rats (Mounnissamy *et al.*, 2008).

Canthium parviflorum Lam. Family: Rubiaceae

Ethanol extract of *Canthium parviflorum* leaves was analyzed by Sathish Kumar *et al.* (2008) for their total antioxidant capacity, reducing power, metal chelating, ABTS+ radical scavenging and hydroxyl radical scavenging activities. The extract at 500µg/ml showed maximum scavenging activity (51.6 per cent) of ABTS radical cation followed by the iron chelation (45.12 per cent) at the same concentration. However, the extract showed only moderate hydroxyl radical scavenging activity (6.42 per cent). Total antioxidant capacity was found to be 12.9mg ascorbic acid equivalents at 500µg/ml extract concentration. There was a positive correlation between the total phenolic content and antioxidant capacity. The results suggest that the phenolics and flavonoids in the leaves provide considerable antioxidant activity.

Canthium setosum Hiern. Family: Rubiaceae

Lagnika *et al.* (2011) evaluated the antioxidant activity of *Canthium setosum* using DPPH radical scavenging assay. The extract showed moderate radical scavenging activity with IC_{50} value of 3.47µg/ml while the standard L-Ascorbic acid had IC_{50} value of 1.1µg/ml.

Capparis atamisguea Kuntze Family: Capparaceae

Borneo *et al.* (2009) analyzed the total phenols content (Folin-Ciocalteau assay) and antioxidant capacity (ferric reducing/antioxidant power – FRAP) of 41 plants from Cordoba (Argentina). *Capparis atamisguea* exhibited the lowest value of total phenols (8.2 mg GAE/g) and antioxidant acapacity (85.2 µmol of Fe(II)/g). A significant linear correlation was found between phenols content and antioxidant capacity.

Capparis cartilaginea Decaisne Family: Capparaceae

Marwah *et al.* (2007) investigated the antioxidant activity of wound-healing herbs using *in vitro* DPPH and phosphomolybdenum assay methods. Out of the 19 plants screened, the aqueous ethanol extract of *Capparis cartilaginea* showed moderate DPPH scavenging activities with per cent IP value 41 per cent, after 15 min of incubation at an effective test concentration of 50µg/ml. The Total Antioxidant Capacity as gallic acid equivalent of 168mg/g of ethanol extract was obtained in the phosphomolybdenum assay. Total phenolic content was 38.3 mg GAE/g of ethanol extract.

Capparis ovata Desf. var. _canescens_ Family: Capparaceae

English Name: Caper

Essential oils of Caper buds and leaves exhibited DPPH by over 70 per cent at the 500µg/ml level. The dichloromethane and methanol extracts from caper buds and leaves exhibited higher antioxidant activities than those of their essential oils in both testing systems (El-Ghorab *et al.*, 2007).

Capparis spinosa L. Family: Capparaceae

English Name: Caper

Souri *et al.* (2004) evaluated the antioxidative activity of *Capparis spinosa* by linoleic acid peroxidation test using 1,3-diethyl-2-thiobarbituric acid as the reagent.

Methanolic extract showed 64.58 per cent inhibition at 40µg concentration with IC_{50} value of 5.84µg.

The antioxidant activity of a caper hydrophilic extract of *Capparis spinosa* was assessed in a number of assays. The extract at 3.5 and 7.0 µM GAE exhibited a dose-dependent peroxyl radical scavenging activity in a methyl lineolate methanol solution oxidized by azo initiator, and reduced hypervalent iron myoglobin species formed from met-Mb an H_2O_2, at 180µM GAE. The hydrophilic extract, at 70-280 µM GAE, caused a dose-dependent inhibition of lipid autooxidation in heated red meat, incubated with simulated gastric fluid for 180 min (Tosoriere *et al.*, 2007).

The evaluation of the antioxidant activities of essential oils from *Capparis spinosa* from Dalmatia (Croatia) by different methods showed that using TBARS method, the oils presented at 1 g/l (the highest concentration tested) lower ablity to inhibit lipid peroxidation than BHA, but close to that of BHT (Kulisic-Bilusic *et al.*, 2010). They also found that the activity was dose-dependent. The essential oil was predominantly constituted by methyl isothiocyanate.

Capsella bursa-pastoris (L.) Medic. Family: Brassicaceae

Gan *et al.* (2010) evaluated the antioxidant activity of *Capsella bursa-pastoris* using the FRAP and TEAC assays, and its total phenolic content was measured by the Folin-Ciocalteu method. For FRAP assay the value was 69.99 µmol Fe(II)/g dry weight. For TEAC assay, the value was 41.37µmol Trolox/g dry weight. Total phenolic content was 4.35 mg gallic acid equivalent/g dry weight of plant material.

Grosso *et al.* (2011) investigated the phytochemical composition and antiradical activity of *Capsella bursa-pastoris*. The vegetal material was rich in kaempferol-3-*O*-rutinoside (mean value 2247.09 mg/kg of dry plant), quinic acid (95628.00 mg/kg of dry plant), arginine (mean value of 1.18 mg/kg of dry plant), palmitic acid (284.48 mg/kg) and β-sitosterol (28 per cent). The extracts presented a concentration-dependent antiradical activity (against DPPH$^{\bullet}$, $O_2^{\bullet -}$ and LOO$^{\bullet}$), being most effective against $^{\bullet}$NO (EC^{25} 0.20 µg/ml).

Capsicum annuum L. Family: Solanaceae

English Name: Red pepper, Green pepper, Sweet pepper, Chillies; Japanese: Akapiiman, Shishitou.

A new capsaicin derivative 6″,7″-dihydro-5′,5″-di-capsaicin isolated from the fruit of *Capsicum annuum* showed almost the same antioxidant activity as capsaicin, but did not have a pungent taste (Ochi *et al.*, 2003).

Maeda *et al.* (2006) examined the phenolic contents of flesh and whole fruit of red pepper and its radical scavenging activities for DPPH radical. The polyphenol content of fruit flesh was 100.5 mg-gallic acid equivalent/100 g and DPPH scavenging activity was 846.7 µmol-Trolox eq./100 g. In the whole fruit the polyphenol content was 101.5 mg-gallic acid equivalent/100 g and DPPH scavenging activity was 525.7 µmol-Trolox eq./100 g).

Antioxidant activity of aroma extracts isolated from *Capsicum annuum* were evaluated by Jang *et al.* (2008). The aroma extracts isolated from green pepper inhibited the oxidation of hexanal by 51 per cent for 1 month at a level of 500μg/ml; these extracts inhibited malonaldehyde formation from cod liver oil by 49 per cent, at a level of 500μg/ml. Its antioxidant activities were comparable to that of the natural antioxidant α-tocopherol.

Pepper has long been recognized as an excellent source of antioxidants, being rich in ascorbic acid and other phytochemicals. Chuah *et al.* (2008) conducted a study to investigate the effect of different cooking methods on the antioxidant properties of coloured peppers. It is concluded that microwave heating and stir-frying without using water are more suitable cooking methods for pepper, to ensure the maximum retention of antioxidant molecules.

Iorizzi *et al.* (2001) investigated the antioxidant activity of some compounds isolated from *Capsicum* sp. Investigation of polar extracts from hot ripe fruits of *Capsicum annuum* L. var. *acuminatum*, yielded three new glycosides named capsoside A, B, capsianoside VII and seven known glycosides as quercetin-3-O-rhamnoside, phenolic glucosides and the lignanglucoside Icariside E5 (1). Icariside E5 shares a vanillyl moiety which is essentially for the biological activity detected in capsaicin and related compounds. This finding prompted Ioriizzi *et al.* (2001) to evaluate the activity of 1 on VR1 on induction of apoptosis by a vanilloid receptor independent pathway and for its antioxidant activity. Icariside E5 did not show capsaicin like activities but significantly prevents the apoptosis induced by oxidative stress mediated by serum withdrawal in Jurkat cells, indicating a potential antioxidant role in cultured cells. These results suggest that icariside E5 may have antioxidant properties that strengthen the importance of peppers in the mediterranean diet (Iorizzi *et al.*, 2001; De Marino *et al.*, 2008).

Serrano *et al.* (2010) reported that antioxidant and nutritive constituents during sweet pepper development and ripening are enhanced by nitrophenolate treatments. Glucose, fructose, ascorbic acid, citric acid, total antioxidant activity and total phenolics increased during pepper development, and their levels were significantly enhanced by nitrophenolate applications. According to them it is advisable to consume peppers at the full red stage in order to achieve the maximum health-beneficial effects by consumers.

Capsicum annuum L. var. _grossam_ (Willd.) Sendt. Family: Solanaceae

Antioxidant potential of selected vegetables commonly used in diet in Asian subcontinent, including *Capsicum annuum* L. var. *grossam* fruits has been evaluated by Gacche *et al.* (2010). The DPPH radical scavenging activity was 25.0 per cent, inhibition of lipid peroxidation was 33.6 per cent, Vitamin C content was 20.48 mg/100g and total phenols was 4.20 mg/g.

Capsicum frutescens L. Family: Solanaceae

English Name: Pepper.

Odukoya *et al.* (2005) evaluated the antioxidant activity of Nigerian dietary spices including *Capsicum frutescens*. The antioxidant activity (expressed as per cent inhibition of oxidation) was 52.96 per cent, reducing power was 10.99 per cent and Total phenolic content was 89.36 mg/100g Tannic acid Equivalent. Antioxidant activity correlated significantly and positively with total phenolics while there was no linear correlation between total antioxidant activity and reducing power neither between reducing power and total phenolic content.

Maeda *et al.* (2006) examined the phenolic contents of leaf of *Capsicum frutescens* and its radical scavenging activities for DPPH radical. The polyphenol content was 21.7 mg-gallic acid equivalent/100 g and DPPH scavenging activity was 98.9 µmol-Trolox eq./100 g).

Caragana tibetica Kom. Family: Fabaceae

70 per cent acetone extract of the stems of *Caragana tibetica* was found to have a potent superoxide anion scavenging activity (Xiang *et al.*, 2005). Tibeticanol, a new piceatannol dimer possessing antioxidant activity, was isolated along with eleven known aromatic compounds. Enzyme oxidation of monomeric stilbene, piceatannol, with horseradish peroxidase and hydrogen peroxide yielded crassigarol E and G as major products. Most of the isolated compounds exhibited superoxide anion scavenging activity (Xiang *et al.*, 2005).

Caralluma adscendens (Roxb.) Haw. Family: Asclepiadaceae

Antioxidant and hypolipidemic effects of *Caralluma adscendens* were evaluated by Tatiya *et al.* (2010) using DPPH *in vitro* lipid peroxidation and reductive ability methods. The treatment showed significant decrease in total cholesterol, LDL, triglyceride and TC/HDL and an increase in HDL in the treated diabetic animal group. *In vitro* antioxidant activity showed that the butanolic extract exhibited potent free radical scavenging effects (Tatiya *et al.*, 2010).

Caralluma edulis (Edgew.) Hook. f. Family: Asclepiadaceae

Ansari *et al.* (2005) evaluated the antioxidant activity of five vegetables traditionally consumed by South-Asian migrants in Bradford, Yorkshire, UK in the DPPH scavenging assay. And *in vitro* non-enzymatic inhibition of bovine brain lipid peroxidation. In both antioxidant assays a strong activity was shown by extracts derived from *Caralluma edulis*.

Caralluma flava N.E.Br. Family: Asclepiadaceae

Marwah *et al.* (2007) investigated the antioxidant activity of wound-healing herbs using *in vitro* DPPH and phosphomolybdenum assay methods. Out of the 19 plants screened, the aqueous ethanol extract of *Caralluma flava* showed the best DPPH scavenging activities with per cent IP value 31.5 per cent, after 15 min of incubation at an effective test concentration of 50µg/ml. Although *Caralluma flava* showed less antioxidant activity in the DPPH assay, very strong antioxidant activity had been demonstrated in the phosphomolybdenum assay. The Total Antioxidant Capacity as gallic acid equivalent of 335mg/g of ethanol extract was obtained in the phosphomolybdenum assay. Total phenolic content was 23.6 mg GAE/g of ethanol extract.

Caralluma quadrangula (Forssk.) N.Br. Family: Asclepiadaceae

Marwah *et al.* (2007) investigated the antioxidant activity of wound-healing herbs using *in vitro* DPPH and phosphomolybdenum assay methods. Out of the 19 plants screened, the aqueous ethanol extract of *Caralluma quadrangula* showed very low DPPH scavenging activities with per cent IP value 14.5 per cent, after 15 min of incubation at an effective test concentration of 50μg/ml. Somewhat surprising is that *C. quadrangula*, a member of succulent asclepiads, lacks antioxidant activity in the DPPH assay. However, *C.quadrangula* demonstrated very strong antioxidant activity in the phosphomolybdenum assay. The Total Antioxidant capacity as gallic acid equivalent of 899mg/g of ethanol extract was obtained in the phosphomolybdenum assay. Total phenolic content was 18.7 mg GAE/g of ethanol extract.

Cardiospermum halicacabum L. Family: Sapindaceae

Kumaran and Karunakaran (2006b) reported that ethanol extract of *Cardiospermum halicacabum* has better antioxidant activity than that of aqueous extract. Sheeba and Asha (2006) reported that ethanol extract exhibited potent *in vitro* hydroxyl radical scavenging and inhibition of lipid peroxidation activities.

Huang *et al.* (2011) investigated the antioxidant properties of *Cardiospermum halicacabum* and its reference compounds *ex vivo* and *in vivo*. Ethanolic extract had better antioxidant activity than that of aqueous extract.

Carduus nutans L. Family: Asteraceae (Compositae)

The antioxidant activity of the water and methanolic extracts obtained from the aerial parts of *Carduus nutans* as well as hexane, chloroform, ethyl acetate, n-butanol and remaining water fractions of the later extract have been investigated by Deliorman Orhan *et al.* (2004). For the antioxidant activity assessment, effects of the extracts and fractions on superoxide anion radical generated by HPX-XOD system and the stable free radical DPPH were studied. Methanol extract and chloroform and ethyl acetate fractions showed a potent free radical scavenging activity, while ethyl acetate and butanol fractions showed significant SOD like activity.

Careya arborea Roxb. Family: Barringtoniaceae

Sambathkumar *et al.* (2006) evaluated the antioxidant activity of methanol extract of *Careya arborea* (MECA) stem bark in various *in-vitro* systems. Antioxidant and free radical scavenging activities of MECA was assessed by using DPPH, superoxide anion radical, nitric oxide radical and hydroxyl radical scavenging assays. The antioxidant activity of MECA increased in a concentration dependent manner. In DPPH radical scavenging assay the IC_{50} value of the extract was found to be 132.25 μg/ml. MECA was found to inhibit the nitric oxide radicals generated from sodium nitroprusside. The IC_{50} value was found to be 72.54 μg/ml, whereas the IC_{50} value of curcumin was 20.4 μg/ml. Moreover, the MECA was found to scavenge the superoxide generated by phenazine methosulphate (PMS)/nicotinamide adenine dinucleotide (NADH)-nitroblue tetrazolium (NBT) system. MECA was also found to inhibit the hydroxyl radical generated by Fenton's reaction, where the IC_{50} value of MECA was found to be more than 1000 μg/ml and for catechin the IC_{50} value was found to be 5 μg/ml.

The methanol and aqueous extracts of *Careya arborea* bark possessed strong antioxidant activity against many oxidants in the *in vitro* antioxidant activity (Senthilkumar *et al.*, 2007, 2008).

Careya sphaerica Roxb. Family: Barringtoniaceae

Maisuthisakul *et al.* (2008) reported that extracts of *Careya sphaerica* possessed strong antioxidant activity, high total phenolic and flavonoid contents. They are of the opinion that plants with high antioxidant activities also have high total phenolic and flavonoid content.

Carica papaya L. Family: Caricaceae

English Name: Papaya; Japanese: Papaya.

Maeda *et al.* (2006) examined the phenolic contents of fruit of papaya and its radical scavenging activities for DPPH radical. The polyphenol content was 19.5 mg-gallic acid equivalent/100 g) and DPPH scavenging activity was 63.4 µmol-Trolox eq./100 g).

Phytochemical flavonols, carotenoids and the antioxidant capacity of fruits of *Carica papaya* was evaluated by Lako *et al.* (2006). *Carica papaya* fruits have moderate Total antioxidant capacity (100 mg/100g TEAC) and are rich in Total polyphenol content (TPP) (26mg/100g GAE), quercetin (90mg/100g) and β-carotene (13mg/100g). Mehdipour *et al.* (2006) reported that antioxidant potentials of Iranian *Carica papaya* juice *in vitro* and *in vivo* are comparable to alpha-tocopherol.

Alfredo (2007) reported high antioxidant activity in papaya fruits. Ayoola *et al.* (2008) carried phytochemical screening and antioxidant activities of the ethanolic extract of leaves of *Carica papaya*. The crude extract was eleven times less potent (than vitamin C) with a maximum inhibition of 83.8 per cent at 5mg/ml. They are of the opinion that the antioxidant activity of this plant may contribute to their claimed antimalarial activity.

Faller and Fialho (2010) evaluated the polyphenol content and antioxidant capacity in organic and conventional plant foods. The results showed that organic fruits tend to have higher hydrolysable polyphenol contents than conventional ones, with values being 72.6 per cent in papaya peels, higher for hydrolysable polyphenols.

Thirty-eight types of fruits commonly consumed in Singapore were systematically analyzed by Isabelle *et al.* (2010) for their hydrophilic oxygen radical absorbance capacity (H-ORAC), total phenolic content (TPC), ascorbic acid (AA) and various lipophilic antioxidants. Papaya had the highest β-cryptoxanthin, lycopene and β-carotene per gram fresh weight.

Carissa carandas L. Family: Apocynaceae

Hegde and Joshi (2009) reported that ethanolic extract of root of *Carissa carandas* possessed strong antioxidant activity.

Sulaiman *et al.* (2010) tested the antioxidant activities of hexane, chloroform and methanol extracts and fractions of *Carissa carandas* unripe fruits using FTC and TBA methods. The chloroform and methanol extracts of unripe fruits showed better activities than BHT. All fractions also showed strong antioxidant activity.

Carissa opaca Stapf ex Haines Family: Apocynaceae

The chloroform and aqueous fractions of *Carissa opaca* fruit, a traditional medicinal fruit in Pakistan, possessed a high amount of total phenolic and flavonoid contents as compared to other solvent fractions with potent antioxidant activities in scavenging DPPH, superoxides, hydroxyl, hydrogen peroxide, ABTS radicals, and had strong iron chelating activity (Sahreen *et al.*, 2010). On the other hand, the ethyl acetate fraction showed the highest inhibition of β-carotene linoleic acid peroxidation and phosphomolybdate assay. A high correlation coefficient existed between EC_{50} values of DPPH, superoxides, hydroxyl, hydrogen peroxide, ABTS radicals, total phenolics and flavonoids, but a non significant correlation was found in the case of iron chelates, β-carotene and phosphomolybdate assay. This study verified that the chloroform and aqueous fractions have strong antioxidant activities which were correlated with its high level of phenolics and flavonoids.

Investigations of Khan *et al.* (2010) revealed that *Carissa opaca* leaves had significant antioxidant potential (inhibition of lipid peroxidation more than 50 per cent) and DPPH-scavenging potential (scavenging more than 80 per cent). Moreover, total phenolic content was also higher in the extracts: Acetone extract = 940 and n-Butanol extract = 600 mg of gallic acid equivalents (GAE). Application of *Carissa opaca* extracts as natural antioxidants in peanut and sunflower oils significantly reduced lipid peroxidation (inhibition more than 60 per cent). A strong correlation of total phenols was observed with DPPH and total antioxidant capacity assays.

***Carlina acanthifolia* All. Family: Asteraceae (Compositae)**

Dordevic *et al.* (2007) studied the antioxidant activities of *Carlina acanthifolia* root essential oil using TBA assay and DPPH and OH radical scavenging assay. The essential oil exhibited substantial, dose-dependent antioxidant activity with Carlina oxide as the main antioxidant component.

***Carpinus laxiflora* (Siebold et Zucc.) Blume Family: Carpinaceae**

Cytoprotective effect on gamma-ray radiation induced oxidative stress from forty one Korean plant extracts was screened by Jung *et al.* (2006). *Carpinus laxiflora* was found to scavenge DPPH radical and intracellular reactive oxygen species. As a result, extract of this plant reduced cell death of Chinese hamster lung fibroblast cells induced by H_2O_2 treatment. In addition, these extracts scavenged ROS generated by radiation.

***Carthamus oxyacantha* Bieb. Family: Asteraceae (Compositae)**

Souri *et al.* (2004) evaluated the antioxidative activity of *Carthamus oxyacantha* by linoleic acid peroxidation test using 1,3-diethyl-2-thiobarbituric acid as the reagent. Methanolic extract showed 71.27 per cent inhibition at 40µg concentration with IC_{50} value of 9.11µg.

***Carthamus tinctorius* L. Family: Asteraceae (Compositae)**

English Name: Seaflower.

A total of eight flavonoids luteolin (1), quercetin (2), luteolin-7-O-beta-D-glucopyranoside (3), luteolin-7-O-(6″-O-acetyl)-beta-D-glucopyranoside (4), quercetin 7-O-β-D-glucopyranoside (5), quercetin-7-O-(6″-O-acetyl)-β-D-glucopyranoside (6), acacetin 7-O-β-D-glucuronide (7) and apigenin-6-C-β-D-glucopyranosyl-8-C-β-D-glucopyranoside (8) were isolated from the leaves of *Carthamus tinctorius* (Lee *et al.*,

2002). The antioxidative activity of these flavonoids was evaluated against 2-deoxyribose degradation and rat liver microsomal lipid peroxidation induced by hydroxyl radicals generated via a Fenton-type reaction. Among these flavonoids compound 1,2, 3, 4, 5 and 6 showed potent antoxidative activities. Compound 7 and 8 were relatively less active (Lee *et al.*, 2002).

Cai *et al.* (2004, 2006) evaluated the antioxidant activity and phenolic compounds in traditional Chinese medicinal plants associated with anticancer. The improved ABTS.+ method was used to systematically assess the total antioxidant capacity (Trolox equivalent antioxidant capacity, TEAC) of the medicinal extracts. The TEAC values and total phenolic content for methanolic extract of inflorescence of *Carthamus tinctorius* were 318.0 µmol Trolox equivalent/100g dry weight (DW), and 2.12 g of gallic acid equivalent/100g DW respectively. A positive significant linear relationship between antioxidant activity and total phenolic content showed that phenolic compounds were the dominant antioxidant components in the tested herbs. Major types of phenolic compounds in *Carthamus tinctorius* include flavonoids (chalcones, carthamin, carthamone), lignans.

Kruawan and Kangsadalampai (2006) investigated the antioxidant activity and phenolic compound contents of water extract of *Carthamus tinctorius*. The extract exhibited strong scavenging activity against DPPH radicals (96.65 per cent) and their FRAP value was 1140.5 µmol/g. The content of phenolic compound in the extracts was determined using the Folin-Ciocalteu reagent and calculated as gallic acid equivalent (GAE). The herbal extract gave a high phenolic compound content (GAE 2139.98 mg/g).

Gan *et al.* (2010) evaluated the antioxidant activities of 40 medicinal plants associated with prevention and treatment of cardiovascular and cerebrovascular diseases using FRAP and TEAC assays and their total phenolic contents were measured by the Folin-Ciocalteu methods. The antioxidant activity of *Carthamus tinctorius* was 26.05µmol Fe(II)/g in FRAP assay, 97.60µmol Trolox/g and Total phenolic content was 7.26 mg GAE/g. A high correlation between antioxidant capacity and total phenolic content indicated that phenolic compounds could be the main contributor of the antioxidant activity of these plants.

Carum carvi L. Family: Apiaceae (Umbelliferae)

English Name: Caraway.

The antioxidative activity of phenolic extracts from seeds of caraway was examined by autoxidation of methyl linoleate by Kahkonen *et al.* (1999). The total phenolics were very low, 6.5mg/g GAE, but antioxidant activity was very high, 64 per cent at 500 ppm level and 46 per cent at 5000 ppm level.

Hydrodistillated extracts from laurel were assessed for their total phenol content, and antioxidant (iron(III) reduction, inhibition of linoleic acid peroxidation, iron (II) chelation, DPPH radical-scavenging and inhibition of hydroxyl radical-mediated 2-deoxy-D-ribose degradation, site and nonsite-specific) activities by Hinneburg *et al.* (2005). The antioxidant activity was moderate.

Mariutti *et al.* (2008) evaluated the antioxidant activity of caraway employing ABTS and DPPH radical assay. DPPH EC_{50} value was 89g/kg while DPPH-TEAC value was 2.8mM/g. ABTS-TEAC value was 3.9mM/g.

Samojlik *et al.* (2010) assayed the antioxidant and hepatoprotective potential of essential oil of *Carum carvi* against CCl_4 damage. Tested essential oils were able to reduce the stable DPPH in a dose dependent manner and to neutralize H_2O_2, reaching 50 per cent neutralization with IC_{50} values of <2.5µl/ml. Caraway essential oil strongly inhibited LP in both systems of induction.

Carum nigrum Family: Apiaceae (Umbelliferae)

Singh *et al.* (2006) evaluated the chemical composition and antioxidant activities of *Carum nigrum* seed essential oil, oleoresin and their selected components.The results showed that both the essential oil and oleoresin were able to reduce the oxidation rate of the mustard oil in the accelerated condition at 60°C in comparison with synthetic antioxidants such as butylated hydroxyanisole and butylated hydroxytoluene at 0.02 per cent. In addition, individual antioxidant assays such as linoleic acid assay, DPPH scavenging activity, reducing power, hydroxyl radical scavenging, and chelating effects have been used.

***Caryopteris incana* (Thunb. ex Houtt.) Miq. Family: Verbenaceae**

Four phenyl propanoids, isolated from *Caryopteris incana* exhibited potential radical scavenging activity against DPPH, hydroxyl and superoxide anion radicals (Gao *et al.*, 1999). Three new phenylethanoid glycosides, incanoside C, incanoside D and incanoside E were isolated together with one known glycoside from the whole plant of *Caryopteris incana* by Gao *et al.* (2000). The three new phenylethanoid glycosides exhibited radical scavenging activities against DPPH radical and inhibitory activities against the oxidation of linoleic acid.

***Caryota mitis* Lour. Family: Arecaceae (Palmae)**

Fu *et al.* (2010) evaluated the antioxidant capacities and total phenolic contents of 56 wild fruits from South China. *Caryota mitis* showed FRAP value of 174µmol Fe(II)/g.

***Casearia grayi* Jessup Family: Flacourtiaceae**

Mosaddik *et al.* (2004) screened some Australian Flacourtiaceae species for *in vitro* antioxidant activity.The total antioxidant activity has been assessed based on scavenging activity of stable ABTS free radicals. Mature stem extract of *Casearia grayi* was found to have most antioxidant activity (IC_{50} = 2.9 microg/ml).

***Casearia sylvestris* Sw. Family: Flacourtiaceae**

Menezes *et al.* (2004) investigated the antioxidant activity of hydroalcoholic extracts of *Rauvolfia sellowii* by DPPH assay and phosphomolybdenum method. The relative activity in phosphomolybdenum complex was 0.65 while scavenging activity against DPPH in IC_{50} (µg/ml) was 30.27.

Casimiroa edulis L. Lave et Lex. Family: Rutaceae

Awaad *et al.* (2006) evaluated the antioxidant activity of the extracts from the leaves of *Casimiroa edulis*. Among all the tested fractions and compounds, the ethanol extract exhibited the strongest antioxidant activity followed by EtOAc extract, giving 842 and 712 µM Trolox equivalents/g dry weight, respectively. Moderate activities were detected for the other tested materials.

Cassia alata L. Family: Caesalpiniaceae

Phansawan and Poungbangpho (2007) reported antioxidant capacities of *Cassia alata* by ABTS method. Ethanol was the best solvent for extraction to give the highest antioxidant capacity, followed by acetone, methanol, distilled water and acetic acid.

Akinmoladun *et al.* (2010) assessed the phytochemical constituents and antioxidant and free radical scavenging activities of *Cassia alata* using seven different antioxidant assay methods. Amount of total flavonoids found in the leaf extracts was 275.16 µg/ml quercetin equivalent. The results suggest that the methanolic extracts possess significant antioxidant and radical scavenging activity that may be due to the phytochemical content of the plant.

Cassia auriculata L. Family: Fabaceae (Leguminosae)

English Name: Tanner's cassia.

Pari and Latha reported antidiabetic and antioxidant activity of *Cassia auriculata* flowers. Oral administration of aqueous extract resulted in decreased free radical formation in tssues studied. The decrease in lipid peroxides and increase in reduced glutathione (GSH), superoxide dismutase (SOD), catalase (CAT), glutathione peroxidase (GPx) and glutathione S-transferase (GST) clearly show the antioxidant properties of flower extracts of *Cassia auricalata* (Pari and Latha, 2002).

The ethanol and methanol extracts of flowers were screened for antioxidant activity. The antioxidant activity was determined by an improved assay based on the decolorization of the radical moncation of ABTS and DPPH radical scavenging method. The ethanol and methanol extracts of flowers showed antioxidant activity in both assays (Kumaran and Karunakaran, 2007).

Suresh Kumar *et al.* (2008) carried out a study to determine the antioxidant activity of *Cassia auriculata*. The total alkaloid content was 22.3mg/g while flavonoid content was 25.15 mg/g. The inhibition of lipid peroxidation was 89.6 per cent which was slightly lesser than Butylated hydroxyl toluene (BHT) (93.6 per cent).

Jain (2009) investigated the antioxidant activity of ethyl extract of *Cassia auriculata* with the help of two *in vitro* antioxidant models – Nitric-oxide scavenging activity and DPPH method. IC_{50} value was calculated and compared with standard Ascorbic acid. Ethyl acetate extract was found to be extremely effective in scavenging nitric oxide (IC_{50} 51.3 µg/ml). In DPPH radical scavenging activity (IC_{50} 96.6 µg/ml) Ethyl acetate extract showed different levels of antioxidant activities in tested models.

Cassia fistula L. Family: Caesalpiniaceae

Munasinghe *et al.* (2001) evaluated the antiradical and antilipoperoxidative effects of *Cassia fistula* extract used by Sri Lankan traditional medicinal practitioners for cardioprotection. DPPH radical scavenging activity was much lower EC_{50} = 59μg/ ml.

Anti-inflammatory and antioxidant activities of the aqueous (CFA) and methanolic extracts of *Cassia fistula* bark were assayed in wistar albino rats by Ilavarasan *et al.* (2005). The extracts showed significant radical scavenging by inhibiting lipid peroxidation initiated by CCl_4 and $FeSO_4$ in rat liver and kidney homogenates. Both extracts exhibited significant antioxidant activity in DPPH, nitric oxide and hydroxyl radical induced *in vitro* assay methods. Both extracts showed dose-dependent protective effect against lipid peroxidation and free radical generation in liver and kidney homogenates.

Aqueous extract of *Cassia fistula* flowers (ACF) was screened by Manonmani *et al.* (2005) for its antioxidant effect in alloxan induced diabetic rats. An appreciable decrease in peroxidation products *viz.* thiobarbituric acid reactive substances, conjugated dienes, hydroperoxides was observed in heart tissues of ACF treated diabetic rats. The decreased activities of key antioxidant enzymes such as superoxide dismutase, catalase, glutathione peroxidase, glutathione reductase and glutathione in diabetic rats were brought back to near normal range upon ACF treatment.

Bahorun *et al.* (2005) investigated the phytochemical constituents and reviewed the antioxidant activity of *Cassia fistula*. FRAP, TEAC, antioxidant activities based on thiocyanate method, oil and water emulsion systems, lipid peroxidation, superoxide scavenging, DPPH radical scavenging assays have been used to assess the free radical scavenging capacities and the reducing potentials of the antioxidant constituents of *Cassia fistula* extracts (Chaminda *et al.*, 2001; Siddhraju *et al.*, 2002; Luximon-Ramma *et al.*, 2002). TEAC and FRAP assays show that *Cassia fistula* vegetative organs, the bark had the highest antioxidant potential followed by the old leaves, the young leaves and twigs. The reproductive organs had the highest antioxidant properties compared to the vegetative parts. In the reproductive organs, the order of efficacy was pods>flower buds> flowers. The antioxidant capacities of the reproductive organs parallel their contents of high levels of catechins, oligomeric and polymeric proanthocyanidins (Luximon-Ramma *et al.*, 2002). Further studies based on lipid peroxidation, DPPH radical and O_2 scavenging (Siddhuraju *et al.*, 2002) also underline the antioxidant of *Cassia fistula* extracts. Here too the antioxidant capacities were directly related to the total phenolic contents of the extracts where the presence of established antioxidants such as xanthones, flavans, flavonolss and di-anthraquinones are potentially responsible for the activities (Yen *et al.*, 2000). In a study highlighting the antiradical and antiperoxidative effects of some plant extracts used by Sri Lankan traditional medicinal practitioners for cardiac-protection, *Cassia fistula* root extracts were showing to exhibit potent DPPH radical scavenging and lipid peroxidation actions, and to afford a 30 per cent protection against deoxyribose damage (Chaminda *et al.*, 2001).

Total phenol, antioxidant and free radical scavenging activities of some medicinal plants, including *Cassia fistula* fruits were studied by Prakash *et al.* (2007). Fruits of *C. fistula* with high phenols (107.8 mg/g) showed poor reducing power (5.9 ASE/ml) and ARP (4.7).

Suresh Kumar *et al.* (2008) carried out a study to determine the antioxidant activity of *Cassia fistula*. The total alkaloid content was 24.6mg/g while flavonoid content was 36.2 mg/g. The inhibition of lipid peroxidation was 89.3 per cent which was higher than Butylated hydroxy toluene (BHT) (93.6 per cent).

Cassia javanica L. Family: Caesalpiniaceae

Kaur and Arora (2011) evaluated the antioxidant potential of hexane, chloroform, ethyl acetate, methanol, and water extract of bark and leaves of *Cassia javanica* by superoxide anion radical scavenging assay. The different extracts showed significant inhibition of superoxide radicals in a dose-dependent manner. The water extract of bark of *C. javanica* showed strong antioxidant potential of 55 per cent at 1000 µg/ml concentration. The methanol leaf extract of *C. javanica* showed strong antioxidant potential of 50.4 per cent at 300 µg/ml concentration.

Cassia occidentalis L. Family: Caesalpiniaceae

Kouakou-Siransy *et al.* (2010) evaluated the antioxidant properties of *Cassia occidentalis* leaves through a scavenger effect on reactive oxygen species (ROS), such as hydrogen peroxide and hypochlorous acid. The plant showed dose-dependent antioxidant activity. Rates of total phenolic, flavonoid and proanthocyaninidin contents were evaluated. The highest rates were to be found in the most active extracts, indicating that antioxidant activity could be influenced by these phytochemical groups.

Mehta *et al.* (2010) determined *in vitro* antioxidant activity of *Cassia occidentalis* by DPPH free radical scavenging, FRAP, lipid peroxidation by TBARS assay methods. The analysis had shown maximum percentage inhibition in case of DPPH method as 66.53 per cent at 160µg/ml and 61.07 per cent in lipid peroxidation at 1000µg/ml. Further, reducing power of methanol extract of seeds increased in concentration dependent manner. Total phenolic content estimation done by using Folin-Ciocalteu reagent and found to be 0.75 per cent w/w.

Cassia siamea Lam. Family: Caesalpiniaceae

Chanwitheesuk *et al.* (2005) investigated the antioxidant activity of *Cassia siamea* using a β-carotene bleaching method. The contents of plant chemicals such as vitamin C, Vitamin E, carotenoids, tannin and total phenolics, were also determined. Methanolic extract showed antioxidant activity. *Cassia siamea* has antioxidant index of 1.92, Vitamin C 48.4 mg per cent, Vitamin E 0.0204mg per cent, Total carotenes 1.92 mg per cent, Total xanthophylls 1.59mg per cent, Tannins 110mg per cent and Total phenolics of 384mg per cent.

Kaur *et al.* (2006) evaluated the antioxidant activity of *Cassia siamea* flowers. The extract was found to contain a large amount of polyphenols and also exhibited an immense reducing ability. At a concentration of 250 µg/ml, 96 per cent of DPPH radicals and at 500 µg/ml, 42.7, 32.7 and 64.4 per cent of O_2-, H_2O_2 and NO respectively could be scavenged by *C. siamea* flower extract. The extract also inhibited OH radical induced oxidation of protein (BSA) and LPO in murine hepatic microsomes. The determination of metal chelating capacity of the extract indicated chelating of metal ions (Fe2+) to be a putative mechanism implicated in the inhibition of OH radical induced BSA oxidation and LPO. *C. siamea* flower extract also exhibited a significant antioxidant activity in acute oxidative tissue injury animal model constituted by CCl_4 induced hepatotoxicity. Oral administration of the extract at a dose of 50-150 mg/kg of body weight significantly protected from CCl_4 induced elevation in AST and ALT in serum, elevation in hepatic LPO, depletion of hepatic GSH and decrease in the activities of antioxidant enzymes: SOD, CAT and GPX.

Kaur and Arora (2011) evaluated the antioxidant activity of ethyl acetate, methanol, and water extract of bark and leaves of *Cassia siamea* by superoxide anion radical scavenging assay. The different extracts showed significant inhibition of superoxide radicals in a dose-dependent manner. Among all the bark extracts of *C. siamea,* the methanol extract showed the maximum inhibition of 60.5 per cent at 800 µg/ml concentration and water extract also showed strong antioxidant potential of 51.3 per cent at 1000 µg/ml concentration. The various leaf extracts of *C. siamea* showed moderate antioxidant potential of 25–50 per cent at 1000 µg/ml.

Cassia sophera L. Family: Caesalpiniaceae

The free radical scavenging activity (FRSA) of the methanol extract and various fractions of methanol extract of *Cassia sophera* leaves was determined by Rahman *et al.* (2008) by using 1,1-diphenyl-2-picrylhydrazyl (DPPH) assay. The FRSA of ethyl acetate fraction was superior to all other fractions (IC_{50} = 15.42 g/ml), which was higher than synthetic antioxidant butylated hydroxyanisole, BHA (18.25 µg/ml). The amount of total phenolic compounds was further determined. The total phenolic content in EtOAc fraction (13.25 per cent) was the highest as compared to other extracts.

Cassia surattensis Burm. Family: Caesalpinaceae

Liu *et al.* (2008) evaluated the antioxidant activity of water, 50 per cent ethanol and 70 per cent acetone extracts of flowers of 15 plants of Hainan. All the extracts of *Cassia surattensis* showed good antioxidant activity in eliminating DPPH, O_2 and SOH radicals.

Cassia tora L. Family: Caesalpiniaceae

Cai *et al.* (2004) evaluated the antioxidant activity and phenolic compounds in traditional Chinese medicinal plants associated with anticancer. The improved ABTS.+ method was used to systematically assess the total antioxidant capacity (Trolox equivalent antioxidant capacity, TEAC) of the medicinal extracts. The TEAC values and total phenolic content for methanolic extract of seed of *Cassia tora* were 290.7 µmol Trolox equivalent/100g dry weight (DW), and 1.20 g of gallic acid equivalent/100g DW respectively. A positive significant linear relationship between antioxidant activity and total phenolic content showed that phenolic compounds were the dominant antioxidant components in the tested herbs. Major types of phenolic compounds in *Cassia tora* include anthraquinones (chrysophanol, emodin, aloe-emodin, rhein).

The antioxidant activity of three different fractions of *Cassia tora* was evaluated by Zhenbao *et al.* (2007) using different antioxidant tests, including inhibition of peroxidation of linoleic acid, reducing power, free-radical and ferrous ions chelating activity. In the above four assays, all the fractions showed antioxidant potential to varying degrees. Among these fractions, ethyl acetate fraction exhibited more antioxidant potency than other fractions. Furthermore, ethyl acetate fraction was found to be more effective in protecting LDL against oxidation in a concentration-dependent manner.

Patnibul *et al.* (2008) screened 18 vegetables for antioxidant activity using silica gel Thin Layer chromatography followed by spraying with DPPH. *Cassia tora* exhibited strong antioxidant activity.

Lobo *et al.* (2010) investigated the antoxidant and antiradical activities of *Cassia tora*. The study was designed to evaluate and compare the antioxidant activity, total phenolics, flavonoids content of aqueous (AETB) and ethanolic (EETB) extracts of fruits. The antioxidant activity was assessed by DPPH, ABTS, nitric oxide, superoxide and hydroxyl radical scavenging assay, FRAP, reducing power and TAC. Two different aqueous (AECT) and ethanolic (EECT) extracts were used for this study. Results showed that AECT possessed stronger antioxidant and antiradical activities than EECT in nitric oxide, DPPH, ABTS radical scavenging assay with IC_{50} values 769.23, 1059.78, 490.08 in AETC and 1000.51, 1799.03 and 943.23µg/ml in EETC respectively. Like antioxidant activity the reducing power increases in a dose dependent manner showing higher absorbance at 700 nm for AECT, *i.e.*, 0.18 as compared to EECT, *i.e.*, 0.14 at 500µg/ml. The FRAP values were found for AECT 0.85, EECT 0.77 and TAC values were found for AETC 0.38 and EECT 0.89. The amount of total phenolic and flavonoid content were also determined.

Arya and Yadav (2010) investigated the antioxidant potency of sequential organic (petroleum ether, benzene, chloroform, methanol) and aqueous extracts of *Cassia tora* by using nitric oxide scavenging activity, β-carotene linoleic acid model system, hydroxyl radical scavenging activity, reducing power, metal chelating activity and superoxide radical scavenging activity. The methanolic extract of the leaves of *Cassia tora* was found to be more effective against free radicals followed by aqueous and other organic extracts respectively. A positive and significant correlation was observed between various test systems and total phenolics showed that the phenolic compounds were responsible for the antioxidant activity of the extracts.

Castanopsis cuspidata (Thunb.) Schott. Family: Fagaceae

Cytoprotective effect on gamma-ray radiation induced oxidative stress from forty one Korean plant extracts was screened by Jung *et al.* (2006). *Castanopsis cuspidata* was found to scavenge DPPH radical and intracellular reactive oxygen species. As a result, extract of this plant reduced cell death of Chinese hamster lung fibroblast cells induced by H_2O_2 treatment. In addition, these extracts scavenged ROS generated by radiation.

Casuarina equisetifolia L. Family: Casuarinaceae

Total phenol, antioxidant and free radical scavenging activities of some medicinal plants, including *Casuarina equisetifolia* bark were studied by Prakash *et al.* (2007). Bark of *Casuarina equisetifolia* had comparatively lower phenols (72.1 mg/g) but exhibited good ARP (16.2) and reducing power (0.7ASE/ml).

Catalpa bignonioides Walt. Family: Bignoniaceae

Antioxidant activity of methanolic extracts from inflorescence rachises, corollas, calyses, leaves, valves of capsules and hypertrophied placenta by DPPH and tyrosine nitration inhibition induced by peroxynitrite was tested by Dvorska *et al.* (2007). The highest total content of phenols, flavones and flavonoles was determined in the leaves. These also show the highest antioxidant activity.

Catharanthus roseus (L.) G.Don Family: Apocynaceae

Aqueous extract of *Catharanthus roseus* (=*Vinca rosea*) flowers were administered orally to alloxan induced diabetic rats once a day for 4 weeks (Kaleem *et al.*, 2005). These treatments led to significant lowering of blood sugar level and reduction in serum lipids. The levels of antioxidant enzymes, catalase and glutathione peroxidase decreased in alloxan induced diabetic rats, however these levels returned to normal in insulin, *P. nigrum* treated rats. There was no significant difference in superoxide dismutase activity compared to controls. Lipid peroxidation level was significantly higher in diabetic rats and it was slightly increased in insulin and *Catharanthus roseus* treated rats as compared to control rat. The results of Kaleem *et al.* (2005) suggest that the treatment with *Catharanthus roseus* is useful in controlling not only the glucose and lipid levels but also in strengthening the antioxidant potential.

Pareira *et al.* (2010) investigated phenolics and organic acids in *Catharanthus roseus*. Neither phenolics nor flavonoids could be identified but organic acid profile composed of oxalic, cis-aconitic, citric, pyruvic, malic and fumaric acids was found. Roots proved to be much richer in these compounds than were other plant parts. Antioxidant activity of the roots was evaluated by Pareira *et al.* (2010) using representative radicals: DPPH, superoxide radical and nitric oxide radical. The results indicate a strong anti-radical activity surpassing that of leaves, petals, stems and seeds. Synergistic studies revealed the contribution of organic acids to the displayed effect, although other compounds, not determined herein, might be responsible for part of the activity (Pareira *et al.*, 2010).

Biswas *et al.* (2010) analyzed antioxidant activity of the aqueous extract of leaves of *Catharanthus roseus* in different systems of assay. Aqueous extract showed DPPH radical scavenging activity with IC_{50} = 241.13µg/ml; Superoxide radical scavenging activity with IC_{50} = 104.04µg/ml; Hydroxyl radical scavenging IC_{50} = 6826.38µg.ml; Lipid peroxidation IC_{50} = 1592.18µg/ml.

Cecropia pachystachya Trec. Family: Cecropiaceae

The antioxidant property of *Cecropia polystachya* was studied using free radical-generating systems (Velazquez *et al.*, 2003). The methanol extract protected against enzymatic and non-enzymatic lipid peroxidation in microsomal membranes of rat.The extract showed the highest scavenging activity on the superoxide and DPPH radicals.

Celastrus hindsii Benth. Family: Celastraceae

Antioxidative compounds were isolated by Ly *et al.* (2006) from the methanol extract of dried leaves of *Celastrus hindsii*. Eight phenolic compounds, rutin, kaempferol 3-rutinoside, rosmarinic acid, lithospermic acid and lithospermic acid B and three novel oligomers of rosmarinic acid, a dimer and two trimers were isolated. These phenolic compounds were shown to have antioxidative active against autoxidation of methyl linoleate in bulk phase and the radical-initiated peroxidation of soybean phosphatidylcholine in liposomes. In the liposomal peroxidation, the number of phenolic hydroxyl group in each molecule was correlated with the effectiveness of antioxidative activity.

Celastrus orbiculatus Lam. Family: Celastraceae

A new flavan-3-ol glycoside, (-)-epicatechin-5-O- β-D-glucosyl-3-benzoate (1), and two known compounds, (-)-epicatechin and (-)-epiafzelechin, were isolated from EtOAc extract of *Celastrus orbiculatus* aerial parts that exhibited significant antioxidant effect in DPPH free radical assay (Hwang *et al.*, 2001).

Celastrus paniculatus Willd. Family: Celastraceae

The effects of aqueous extracts of *Celastrus paniculatus* seeds were shown to have antioxidant properties in rats. Godkar *et al.* (2003) investigated the free radical scavenging capacity of three aqueous extracts obtained from *Celastrus paniculatus* seeds. All the water soluble extracts exhibited a dose-dependent free radical scavenging capacity for DPPH and also for superoxide-generated assays (*in vitro* assays). All the water soluble extracts significantly attenuated H_2O_2-induced neuronal death, and acid extract was the most effective in protecting the neuronal cells against oxidative injury caused by H_2O_2. The data presented by Godkar *et al.* (2003) suggest that *Celastrus paniculatus* seed water soluble extract protected neuronal cells in part by their free radical scavenging properties, by reducing lipid peroxidation, and also by their ability to induce the antioxidant enzyme catalase.

Ramadan *et al.* (2009) determined the lipid composition and characteristics of *Celastrus paniculatus* seed oil (Malkanguni oil). The seeds are a rich source of oil (40 per cent). The amounts of neutral lipids in the oil were the highest, followed by glycolipids and phospholipids. Oleic acid followed by palmitic and linoleic were the major fatty acids in seed oil and its lipid classes. The seed oil being characterized by a relatively high amount of phytosterols, where in the sterol marker was β-sitosterol followed by campesterol and stigmasterol. γ-Tocopherol was the major tocopherol isomer while the rest being α-tocopherol. When *C. paniculatus* oil and extra virgin olive oil were compared upon their radical scavenging activity (RSA) toward the stable 1,1-diphenyl-2-picrylhydrazyl (DPPH) radical *C. paniculatus* oil exhibited stronger RSA.

Zohera *et al.* (2010) evaluated the antioxidant potential of methyl, ethyl acetate, pet ether and water extracts of *Celastrus paniculatus* seed using DPPH free radical assay, reducing power assessment, nitric oxide scavenging assay and cupric ion reducing capacity assay. The extracts showed moderate antioxidant activity in a dose dependent manner. These extracts were found to contain phenolics and flavonoid compounds. In DPPH radical scavenging assay, ethyl acetate extract had the lowest IC_{50} value (585.58µg/ml) compared to ascorbic acid. In nitric oxide scavenging assay IC_{50} value was found to be 122.99 µg/ml, 320.54µg/ml, 601.81µg/ml and 206.37µg/ml respectively for the water, methanol, ethyl acetate and pet ether extracts. Compared to 6.83μg/ml which was the IC_{50} value for the reference ascorbic acid. The extracts also showed good reducing power.

Celosia argentea L. Family: Amaranthaceae

Li *et al.* (2008) evaluated the antioxidant properties of 45 medicinal plants including *Celosia argentea* using FRAP and TEAC assays and the total phenolic contents of these plants were measured. *Celosia argentea* showed very low antioxidant activity with 1.44 μmol Fe(II)/g FRAP value, 1.45μmol Trolox/g of TEAC value and 1.73 mg GAE/g of Phenolic content.

Celosia argentea L. var. cristata L. (Syn: Celosia cristata L.) Family: Amaranthaceae

Odukoya *et al.* (2007) evaluated the antioxidant activity and total phenolic contents of Nigerian green leafy vegetables including *Celosia argentea* var. *cristata*. Potential free radical scavenging activity of these vegetables was confirmed by spraying spots of the extracts with DPPH (yellow color on purple background). Antioxidant activity was assayed in linoleic acid model system. Total polyphenols as Tannic Acid Equivalent (TAE) and Ascorbic Acid. Antioxidant activity in *Amaranthus hybridus* was very high 68.41 per cent (inhibition of lipid peroxidation). There was low correlation between AA/TAE, AA/ASC and TAE/ASC.

Gan *et al.* (2010) evaluated the antioxidant activities of 40 medicinal plants associated with prevention and treatment of cardiovascular and cerebrovascular diseases using FRAP and TEAC assays and their total phenolic contents were measured by the Folin-Ciocalteu methods. The antioxidant activity of *Celosia cristata* was 35.13μmol Fe(II)/g in FRAP assay, 34.83μmol Trolox/g and Total phenolic content was 2.97 mg GAE/g. A high correlation between antioxidant capacity and total phenolic content indicated that phenolic compounds could be the main contributor of the antioxidant activity of these plants.

Centaurea americana Nutt. Family: Asteraceae (Compositae)

The antioxidant properties of extracts and Americanin, isolated from the seeds of *Centaurea americana* were screened for antioxidant properties (Shoeb *et al.*, 2007).

Centaurea bruguieriana (DC.) Hand-Mzt. subsp. belangerana (DC.) Bornm. Family: Asteraceae (Compositae)

Souri *et al.* (2004) evaluated the antioxidative activity of *Centaurea bruguieriana* subsp. *belangerana* by linoleic acid peroxidation test using 1,3-diethyl-2-thiobarbituric acid as the reagent. Methanolic extract showed 91.00 per cent inhibition at 40μg concentration with IC_{50} value of 1.74μg which is approximately in the range of a-tocopherol (IC_{50} = 0.60μg).

Centaurea calçitrapa L. Family: Asteraceae (Compositae)

A total of 27 extracts from non-cultivated and weedy vegetables traditionally consumed by ethnic Albanians in southern Italy were tested for their free radical scavenging activity (FRSA) in the DPPH screening assay, for their *in vitro* non-enzymatic inhibition of bovine brain lipid peroxidation and for their inhibition of xanthine oxidase. In both antioxidant assays strong activity was shown for *Centaurea calcitrapa* young whorls. The activity was comparable to quercetin and *Rhodiola rosea* extract (Pieroni *et al.*, 2002).

Centaurea depressa M.Bieb. Family: Asteraceae (Compositae)

Extracts of *Centaurea depressa* were investigated by Hosseinimehr *et al.* (2007) for their total flavonoids, phenol contents and their radical scavenging activity using DPPH assays. Quercetin and butylated hydroxytoluene were used as standard reference with well documented antioxidant activity. Methanolic extract sample showed free radical scavenging activity. A correlation between radical scavenging capacitiy of the extract with total phenolic compounds content was observed.

Centaurea ensiformis Hub-Mor. Family: Asteraceae (Compositae)

The n-hexane, chloroform, ethyl acetate and ethyl alcohol extracts of aerial parts of *Centaurea ensiformis* have been assessed for antioxidant activities by Ugur *et al.* (2009). The antioxidant activities were determined by using four complementary *in vitro* assays: inhibition of DPPH, total antioxidant activity, and the total amount of phenolic and flavonoid compounds. The total antioxidant activity increased with the increasing amount of the extracts (20, 40, 80 and 160 µg) which contained linoleic acid emulsion. The major compounds of the hexane extract of the plant were caryophellene oxide (28.72 per cent), spathulenol (17.81 per cent), eudesmol (13.03 per cent) and β-bourbonene (8.51 per cent).

Centaurea mucronifera DC. Family: Asteraceae (Compositae)

Tepe *et al.* (2006) screened the antioxidative properties of the methanolic extracts of *Centaurea mucronifera* by two complementary test systems, namely DPPH free radical-scavenging and β-carotene/linoleic acid. The extract showed good antioxidant activity.

Centaurium erythraea Rabin. Family: Gentianaceae

English Name: Small centaury.

Valentão *et al.* (2003) evaluated the hydroxyl and hypochlorous acid scavenging activity of *Centaurium erythraea* and compared with that of green tea (*Camellia sinensis*). The obtained results demonstrate that small centaury infusion exhibits interesting antioxidant properties, expressed both by its capacity to effectively scavenge hydroxyl radical and hypochlorous acid, although with a lower activity against the second than that observed for green tea.

Centaurium spicatum (L.) Janch. Family: Gentianaceae

Shahat *et al.* (2003) isolated three known acetylated flavonol glycosides and a new pentacetylated flavonoid glycoside from *Centaurium spicatum*. The anticomplement and antioxidant activities of these compounds were evaluated. The triacetylated flavonoid glycoside showed the highest activity in the two assays.

Centella asiatica (L.) Urban. Family: Apiaceae (Umbelliferae)

Chanwitheesuk *et al.* (2005) investigated the antioxidant activity of leaves of *Centella asiatica* using a β-carotene bleaching method. The contents of plant chemicals such as vitamin C, Vitamin E, carotenoids, tannin and total phenolics, were also determined. Methanolic extract showed antioxidant activity. *Centella asiatica* has antioxidant index of 7.10, Vitamin C 3.28 mg per cent, Vitamin E 0.0031mg per cent, Total carotenes 12.8 mg per cent, Total xanthophylls 10.6mg per cent, Tannins 24.3mg per cent and Total phenolics of 98.5mg per cent.

In order to find their potential sources, the crude extracts of different parts of some selected food and medicinal plants, including *Centella asiatica*, were studied by Bajpai *et al.* (2005) for total phenolic contents (TPC) and antioxidant activity (AOA). AOA was assayed by auto-oxidation of β-carotene and linoleic acid and expressed as the per cent of inhibition relative to control. TPC in aerial parts of *Centella asiatica* was 23.1 mg/g (GAE) while antioxidant activity was 71.2 per cent.

Wong *et al.* (2006) investigated the antioxidant properties of 25 edible tropical plants including *Centella asiatica* using DPPH and FRAP assays. Total phenolic content was also estimated. *Centella asiatica* showed good antioxidant activity.

Wojcikowski *et al.* (2007) evaluated the antioxidant activity of aerial parts of *Centella asiatica.* Oxygen radical absorbance capacity of the methanol extract and aqueous extracts were 496.84 and 202.94µmol Trolox equivalent per g of dried starting material, respectively.

Subhasree *et al.* (2009) carried out the antioxidant activity of leaf of *Centella asiatica* by measuring the ability of methanol extracts to scavenge radicals generated by *in vitro* systems and by their ability to inhibit lipid peroxidation. The levels of non-enzymatic antioxidants were also determined by standard spectrophotometric methods. Correlation and regression analysis established a positive correlation between some of these antioxidants and the *in vitro* free radical-scavenging activity of the plant extracts. The conclusions drawn from this study indicate that *in vivo* studies, isolation and analysis of individual bioactive compounds will reveal the crucial role that this plant may play in several therapeutic formulations.

Jain and Agrawal (2009) reported antioxidant activity of methanolic extract of *Centella asiatica.*

Huda-Faujan *et al.* (2007, 2009) reported antioxidative activities of aqueous and methanolic extracts of *Centella asiatica* and four other plants. The analysis carried out was total phenolic content, ferric reducing power, ferric thiocyanate (FTC) and TBA tests. The phenolic content was 3.72 mg (aqueous extract) and 7.79 mg (methanolic extract) TAE/100 g fresh weight and antioxidant activity of 53.1 per cent (aqueous extract) and 66.17 per cent (methanolic extract) using FTC method. Increasing the concentration of the extracts resulted in increased ferric reducing antioxidant power. Total phenolic contet has positive correlation with antioxidant capacity.

Centipeda minima (L.) A.Br. et Aeschers. Family: Asteraceae (Compositae)

Cai *et al.* (2004) evaluated the antioxidant activity and phenolic compounds in traditional Chinese medicinal plants associated with anticancer. The improved ABTS.+ method was used to systematically assess the total antioxidant capacity (Trolox equivalent antioxidant capacity, TEAC) of the medicinal extracts. The TEAC values and total phenolic content for methanolic extract of aerial parts of *Centipeda minima* were 72.8 μmol Trolox equivalent/100g dry weight (DW), and 0.29 g of gallic acid equivalent/100g DW respectively. A positive significant linear relationship between antioxidant activity and total phenolic content showed that phenolic compounds were the dominant antioxidant components in the tested herbs. Major types of phenolic compounds in *Centipeda minima* include phenolic terpenes (triptophenolide, triptophenolide methyl ether).

Antioxidant capacities of 56 selected Chinese medicinal plants, including *Centipeda minima* were evaluated by Song *et al.* (2010) using the TEAC and FRAP assays, and their total phenolic content was measured by the Folin-Ciocaleteu method. The TEAC value was 16.87μmol Trolox/g, FRAP value was 13.31μmol Fe^{2+}/g and phenolic content was 2.34mg GAE/g.

Cephalanoplos segetum (Bge.) Kitam. Family: Asteraceae (Compositae)

Gan *et al.* (2010) evaluated the antioxidant activities of 40 medicinal plants associated with prevention and treatment of cardiovascular and cerebrovascular diseases using FRAP and TEAC assays and their total phenolic contents were measured by the Folin-Ciocalteu methods. The antioxidant activity of *Cephalanoplos segetum* was 31.55μmol Fe(II)/g in FRAP assay, 32.52μmol Trolox/g and Total phenolic content was 1.78 mg GAE/g. A high correlation between antioxidant capacity and total phenolic content indicated that phenolic compounds could be the main contributor of the antioxidant activity of these plants.

***Cerasus mahaleb* (L.) Miller var. *mahaleb* Family: Rosaceae**

Serteser *et al.* (2009) studied the antioxidant activity of 50 per cent aqueous methanol extracts of *Cerasus mahaleb* var. *mahaleb* including free radical scavenging, hydrogen peroxide scavenging and metal (Fe^{2+}) chelating activities.The DPPH radical scavenging effects of fruit extract was 0.608. Fe^{2+} chelating activity (per cent) of fruit extracts showed 35.49 per cent inhibition. Hydrogen peroxide inhibition activity of fruit extracts was 31.53 per cent.

***Ceratonia siliqua* L. Family: Fabaceae (Leguminosae)**

English Name: Carob.

Kumazawa *et al.* (2002) evaluated the antioxidant activity of polyphenols from pods of *Ceratonia siliqua.* The total polyphenol content in CPP determined by Folin-Ciocalteu method was 19.2 per cent. The condensed tannin content determined by vanillin and proanthocyanidin assay systems was 4.37 per cent and 1.36 per cent, respectively. β-carotene bleaching, DPPH free radical scavenging, inhibition of lipid peroxidation by the erythrocyte ghost, and microsomal assay systems were used to evaluate the antioxidant activity. CPP showed a stronger inhibitory effect against the discoloration of β-carotene than other polyphenol compounds such as catechins and procyanidins. CPP had weaker antioxidant activity in the DPPH free radical scavenging, the erythrocyte ghost, and microsomal systems than authentic polyphenol compounds at the same concentrations. The activity adjusted by the polyphenol concentration was, however, comparable to that of authentic polyphenol compounds.

El Hajaji *et al.* (2011) evaluated the *in vitro* antioxidant property and phytochemical constituents of the crude ethyl acetate and methanol extracts of the three genders of carob tree barks. The highest antioxidant activity and the higher amounts of total phenols were shown in methanol crude bark extract for the three genders. Variety significantly affected the phenol content and the antioxidant activity, with the spontaneous male variety globally showed a higher polyphenol concentration and antioxidant activity than the grafted female and spontaneous female.

Ceriops decandra (Griff.) Ding Hou Family: Rhizophoraceae

Vadlapudi and Naidu (2009) evaluated the antioxidant potential of *Ceriops decandra*, a mangrove plant. The SOD was 1.49U/mg, Catalase was 0.63U/mg, Ascorbic acid was 150mg/100g, DPPH radical inhibition was 26 per cent and FRAP units were 430.

Chaenomeles speciosa (Sweet.) Nakai Family: Rosaceae

Gan *et al.* (2010) evaluated the antioxidant activity of *Chaenomeles speciosa* using the FRAP and TEAC assays, and its total phenolic content was measured by the Folin-Ciocalteu method. For FRAP assay the value was 195.15 µmol Fe(II)/g dry weight. For TEAC assay, the value was 107.61µmol Trolox/g dry weight. Total phenolic content was 13.58 mg gallic acid equivalent/g dry weight of plant material.

Chaerophyllum hirsutum L. Family: Apiaceae (Umbelliferae)

The methanol extract from aerial parts showed antioxidant activity based on the DPPH test (Dall'Acqua and Innocenti, 2004). Phytochemical investigations showed the presence of caffeoyl derivatives as well as flavonoid compounds, which might be the active principles responsible for the antioxidant activity of the methanol extract. The activities of flavonoids were comparable to that of rutin. These results were in agreement with the known importance of the catechol group for antioxidant activity.

Chaerophyllum libanoticum Boiss. et Kotschy. Family: Apiaceae (Umbelliferae)

The essential oil obtained by hydrodistillation from the crushed fruits was evaluated using an *in vitro* radical scavenging activity test. The essential oil scavenged DPPH resulting in IC_{50} >30mg/ml. In addition, the effect on inhibition of lipid peroxidation of the essential oil was assayed using β-carotene bleaching and haemoglobin induced linoleic acid peroxidation methods resulting in 16 per cent antioxidant activity (Demirci *et al.*, 2007).

Chamaenerion angustifolium (L.) Scop. Family: Onagraceae

Myagmar and Aniya (2000) evaluated the free radical scavenging action of some medicinal herbs growing in Mongolia. The aqueous extract of nine herbs including *Chamaenerion angustifolium* were used. The free radical scavenging action was determined *in vitro* and *ex vivo* by using ESR spectrometer and chemiluminiscence analyzer. The results showed that *Chamaenerion angustifolium* extract possess strong scavenging action of DPPH, superoxide and hydroxyl radicals. *Chamaenerion angustifolium* also depressed reactive oxygen production from polymorphonuclear leukocytes stimulated by phorbol-12-myristate *ex vivo*.

Chamomilla recutita (L.) Rausch. Family: Asteraceae (Compositae)

The antioxidative efficacy of the different flavones in several model lipid peroxidation systems was determined by Nissen *et al.* (1996). *In vivo* studies on human skin show that *Chamomilla recutita* extracts have a preventive effect on the UV-induced erythema. It was proposed that the flavones are effective scavengers of reactive oxygen species and good chelators of iron. Some flavones seem to have an inhibitory effect on the enzyme ornithine decarboxylase.

Changium smyrnoides Wolff Family: Apiaceae (Umbelliferae)

Antioxidant capacities of 56 selected Chinese medicinal plants, including *Changium smyrnoides* were evaluated by Song *et al.* (2010) using the TEAC and FRAP assays, and their total phenolic content was measured by the Folin-Ciocaleteu method. The TEAC value was 2.07μmol Trolox/g, FRAP value was 0.35μmol Fe^{2+}/g and phenolic content was 0.50mg GAE/g.

Chassalia chartacea Craib Family: Rubiaceae

Methanol exteracts of seven Malaysian medicinal plants, including *Chasalia chartacea* were screened by Saha *et al.* (2004) for antioxidant and nitric oxide inhibitory activities. Antioxidant activity was measured by using FTC, TBA and DPPH free radical scavenging methods and Greiss assay was used for the measurement of nitric oxide inhibition in LPS and interferon-γ-treated RAW264.7 cells. All the extracts showed strong antioxidant activity comparable to or higher than that of α-tocopherol, BHT and quercetin in FTC and TBA methods. In the DPPH radical scavenging assay the extracts were weak. In the Greiss assay *Chasalia chartacea* showed strong inhibitory activity on nitric oxide production in LPS and IFN-γ-induced RAW 264.7 cells.

Chassalia minor Ridl. Family: Rubiaceae

Ahmad *et al.* (2010) evaluated the antioxidant potential of 22 species of medicinal plants from Malaysian Rubiaceae including *Chassalia minor*. FTC, TBA, TPC and the DPPH assays were employed. The tested extracts showed strong antioxidant potential with percent inhibition of 96.2 per cent in the FTC, and 95.3 per cent in TBA assays. The TPC of the extract was 15.77mg GAE/g. A good correlation was observed between total phenolic content and radical-scavenging activities.

Chassalia pubescens Ridl. Family: Rubiaceae

Ahmad *et al.* (2010) evaluated the antioxidant potential of 22 species of medicinal plants from Malaysian Rubiaceae including *Chassalia pubescens*. FTC, TBA, TPC and the DPPH assays were employed. The tested extracts showed strong antioxidant potential with percent inhibition of 98.5 per cent in the FTC, and 93.8 per cent in TBA assays. The TPC of the extract was 12.54mg GAE/g. A good correlation was observed between total phenolic content and radical-scavenging activities.

Cheiloclinium cognatum (Miers) A.C.Sm. Family: Celastraceae

Triterpenes isolated from the root bark were investigated for their radical scavenging abilities involving DPPH assay (Jeller *et al.*, 2004).

Chelidonium majus L. Family: Papaveraceae

English Name: Greater Celandine

Antioxidant activity in the alcoholic extracts (20 and 40 per cent) of the greater celandine herb was investigated by Then *et al.* (2003) by FRAP method. Since the antioxidant activity of the extracts greatly depends on the quality of compounds, the phytochemical examination for alkaloid and element content were also examined. According to the results the antioxidant activity does not depend on the alkaloid content of the drug during the vegetative period and on the alkaloid content of the alcoholic extracts. It seems that the antioxidant activity of the extracts is also independent from the transition metal element content.

Chenopodium album L. Family: Chenopodiaceae

Antioxidant potential of *Chenopodium album* has been investigated by Singh *et al.* (2009) for their ability to scavenge DPPH, reducing power assay, nitric oxide radical, along with Hydrogen peroxide radical scavenging activity. Antioxidant activity was assessed using two different extracts of methanol and petroleum ether using ascorbic acid as a reference compound. Methanol extract in comparison with the petroleum ether had been found to be more efficient. The results obtained in this indicate that the whole parts of the herb are a potential source of natural antioxidant.

Chenopodium botrys L. Family: Chenopodiaceae

Souri *et al.* (2004) evaluated the antioxidative activity of *Chenopodium botrys* by linoleic acid peroxidation test using 1,3-diethyl-2-thiobarbituric acid as the reagent. Methanolic extract showed 51.59 per cent inhibition at 40µg concentration with IC_{50} value of 45.76µg.

Chenopodium foliosum (Moench) Asch Family: Chenopodiaceae

Serteser *et al.* (2009) studied the antioxidant activity of 50 per cent aqueous methanol extracts of *Chenopodium foliosum* including free radical scavenging, hydrogen peroxide scavenging and metal (Fe^{2+}) chelating activities.The DPPH radical scavenging effects of fruit extract was 0.686. Fe^{2+} chelating activity (per cent) of fruit extracts showed 43.76 per cent inhibition. Hydrogen peroxide inhibition activity of fruit extracts was 59.39 per cent.

Chenopodium pallidecaule Aellen Family: Chenopodiaceae

Penarrieeta *et al.* (2005) measured the Total Antioxidant Capacity (TAC) in some Andean foods. Eight Andean foods, including *Chenopodium pallidecaule* were analyzed by two methods ABTS and FRAP to assess TAC. The highest TAC value was observed in *Chenopodium pallidecaule* both water-soluble fraction and water-insoluble fraction.

Chenopodium quinoa Willd. Family: Chenopodiaceae

English Name: Quinoa.

Penarrieeta *et al.* (2005) measured the Total Antioxidant Capacity (TAC) in some Andean foods. Eight Andean foods, including *Chenopodium quinoa* were analyzed by two methods ABTS and FRAP to assess TAC. TAC value of *Arracacia xanthorhiza* was good in both water-soluble fraction and water-insoluble fraction.

Nsimba *et al.* (2008) evaluated the antioxidant activity of various extracts and fractions of *Chenopodium quinoa* employing DPPH assay and FRAP assay. The antioxidant activity was less correlated to the phenolics content suggesting that non-phenolic compounds might play major free radicals scavenging activity.

Alvarez-Jubete *et al.* (2010) examined the polyphenol composition and antioxidant properties of methanolic extracts from quinoa, and evaluated how these properties were affected following two types of processing: sprouting and baking. The total phenol content amongst the seed extracts was significantly higher in quinoa. Antioxidant capacity, measured by the DPPH scavenging capacity and the ferric ion reducing antioxidant power assays was also higher for quinoa extract. The total phenol content and antioxidant activity was generally found to increase with sprouting and decrease in levels was observed following bread making. Analysis by liquid chromatography coupled with diode array detector revealed the presence of phenolic acids, catechins, flavonol, flavone and flavonol glycosides.

Chimarrhis turbinata DC. Family: Rubiaceae

Glucoalkaloid cordifoline isolated from *Chimarrhis turbinata* showed strong free radical scavenging activity against DPPH as well as pronounced antioxidant activity evidenced by redox properties measured by EICD-HPLC (Cardoso *et al.*, 2004). In an *in vitro* rat brain assay, turbinatine showed moderate antioxidant activity (IC_{50} 1.86 μM), compared to the standard compound, galanthamine (IC_{50} 0.92 μM). An antioxidant 1-(3′4′-dihydroxycinnamoyl) cyclopentane-2,3-diol and two known trns and cis chlorogenic acid methyl esters were isolated from the ethanolic extract of the leaves of *Chimarrhis turbinata* (Cardoso *et al.*, 2006).

Chionanthus virginicus L. Family: Oleaceae

The potential antioxidant activities of two secoiridoids from root bark of *Chionanthus virginicus* were investigated to evaluate their potential value as the natural products for food or cosmetic applications. These secoiridioids, as antioxidants neutralized the activities of radicals and inhibited the peroxidation reactions of linoleic acid emulsion. Ligustroside and oleuropein showed 71.9, 82.4, 80.7 and 90.4 per cent inhibition on lipid peroxidation of linoleic acid emulsion at the concentration of 10 and 20 μg/ml. In addition, ligustroside and oleuropein had effective DPPH, ABTS and superoxide anion radical scavenging, hydrogen peroxide scavenging, total reducing power and metal chelating on ferrous ion activities (Gülçin, 2009).

Chlorophytum borivilianum Santapau et Fernandes Family: Liliaceae

Govindarajan *et al.* (2005) screened *Chlorophytum borivilianum* to determine its antioxidant activity, isolation of the sapogenins and standardization of the isolated sapogenin fraction. Potent antioxidant activity of ethanolic extract was found by their ability to scavenge DPPH (84.51 per cent), hydroxyl radical (48.95 per cent), feryl bi-pyrdyl complex (84.53 per cent) along with the inhibition of lipid peroxidation (67.17 per cent) at 100 μg/ml concentration. The ethanolic extract also exhibited significant inhibition of superoxide anion radical generated by photochemiluminiscence.

Under *in vitro* DPPH free radical scavenging assay and lipid peroxidation assay the aqueous extract of *Chlorophytum borivilianum* considerably inhibited, in a dose-dependent manner, the levels of DPPH free radicals and thiobarbituric acid reactive substances, respectively thus showing significant antioxidant property (Kenjale *et al.*, 2007).

Chlorophytum tuberosum (Roxb.) Baker Family: Liliaceae

In order to find their potential sources, the crude extracts of different parts of some selected food and medicinal plants, including *Chlorophytum tuberosum*, were studied by Bajpai *et al.* (2005) for total phenolic contents (TPC) and antioxidant activity (AOA). AOA was assayed by auto-oxidation of β-carotene and linoleic acid and expressed as the per cent of inhibition relative to control. TPC in tubers of *Chlorophytum tuberosum* was 4.6 mg/g (GAE) while antioxidant activity was 52.0 per cent.

Antioxidant potential of *Chlorophytun tuberosum* has been investigated by Narasimhan *et al.* (2006) for their ability to scavenge DPPH, nitric oxide radical along with their capacity to reduce lipid peroxidation in rat liver homogenate, chelation of ferrous ion, radical scavenging potential using chemiluminiscence and their total antioxidant capacity. The extract has been found to possess antioxidant activity in all the models tested as evident by IC_{50} values being 225.31, 888.44, 809.22 and 422.97 µg/ml for scavenging of DPPH, nitric oxide, lipid peroxidation and ferry bi-pyridyl complex, respectively, along with a integral antioxidant activity of 2.986 mmol ascorbic acid/g equivalents in photochemiluminiscence assay (Narasimhan *et al.*, 2006).

Chondrilla juncea L. Family: Asteraceae (Compositae)

A total of 27 extracts from non-cultivated and weedy vegetables traditionally consumed by ethnic Albanians in southern Italy were tested for their free radical scavenging activity (FRSA) in the DPPH screening assay, for their *in vitro* non-enzymatic inhibition of bovine brain lipid peroxidation and for their inhibition of xanthine oxidase. Extracts from *Chondrilla juncea* showed strong *in vitro* inhibition of xanthine oxidase, with an activity higher than that of a reference extract from *Ledum groenlandicum* (Pieroni *et al.*, 2002).

Chromolaena odorata (L.) King et Robins. Family: Asteraceae (Compositae)

Akinmoladun *et al.* (2010) assessed the phytochemical constituents and antioxidant and free radical scavenging activities of *Chromolaena odorata* using seven different antioxidant assay methods. Total flavonoids was 272.12 µg/ml quercetin equivalent. The extract demonstrated high lipid peroxidation inhibitory activity. The results suggest that the methanolic extracts possess significant antioxidant and radical scavenging activity that may be due to the phytochemical content of the plant.

Rao *et al.* (2010) assessed the *in vitro* antioxidant activity and total phenolic content of chloroform extract of *Chromolaena odorata* leaves. The DPPH activity of the extract (0.1-5mg/ml) was increased in a dose dependent manner, which was found in the range of 23.48-91.61 per cent as compared to ascorbic acid (33.69-94.10 per cent). The IC_{50} values of chloroform extract in DPPH radical, hydroxyl radical, nitric oxide, ABTS radical were obtained to be 0.31, 0.43, 0.28 and 1.32 mg/ml, respectively. However the IC_{50} values for the standard ascorbic acid were noted to be 0.24, 0.41,

0.23 and 1 mg/ml, respectively. Measurement of total phenolic content of the chloroform extract was achieved using Folin-Ciocalteau reagent containing 242.2 mg/g of phenolic content which was found significantly higher when compared to reference standard gallic acid.

Povichit *et al.* (2010) investigated phenolic content and free radical scavenging effect of *Chromolaena odorata* by DPPH and ABTS assays, anti-lipid peroxidation activity by TBARS and for antiglycation activity. The results revealed that the total phenolic content showed good correlation with free radical scavenging by ABTS and anti-lipid peroxidation by TBARS, but showed no correlation with antiglycation. The extract of *Chromolaena odorata* moderate antioxidant effect and also showed a promising antiglycation effect. The IC_{50} were 1.33mg/ml for the DPPH method, TEAC value of 0.29 mg/Trolox/mg sample for the ABTS method, IC_{50} of 10.61 mg/ml for the TBARS method and IC_{50} of 0.60 for the antiglycation method.

Chrozophora tinctoria (L.) Rabin. Family: Euphorbiaceae

Methanolic extract as well as isolated compounds of *Chrozophora tinctoria* have been found to have antioxidant effects, shown by the inhibition of lipid peroxidation and DPPH assay (Delazar *et al.*, 2006).

Antioxidant activity of different extracts of *Chrozophora tinctoria* were evaluated by Shahwar *et al.* (2010) through DPPH radical scavenging activity, phosphomolybdate and ferric thiocyanate methods. The DPPH radical scavenging activity of ethyl acetate extract was 32.4 per cent, total antioxidant activity was 0.331 and inhibition of lipid peroxidation was 59.5 per cent while total phenols was 353.5 mg GAE/g of crude extract.

Chrysanthellum indicum DC. ssp. *afroamericanum* Family: Asteraceae (Compositae)

Borneo *et al.* (2008) evaluated the antioxidant activity of ethanolic extract of *Chrysanthellum indicum* ssp. *afroamericanum*. The FRAP value was 264.3µmol of Fe(II)/g and DPPH radical scavenging activity was EC_{50} = 349.2.

Chrysanthemum cinerariaefolium (Trev.) Vis. Family: Asteraceae (Compositae)

Souri *et al.* (2004) evaluated the antioxidative activity of *Chrysanthemum cinerariaefolium* by linoleic acid peroxidation test using 1,3-diethyl-2-thiobarbituric acid as the reagent. Methanolic extract showed 85.20 per cent inhibition at 40µg concentration with IC_{50} value of 0.57µg which is approximately in the range of α-tocopherol (IC_{50} = 0.60µg).

Chrysanthemum coronarium M.J.Roem. Family: Asteraceae (Compositae)

English Name: Garland chrysanthemum; Japanese: Shungoku.

Maeda *et al.* (2006) examined the phenolic contents of leaf of chrysanthemum and its radical scavenging activities for DPPH radical. The polyphenol content was 67.1 mg-gallic acid equivalent/100 g) and DPPH scavenging activity was 335 µmol-Trolox eq./100 g).

Chrysanthemum frutescens L. Family: Asteraceae (Compositae)

The chloroform and methanolic extracts of 124 Egyptian plant species belonging to 56 families were investigated by Moussa *et al.* (2011) and compared their antioxidant activity by DPPH scavenging assay. Among the 124 plant species tested 18 exhibited extremely high antiradical activity (>80 per cent inhibition). The IC_{50} value of DPPH radical scavenging of *Chrysanthemum frutescens* was 27.83µg/ml, while total phenolic and flavonoid contents were 69.06 mg Tannic acid equivalent/g extract and 208.66 mg Rutin equivalent/g extract, respectively. Correlation coefficient between DPPH radical scavenging activity and the total phenolic and flavonoid contents suggest that phenolics and flavonoids in the extracts were partly responsible for the antiradical activities.

Chrysanthemum indicum L. Family: Asteraceae (Compositae)

Cai *et al.* (2004) evaluated the antioxidant activity and phenolic compounds in traditional Chinese medicinal plants associated with anticancer. The improved ABTS.+ method was used to systematically assess the total antioxidant capacity (Trolox equivalent antioxidant capacity, TEAC) of the medicinal extracts. The TEAC values and total phenolic content for methanolic extract of inflorescence of *Chrysanthemum indicum* were 303.6 µmol Trolox equivalent/100g dry weight (DW), and 2.01 g of gallic acid equivalent/100g DW respectively. A positive significant linear relationship between antioxidant activity and total phenolic content showed that phenolic compounds were the dominant antioxidant components in the tested herbs. Major types of phenolic compounds in *Chrysanthemum indicum* include phenolic terpenes (triptophenolide, triptophenolide methyl ether).

Kruawan and Kangsadalampai (2006) investigated the antioxidant activity and phenolic compound contents of water extract of *Chrysanthemum indicum*. The extract exhibited strong scavenging activity against DPPH radicals (93.33 per cent) and their FRAP value was 1626 µmol/g. The content of phenolic compound in the extracts was determined using the Folin-Ciocalteu reagent and calculated as gallic acid equivalent (GAE). The herbal extract gave a high phenolic compound content (GAE 214.34 mg/g).

Nikolova and Dzhurmanski (2009) evaluated the free radical scavenging activity of total methanol extract of *Chrysanthemum indicum*. The extract exhibited strong antioxidant activity and its IC_{50} value for DPPH radical was 87.64 µg/ml.

Antioxidant capacities of 56 selected Chinese medicinal plants, including *Chrysanthemum indicum* were evaluated by Song *et al.* (2010) using the TEAC and FRAP assays, and their total phenolic content was measured by the Folin-Ciocalteu method. The TEAC value was 51.91µmol Trolox/g, FRAP value was 72.16µmol Fe^{2+}/g and phenolic content was 11.28mg GAE/g.

Chrysanthemum morifolium Ramat. Family: Asteraceae (Compositae)

The antioxidative activity of water extracts of three herbs, including the flower of *Chrysanthemum morifolium* were investigsted by Osawa and Namiki (1981). The extracts showed marked antioxidative activity, not only in linoleic acid but also in liposome model systems, indicating that the three herbal water extracts may protect the cell from damage by peroxidation. The extracts possessed high contents of phenolic compounds and exhibited reducing power, revealing that these herbal extracts may contain reductones. The water extracts of three herbs also showed good hydrogen-donating abilities, indicating that they had effective activities as radical scavengers.

Cai *et al.* (2004) evaluated the antioxidant activity and phenolic compounds in traditional Chinese medicinal plants associated with anticancer. The improved ABTS.+ method was used to systematically assess the total antioxidant capacity (Trolox equivalent antioxidant capacity, TEAC) of the medicinal extracts. The TEAC values and total phenolic content for methanolic extract of floral bud of *Chrysanthemum morifolium* were 731.6 μmol Trolox equivalent/100g dry weight (DW), and 4.76 g of gallic acid equivalent/100g DW respectively. A positive significant linear relationship between antioxidant activity and total phenolic content showed that phenolic compounds were the dominant antioxidant components in the tested herbs. Major types of phenolic compounds in *Chrysanthemum morifolium* include flavones (acacetin, luteolin-7-glucoside).

Two new dicaffeoylquinic acids, 3,5-dicaffeoyl-epi-quinic acid (1) and 1,3-dicaffeoyl-epi-quinic acid (2) were isolated from *Chryanthemum morifolium* together with six known di-caffeoylquinic acid derivatives and three flavonoids (Kim and Lee, 2005). These compounds were assessed for antioxidant activity in the DPPH radical and superoxide anion radical scavenging systems. Two new compounds showed potent superoxide anion radical scavenging activity (IC_{50}=2.9 and 2.6μg/ml respectively) in the xanthine/xanthine oxidase system as compared to quercetin and also showed potent DPPH radical scavenging activity (IC_{50} = 5.6 and 5.8 μg/ml respectively).

Antioxidant capacities of 56 selected Chinese medicinal plants, including *Chrysanthemum morifolium* were evaluated by Song *et al.* (2010) using the TEAC and FRAP assays, and their total phenolic content was measured by the Folin-Ciocalteu method. The TEAC value was 80.04μmol Trolox/g, FRAP value was 149.24μmol Fe^{2+}/g and phenolic content was 14.79mg GAE/g.

Chrysanthemum multiflorum Family: Asteraceae (Compositae)

The antioxidant activity of *Chrysanthemum multiflorum* has been investigated by Chen *et al.* (2005) using both enzymatic and non-enzymatic *in vitro* antioxidant assays. Extracts inhibited xanthine oxidase and lipoxygenase activities and were scavengers of the ABTS radical cation using the Trolox equivalent antioxidant capacity assay.

Chrysophyllum cainito L. Family: Sapotaceae

Eng. Name: Star apple.

Luo *et al.* (2002) reported polyphenolic antioxidants from the fruits of Star apple.

Chrysophyllum inornatum Family: Sapotaceae

Santos *et al.* (2009) evaluated the antioxidant activity of extracts from nine plant species belonging to the Brazilian flora acting on the free radicals DPPH and TEMPOL (4-hydroxy-2,2,6,6-tetramethyl-1-piperidinyloxy-1-oxyl) by electron paramagnetic resonance (EPR), and acting on the hydroxyl radical by the spin trapping technique generated by a Fenton reaction. Results showed that the extracts of *Chrysophyllum inornatum* display the strongest antioxidant activities with DPPH scavenging IC_{50} value of 0.077 mg/ml while Hydroxyl scavenging IC_{50} value was 13.945 mg/ml.

***Chukrassia tabularis* A.Juss. Family: Meliaceae**

Kaur *et al.* (2008) evaluated the antioxidant potential of chloroform, ethyl acetate n-butanol and water fractions of 80 per cent methanol extracts of leaves of *Chukrassia tabularis* by DPPH, deoxyribosee degradation, reducing power and DNA nicking assays. The different fractions showed significant activities in all the free radical scavenging tests and these findings have also shown direct relationship between antioxidant activity and phenolic content. Among the fractions, ethyl acetate fraction exhibited highest inhibition of 93.14 per cent, 89.99 per cent, 87.04 per cent in DPPH, non-site specific and site-specific deoxyribose degradation assays, respectively and 91.20 per cent reduction of ferricyanide.

***Cicer arietinum* L. Family: Fabaceae (Papilionaceae)**

English Name: Chick pea; Bengal gram.

Sreeramulu *et al.* (2009) evaluated the antioxidant activity of Bengal gram by DPPH scavenging assay, FRAP assay and reducing power. Phenolic content was 194.90 mg/100g. DPPH radical scavenging activity expressed as trolox equivalent, FRAP (µmol/g) and reducing power (mg/g) were 0.53, 128 and 1.88, respectively in Bengal gram (whole). In Bengal gram dhal phenolic content was 149.37 mg/100g). For Bengal gram dhal the DPPH radical scavenging activity expressed as trolox equivalent, FRAP (µmol/g) and reducing power (mg/g) were 0.49, 84 and 1.50, respectively.

***Cichorium intybus* L. var. *foliosum* Hegy Family: Asteraceae (Compositae)**

English Name: Chicory.

Sultana *et al.* (1995) reported that the presence of plant extract *Cichorium intybus* in the reaction mixture containing calf thymus DNA and free radical generating system protect DNA against oxidative damage to its deoxyribose moiety. The effect was dependent on the concentration of plant extracts.

A total of 27 extracts from non-cultivated and weedy vegetables traditionally consumed by ethnic Albanians in southern Italy were tested for their free radical scavenging activity (FRSA) in the DPPH screening assay, for their *in vitro* non-enzymatic inhibition of bovine brain lipid peroxidation and for their inhibition of xanthine oxidase. Extracts from *Cichorium intybus* showed strong *in vitro* inhibition of xanthine oxidase, with an activity higher than that of a reference extract from *Ledum groenlandicum*.

Souri *et al.* (2004) investigated the antioxidant activity of methanol extracts of chicory, used as vegetable in Iranian diet against linoleic acid peroxidation. The extracts showed good antioxidant activity.

Antioxidant and DPPH radical scavenging activities, reducing powers and the amount of total phenolic compounds of some medicinal Asteraceae species were studied by Özgen *et al.* (2004). Water extract of *Cichorium intybus* showed good antioxidant activity with 51.6 per cent inhibition at 500μg/ml concentration. The DPPH radical scavenging activity was 84.9 per cent at 500μg/ml concentration and reducing power was 19.3 μg/ml Ascorbic acid equivalent. Total phenolic amount was 25.2μg/ml Gallic acid equivalent.

Jeong *et al.* (2010) reported that Lunasin peptide purified from *Cichorium intybus* protects DNA from oxidative damage by suppressing the generation of hydroxyl radical via blocking fenton reaction between Fe(2+) and H_2O_2 by chelating Fe(2+).

Cienfuegosia digitata Cav. Family: Malvaceae

Polyphenol contents, antioxidant and anti-inflammatory activities of *Cienfuegosia digitata* were investigated by Kiessoun *et al.* (2010). Total phenolic content was 53.02 mg GAE/100g extract, total flavonoid content was 5.89 quercetin equivalent/100mg extract, total flavonol content was 3.14mg quercetin equivalent/100 mg extract and total tannin content was 45.33 mg TAE/100 mg extract. Antioxidant activity in DPPH assay was 10.2 mmol Ascorbic acid equivalent (AAE)/g extract, in FRAP assay 4.33mmol AAE/g extract and in ABTS assay 1.91 mmol AAE/g extract.

Cimcifuga dahurica (Thurcz.) Maxim. Family: Ranunculaceae

Cai *et al.* (2004) evaluated the antioxidant activity and phenolic compounds in traditional Chinese medicinal plants associated with anticancer. The improved ABTS.+ method was used to systematically assess the total antioxidant capacity (Trolox equivalent antioxidant capacity, TEAC) of the medicinal extracts. The TEAC values and total phenolic content for methanolic extract of root of *Cimcifuga dahurica* were 191.4 μmol Trolox equivalent/100g dry weight (DW), and 0.76 g of gallic acid equivalent/100g DW respectively. A positive significant linear relationship between antioxidant activity and total phenolic content showed that phenolic compounds were the dominant antioxidant components in the tested herbs. Major types of phenolic compounds in *Cimcifuga dahurica* include phenolic acids (ferulic acid, caffeic acid), tannins.

Cimcifuga foetida L. Family: Ranunculaceae

Antioxidant capacities of 56 selected Chinese medicinal plants, including *Cimcifuga foetida* were evaluated by Song *et al.* (2010) using the TEAC and FRAP assays, and their total phenolic content was measured by the Folin-Ciocaleteu method. The TEAC value was 119.50μmol Trolox/g, FRAP value was 199.08μmol Fe^{2+}/g and phenolic content was 12.57mg GAE/g. The strong correlation between TEAC value and FRAP value suggested that the antioxidants in these plants possess free radical scavenging activity and oxidant reducing power, and the high positive correlation between antioxidant capacities and total phenolic content implied that phenolic compounds are a major contributor to the antioxidant activity of these plants.

Cimcifuga racemosa (L.) Nutt. Family: Ranunculaceae

Wojcikowski *et al.* (2007) evaluated the antioxidant activity of aerial parts of *Cimcifuga racemosa*. Oxygen radical absorbance capacity of the methanol extract and aqueous extracts were 42.59 and 1222.36μmol Trolox equivalent per g of dried starting material, respectively.

Cinchona ledgeriana Moens Family: Rubiaceae

Al-Mustafa and Al-Thunibat (2008) evaluated the antioxidant activity of *Cinchona ledgeriana* leaves. The level of antioxidant activity was determined by DPPH and ABTS assay in relation to total phenolic contents of the medicinally used part. They concluded that *Cinchona ledgeriana* leaves have high antioxidant capacity (DPPH-TEAC> or = 80mg/g).

Cinchona pubescens Hort. Family: Rubiaceae

Mosquera *et al.* (2007) studied the antioxidant activity of the extracts using DPPH free-radical scavenging method. The percentage of the antioxidant activity of the n-Hexane fraction was 10.884, Dichloromethane fraction was 23.78 and methanol fraction was 36.385.

Cinnamomum burmanni Blume Family: Lauraceae

Prasad *et al.* (2009) evaluated the flavonoid contents and antioxidant activities from five species of *Cinnmomum* including *Cinnamomum burmanni. Cinnamomum burmanni* had the highest flavonoid content among the five species. *Cinnamomum burmannii* showed good antioxidant capability in DPPH radical scavenging activity, reducing power and superoxide anion scavenging activity at various concentrations. Three flavonoid compounds namely quercetin, kaempferol and quercitrin were identified.

Cinnamomum camphora (L.) Presl. Family: Lauraceae

Ethyl acetate and Butanol extracts of *Cinnamomum camphora* displayed strong anti-oxidant activity with IC_{50} values of 14 and 15μg, respectively, when tested by the DPPH and Xanthine oxide assays (Lee *et al.*, 2006).

Cinnamomum cassia Presl. Family: Lauraceae

The antioxidant activities of hot water extracts (HWECC) and ethanol extracts (EECC) from the dry bark of *Cinnamomum cassia* were evaluated by Lin *et al.* (2003). Results showed that at 1.0mg/ml, the ethanol extracts of *Cinnamomum cassia* (96.30 per cent) exhibited a greater inhibition than the α-tocopherol (93.74 per cent) on $FeCl_2$-ascorbic acid induced lipid peroxidation of rat liver homogenate *in vitro*. The same extract showed an excellent antioxidant activity in enzymatic and nonenzymatic liver tissue systems. EECC revealed the strongest antioxidant activity followed by α-tocopherol and HWECC.

Methanol extract of *Cinnamomum cassia* showed relatively high levels of DPPH radical scavenging activity (IC_{50} <6.0 μg/ml) (Lee *et al.*, 2003). The activities of superoxide dismutase, catalase and glutathione peroxidase were dose-dependently enhanced in V79-4 cells treated with the plant extract.

Cai *et al.* (2004) evaluated the antioxidant activity and phenolic compounds in traditional Chinese medicinal plants associated with anticancer. The improved ABTS.+ method was used to systematically assess the total antioxidant capacity

(Trolox equivalent antioxidant capacity, TEAC) of the medicinal extracts. The TEAC values and total phenolic content for methanolic extract of bark of *Cinnamomum cassia* were 570.6 μmol Trolox equivalent/100g dry weight (DW), and 2.83 g of gallic acid equivalent/100g DW respectively. A positive significant linear relationship between antioxidant activity and total phenolic content showed that phenolic compounds were the dominant antioxidant components in the tested herbs. Major types of phenolic compounds in *Cinnamomum cassia* include phenolic acids (cinnamic acid, protocatechuic acid), coumarin, tannins.

Kim *et al.* (2008) reported that *Cinnamomum cassia* extract significantly decreased on blood glucose activities of glutathione reductase, glutathione S-transferase, glutathione peroxidase, treatment also significantly decreased lipid peroxidation.

8-methoxyl-9-hydroxy-3'4'-methylenedioxy-3S,4R-diphenyl butyrolactone, named cinnamomumolide, isolated from the dried tender stems of *Cinnamomum cassia* by Liu *et al.* (2009) exhibited antioxidant activity in protecting against the injury caused by hydrogen peroxide oxidation on human umbilical vein endothelial cells with EC_{50} value of 10.7 μmol.

Prasad *et al.* (2009) evaluated the flavonoid contents and antioxidant activities from five species of *Cinnmomum* including *Cinnamomum cassia. Cinnamomum cassia* showed good antioxidant capability in DPPH radical scavenging activity, reducing power and superoixde anion scavenging activity at various concentrations. Three flavonoid compounds namely quercetin, kaempferol and quercitrin were identified.

Antioxidant capacities of 56 selected Chinese medicinal plants, including *Cinnamomum cassia* were evaluated by Song *et al.* (2010) using the TEAC and FRAP assays, and their total phenolic content was measured by the Folin-Ciocalteu method. The TEAC value was 52.75μmol Trolox/g, FRAP value was 35.59μmol Fe^{2+}/g and phenolic content was 9.71mg GAE/g.

Cinnamomum insulari-montanum Hayata Family: Lauraceae

Hou *et al.* (2003) evaluated the antioxidant activity of 70 per cent aqueous acetone extract of *Cinnamomum insulari-montanum* by various assays including DPPH, hydroxyl radicals and reducing power assay. The extract exhibited moderate activity against DPPH radicals (IC_{50} = 24.9 μg/ml) and inhibited formation of OH generated in the Fenton reaction system (1.1 μg/ml). The total phenolics were 3.0 mg of GAE/g.

Cinnamomum kanehirai Hayata Family: Lauraceae

Lee *et al.* (2006) investigated the antioxidant activity of ethanol extract of leaves using DPPH, hydroxyl and superoxide radical scavenging and reducing power activities as well as the induction of heme oxygenase-1 (HO-1). The IC_{50} values against DPPH radical was 6.4μg/ml, OH radical 0.16μg, superoxide radical 6835μg/ml and total phenolics was 32.1 mg of GAE/g.

Cinnamomum pauciflorum Nees Family: Lauraceae

Prasad *et al.* (2009) evaluated the flavonoid contents and antioxidant activities from five species of *Cinnmomum* including *Cinnamomum pauciflorum. Cinnamomum pauciflorum* showed good antioxidant capability in DPPH radical scavenging activity, reducing power and superoixde anion scavenging activity at various concentrations. Three flavonoid compounds namely quercetin, kaempferol and quercitrin were identified.

Cinnamomum tamala Nees et Eberm. Family: Lauraceae

In order to find their potential sources, the crude extracts of different parts of some selected food and medicinal plants, including *Cinnamomum tamala*, were studied by Bajpai *et al.* (2005) for total phenolic contents (TPC) and antioxidant activity (AOA). AOA was assayed by auto-oxidation of β-carotene and linoleic acid and expressed as the per cent of inhibition relative to control. TPC in leaves of *Cinnamomum tamala* was 12.5 mg/g (GAE) while antioxidant activity was 52.3 per cent.

The antioxidant activity of essential oil and oleoresins were evaluated against mustard oil by peroxide, *p*-aniside, thiobarbituric acid and total carbonyl value method. In addition, their inhibitory action by FTC method, scavenging capacity by DPPH method was also studied by Kapoor *et al.* (2009). Results showed that both the volatile oil and oleoresins have effective antioxidant activities.

Prasad *et al.* (2009) evaluated the flavonoid contents and antioxidant activities from five species of *Cinnmomum* including *Cinnamomum tamala*. *Cinnamomum tamala* showed good antioxidant capability in DPPH radical scavenging activity, reducing power and highest superoixde anion scavenging activity at various concentrations. Three flavonoid compounds namely quercetin, kaempferol and quercitrin were identified.

Antioxidant activity of different extracts of *Cinnamomum tamala* were evaluated by Shahwar *et al.* (2010) through DPPH radical scavenging activity, phosphomolybdate and ferric thiocyanate methods. The DPPH radical scavenging activity of petroleum ether extract was 94.3 per cent, total antioxidant activity methanol extract was 1.131 and inhibition of lipid peroxidation was 54.4 per cent while total phenols was 289.3 mg GAE/g of crude extract.

Cinnamomum verum Presl. Family: Lauraceae

In order to gain insight into the antioxidant effect of *Cinnamomum verum* hepatic and cardiac antioxidant enzymes, GSH content and lipid conjugated dienes were studied in rats fed high fat diet along with *Cinnamomum verum*. The antioxidant enzyme activities were found to be significantly enhanced whereas GSH content was markedly restored in rats fed a fat diet with spices. In addition, this spice partially counteracted increase in lipid conjugated dienes and hydroperoxides, the primary producers of lipid peroxidation (Dhuley, 1999).

Cinnamomum zeylanicum Blume Family: Lauraceae

English Name: Cinnamon

The antioxidant activity of ether, methanol and water extracts of *Cinnamomum zeylanicum* was investigated using β-carotene/linoleic acid system. The etheric (0.69 mg), methanolic (0.88 mg) and aqueous (0.44 mg) cinnamon extracts, inhibited the oxidative process in 68 per cent, 95.5 per cent and 87.5 per cent respectively. The BHT control inhibited 80 per cent oxidation. The spray reagents (1) β-carotene/linoleic acid and (2) $FeCl_3/K_3Fe(CN)_4$ 1 per cent sol, showed spots in TLC with antioxidant activity (1) and blue color (2), indicating the presence of phenolic compounds with Rf values of 0.50. Five fractions were obtained by column partition with antioxidant activity and the presence of phenolic compounds.

Jayathilakan *et al.* (2007) evaluated antioxidant potential of synthetic and natural antioxidants and its effect on warmed-over-flavour in different species of meat. *Cinnamomum zeylanicum* showed good antioxidant activity.

Prasad *et al.* (2009) evaluated the flavonoid contents and antioxidant activities from five species of *Cinnamomum* including *Cinnamomum zeylanicum. Cinnamomum zeylanicum* exhibited the highest total phenolic content among the five species. *Cinnamomum zeylanicum* showed the highest DPPH radical scavenging activity and reducing power at various concentrations. Three flavonoid compounds namely quercetin, kaempferol and quercitrin were identified.

Twenty-five essential oils, including cinnamon leaf oil, were tested by Wei and Shibamoto (2010) for antioxidant activities using a conjugated diene assay, the aldehyde/carboxylic acid assay, the DPPH free radical scavenging assay, and the malonaldehyde/gas chromatography (MA/GC) assay. They were also tested for lipoxygenase inhibitory activities using the lipoxygenase inhibitor-screening assay. Cinnamon leaf oil showed activities in the aldehyde/carboxylic acid assay (100 per cent) and DPPH assay (84 per cent) but only moderate activities in the conjugated diene assay (24 per cent) and MA/GC assay (48 per cent).

Antioxidant activity of different extracts of *Cinnamomum zeylanicum* were evaluated by Shahwar *et al.* (2010) through DPPH radical scavenging activity, phosphomolybdate and ferric thiocyanate methods. The DPPH radical scavenging activity of methanol extract was 96.8 per cent, total antioxidant activity was 1.103 and inhibition of lipid peroxidation was 69.7 per cent while total phenols was 398.8 mg GAE/g of crude extract.

Cirsium arvense (L.) Scop., *C. oleraceum* (L.) Scop., *C. palustre* (L.) Coss. ex Scop., *C. rivulare* (Jacq.) All. and *C. vulgare* (Savi) Ten. Family: Asteraceae (Compositae)

Methanolic extracts from inflorescences and leaves of five *Cirsium* species and $CHCl_3$, Et_2O, EtOAc, n-BuOH fractions obtained from those extracts were investigated for their antioxidant activity. The antioxidant activity measurement, expressed as Total Antioxidant Status, ranged from 0.98 to 2.71 mM/L. There was a positive correlation between the total phenolic content and antioxidant activity for MeOH extracts and EtOAc fractions (Nazaruk, 2007).

Cirsium japonicum DC. Family: Asteraceae (Compositae)

Gan *et al.* (2010) evaluated the antioxidant activities of 40 medicinal plants associated with prevention and treatment of cardiovascular and cerebrovascular diseases using FRAP and TEAC assays and their total phenolic contents were measured by the Folin-Ciocalteu methods. The antioxidant activity of *Cirsium japonicum* was 21.49μmol Fe(II)/g in FRAP assay, 24.86μmol Trolox/g and Total phenolic content was 1.58 mg GAE/g. A high correlation between antioxidant capacity and total phenolic content indicated that phenolic compounds could be the main contributor of the antioxidant activity of these plants.

Cirsium setosum (Willd.) Kitam. Family: Asteraceae (Compositae)

Cai *et al.* (2004) evaluated the antioxidant activity and phenolic compounds in traditional Chinese medicinal plants associated with anticancer. The improved ABTS.+ method was used to systematically assess the total antioxidant capacity (Trolox equivalent antioxidant capacity, TEAC) of the medicinal extracts. The TEAC values and total phenolic content for methanolic extract of floral bud of *Cirsium setosum* were 257.8 μmol Trolox equivalent/100g dry weight (DW), and 0.71 g of gallic acid equivalent/100g DW respectively. A positive significant linear relationship between antioxidant activity and total phenolic content showed that phenolic compounds were the dominant antioxidant components in the tested herbs. Major types of phenolic compounds in *Cirsium setosum* include flavonols (rutin), phenolic acids (caffeic acid, chlorogenic acid).

Cissampelos owarensis P. Beauv. ex. DC. Family: Menispermaceae

Habila *et al.* (2011) evaluated the phytochemicals and antioxidant activity of four Nigerian medicinal plants, including *Cissampelos owarensis*. The percentage of antioxidant activity (AA per cent) of the plants extract showed a dose dependent increase. *C. owarensis* had the highest percentage antioxidant activity at 125 μg/ml (91 per cent). The results of reducing potential and Total Phenolic contents expressed in terms of Gallic acid equivalent showed that *Cissampelos owarensis* possessed the highest reducing potential and total phenolic content (0.703nm, 12.4mg) when compared with Gallic acid standard (1.268nm).

Cissampelos pareira L. Family: Menispermaceae

The alkaloidal fraction (AFCP) of roots of *Cissampelos pareira* was screened for *in vitro* antioxidant activity and immunomodulatory activity in mice. AFCP possess strong antioxidant activity which was revealed by its stability to scavenge the stable free radical DPPH, superoxide ion and to inhibit lipid peroxidation in rat liver homogenate induced by iron/ADP/Ascorbate complex. The HPTLC fingerprint profile was also established for the identification of AFCP which was found to contain 0.176 per cent of berberine. Tetrandrine and berberine were isolated from roots (Bafna and Mishra, 2009).

Amresh *et al.* (2007) evaluated the antioxidant activity of *Cissampelos pareira*. The roots of *C. pareira* (50 per cent ethanol extract) were found to contain a large amount of polyphenols and exhibit significant and dose-dependent reducing ability, indicative of potent antioxidant ability both *in vitro* and *in vivo*. The *C. pareira* extract (CPE) showeed significant antioxidant activity in the DPPH assay. The ability of CPE to scavenge reactive oxygen and nitrogen species was tested; and it was found to significantly scavenge superoxide, hydrogen peroxide, hydroxyl radicals, and nitric oxide at a dose regimen of 50 to 400μg/kg *in vitro*. The CPE also inhibited hydroxyl radical-induced oxidation of proteins *in vitro*. The efficacy of CPE was further tested *in vivo*. It was found to exhibit a potent protective activity in an acute oxidative tissue injury animal model: benzo(a)pyrene-induced gastric toxicity in mice. Treatment with CPE at a dose regimen of 50 and 100 mg/kg body weight in mice, significantly and dose-dependently protected the gastric mucosa against benzo(a)pyrene-induced oxidative stresss. The CPE, significantly and dose-dependently also afforded protection against gastric lipid peroxidation glutathione level, and activities of other antioxidant enzymes namely SOD, CAT and glutathione peroxidase and glutathione S-transferase.

Cissus quadrangularis L. Family: Vitaceae

Protective effect of *Cissus quadrangularis* on neutrophil mediated tissue injury induced by aspirin in rats was investigated by Jainu *et al.* (2006). The findings suggest that the gastroprotective activity of ethanol extract of *Cissus quadrangularis* could be mediated possibly through its antioxidant effect as well as by the attenuation of the oxidative mechanism and neutrophil infiltration.

Cistanche deserticola Y.C.Ma Family: Orobanchaceae

The acetone-aqueous (9:1) extract from the stem of *Cistanche deserticola* showed a strong free radical scavenging activity (Xiong *et al.*, 1996). Nine major phenylethanoid compounds were isolated from this extract. All of these compounds showed stronger free radical scavenging activities than α-tocopherol on DPPH radical and xanthine/xanthine oxidase generated superoxide anion radical. Among the nine compounds, isoacetoside and tubuloside B, whose caffeoyl moiety is at 6′-position of the glucose, showed an inhibitory effect on XOD (Xiong *et al.*, 1996).

Cistanche salsa (C.A.Mey) Beck. Family: Orobanchaceae

Structure-activity relationships of phenylethanoid glycosides in plants of *Cistanche salsa* on antioxidative activity by DPPH method was studied by Yang *et al.* (2009). The antioxidative activity of phenylethanoid glycosides is related to the number of phenolic hydroxyl, steric hindrance, 2-acetyl on the middle glucopyranose and the location of phenolic hydroxyl. Additionally, it may be related to the alpha, beta-unsaturated ketone phenyl-2-propenoyl (Yang *et al.*, 2009).

Cistus ladanifer L. Family: Cistaceae

Phenolic and tannin content of *Cistus ladanifer* leaves aqueous extract was determined and its antioxidant and antimicrobial activity were fully studied by several *in vitro* assays (Barrajon-Catalan *et al.*, 2010). *C. ladanifer* extract was effective to inhibit peroxy radicals. The major compounds are ellagitannins, especially punicalagins derivatives. The leaves of this plant is a source for water-soluble ellagitannins-enriched polyphenolic extracts with antioxidant and antimicrobial activities.

Cistus laurifolius L. Family: Cistaceae

Serteser *et al.* (2009) studied the antioxidant activity of 50 per cent aqueous methanol extracts of *Cistus laurifolius* including free radical scavenging, hydrogen peroxide scavenging and metal (Fe^{2+}) chelating activities.The DPPH radical scavenging effects of leaves extract was 0.345 while fruit extract was 0.580. Fe^{2+} chelating activity (per cent) of leaves extracts was 28.43 while fruit extracts showed

36.27 per cent inhibition. Hydrogen peroxide inhibition activity of leaves extracts was 41.54 per cent and fruit extracts was 32.79 per cent.

Cistus populifolius L. Family: Cistaceae

Phenolic and tannin content of *Cistus populifolius* leaves aqueous extract was determined and its antioxidant and antimicrobial activity were fully studied by several *in vitro* assays (Barrajon-Catalan *et al.*, 2010). *C. populifolius* extract was effective to inhibit peroxy radicals. The major compounds are ellagitannins, especially punicalagins derivatives. The leaves of this plant is a source for water-soluble ellagitannins-enriched polyphenolic extracts with antioxidant and antimicrobial activities.

Citrullus colocynthis (L.) Schrad. Family: Cucurbitaceae

English Name: Bitter apple.

Souri *et al.* (2008) investigated the antioxidant activity and radical scavenging activity of methanolic extract of fruit. The antioxidant activity of this plant extract against linoleic acid peroxidation expressed as IC_{50} was 24.851ng/ml and radical scavenging activity against DPPH expressed as IC_{50} was 74.41 µg/ml.

Citrullus lanatus (Thunb.) Matsumura et Nakai Family: Cucurbitaceae

English Name: Watermelon.

The total antioxidant activity of melon was measured by Wang *et al.* (1996) using automated oxygen radical absorbance capacity (ORAC) assay. On the basis of the wet weight of the fruits (edible portion), melon had the highest ORAC activity (0.97micromoles of Trolox equivalents per gram). On the basis of dry weight of the fruits melon had the highest ORAC activity (12.9µmoles/gTrolox equivalent). The contribution of the fruit pulp fraction (extracted with acetone) to the total ORAC activity of a fruit was usually less than 10 per cent.

Thirty-eight types of fruits commonly consumed in Singapore were systematically analyzed by Isabelle *et al.* (2010) for their hydrophilic oxygen radical absorbance capacity (H-ORAC), total phenolic content (TPC), ascorbic acid (AA) and various lipophilic antioxidants. Red watermelon had the highest β-cryptoxanthin, lycopene and β-carotene per gram fresh weight.

Tlili *et al.* (2010) reported that the amount of each specific antioxidant, as well as the HAA and LAA, were both influenced by genotype and sampling area, emphasizing the need to evaluate watermelon biodiversity in order to improve its nutritional value.

Citrus aurantifolia (Christm. et Panz.) Swingle Family: Rutaceae

Ghafar *et al.* (2010) investigated the total phenolic, flavonoid and hesperidine contents of four *Citrus* species including *C.aurantifolia*. Additionally, the antioxidant activities were also determined by FRAP and DPPH radical scavenging activity. The Total phenolic content was 211.7 mg of gallic acid equivalent and Flavonoid content was 10.67 mg/100 ml of juice. The antioxidant activity was DPPH 80 per cent of EC_{50} value while FRAP was 80µmol Fe^{2+} equivalent/100ml of juice.

Citrus aurantium Franch. Family: Ranunculaceae

Methanol aqueous extract of *Citrus aurantium* was screened by Kim *et al.* (1997) for free radical scavenging activity using the DPPH free radical generating system. The extract showed free radical scavenging activity.

Cai *et al.* (2004) evaluated the antioxidant activity and phenolic compounds in traditional Chinese medicinal plants associated with anticancer. The improved ABTS.+ method was used to systematically assess the total antioxidant capacity (Trolox equivalent antioxidant capacity, TEAC) of the medicinal extracts. The TEAC values and total phenolic content for methanolic extract of immature fruit of *Citrus aurantium* were 703.6 μmol Trolox equivalent/100g dry weight (DW), and 5.07 g of gallic acid equivalent/100g DW respectively. A positive significant linear relationship between antioxidant activity and total phenolic content showed that phenolic compounds were the dominant antioxidant components in the tested herbs. Major types of phenolic compounds in *Citrus aurantium* include flavonones (hesperidin, naringenin, naringin).

Citrus bergamia Risso et Poiteau Family: Rutaceae

English Name: Bergamot.

Trovato *et al.* (2009) evaluated the protective effect of treatment with *Citrus bergamia* juice (1ml/day, for 30 days) against hypercholesterolemic diet-induced renal injury in rat. The antioxidant potential of *C.bergamia* juice was examined in two *in vitro* systems: in the DPPH test the juice showed a noticeable effect on scavenging free radicals (IC_{50}=25.01); in the reducing power assay it showed a strong activity too (1.44 ASE/ml). These findings suggest that *C. bergamia* juice has a protective role in hypercholesterolemic diet-induced renal damage, which may be attributed to its antioxidant properties.

Twenty-five essential oils, including Bergamot oil, were tested by Wei and Shibamoto (2010) for antioxidant activities using a conjugated diene assay, the aldehyde/carboxylic acid assay, the DPPH free radical scavenging assay, and the malonaldehyde/gas chromatography (MA/GC) assay. They were also tested for lipoxygenase inhibitory activities using the lipoxygenase inhibitor-screening assay. Bergamot oil exhibited 100 per cent antioxidant activity in the aldehyde/carboxylic acid assay.

Citrus grandis Osbeck Family: Rutaceae

English Name: Buntan Pumelo.

Antioxidant activity of young fruits was investigated using DPPH assay. Radical scavenging activity was highest in water and butanol extracts of the fruit. Both of these fractions were found to significantly reduce H_2O_2 induced cytotoxicity in HepG2 cells in a dose dependent manner. Pretreatment of the cells with either extract reduced generation of reactive oxygen species. The butanol fraction inhibited lipid oxidation to a greater extent than the aqueous fraction in the same cells.

Dangyuja (*Citrus grandis*) extract treated with *Aspergillus saitoi* had significantly higher antioxidant and antidiabetic activity than both its glycosides and aglycones in ORAC and supercoiled DNA strand scission assay, rat intestinal α-glucosidase and pancreatic α-amylase inhibition, suggesting that the fermentation of flavanone aglycones of Dangyuja extract with *A. saitoi* can increase antioxidant antidiabetic activity compared to their glycosides and aglycones (Kim *et al.*, 2009).

The antioxidant potentials of buntan pumelo and its fermented products were determined by Jang *et al.* (2010). The essential oil from peel had higher total phenolic (214mM) and flavonoid (134mg QE/g of dried material) contents than those of different solvent extracts from fruit pulp. In addition, DPPH free radical scavenging activity and ferric-reducing antioxidant power valued determined for the essential oil were 26.1 per cent and 2.3mM, respectively, which were significantly higher than those of various fruit pulp extracts. The ethanol-fermented products of pummel juice had higher total phenolic and flavonoid contents than those of fresh juice. For maintenance of the substantial antioxidant properties of pumelo products, ethanol fermented juice rather than fresh or acetate-fermented juice was recommended by Jang *et al.* (2010). Through correlation analysis, the phenolic compounds in the fermented pumelo products were found to be the major contributors to the free radical-scavenging activity and ferric-reducing power.

Citrus hystrix DC. Family: Rutaceae

Wong *et al.* (2006) investigated the antioxidant properties of 25 edible tropical plants including *Citrus hystrix* using DPPH and FRAP assays. Total phenolic content was also estimated. *Citrus hystrix* showed moderate antioxidant activity.

Butryee *et al.* (2009) investigated the effect of processing on the flavonoid content and antioxidant capacity of *Citrus hystrix* leaf. The results suggest that method of processing can significantly affect the content of flavonoids and their TAC values.

Ghąfar *et al.* (2010) investigated the total phenolic, flavonoid and hesperidine contents of four *Citrus* species including *C. hystirx*. Additionally, the antioxidant activities were also determined by FRAP and DPPH radical scavenging activity. *C. hystix* had the highest Total phenolic content (490.74mg of gallic acid equivalent) and Flavonoid content (22.25 mg/100 ml of juice). The antioxidant activity was DPPH 35mg of EC_{50} value while FRAP was 89µmol Fe^{2+} equivalent/100ml of juice.

Citrus indica Tanaka Family: Rutaceae

Povichit *et al.* (2010) investigated phenolic content and free radical scavenging effect of fruit of *Citrus indica* by DPPH and ABTS assays, anti-lipid peroxidation activity by TBARS and for antiglycation activity. The results revealed that the total phenolic content showed good correlation with free radical scavenging by ABTS and anti-lipid peroxidation by TBARS, but showed no correlation with antiglycation. The extract of *Citrus indica* showed moderate antioxidant effect and also showed a promising antiglycation effect. The IC_{50} were 5.20mg/ml for the DPPH method, TEAC value of 0.13 mg/Trolox/mg sample for the ABTS method, IC_{50} of 7.60 mg/ml for the TBARS method and IC_{50} of 17.64 for the antiglycation method.

Citrus limetta Risco Family: Rutaceae

Kundu Sen *et al.* (2010) evaluated *in vitro* antioxidant activity of methanol extract of *Citrus limetta* fruit peel on reactive oxygen and nitrogen species. The extract had shown good free radical scavenging activity in a dose dependent manner. The IC_{50} values for DPPH, superoxide radical, nitric oxide radical inhibition and lipid peroxidation assays of *Citrus limetta* were 7.36 µg/ml, 9.36µg/ml, 136.45µg/ml and 111µg/ml, respectively. Measurement of total phenolic compounds by Folin-Ciocalteu's reagent indicated that 1 mg extract contains 126.85µg equivalent of pyrocatechol.

Citrus limon **(L.) Burm.f. Family: Rutaceae**

Citrus limon and some other Rutaceae members are rich source of antioxidants (Oza and Pandya, 2007). Yang *et al.* (2010) reported antioxidant activity of the essential oil from *Citrus limon*.

Citrus maxima **Merr. Family: Rutaceae**

English Name: Shaddock, Squash, Grapefruit.

Huang *et al.* (2004) evaluated the antioxidant activities of various fruits and vegetables produced in Taiwan. Squash seeds showed the highest antioxidant activity.

Gorinstein *et al.* (2003, 2004) investigated the antioxidant properties of Grapefruit juice. The contents of total polyphenols, flavonoids and anthocyanins in fresh Grapefruit juice were 906.9, 44.8 and 68.7µg/ml, respectively. The antioxidant potential measured by scavenging activity against nitric oxide, the β-carotene-linoleate model system (β-carotene) and the DPPH and ABTS assays was good. A high level of correlation between contents of total polyphenols and flavonoids and antioxidant potential values of the juice was found. Diets supplemented with orange juice improved plasma lipid metabolism only in rats fed added cholesterol.

Mathur *et al.* (2010) evaluated the antioxidant activity of peel and pulp of *Citrus maxima* by superoxide anion radical scavenging assay, scavenging of Hydrogen peroxide, inhibition of lipid peroxide formation. Ascorbic acid content was 18mg/g in fruit. Total phenolic content in peel was 28µg/g gallic acid equivalent while in pulp it is 40 µg/g gallic acid equivalent. Shaddock ethanolic fraction showed maximum superoxide anion radical scavenging activity (90 per cent), preceded by chloroform fraction (75 per cent) and aqueous fraction (50 per cent). Hydrogen peroxide scavenging activity of Shaddock was 92 per cent in ethanolic fraction of pulp followed by chloroform fraction (75 per cent) and aqueous fraction (50 per cent). Inhibition of Lipid peroxide formation of pulp was 85 per cent (ethanolic fraction), 75 per cent (chloroform fraction), and 45 per cent (aqueous fraction).

Singh *et al.* (2010) reported considerable antioxidant activities of the essential oil of *Citrus maxima*. Several compounds could be identified in significant amounts in the oil (limonene, E-citral, Z-citral and 3,3-dimethyl-1-hexene).

Kundu Sen *et al.* (2010) evaluated *in vitro* antioxidant activity of methanol extract of *Citrus maxima* leaf on reactive oxygen and nitrogen species. The extract had shown good free radical scavenging activity in a dose dependent manner. The IC_{50} values for DPPH, superoxide radical, nitric oxide radical inhibition and lipid peroxidation assays of *Citrus limetta* were 5.5 µg/ml, 79.16µg/ml, 174.97µg/ml and 111.5µg/ml, respectively. Measurement of total phenolic compounds by Folin-Ciocalteu's reagent indicated that 1 mg extract contains 116.7µg equivalent of pyrocatechol.

Citrus medica L. Family: Rutaceae

The n-hexane extract of Diamonte citron (*Citrus medica* acv. Diamonte) peel showed significant antioxidant activity that was carried out using different assays (DPPH test, β-carotene bleaching test and bovine brain peroxidation assay) (Conforti *et al.*, 2007).

Citrus microcarpa Bunge Family: Rutaceae

Ghafar *et al.* (2010) investigated the total phenolic, flavonoid and hesperidine contents of four *Citrus* species including *C. microcarpa*. Additionally, the antioxidant activities were also determined by FRAP and DPPH radical scavenging activity. The Total phenolic content was 105.0 mg of gallic acid equivalent and Flavonoid content was 8.70 mg/100 ml of juice. The antioxidant activity was DPPH 120mg of EC_{50} value while FRAP was 48.18µmol Fe2+ equivalent/100ml of juice.

Citrus nobilis Lour. Family: Rutaceae

English Name: King Orange.

Chemical composition and antioxidant activity of the peel oil of *Citrus nobilis* was evaluated by Gursoy *et al.* (2010) in four complementary test systems namely; β-carotene/linoleic acid, DPPH radical scavenging, reducing power and metal chelating activities. In the first system, antioxidant activity increased with the increasing concentration. At 20 mg/ml concentration, antioxidant property of the oil was 96.8 per cent and as strong as the positive controls BHT and α-tocopherol. Scavenging effect of the oil was superior to the positive controls BHT and α-tocopherol at 1.5 mg/ml concentration (96.4 per cent). Reducing power and chelating effect of the essential oil increased with the increasing concentration.

Citrus paradisii Macf. Family: Rutaceae

English Name: Grapefruit, Star ruby red grapefruit.

The total antioxidant activity of grapefruit was measured by Wang *et al.* (1996) using automated oxygen radical absorbance capacity (ORAC) assay. On the basis of the wet weight of the fruits (edible portion), grapefruit had the good ORAC activity (4.83micromoles of Trolox equivalents per gram). On the basis of dry weight of the fruits grapefruit had the highest ORAC activity (48.3µmoles/gTrolox equivalent). The contribution of the fruit pulp fraction (extracted with acetone) to the total ORAC activity of a fruit was usually less than 10 per cent.

Vinson *et al.* (2002) reported antioxidant activity of grapefruit juice. Ascorbate contributed 66 to 77 per cent of the antioxidant content of grapefruit juices.

The antioxidant activity of *Citrus paradisii* seeds glyceric extract dissolved in ethanol and in aqueous media was evaluated using three different methods: evaluation by DPPH assay, by 5-lipoxygenase assay and by luminal/xanthine/xanthin oxidase chemiluminescence assay. The total phenolic content was determined by the Prussian Blue method opportunely modified. The grapefruit seeds glyceric extract utilized as aqueous solutions demonstrated antioxidant properties better than those displayed by alcoholic solutions (Giamperi *et al.*, 2004).

Eleven known coumarins: osthol, xanthotoxin, bergapten, isopinellin, scoparone, limettin, aurapten, auraptene, scopoletin, bergaptol and umbelliferone were isolated for the first time from n-hexane fraction of peels of star ruby red grapefruit Egyptian cultivar by Mohamed (2004). In addition two steroids: β-sitosterol and stigmasterol were isolated. Three flavonoid glycosides were isolated fom n-butanol fraction. Nobiletin and tangerine were isolated from the dichloromethane fractions. Alcoholic extract showed the highest antioxidant activity.

Deyhim *et al.* (2006) reported that drinking of Grapefruit juice prevented oxidative stress by enhancing total antioxidant capacity and elevating liver antioxidant enzymes while modulating cardiovascular risk factors. Yang *et al.* (2010) reported antioxidant activity of the essential oil from *Citrus paradisii.*

Citrus ponki (Hayata) Hort. ex Tanaka Family: Rutaceae

Chen *et al.* (2009) studied the antioxidant activity of fruit peel extract towards xanthine oxidase, lipoxygenase and tyrosinase activities. The inhibitory effect of acetone extract on xanthine oxidase activity was nil, tyrosine activity was more than 20 per cent and lipoxygenase activity was about 50 per cent.

Citrus reticulata Blanco Family: Rutaceae

English Name: Kinnow.

The antioxidative activities of twenty types of Citrus fruits were investigated by Tanizawa *et al.* (1992) with a screening method which was based on rat liver microsomal lipid peroxidation induced by NADPH and ADP. The activities of the exocarp were greater than those of sarcocarp and the activities from immature fruits were greater than those from mature fruits. The strongest antioxidative activity was found in ponkan (*Citrus reticulata*) collected in July.

The total antioxidant activity of orange was measured by Wang *et al.* (1996) using automated oxygen radical absorbance capacity (ORAC) assay. On the basis of the wet weight of the fruits (edible portion), orange had good ORAC activity (7.50micromoles of Trolox equivalents per gram). On the basis of dry weight of the fruits strawberry had the highest ORAC activity (51.7μmoles/gTrolox equivalent). The contribution of the fruit pulp fraction (extracted with acetone) to the total ORAC activity of a fruit was usually less than 10 per cent.

Nectars of orange in Samsun markets in Turkey were analyzed by Tosun and Ustun (2003) for total antioxidant capacity, total phenolic content, ascorbic acid content and total carotenoids. Ferric reducing/antioxidant power (FRAP) assay was used to measure the total antioxidant power. Orange nectar had the highest concentration of ascorbic acid (589.8 mg/kg). Total phenols were 194.20 mg/l and total carotenoids 3.43 mg/l.

Faller and Fialho (2010) evaluated the polyphenol content and antioxidant capacity in organic and conventional plant foods. The results showed that organic fruits tend to have higher hydrolysable polyphenol contants than conventional ones, with values being 11.5 per cent in orange peels, higher for hydrolysable polyphenols.

Mathur *et al.* (2010) evaluated the antioxidant activity of peel and pulp of *Citrus reticulata* by superoxide anion radical scavenging assay, scavenging of Hydrogen peroxide, inhibition of lipid peroxide formation. Ascorbic content was 6.06mg/g in fruit. Total phenolic content in peel was 148µg/g gallic acid equivalent while in pulp it is 76 µg/g gallic acid equivalent. Kinnow ethanolic fraction showed maximum superoxide anion radical scavenging activity (87 per cent), preceded by chloroform fraction (25 per cent) and aqueous fraction (5 per cent). Inhibition of lipid peroxide formation was 83 per cent (ethanolic fraction), 24 per cent (chloroform fraction) and 10 per cent (aqueous fraction).

Citrus sinensis (L.) Osbeck Family: Rutaceae

English Name: Orange, Blood sweet orange, Blood orange.

Total Antioxidant Activities (TAA) of freshly squeezed and processed blood and blond orange juices were measured by Arena *et al.* (2001) using the ABTS radical-cation method. Blood juices have TAA values higher than blond juices, and freshly-squeezed juices were higher than processed. Ascorbic acid was the main contributor for the antioxidant activity followed by hydroxycinnamic acids and anthocyanins.

Vinson *et al.* (2002) reported antioxidant activity of orange juice. Ascorbate contributed 56 to 77 per cent of the antioxidant content of orange juices. In orange juice polyphenols accounted for 84-85 per cent of antioxidant quality.

Cai *et al.* (2004) evaluated the antioxidant activity and phenolic compounds in traditional Chinese medicinal plants and fruits associated with anticancer. The improved ABTS.+ method was used to systematically assess the total antioxidant capacity (TEAC) of the medicinal extracts. The TEAC values and total phenolic content for methanolic extract of fruit of *Citrus sinensis* were 67.2 µmol Trolox equivalent/100g dry weight (DW), and 0.51 g of gallic acid equivalent/100g DW respectively.

Six coumarins isoimpratorin, osthol, bergapten, scoparone, limettin and bergaptol were isolated from n-hexane fraction of peels of blood sweet orange Egyptian culticar (Mohamed, 2004). In addition two steroids: β-sitosterol and stigmasterol were isolated. Three flavonoid glycosides were isolated fom n-butanol fraction. Nobiletin and tangerine were isolated from the dichloromethane fractions. Alcoholic extract showed the highest antioxidant activity. Effect of thermal treatments on antioxidant and antiradical activity of blood orange juice was studied by Lo Scalzo *et al.* (2004). Thermal treatment generally induced an increase in the main phenolic substances of orange juice, such as anthocyanins and total cinnamates, while ascorbic acid underwent a decrease. Antioxidant properties, evaluated in a lipoxygenase-linoleic acid system, were higher in thermally treated samples, while free radical scavenging activity, evaluated by ESR spin trapping of hydroxyl radicals and DPPH quenching were enhanced in untreated juices.

Gorinstein *et al.* (2003, 2004) investigated the antioxidant properties of orange fruit juices. The contents of total polyphenols, flavonoids and anthocyanins in fresh orange juice were 962.1, 50.1 and 69.9µg/ml, respectively. The antioxidant potential measured by scavenging activity against nitric oxide, the β-carotene-linoleate model system (β-carotene) and the DPPH and ABTS assays was good. A high level of correlation between contents of total polyphenols and flavonoids and antioxidant potential values of the juice was found. Diets supplemented with orange juice improved plasma lipid metabolism only in rats fed added cholesterol.

Seven different extracts, fractions and residues of Navel sweet orange peel were evaluated for their radical scavenging activity by DPPH and luminal induced chemiluminiscence methods. Also, the Folin-Ciocalteu method was used to determine the total phenolic content. High phenolic content and radical scavenging activities were found for the ethyl acetate fraction. The radical scavenging activity of the ethyl acetate fraction approached the activity of the standards. Total phenolic content showed a small relation with radical scavenging activity. Among the various classes of flavonoids: flavone glycosides, flavones and flavonols seem to prevail.

Odukoya *et al.* (2005) evaluated the antioxidant activity of Nigerian dietary spices including *Citrus sinensis*. The antioxidant activity (expressed as per cent inhibition of oxidation) was 21.37 per cent, reducing power was 67.81 per cent and Total phenolic content was 38.41 mg/100g Tannic acid Equivalent. Antioxidant activity correlated significantly and positively with total phenolics while there was no linear correlation between total antioxidant activity and reducing power neither between reducing power and total phenolic content.

Deyhim *et al.* (2006) reported that drinking of orange juice prevented oxidative stress by enhancing total antioxidant capacity and elevating liver antioxidant enzymes while modulating cardiovascular risk factors. The phenolic compounds and antioxidant capacities of orange juices obtained from cvs. Moro and Sanguinella (*Citrus sinensis*) were determined by Kelebek *et al.* (2007). A total of 18 phenolic compounds were identified and quantified in Moro and Sanguinello orange juices, including hydroxylbenzoic acids (2), hydroxycinnamic acids (5), flavanones (5), and anthocyanins (6). Ferulic acid was the most dominant hydroxycinnamic acid and cyanidin 3-(6″-malonyl glucoside) and cyanidin-3-glucoside were the most dominant anthocyanins in both cultivars. Antioxidant activities of orange juices were measured using the DPPH method. According to DPPH assays, the antioxidant capacity of Moro juice was found to be higher than that of Sanguinello juice.

Mathur *et al.* (2010) evaluated the antioxidant activity of peel and pulp of *Citrus sinensis* by superoxide anion radical scavenging assay, scavenging of Hydrogen peroxide, inhibition of lipid peroxide formation. Ascorbic content was 4.3mg/g in fruit. Total phenolic content in peel was 134µg/g gallic acid equivalent while in pulp it was 28 µg/g gallic acid equivalent. Superoxide anion radical scavenging activity was low (for peel 7.25 per cent in chloroform extract and 25 per cent for the pulp.) The antioxidant activities were also low compared to other citrus fruits.

Ghafar *et al.* (2010) investigated the total phenolic, flavonoid and hesperidine contents of four *Citrus* species including *C. sinensis*. Additionally, the antioxidant activities were also determined by FRAP and DPPH radical scavenging activity. The Total phenolic content was 135.3 mg of gallic acid equivalent and Flavonoid content was 2.99 mg/100 ml of juice. The antioxidant activity was DPPH 95mg of EC_{50} value while FRAP was 60µmol Fe^{2+} equivalent/100ml of juice.

Singh *et al.* (2010) reported considerable antioxidant activities of the essential oil of *Citrus sinensis.* Limonene is predominant in the essential oil.

Citrus x tangerina Tanaka Family: Rutaceae

English Name: Tangerine

Vinson *et al.* (2002) reported antioxidant activity of tangerine juice. Ascorbate contributed 66 to 77 per cent of the antioxidant content of tangerine juices.

Faller and Fialho (2010) evaluated the polyphenol content and antioxidant capacity in organic and conventional plant foods. The results showed that organic fruits tend to have higher hydrolysable polyphenol contents than conventional ones. Fruit peels also showed higher concentration of polyphenols than pulp, reaching, for tangerines, twice the amount found in pulps, which reflected in higher antioxidant capacity.

Citrus unshiu Marc. Family: Rutaceae

English Name: Satsuma mandarin.

Suguira *et al.* (2006) reported that chronic administration of *Citrus unshiu* fruit improves oxidative stress in streptozotocin-induced diabetic rat liver. After a ten-week administration of fruits of Satsuma mandarin were examined on an antioxidant enzymes and glutathione levels in the liver were evaluated. The superoxide dismutase (SOD), catalase, and glutathione-peroxidase activities and glutathione level in diabetic rats liver decreased significantly compared with those in normal rats. In contrast, the SOD, GR and glutathione levels in the Satsuma mandarin (1 or 3 per cent) diet-fed rat livers were signifantly higher than those in the normal diet-fed rat livers.

Clausena lansium (Lour.) Skeels Family: Rutaceae

English Name: Wampee.

Adebajo *et al.* (2009) reported that the methanolic extract of *Clausena lansium* stem bark exhibited antioxidant activities.

Nagendra Prasad *et al.* (2010) reported antioxidant and anticancer activities of 8-hydroxypsoralen isolated from *Clausena lansium* peel. This isolate (8-hydroxypsoralen) exhibited good scavenging activities against DPPH radical and superoxide anion as well as significant reducing power.

Clematis chinensis Osbeck Family: Ranunculaceae

Cai *et al.* (2004) evaluated the antioxidant activity and phenolic compounds in traditional Chinese medicinal plants associated with anticancer. The improved ABTS.+ method was used to systematically assess the total antioxidant capacity (Trolox equivalent antioxidant capacity, TEAC) of the medicinal extracts. The TEAC values and total phenolic content for methanolic extract of root of *Clematis chinensis* were 95.8 μmol Trolox equivalent/100g dry weight (DW), and 0.55 g of gallic acid equivalent/100g DW respectively. A positive significant linear relationship between antioxidant activity and total phenolic content showed that phenolic compounds were the dominant antioxidant components in the tested herbs. Major types of phenolic compounds in *Clematis chinensis* include coumarins (pyranocoumarin) other phenolics (syringaresinol).

Gan *et al.* (2010) evaluated the antioxidant activity of *Clematis chinensis* using the FRAP and TEAC assays, and its total phenolic content was measured by the Folin-Ciocalteu method. For FRAP assay the value was 82.27 μmol Fe(II)/g dry weight. For TEAC assay, the value was 45.04μmol Trolox/g dry weight. Total phenolic content was 7.85 mg gallic acid equivalent/g dry weight of plant material.

Cleome arabica L. Family: Capparaceae

Selloum *et al.* (1998) reported the antioxidant potency of *Cleome arabica* leaves of Algerian origin.

Cleome gynandra L. (Syn.: *Gynandropsis gynandra*) Family: Capparaceae

The total antioxidant activity, as evaluated by determining the ability of the leaf extract (70 per cent ethanol) to scavenge the stable 2,2′-azinobis (3-ethylbenzothiazoline)-6-sulphonate radical cationic chromophore ($ABTS^+$) is found to be 523.67±4.16 mg vitamin C equivalents/100g (Anbazhagi *et al.*, 2009). Polyphenolic phytochemicals especially the flavonoids, vitamin C and essential metal ions accumulate in leaves may supply substantial antioxidants, which in turn may inhibit, prevent or retard the development of several chronic diseases.

Using Trolox assay, Cook *et al.* (1998) estimated the antioxidant activity of 17 wild edible plants of Niger Republic used for food and traditional medicine. They observed that *Gynandropsis gynandra* possessed strong antioxidant activity.

Cleome icosandra L. Family: Capparaceae

Kalim *et al.* (2010) evaluated the antioxidant activity of Methanol (50 per cent) extract of *Cleome icosandra*. IC_{50} values for DPPH, ABTS, NO, OH, O_2 and ONOO were 7.28, 2.54, 20.13, 152.23, 30.96 and 532.85μg/ml. The extract contained significant amount of polyphenols.

Cleome viscosa L. Family: Capparaceae

Vadlamudi and Naidu (2010) evaluated the antioxidant activity of ethanolic extracts of *Cleome viscosa* leaves. SOD activity was 0.87U/mg, Catalase activity was 2.3U/mg while vitamin C content was 100 mg/100g.

Clerodendron cyrtophyllum Turcz. Family: Verbenaceae

Li *et al.* (2008) evaluated the antioxidant properties of 45 medicinal plants including *Clerodendron cyrtophyllum* using FRAP and TEAC assays and the total phenolic contents of these plants were measured. *Clerodendron cyrtophyllum* showed 67.97 μmol Fe(II)/g FRAP value, 38.88μmol Trolox/g of TEAC value and 11.85 mg GAE/g of Phenolic content.

Clerodendrum infortunatum L. Family: Verbenaceae

Three successive extract of *Clerodendrum infortunatum* leaves have been studied by Gouthamchandra *et al.* (2010) for their potential as antioxidants in DPPH model. The scavenging activity of ethanol extract was found to be high when compared to petroleum ether and chloroform extracts. The antioxidant and its protective effects against CCl_4 induced oxidative stress in rats were significantly high. The presence of high antioxidant and pharmacological properties correlates to the total phenolic contents in the plant.

Clerodendrum trichotomum Thunb. Family: Verbenaceae

The antioxidant properties of jionoside D and isoacteoside, isolated from *Clerodendrum trichotomum* were investigated by Chae *et al.* (2004, 2005). This compound scavenged intracellular reactive oxygen species (ROS) and DPPH radical, and prevented lipid peroxidation. This radical scavenging activity of isoacteoside protected cell viability of Chinese hamster lung fibroblast (V79-4) cells exposed to hydrogen peroxide (H_2O_2). Further more isoacteoside reduced the apoptotic cells formation induced by H_2O_2, as demonstrated by the decreased number of sub-G1 hypo-diploid cells and apoptotic cell body formation. However, isoacteoside increased the activities of cellular antioxidant enzymes, superoxide dismutase (SOD) and catalase (CAT). Taken together, these findings suggest that isoaceteoside, isolated from *C. trichotomum*, possesses antioxidant properties.

Cleyera japonica Thunb.var. _morii_ (Yamamoto) Masamune Family: Theaceae

Hou *et al.* (2003) evaluated the antioxidant activity of 70 per cent aqueous acetone extract of *Cleyera japonica* var. *mori* by various assays including DPPH, hydroxyl radicals and reducing power assay. The extract exhibited stronger activity against DPPH radicals (IC_{50} = 5.4μg/ml) and inhibited formation of OH generated in the Fenton reaction system (0.5μg/ml).The quantity of phenolics was 50.3 mg of GAE/g).

Clibadium funzike Family: Asteraceae (Compositae)

Mosquera *et al.* (2007) studied the antioxidant activity of the extracts of *Clibadium funzike* using DPPH free-radical scavenging method. The percentage of the antioxidant activity of the n-Hexane fraction was 11.34, Dichloromethane fraction was 45.06 and methanol fraction was 46.714.

Clinopodium vulgare L. Family: Lamiaceae (Labiatae)

Tepe *et al.* (2007) examined the chemical composition and *in vitro* antioxidant activity of the essential oil of *Clinopodium vulgare* by using DPPH and β-carotene-linoleic acid assays. In the first case, IC_{50} value of *C. vlugare* essential oil was determined as 63μg/ml. IC_{50} value of thymol and γ-terpinene, the major compounds of the oil, was determined as 161μg/ml. and 122μg/ml., respectively, whereas p-cymene did not show antioxidant activity. In β-carotene-linoleic acid system, *C. vulgare* essential oil exhibited 52.3 per cent inhibition against linoleic acid oxidation.

Clitoria ternatea L. Family: Fabaceae

English Name: Butterfly pea

Kamken and Wulkinson (2009) investigated the potential antioxidant activity of flower extracts of *Clitoria ternatea* and an extract containing eye gel formulation. Aqueous extracts were shown to have stronger antioxidant activity (as measured by

DPPH scavenging activity) than ethanol extracts) IC_{50} values were 1 mg/ml and 4mg/ml, respectively). Aqueous extracts incorporated into an eye gel formulation were also shown to retain this activity however, it was significantly less than a commercial antiwrinkle cream included for comparison. The total phenolic content was 1.9 mg/g extract as gallic acid equivalents. The data from this study support the use of *C. ternatea* extract as antioxidant inclusions in cosmetic products.

Vadlamudi and Naidu (2010) evaluated the antioxidant activity of ethanolic extracts of *Clitoria ternatea* leaves. SOD activity was 1.3U/mg, Catalase activity was 2.9U/mg while vitamin C content was 210 mg/100g.

Narayanaswamy and Balakrishnan (2011) evaluated the antioxidant activity of 13 medicinal plants including leaf and flower of *Clitoria ternatea*. The aqueous extract showed highest 78 per cent inhibition of DPPH radical while ethanolic extract showed 82 per cent inhibition of DPPH radical. The total phenolic contents of herbs were also determined by Folin-Ciocalteu method. *Clitoria ternatea* contained maximum amount of phenols (10.64 mg catechol/g of plant tissue)

Cnestis ferruginea DC. Family: Connaraceae

Oke and Hamburger (2002) studied the antioxidant activity of ethanol extract of leaf and stem using DPPH assay. The extract showed high antioxidant activity.

Cnidium officinale Makino Family: Apiaceae (Umbelliferae)

Jeong *et al.* (2009a,b) investigated the DPPH radical, hydroxyl radical and intracellular ROS scavenging capacity, and Fe^{2+} chelating activity of the extracts from *Cnidium officinale*. These extracts showed a dose-dependent free-radical scavenging capacity and a protective effect on DNA damage and the lipid peroxidation causing the cell damage by ROS. Essential oils showed good antioxidant properties in that IC_{50} value in DPPH and ABTS showed 6.79 and 7.33µg/ml (Jeong *et al.*, 2009a).

Cnidoscolus aconitifolius (Miller) M.Johnston Family: Euphorbiaceae

English Name: Tree spinach.

Total phenolic content and antioxidant capacity of tree spinach species *Cnidoscolus aconitifolius* were determined in raw and cooked leaf extracts. Antioxidant capacity was assessed by the oxygen radical absorbance capacity (ORAC) assay, and flavonoid glycoside composition was quantified by HPLC and identified by GC. Total phenolics and antioxidant capacity were higher than in cooked leaf extracts. The ORAC values were strongly correlated with total phenolic content in all leaf extracts. The major flavonoids isolated from the leaf extracts were kaempferol-3-O-glycosides and quercetin-3-O-glycosides. Cooking reduced antioxidant activity and phenolic content and resulted in losses of some kaempferol glycoside and quercetin glycoside residues in leaf extracts. The results of this indicate that tree spinach leaves are a rich source of natural antioxidant for foods (Kuti and Kunuru, 2004).

Oloyede *et al.* (2010) evaluated the antioxidant activity of the hexane, ethyl acetate, butanol and aqueous extracts of the aerial parts of *Cnidoscolus aconitifolius*. Antioxidant activity determined by ferric thiocyanate method showed that the species provided atleast one fraction with highly promising antioxidant activity. Hexane, ethyl acetate, butanol and water extract at various concentrations (50, 100, 200 and 500 μg/ml) showed antioxidant activities in a concentration dependent manner, the butanol extract however at the concentration of 500μg/ml showed 70-78 per cent antioxidant activity which was very close to that of 500μg/ml of α-tocopherol (82 per cent), the reference compound.

Adefegha and Oboh (2011) reported that cooking decreases the vitamin C contents in the leafy vegetable *Cnidoscolus aconitifolius*, while it increased the phenolic content and antioxidant activities.

Cnidoscolus chayamansa McVaugh Family: Euphorbiaceae

Total phenolic content and antioxidant capacity of tree spinach species *Cnidoscolus chayamansa* were determined in raw and cooked leaf extracts. Antioxidant capacity was assessed by the oxygen radical absorbance capacity (ORAC) assay, and flaonoid glycoside composition was quantified by HPLC and identified by GC. Total phenolics and antioxidant capacity were higher than in cooked leaf extracts. The ORAC values were strongly correlated with total phenolic content in all leaf extracts. The major flavonoids isolated from the leaf extracts were kaempferol-3-O-glycosides and quercetin-3-O-glycosides. Cooking reduced antioxidant activity and phenolic content and resulted in losses of some kaempferol glycoside and quercetin glycoside residues in leaf extracts. The results of this indicate that tree spinach leaves are a rich source of natural antioxidant for foods (Kuti and Kunuru, 2004).

Coccinia indica Wight et Arn. (Syn. *Coccinia grandis* (L.) Voigt) Family: Cucurbitaceae

It is cultivated in India, wild variety grows in India, Tropical Africa, Australia, Fiji and Oriental countries. Leaf extract possesses antioxidant activity, which may be attributed to its protective action on lipid peroxidation and to the enhancing effect on cellular antioxidant defence contributing to the protection against oxidative damage in streptozotocin diabetes (Venkateswaran and Pari, 2002).

Chanwitheesuk *et al.* (2005) investigated the antioxidant activity of *Coccinea grandis* using a β-carotene bleaching method. The contents of plant chemicals such as vitamin C, Vitamin E, carotenoids, tannin and total phenolics, were also determined. Methanolic extract showed antioxidant activity. *Coccinea grandis* has antioxidant index of 2.87, Vitamin C 16.6 mg per cent, Vitamin E 0.0070mg per cent, Total carotenes 1.94 mg per cent, Total xanthophylls 2.65mg per cent, Tannins 17.7mg per cent and Total phenolics of 74.7 mg per cent.

Mazumder *et al.* (2008) investigated the effect of leaves powder, its methanol and aqueous extracts on aspirin-induced gastric ulcer model in rats. The leaf powder showed decrease in level of Lipid Peroxidation and Superoxide dismutase (SOD) activity.

Biswas *et al.* (2010) analyzed antioxidant activity of the aqueous extract of leaves of *Coccinia indica* in different systems of assay. Aqueous extract showed Superoxide radical scavenging activity with IC_{50} = 418.16μg/ml; Hydroxyl radical scavenging IC_{50} = 552.93μg.ml; Lipid peroxidation IC_{50} = 1637.04μg/ml.

Cocos nucifera L. Family: Arecaceae (Palmae)

English Name: Coconut.

Tender *Cocos nucifera* water, variety Chandrasekhara, was tested for its ability to scavenge free radicals, inhibit lipid peroxidation and protect haemoglobin from nitrite-induced oxidation (Mantena *et al.*, 2003). Fresh sample of coconut water scavenged DPPH, ABTS and superoxide radicals but promoted the production of hydroxyl radicals and increased lipid peroxidation. The activity was most significant for fresh samples of coconut water and diminished significantly upon heat, acid or alkali treatment or dialysis. Maturity of coconut drastically decreased the scavenging ability of coconut water against DPPH, ABTS and superoxide radicals. Coconut water protected hemoglobin from nitrite-induced oxidation to methanoglobin when added before the autocatalytic stage of the oxidation.

Alviano *et al.* (2008) evaluated the antioxidant activity of aqueous extract of the husk fiber of *Cocos nucifera* by DPPH photometric assay. The extract showed good antioxidant activity. The antioxidant activity EC_{50} vlaue of the extract was 10μg/ml.

Hydromethanol extracts of 15 Bangladeshi medicinal plants, including *Cocos nucifera*, were evaluated by Hasan *et al.* (2009) for antioxidant potential using DPPH radical scavenging assay. *Cocos nucifera* developing kernel exhibited radical scavenging activity with an IC_{50} value of 13.67 μg/ml compared to the IC_{50} value of 5.15μg/ml as shown by the reference antioxidant ascorbic acid, in a dose dependent fashion.

Codonopsis pilosula (Franch.) Mannb. Family: Campanulaceae

Ng *et al.* (2004) investigated the antioxidant effect of aqueous and organic extracts of roots of *Codonopsis pilosula*. The aqueous extracts were the most potent in inhibiting erythrocyte hemolysis. The organic extracts had weaker lipid peroxidation.

Coffea arabica L. Family: Rubiaceae

English Name: Coffee.

Caffeine (1,3,7-trimethyl xanthine), an ingredient of *Coffea arabica*, has been investigated by Devasagayam *et al.* (1996) for its potential antioxidant activity against oxidative damage to rat liver microsomes. The results obtained showed that caffeine was an effective inhibitor of lipid peroxidation, at millimolar concentrations, against hydroxyl radical, peroxyl radical and singlet oxygen. The extent of inhibition was high against peroxidation induced by OH medium against singlet oxygen and low against peroxyl radical. In general, the antioxidant ability of caffeine was similar to that of the established biological antioxidant glutathione and significantly higher than ascorbic acid.

The antioxidative activity of column chromatographic fractions obtained from brewed coffee was investigated by Yanagimoto *et al.* (2004) to find antioxidants and to assess the benefit of coffee drinking. The results indicate that brewed coffee contains many antioxidant and consumption of antioxidant-rich brewed coffee may inhibit diseases caused by oxidative damages.

In vitro methods such as the ABTS+ decolorization assay and the oxygen radical absorbance capacity assay (ORAC) were used to assess the capacity of coffee compounds to scavenge free radicals (Gomez Ruiz *et al.*, 2007). Colonic metabolites such as m-coumaric acid and dihydroferulic acid showed high antioxidant activity.

Green coffee conserves were prepared from *Coffea arabica* by flaking, powdering and extraction with solvent mixtures of isopropanol and water in different ratios and their antioxidant properties were investigated by Naidu *et al.* (2008). The yields of conserves were highest 27 per cent when isopropanol and water in ratio of 60:40 was employed. The total polyphenol content was determined and found to be higher (31.7-32.2 per cent) in these conserves. At a concentration of 200 ppm, coffee conserves exhibited 92 per cent antioxidant activity in comparison to 95 per cent for BHA. The conserves were analyzed by HPLC and three phenolic compounds could be identified. Results indicated that the conserves from green coffee possess potential antioxidant activity and could be used as nutraceuticals as well as preservatives in food formulations (Naidu *et al.*, 2008).

Ramalakshmi *et al.* (2008) have shown that the high antioxidant potential of the methanol extract of low-grade coffee beans is due the presence of phenolic compounds including chlorogenic acids, which make them more suitable as a source of natural antioxidant and their utility can be explored in food industry.

Coffea robusta Linden Family: Rubiaceae

Green coffee conserves were prepared from *Coffea robusta* by flaking, powdering and extraction with solvent mixtures of isopropanol and water in different ratios and their antioxidant properties were investigated by Naidu *et al.* (2008). The yields of conserves were highest 29 per cent when isopropanol and water in ratio of 60:40 was employed. The total polyphenol content was determined and found to be higher (31.7-32.2 per cent) in these conserves. At a concentration of 200 ppm, coffee conserves exhibited 88 per cent antioxidant activity in comparison to 95 per cent for BHA. The conserves were analyzed by HPLC and three phenolic compounds could be identified. Results indicated that the conserves from green coffee possess potential antioxidant activity and could be used as nutraceuticals as well as preservatives in food formulations (Naidu *et al.*, 2008).

Coix lacryma-jobi L. Family: Poaceae (Graminae)

English Name: Adlay.

An activity-directed fractionation and purification process was used by Kuo *et al.* (2002) to identify the antioxidative components of adlay hulls. Hulls of adlay were extracted with methanol and then separated into water, 1-butanol, ethyl acette and hexane fractions. The 1-butanol-soluble fraction exhibited greater capacity to scavenge DPPH radicals when compared with fractions soluble in water, ethyl acetate and hexane phases. Six compounds showing strong antioxidant activity were identified. These are coniferylalcohol, syringic acid, ferulic acid, syringaresinol, 4-ketopinoresinol and a new lignan, mayuenolide.

Gan *et al.* (2010) evaluated the antioxidant activity of *Coix lacryma-jobi* using the FRAP and TEAC assays, and its total phenolic content was measured by the Folin-

Ciocalteu method. For FRAP assay the value was 7.75 μmol Fe(II)/g dry weight. For TEAC assay, the value was 4.69μmol Trolox/g dry weight. Total phenolic content was 2.34 mg gallic acid equivalent/g dry weight of plant material.

Cola nitida (Vent.) Schott et Endl. Family: Sterculiaceae

An extract containing oligomeric proanthocyanidins (OPC) from *Cola nitida* was prepared by maceration in acetone/water and evaluated by Rakotoarison *et al.* (1996) for their capacity to scavenge oxygen free radical. Results expressed in inhibitory concentrations IC_{50} in mg/l, showed effective antioxidant activities of cola extract. Indeed by comparison with known antioxidant activities of molecules such as mesna, N-acetyl-cystein or glutathione, OPC extract presented better efficacy. The OPC are known to contain proanthocyanidins B1, B2, B3 and B4 with the predominance of B1 and B2 dimers. Thus, it is concluded that these molecules protect tissues against free oxygen radicals (Rakotoarison *et al.*, 2004).

Colchicum balansae Planchon Family: Colchicaceae

Mammadov *et al.* (2009) evaluated the antioxidant activity of extracts from tubers and leaves of *Colchicum balansae*. Result of this study show that leaves extracts exhibited higher antioxidant activity than tuber extracts with all types of solvent. The highest antioxidant activity efficiency was determined in extract leaf-ethanol (64 per cent) and the least efficiency in extract tuber-benzine (14.5 per cent). All extracts of CB tubers and leaves have effective free radical scavenging and reducing power. The highest free radical scavenging activity was determined in extract leaf-benzene (68.35 per cent), this activity was followed by acetone (61.23 per cent), methanol (58.67 per cent) and ethanol extracts (54.74 per cent) respectively. The highest radical scavenging activity was determined in extract tuber benzene (61.28 per cent), and the least efficiency in extract tuber-ethanol (20.48 per cent). In addition, all extracts of tubers and leaves were determined as pyrocatechol equivalents.

Colchicum speciosum Stev. Family: Colchicaceae

English Name: Colchicum corm.

Souri *et al.* (2008) investigated the antioxidant activity and radical scavenging activity of methanolic extract of root stock. The antioxidant activity of this plant extract against linoleic acid peroxidation expressed as IC_{50} was 39.77ng/ml and radical scavenging activity against DPPH expressed as IC_{50} was 24.63 μg/ml.

Coleus amboinicus Lour. Family: Lamiaceae (Labiatae)

Chanwitheesuk *et al.* (2005) investigated the antioxidant activity of *Coleus amboinicus* using a β-carotene bleaching method. The contents of plant chemicals such as vitamin C, Vitamin E, carotenoids, tannin and total phenolics, were also determined. Methanolic extract showed antioxidant activity. *Coleus amboinicus* has antioxidant index of 4.22, Vitamin C 10.2 mg per cent, Vitamin E 0.0054mg per cent, Total carotenes 2.54 mg per cent, Total xanthophylls 4.24mg per cent, Tannins 24.3mg per cent and Total phenolics of 54.8 mg per cent.

Coleus aromaticus Benth. Family: Lamiaceae (Labiatae)

The antioxidant activity of freeze-dried aqueous extract of the leaves of *Coleus aromaticus* was investigated by Kumaran and Karunakaran (2006a, 2007a) employing various established *in vitro* systems. The extract showed notable inhibitory activity in the β-carotene-lineolated system. The multiple antioxidant activity of the extract was evident as it showed significant reducing power, superoxide scavenging activity, nitric oxide scavenging activity and also ferrous iron chelating potency. Rosmarinic acid was found as a major component and principally responsible for the radical scavenging activity.

A study was undertaken by Rao *et al.* (2006) to elucidate *in vitro* free radical scavenging potential and inhibition of lipid peroxidation by *C.aromaticus* hydroalcoholic extract (CAE). Anti-clastogenic and radioprotective potential of CAE were studied using micronucleus assay after irradiating Chinese hamster fibroblast (V79) cells. CAE at 10, 20, 40, 60, 80, 100 and 120 µg/ml resulted in a dose-dependent increase in radical scavenging ability against various free radicals *viz.*, DPPH, ABTS, superoxide anion ($O_2^{\bullet -}$), hydroxyl ($OH^{\bullet}$) and nitric oxide ($NO^{\bullet}$) generated *in vitro*. A maximum scavenging potential was noticed at 100 µg/ml and a saturation point was reached thereafter with the increasing doses of CAE. The free radical scavenging potential of the extract was in the order of DPPH > ABTS > Superoxide > Hydroxyl > Nitric oxide. CAE also exhibited a moderate inhibition of lipid peroxidation *in vitro*, with a maximum inhibition at 60 µg/ml (33 per cent), attaining saturation at higher doses. The extract also rendered protection against radiation induced DNA damage, as evidenced by the significant decrease in the percentage of radiation-induced micronucleated cells (MN) and frequency of micronuclei (total). A maximum anticlastogneic effect/radioprotection was noticed at a very low concentration *i.e.*, 5 µg/ml of CAE, treated 1 h prior to 2 Gy of gamma radiation. A significant anticlastogenic/radioprotective effect was also observed when the cells were treated with an optimum dose of CAE (5 µg/ml) 1 h prior to 0.5, 1, 2 and 4 Gy of gamma radiation compared with the respective radiation control groups. Overall, their results established an efficient antioxidant, anticlastogenic and radioprotective potential of CAE, which may be of great pharmacological importance.

Coleus blumei Benth. Family: Lamiaceae (Labiatae)

Zakaria *et al.* (2008) studied the antioxidant properties of the methanol extract of leaves and stems of *Coleus blumei* by DPPH radical scavenging activity. The extracts showed antioxidant activity comparable to butylated hydroxytoulene (BHT).

Colocasia esculenta Schott. Family: Araceae

English Name: Taro.

Maeda *et al.* (2006) examined the phenolic contents of taro and its radical scavenging activities for DPPH radical. The polyphenol content was 87.1 mg-gallic acid equivalent/100 g) and DPPH scavenging activity was 343.9 µmol-Trolox eq./ 100 g).

Phytochemical flavonols, carotenoids and the antioxidant capacity of leaves of *Colocasia esculenta* was evaluated by Lako *et al.* (2006). *Colocasia esculenta* leaves have moderate Total antioxidant capacity (130 mg/100g) and Total polyphenol content (TPP) (120mg/100g) and β-carotene (19mg/100g).

In a screening of South African indigenous food plants for antioxidant activity by testing for inhibition of lipid peroxidation the teas with the tuber of *Colocasia esculenta* had high activity (Lindsey *et al.*, 2006).

Combretum acutifolium; Combretum apiculatum Sond., *C. caffrum* (Eckl. et Zeyh.) Kuntze; *Combretum celastroides* Welw. ex Laws. *Combretum erythrophyllum* (Burch.) Sond.; *Combretum imberbe* Wawra; *Combretum kraussii* Hochst.; *Combretum molle* R.Br.; *Combretum nelsonii* Duemmer; *Combretum petrophilum* Retief; *Combretum padoides* Engl. et Diels Family: Combretaceae

Masoko and Eloff (2007) evaluated the antioxidant activity of leaves of 24 species of *Combretum*.Acetone and methanol extracts of *Combretum caffrum, C. celastroides, C. erythrophyllum, C. imberbe, C. padoides, C. kraussii* showed moderate antioxidant activity.

Combretum collinum Fresen; *Combretum hereroense* Schinz; *Combretum woodii* Duemmer; Family: Combretaceae

Masoko and Eloff (2007) evaluated the antioxidant activity of leaves of 24 species of *Combretum*. Acetone and methanol extracts of *Combretum collinum, Combretum hereroense* and *Combretum woodii* showed strong antioxidant activity.

Combretum edwardsii Exell; *Combretum zeyheri* Sond. Family: Combretaceae

Masoko and Eloff (2007) evaluated the antioxidant activity of leaves of 24 species of *Combretum*. Acetone extract of *Combretum zeyheri* showed moderate antioxidant activity.

Combretum micranthum G.Don Family: Combretaceeae

Karou *et al.* (2005) evaluated the polyphenol content and antioxidant activity of *Combretum micranthum* by ABTS assay. Polyphenols in the lyophilized extract of leaves was 37.80 per cent. The antioxidant activity was 2.08µmol Trolox/µg in the Phosphomolybdenum assay and 16.37 µmol Trolox/µg in ABTS assay. The total phenolic compounds were highly correlated with the antioxidant activities.

Commelina benghalensis L. Family: Commelinaceae

Hydromethanol extracts of 15 Bangladeshi medicinal plants, including *Commelina benghalensis*, were evaluated by Hasan *et al.* (2009) for antioxidant potential using DPPH radical scavenging assay. *Commelina benghalensis* aerial parts exhibited radical scavenging activity with an IC_{50} value of 21.53 µg/ml compared to the IC_{50} value of 5.15µg/ml as shown by the reference antioxidant ascorbic acid, in a dose dependent fashion.

Commelina diffusa Burm.f. Family: Commelinaceae

Extracts of *Commelina diffusa* have been shown to have anti-inflammatory and antioxidant properties due to flavonoids, triterpenoids, diterpenoids and caffeic acid derivatives (Houghton *et al.*, 2005).

Commiphora berryi (Arn.) Engl. Family: Burseraceae

Gowri Shankar *et al.* (2008) reported the hepatoprotective and antioxidant effects of *Commiphora berryi* bark extract against CCl_4-induced oxidative damage in rats. They reported dose-dependent reduction in the serum levels of liver enzymes.

Commiphora caudata (Wight et Arn.) Engl. Family: Burseraceae

Annu *et al.* (2010) investigsted the anti-inflammatory, analgesic and lipid-peroxidation studies on leaves of *Commiphora caudata* leaves. Ethanol extract of leaves showed significant inhibition of $FeCl_2$-ascorbic acid stimulated lipid peroxidation in rat liver homogenate *in vitro* at 50 µg/ml dose. Lower doses studied did not produce significant anti-lipid peroxidation effects *in vitro* and doses higher than 50µg/ml did not significantly increase the anti-lipid peroxidation effect any further.

Commiphora myrrha (Nees) Engl. Family: Burseraceae

The resin exudates and sap content is found to be effective against arthritis, β-sitosterol, campestrol and eugenol in myrrh have antioxidative effect (Mishra *et al.*, 2007).

Commiphora ornifolia (Balf.f.) Gillett Family: Burseraceae

Mothana *et al.* (2009) evaluated the antioxidant activity of 26 Yemeni medicinal plants using DPPH assay. Methanolic extracts of *Commiphora ornifolia* showed good antioxidant potential at low concentrations (more than 80 per cent at 50g/ml).

Mothana *et al.* (2010) investigated the chemical composition and antioxidant activity of essential oil of *Commiphora ornifolia*. The authors found low antioxidant activity even at high concentrations.

Commiphora parvifolia Engl. Family: Burseraceae

Mothana *et al.* (2010) investigated the chemical composition and antioxidant activity of essential oil of *Commiphora parvifolia*. The authors found low antioxidant activity even at high concentrations.

Commiphora wightii (Arn.) Bhandari (Syn: _Commiphora mukul_ Engl. Family: Burseraceae

English Name: Guggul

The gum yielded from the plant is used for treatment of arthritis and lowering blood pressure (Mishra *et al.*, 2007). Al-Burtamani and Fatope (2004) isolated flavonoids, naringenin and 3-hydroxynaringenin from *Commiphora wightii*. The antioxidant activities of these compounds were estimated in DPPH assay. The percent of IP of the crude methanol extract, naringenin, 3-hydroxynaringenin and gallic acid (control) were 54,39, 66 and 88 per cent respectively.

Bioactivity-directed fractionation and purification afforded components of *Commiphora wightii*. This fraction showed moderate scavenging effect against DPPH radicals (Zhu *et al.*, 2001).

Comocladia engleriana Loes. Family: Anacardiaceae

Ruiz-Terán *et al.* (2008) evaluated the antioxidant activity of aerial parts of *Comocladia engleriana* using DPPH and β-carotene-linoleic acid bleaching assay. The extract displayed antioxidant activity comparable to the commercial antioxidant BHA. For the methanolic extract analysed, a clear relation between the total phenolic content of the extracts and their antioxidant activity was found.

Conyza canadensis (L.) Cronquist Family: Asteraceae (Compositae)

Antibacterial, antioxidant and cytotoxic activities of petroleum ether, ethyl acetate and methanol extracts of *Conyza canadensis* were investigated by Hayat *et al.* (2009). Antioxidant activity was measured by the DPPH method. Among the tested extracts methanolic extract showed good antioxidant activity with IC_{50} of 120μg/ml.

Copaifera langsdorffii Desf. Family: Myristichaceae

Santos *et al.* (2009) evaluated the antioxidant activity of extracts from nine plant species belonging to the Brazilian flora acting on the free radicals DPPH and TEMPOL (4-hydroxy-2,2,6,6-tetramethyl-1-piperidinyloxy-1-oxyl) by electron paramagnetic resonance (EPR), and acting on the hydroxyl radical by the spin trapping technique generated by a Fenton reaction. Results showed that the extracts of *Copaifera langsdorffii* display strong antioxidant activities with DPPH scavenging IC_{50} value of 0.057 mg/ml while Hydroxyl scavenging IC_{50} value was 1.772 mg/ml.

Copernicia cerifera Mart. Family: Arecaceae (Palmae)

Silva *et al.* (2005) evaluated the antioxidant activity of ethanol and butanol extracts from leaves and mesocarp of fruit of *Copernicia cerifera*. They evaluated the capacity of the extracts to inhibit the reduction of free radical DPPH and to protect *Saccharomyces cerevisae* cells against the lethal oxidative stress caused by tert-butylhydroperoxide (TBH). Ethanol and butanol extract of leaves and mesocarp of fruits of *Copernicia cerifera* were able to increase the tolerance of *Saccharomyces cerevisae* to TBH and showed to be active as DPPH radical scavengers, thus indicating that this plant extracts could be considered as potential sources of antioxidants. The extracts exhibited a DPPH radical scavenging activity higher than that obtained from *Ginkgo biloba*, a reference plant with well documented antioxidant activity.

Coptis chinensis Franch. Family: Ranunculaceae

The antioxidant properties of twenty medicinal herbs used in the traditional Mediterranean and Chinese medicine including *Coptis chinensis* were studied by Schinella *et al.* (2002). All the extract were assayed at a concentration of 100μg/ml. Most of the extracts were weak scavengers of the hydroxyl radical and *Coptis chinensis* exhibited the highest scavenging activity.

Cai *et al.* (2004) evaluated the antioxidant activity and phenolic compounds in traditional Chinese medicinal plants associated with anticancer. The improved ABTS.+ method was used to systematically assess the total antioxidant capacity (Trolox equivalent antioxidant capacity, TEAC) of the medicinal extracts. The TEAC values and total phenolic content for methanolic extract of root of *Coptis chinensis* were 325.4 μmol Trolox equivalent/100g dry weight (DW), and 1.48 g of gallic acid equivalent/100g DW respectively. A positive significant linear relationship between antioxidant activity and total phenolic content showed that phenolic compounds were the dominant antioxidant components in the tested herbs. Major types of phenolic compounds in *Coptis chinensis* include phenolic acids (ferulic acid, chlorogenic acid), phenolic alkaloids (magnoflorine).

The antioxidant activity of magnoflorine, an alkaloid isolated from the rhizome of *Coptis chinensis* was examined against the oxidation of low density lipoprotein and modified LDL (Hung *et al.*, 2007). Magnoflorine was found to inhibit the copper-mediated oxidation of LDL, as well as glycated and glycoxidated LDL by increasing the lag time of conjugated diene formation and preventing the generation of TBARS. In addition, the results from the fluorescence emission spectra of tryptophan supported that the antioxidant activity of magnoflorine could be associated with the protective effect on the structural modification of apolipoprotein B required for LDL oxidation. Results suggested that magnoflorine may be useful for preventing the oxidation of various LDL forms.

Li *et al.* (2008) evaluated the antioxidant properties of 45 medicinal plants including *Coptis chinensis* using FRAP and TEAC assays and the total phenolic contents of these plants were measured. *Coptis chinensis* showed 112.2 μmol Fe(II)/g FRAP value, 61.66μmol Trolox/g of TEAC value and 12.89 mg GAE/g of phenolic content. A high correlation between antioxidant capacities and their phenolic contents indicated that phenolic compounds were a major contributor of antioxidant activity of these plants.

Corchorus capsularis L. Family: Tiliaceae

Zakaria (2007) evaluated the free radical scavenging properties of *Corchorus capsularis* using DPPH and superoxide anion radical scavenging assays. The extract was found to show remarkable antioxidant activity in both assays. The DPPH radical scavenging activity was 97.50 per cent and superoxide scavenging activity was 94.90 per cent. Phytochemicals screening of this plant demonstrated the presence of flavonols, triterpenes, saponins and steroids.

Corchorus olitorius L. Family: Tiliaceae

English Name: Jew's mallow; Japanese: Morohaiya.

Oboh and Akindahunsi (2004) investigated the effect of preservation of green leafy vegetables of Nigeria on the antioxidant phytoconstituents (Vitamin C and Total phenol) and activity (reducing property and free radical scavenging ability) including *Corchorus olitorius*. The result of the study revealed that sun-drying of green leafy vegetables cause a significant decrease in the vitamin C content (16.67-64.68 per cent loss). Conversely it leads to a significant increase in the total phenol content (6.45 -223.08 per cent grain) reducing property (16.00-362.50 per cent grain) and free radical scavenging ability (126.00-5757.00 per cent gain) of green leafy vegetables.

Maeda *et al.* (2006) examined the phenolic contents of leaf of *Corchorus olitorius* and their radical scavenging activities by using DPPH assay. High contents of phenolic compounds and strong DPPH scavenging activities were observed in extracts from this species. They also showed radical scavenging activities for the superoxide radical and tert-butylhydroperoxyl radical (t-BuOO·).

Odukoya *et al.* (2007) evaluated the antioxidant activity and total phenolic contents of Nigerian green leafy vegetables including *Corchorus olitorius*. Potential free radical scavenging activity of these vegetables was confirmed by spraying spots of the extracts with DPPH (yellow color on purple background). Antioxidant activity was assayed in linoleic acid model system. Total polyphenols as Tannic Acid Equivalent (TAE) and Ascorbic Acid. In terms of TAE Odukoya *et al.* (2007) reported 153.63 mg AA/100g. There was low correlation between AA/TAE, AA/ASC and TAE/ASC.

Antioxidant activities of eight leafy vegetables of Ghana including *Corchorus olitorius* were assessed by Morrison and Twumasi (2010). The total antioxidant capacity (TAC) and Total Phenol content (TPC) in the methanol extracts (METE) and hydro-ethanol extracts (HETE) were measured. The TPC of the methanol extract of *Corchorus olitorius* was 0.292 mg/ml TAE and TAE was 0.290 mg/ml Ascorbic acid equivalent (AAE) while TPC of hydro-ethanol extract was 0.432 mg/ml TAE and TAE was 0.421 mg/ml Ascorbic acid equivalent (AAE). Radical scavenging activity of methanol extract was EC_{50} = 0.1635g/ml and HETE was 0.1452 mg/ml while Fe^{3+} reducing potential was EC_{50} = 0.1452 (METE) and 0.1452 (HETE).

Salawu *et al.* (2011) evaluated the cellular antioxidant activities of green leafy vegetables including *Corchorus olitorius* against peroxyl radical-induced oxidation in HepG2 cells. HPLC/DAD/MS investigation of the ethanolic extract of the vegetal matter revealed the presence of 5 phenolic compounds. At a concentration of 1 mg/ml the quenching of peroxyl radical in HepG2 cells revealed that *Corchorus olitorius* has moderate antioxidant activity.

Cordia gilletii De Wild. Family: Boraginaceae

The methanol and dichloromethane extracts of *Cordia gilletii* showed the greater antioxidant activity by scavenging the free radical DPPH with IC_{50} values of 3.2 and 8.1 µg/ml, respectively (Okusa and Penge, 2007).

Cordia multispicata Cham. Family: Boraginaceae

The total phenolic content and antioxidant activity of extracts and four flavonoids isolated from leaves of *Cordia multispicata* were evaluated using Folin-Ciocalteu reagent, DPPH free radical scavenging and inhibition of peroxidation of linoleic acid by FTC method. Total phenolic content was 66.1 to 231 mg/g, inhibition of lipid peroxide formation was IC_{50} = 51.2 to 89 mg/l (da Silva *et al.*, 2010).

Cordia myxa L. Family: Boraginaceae

English Name: Sebesten plums.

Souri *et al.* (2008) evaluated the antioxidant activity against linoleic acid peroxidation using 1,3-diethyl-2-thiobarbituric acid as reagent. Antioxidant activity (IC_{50}) against peroxidation of linoleic acid (2mg/ml) was 132.53 and phenolic content was 373.91 mg/100g dry weight. The results of this study showed that there is no significant correlation between antioxidant activity and phenolic content of the studied plant materials and phenolic content could not be a good indicator of antioxidant capacity.

Cordia perrottettii A.DC. Family: Boraginaceae

Marwah *et al.* (2007) investigated the antioxidant activity of wound-healing herbs using *in vitro* DPPH and phosphomolybdenum assay methods. Out of the 19 plants screened, the aqueous ethanol extract of *Cordia perrottettii* showed the best DPPH scavenging activities with per cent IP value 85.4 per cent, after 15 min of incubation at an effective test concentration of 50µg/ml. The Total Antioxidant Capacity as gallic acid equivalent of 646mg/g of ethanol extract was obtained in the phosphomolybdenum assay. Total phenolic content was 145 mg GAE/g of ethanol extract.

Coriandrum sativum L. Family: Umbelliferae (Apiaceae)

English Name: Coriander.

Coriander seeds have been shown to have antiperoxidative properties (Chitra and Leelamma, 1999). The activity of polyphenolic compounds from coriander seeds on oxidative damage induced by H_2O_2 in human lymphocytes has been shown (Hashim *et al.*, 2005). Quercetin 3-glucuronide, isoquercitrin and rutin have been separated and identified in coriander fruits (Kunzemann and Hermann, 1977). Guerra *et al.* (2005) determined the antioxidant activity of the fractions containing approximately 1.5 mg of carotenoids by the coupled oxidation of the β-carotene and the linoleic acid. While Wangensteen *et al.* (2004) and Wong and Kitts (2006) confirmed antioxidant property of *C. sativum* using DPPH assay.

Sabu and Kuttan (2003) investigated the antioxidant activity of Indian herbal drugs, including *Coriandrum sativum,* in rats with alloxan-induced diabetes. The extract of *Coriandrum sativum* showed antidiabetic and antioxidant activity.

Souri *et al.* (2004) investigated the antioxidant activity of methanol extracts of leaves and twigs of coriander, used as vegetable in Iranian diet against linoleic acid peroxidation. The extracts showed good antioxidant activity.

Chanwitheesuk *et al.* (2005) investigated the antioxidant activity of leaves of *Coriandrum sativum* using a β-carotene bleaching method. The contents of plant chemicals such as vitamin C, Vitamin E, carotenoids, tannin and total phenolics, were also determined. Methanolic extract showed antioxidant activity. *Coriandrum sativum* has antioxidant index of 1.41, Vitamin C 11 mg per cent, Vitamin E 0.0038mg per cent, Total carotenes 2.52 mg per cent, Total xanthophylls 1.05mg per cent, Tannins 24.3mg per cent and Total phenolics of 33mg per cent.

Singh *et al.* (2006) reported that the oleoresin from *Coriandrum sativum* had a better antioxidant activity for sunflower oil than essential oil and synthetic antioxidants, BHA and BHT.

Wong *et al.* (2006) investigated the antioxidant properties of 25 edible tropical plants including *Coriandrum sativum* using DPPH and FRAP assays. Total phenolic content was also estimated. *Coriandrum sativum* showed good antioxidant activity.

Souri *et al.* (2008) evaluated the antioxidant activity against linoleic acid peroxidation using 1,3-diethyl-2-thiobarbituric acid as reagent. Antioxidant activity (IC_{50}) against peroxidation of linoleic acid (2mg/ml) was 41.25 and phenolic content was 227.69 mg/100g dry weight.

Mariutti *et al.* (2008) evaluated the antioxidant activity of coriander employing ABTS and DPPH radical assay. DPPH EC_{50} value was 235g/kg while DPPH-TEAC value was 1.1mM/g. ABTS-TEAC value was 2.21mM/g.

Antioxidant activities of ethanol extracts from seven Apiaceae fruits have been studied by Nickavar and Abolhasani (2009) by the DPPH radical scavenging test. All the studied extracts, including *Coriandrum sativum*, showed antioxidant capability. For *Coriandrum sativum* IC_{50} value of the DPPH scavenging activity was 222.40µg/ml and total flavonoid content was 9.81µg/mg. A positive correlation was found between the antioxidant potency and flavonoid content of the fractions.

Antioxidant potential of selected vegetables commonly used in diet in Asian subcontinent, including *Coriandrum sativum* leaves has been evaluated by Gacche *et al.* (2010). The DPPH radical scavenging activity was 62.4 per cent, inhibition of lipid peroxidation was 54.9 per cent, Vitamin C content was 20.48 mg/100g and total phenols was 2.20 mg/g.

Samojlik *et al.* (2010) assayed the antioxidant and hepatoprotective potential of essential oil of *Coriandrum sativum* against CCl_4 damage. Tested essential oils were able to reduce the stable DPPH in a dose dependent manner and to neutralize H_2O_2, reaching 50 per cent neutralization with IC_{50} values of <4.05µl/ml. Coriander essential oil exhibited prooxidant activity in LP in both systems of induction. *In vivo* investigation conferred leak of antioxidative capacity of coriander essential oil.

Coridothymus capitatus (L.) Reichb.f. Family: Lamiaceae (Labiatae)

English Name: Spanish oregano.

Demo *et al.* (1999) reported antioxidant α-tocopherol in the Hexane extract of *Coridothymus capitatus*. Kosar *et al.* (2003) prepared water extracts from steam distilled essential oil-extracted *Coridothymus capitatus*. The HPLC-DPPH on-line method was applied to the qualitative and quantitative analysis of this plant extract. There was a strong correlation between the scavenging (negative) peak area and the concentration of the radical scavenging reference substances used. The radical scavenging compounds within the extracts were determined as benzoic acid and hydroxycinnamic acid derivatives, flavonoids and diterpenoids according to their retention time and UV spectral data. Rosmarinic acid and carnosic acid were identified as the dominant radical scavengers in these extracts by this method.

Water-soluble extract from *Coridothymus capitatus* was screened by Dorman *et al.* (2004) for antioxidant properties in a battery of six *in vitro* assays. The extract demonstrated varying degrees of efficacy in each screen. Spanish oregano was the most effective inhibitor of nonsite-specific hydroxyl radical-mediated 2-deoxy-d-ribose degradation.The extract contained Folin-Ciocalteu reagent-reactive substances, which was confirmed by the presence of polar phenolic analytes (*i.e.*, hydroxybenzoates, hydroxycinnamates, and flavonoids).

Cornus mas L. Family: Cornaceae

English Name: Cornelian cherry.

Leskovac *et al.* (2007) reported radioprotective and antioxidant properties of *Cornus mas*. Treatment with 0.1mg/ml showed reduced levels of lipid peroxidation products (50.04 per cent). Pantelidis *et al.* (2007) assayed the antioxidant activity (determined as FRAP and deoxyribose protection), ascorbic acid, phenol and anthocyanin contents. This study outlines that the native Cornelian cherry population is an extremely rich source of antioxidants.

Serteser *et al.* (2009) studied the antioxidant activity of 50 per cent aqueous methanol extracts by various antioxidant assays, including free radical scavenging, hydrogen peroxide scavenging and metal (Fe^{2+}) chelating activities. The methanolic extract examined in the assay showed the strongest activities. The DPPH radical scavenging effects of leaves extract was 0.716 while fruit extract was 0.928. Fe^{2+} chelating activity (per cent) of leaves extracts was 45.72 while fruit extracts showed 54.24 per cent inhibition. Hydrogen peroxide inhibition activity of leaves extracts was 65.42 per cent and fruit extracts was 74.35 per cent.

Tural and Koca (2008) reported that the methanolic extracts of the Cornelian cherry fruits showed EC_{50} (mg/ml) (DPPH reduction) values as 0.52.

Antioxidant capacity, total anthocyanins, total phenols, ascorbic acid and total flavonoids of a number of selected cornelian cherry genotypes were investigated by Hassanpour *et al.* (2011). The range of ascorbic acid content of genotypes was 183.25 – 299.5 mg/100g fresh weight. There were linear relationships between the antioxidant capacities with total phenols, total flavonoid and ascorbic acid. C_{24} genotype had the highest total antioxidant capacity and total phenolics.

Cornus officinalis Sieb. et Zucc. Family: Cornaceae

Methanol aqueous extract of *Cornus officinalis* was screened by Kim *et al.* (1997) for free radical scavenging activity using the DPPH free radical generating system. The extract showed free radical scavenging activity.

Cai *et al.* (2004) evaluated the antioxidant activity and phenolic compounds in traditional Chinese medicinal plants associated with anticancer. The improved ABTS.+ method was used to systematically assess the total antioxidant capacity (Trolox equivalent antioxidant capacity, TEAC) of the medicinal extracts. The TEAC values and total phenolic content for methanolic extract of fruit of *Cornus officinalis* were 651.8 μmol Trolox equivalent/100g dry weight (DW), and 2.30 g of gallic acid equivalent/100g DW respectively. A positive significant linear relationship between antioxidant activity and total phenolic content showed that phenolic compounds were the dominant antioxidant components in the tested herbs. Major types of phenolic compounds in *Cornus officinalis* include pehnolic acids (gallic acid), flavonols (kaempferol, quercetin and their glycosides), tannins (cornus-tannins).

Cornus sanguinea L. subsp. australis (C.A.Mey) Jav. Family: Cornaceae

Serteser *et al.* (2009) studied the antioxidant activity of 50 per cent aqueous methanol extracts by various antioxidant assays, including free radical scavenging, hydrogen peroxide scavenging and metal (Fe^{2+}) chelating activities. The methanolic extract examined in the assay showed the strongest activities. The DPPH radical scavenging effects of leaves extract was 0.693 while fruit extract was 0.865. Fe^{2+} chelating activity (per cent) of leaves extracts was 44.64 while fruit extracts showed 52.34 per cent inhibition. Hydrogen peroxide inhibition activity of leaves extracts was 62.32 per cent and fruit extracts was 71.23 per cent.

Cornus sanguinea L. subsp. _sanguinea_ Family: Cornaceae

Serteser *et al.* (2009) studied the antioxidant activity of 50 per cent aqueous methanol extracts by various antioxidant assays, including free radical scavenging, hydrogen peroxide scavenging and metal (Fe^{2+}) chelating activities. The methanolic extract examined in the assay showed the strongest activities. The DPPH radical scavenging effects of leaves extract was 0.672 while fruit extract was 0.829. Fe^{2+} chelating activity (per cent) of leaves extracts was 44.92 while fruit extracts showed 51.24 per cent inhibition. Hydrogen peroxide inhibition activity of leaves extracts was 61.28 per cent and fruit extracts was 69.03 per cent.

Cornus stolonifera Michx. Family: Cornaceae

McCune (1999) evaluated antioxidant activities of indigenous medicinal plants of Canadian boreal forest by three different assays (DPPH, NBT/xanthine oxidase and DCF/APPH). *Cornus stolonifera* showed antioxidant activity similar to vitamin, and green tea in all the assays.

Coronopus didymus (L.) Sm. Family: Brassicaceae (Cruciferae)

The whole plant aqueous extract of *Coronopus didymus* was fractionated on the basis of polarity (Prabhakar *et al.*, 2006). The most non-polar fraction was found to be more active than other fractions in scavenging DPPH, ABTS, nitric oxide and hydroxyl radicals in steady-state conditions. Stop-flow spectrometric studies showed 58.13 per cent inhibition of 100μM DPPH at a concentration of 150μg/ml of non-polar fraction in 1000s and 32.31 per cent scavenging of 960μM ABTS at a concentration of 300μg/ml of CDF in 100s (Prabhakar *et al.*, 2006).

Corylus avellana L. Family: Corylaceae

English Name: Hazelnut.

Three hazelnut cultivars produced in Portugal were characterized by Oliveira *et al.* (2007, 2008) in respect of their chemical composition, antioxidant potential and antimicrobial activity. Antioxidant potential was assessed by the reducing power assay, the scavenging effect on DPPH radicals and β-carotene lineolate model system. Aqueous hazelnut extract presented antioxidant activity in a concentration-dependent way, in general with similar behaviour for all cultivars. Eight phenolic compounds were identified: 3-,4- and 5-caffeoylquinic acids, caffeoyltartaric acid, p-coumaroyltartaric acid, myricetin-rhamnoside, quercetin-3-rhamnoside and kaempferol 3-rhamnoside. The hazel leaves extract presented a high antioxidant activity in a concentration-dependent way (Oliveira *et al.*, 2007).

Coscinium blumeanum Miers Family: Menispermaceeae

Keawpradub *et al.* (2005) evaluated the antioxidant activities of *Coscinium blumeanum* employing DPPH radical scavenging assay. Methanol and chloroform extracts exhibited moderate antioxidant activity with EC_{50} values of 25-55µg/ml. An ester triacontanyl caffeate was isolated from *C.blumeanum*. Triacontanyl caffeate was mainly responsible for antioxidant activity with an EC_{50} value of 98.0µg/ml.

Coscinium fenestratum (Gaertn.) Colebr. Family: Menispermaceae

It is indigenous to India and Srilanka, it is also found in Cambodia, Vietnam, Indonesia and Malaysia. The stem contains berberine, ceryl alcohol, hentriacontane, sitosterol, palmitic acid, oleic acid and saponin together with some resinous material (Anonymous, 1992). Roots contain tertiary alkaloids, berlambine, dihydroberlambine and noroxyhydrastine (Datta *et al.*, 1988). The antioxidant activity of *C. fenestratum* may be due to the presence of berberine and phenolic compounds (Punitha *et al.*, 2005b).

Effects of aqueous extract of *Coscinium fenestratum* was studied by Punitha *et al.* (2005a) in STZ-nicotinamide induced type-2 diabetic rats, at two dose levels of 250 mg/kg and 500 mg/kg. At the end of the experimental period, diabetic rats treated with aqueous extract at both dose levels showed a significant increase in the levels of enzymatic antioxidants such as glutathione peroxidase, glutathione synthetase, peroxidase, superoxide dismutase and catalase as compared to the untreated control. Similarly, a significant increase was also observed in the levels of the non enzymatic antioxidants ceruloplasmin, ascorbic acid and tocopherol. The results suggest that the aqueous stem extract of *Coscinium fenestratum* prevents type 2 diabetes mellitus induced oxidative stress (Punitha *et al.*, 2005a).

Oral administration of *Coscinium fenestratum* extract (200 mg/kg.b.wt.) to mice resulted in a reduction of radiation induced peroxidation of tissue lipids and restore the tissue antioxidant levels such as glutathione, glutathione peroxidase and Superoxide dismutase (Divyasree and Nair, 2008). The extract showed significant free radical scavenging activity in terms of the reduction of the stable free radical DPPH and protection of deoxyribose against fenton reaction induced hydroxyl radical damage.

Cosmos caudatus H.B.K. Family: Asteraceae (Compositae)

Wong *et al.* (2006) investigated the antioxidant properties of 25 edible tropical plants including *Cosmos caudatus* using DPPH and FRAP assays. Total phenolic content was also estimated. *Cosmos caudatus* showed highest antioxidant activity, 200µmol Trolox of DPPH radical scavenging. A strong correlation between TEAC values obtained for the DPPH assay and those for FRAP assay implied that compounds in the extracts were capable of scavenging DPPH radical and reducing ferric ions. A satisfactory correlation of TPC with TEAC suggested that polyphenols in the extracts were partly responsible for the antioxidant activities while its correlation with CCA was poor, indicating that polyphenols might not be the main cupric ion chelators.

Abbas *et al.* (2006) evaluated the antioxidant activity of *Cosmos caudatus* using DPPH assay. Methanolic extract showed DPPH radical scavenging activity.

Extracts from 11 vegetables of Indonesian origin were screened for flavonoid content, total phenolics, and antioxidant activity by Andarwulan *et al.* (2010). Flavonoid content in *Cosmos caudatus* was 52.19mg/100g fresh weight. *Cosmos caudatus* has the greatest total phenols with 1.52 mg GAE/100g fw. The flavonoid content of the vegetables studied were mainly quercetin and kaempferol. *Cosmos caudatus* had the highest antioxidant activity as measured by ferric cyanide reducing power, DPPH and ABTS scavenging and inhibition of linoleic acid oxidation. They concluded that *Cosmos caudatus* was rich source of dietary flavonoids and antioxidants.

Huda-Faujan *et al.* (2007, 2009) reported antioxidative activities of aqueous and methanolic extracts of *Cosmos caudatus* and four other plants. The analysis carried out was total phenolic content, ferric reducing power, ferric thiocyanate (FTC) and TBA tests. The phenolic content was 21.41 mg (aqueous extract) 18.43 mg (methanolic extract) TAE/100 g fresh weight and antioxidant activity of 59.7 per cent (aqueous extract), 68.67 per cent (methanolic extract) using FTC method. Increasing the concentration of the extracts resulted in increased ferric reducing antioxidant power. TBA analysis showed that *C. caudatus* extract had the highest antioxidant effect. Total phenolic contet had positive correlation with antioxidant capacity.

Costus afer Ker.-Gawl. Family: Costaceae

Phytochemical constituents and antioxidant activities of aqueous and methanol stem extracts of *Costus afer* were evaluated by Anyasor *et al.* (2010). *C. afer* contained flavonoids, phenols, anthraquinones, cardiac glycosides, terpenoids, alkaloids and tannins. The aqueous extract had a higher free radical scavenging activity with IC_{50} value of 64.42μg/ml (inhibition of DPPH radical activity) than methanol extract (92.33μg/ml). Furthermore, total phenolic content in aqueous extract (0.66 mg gallic acid equivalent/g) was significantly higher than methanol extract (0.52 mg gallic acid equivalent/g). In addition, inhibition of lipid peroxidation by aqueous extract (80.60 per cent) was significantly higher than methanol extract (77.00 per cent).

Costus igneus N. E.Br. Family: Costaceae

Leaves powder of *Costus igneus* was analyzed by Vishalakshi Devi and Urooj (2010) for its proximate composition and antioxidant components. Resuts indicate that the leaves are rich in antioxidant components such as ascorbic acid, α-tocopherol, β-carotene, etc.

Costus speciosus (Koen.) Sm. Family: Costaceae

Vijayalakshmi and Sarada (2007) investigated the polyphenol content and antioxidant activity of different parts of *Costus speciosus*.

Leaves powder of *Costus igneus* was analyzed by Vishalakshi Devi and Urooj (2010) for its proximate composition and antioxidant components. Resuts indicate that the leaves are rich in antioxidant components such as ascorbic acid, α-tocopherol, β-carotene etc.

Cotinus coggygria Scop. Family: Anacardiaceae

Ivanova *et al.* (2005) reported high phenolics content and antioxidant properties in *Cotinus coggygria*.

Niiforovi *et al.* (2010) examined the antioxidant activity of *Cotinus coggygria* including total antioxidant capacity, DPPH free-radical scavenging, the inhibitory activity toward lipid peroxidation, Fe(3+)-reducing power, Fe(2+)-chelating ability and hydroxyl scavenging activity. Total phenolic and flavonoid contents were also determined for each alcoholic extract. *Cotinus coggygria* extract contained the highest amount of total phenols (413 mg GAE/g dry extract). Alcoholic extract of *Cotinus coggygria* showed the highest total antioxidant capacity (313 mg AA/g dry extract) as well as DPPH free-radical scavenging (IC_{50} =9μg/ml), inhibitory activity toward lipid peroxidation (IC_{50} = 3μg/ml) and reducing power.

Coulutea persica Boiss. Family: Fabaceae (Leguminosae)

Souri *et al.* (2004) evaluated the antioxidative activity of *Coulutea persica* by linoleic acid peroxidation test using 1,3-diethyl-2-thiobarbituric acid as the reagent. Methanolic extract showed 81.52 per cent inhibition at 40μg concentration with IC_{50} value of 6.57μg.

***Coussarea hydrangeifolia* (Benth.) Benth. et Hook. f. ex Muell.-Arg. Family: Rubiaceae**

Phenyl propanoid glycosides isolated from the leaves of *Coussarea hydrangeifolia* were evaluated for their antioxidant activity by DPPH assay. The antioxidant activities of 1,6-di-O-caffeoyl-β-D-glucopyranoside and 1-O-(E)-caffeoyl-β-D-glucopyranoside (IC_{50} values of 15.0 and 19.2μM) respectively were comparable to that of the standard positive control caffeic acid, whilst rest were only weakly active(Hamerski *et al.*, 2005).

***Craibiodendron henryi* W.W.Smith Family: Ericaceae**

Iridiod glycosides and grayanane diterpenoids isolated from the roots of *Craibiodendron henryi* were evaluated for their antioxidant activity (Huang *et al.*, 2005).

***Crassocephalum crepidioides* (Benth.) S.Moore Family: Asteraceae (Compositae)**

Free radical scavenging and protective actions against chemically induced hepatotoxicity of *Crassocephalum crepidioides* were investigated by Aniya *et al.* (2005). A water extract of *C. crepidioides* strongly scavenged superoxide anion, hydroxyl radical and also stable radical DPPH. Galactosamine (GaIN, 400 mg/kg) and lipopolysaccharide (LPS, 0.5microg/kg) induced hepatotoxicity of rats as seen by an elevation of serum alanine aminotransferase (ALT) and aspartate aminotransferase (AST) and of lipid peroxidation in liver homogenates was significantly depressed when the herbal extract was given intraperitoneally 1 and 15 h before GaIN and LPS treatment. Similarly carbon tetrachloride (CCl_4) induced liver injury, as evidenced by an incrase in AST and ALT activities in serum, was also inhibited by the extract pretreatment. Isochlorogenic acids, quercetin and kaempferol glycosides were identified as active components of *C. crepidioides* with strong free radical scavenging action. These results demonstrate that *C. crepidioides* is a potent antioxidant and protective against GaIN plus LPS-or CCl_4-induced hepatotoxicity (Aniya *et al.*, 2005).

Odukoya *et al.* (2007) evaluated the antioxidant activity and total phenolic contents of Nigerian green leafy vegetables including *Crassocephalum crepidioides*.

Crassocephalum rubens **Family: Asteraceae**

Iwaleva *et al.* (2005) reported antioxidant and cytoprotective activities of boiled, cold and methanolic extracts of nine edible vegetables in South West Nigeria which were evaluated in the DPPH assay and hemeagglutination assay in bovine erythrocytes, respectively. *Crassocephalum rubens* showed the highest antioxidant activity (56.5 per cent).

Crassula cv. himaturi **Family: Crassulaceae**

A new compound 11-O-(4′-O-methylgalloyl)-bergenin isolated from the MeOH extract of *Crassula* cv. *himaturi* showed antioxidant activity (EC_{50} 23.9 μM) equivalent to that of L-ascorbic acid or quercetin.

Crataegus aronia **(L.) Bosc. ex DC. (Syn.** ***Crataegus azarolus*** **L.) Family: Rosaceae**

Water-soluble extracts of *Crataegus aronia* (Syn.: *C. azarolus*) inhibited oxidation of β-carotene, AAPH-induced plasma oxidation and Fe^{2+} induced lipid peroxidation in rat liver homogenates. In addition, the extract is an efficient scavenger of the O_2, increases intracellular GSH levels and is not cytotoxic. Based on these observations Ljubuncic *et al.* (2005) is of the opinon that the therapeutic benefic at leat in part, is attributed to its effective inhibition of oxidative processes, efficient scavenging of O_2 and possible increasing GSH biosynthesis.

Al-Mustafa and Al-Thunibat (2008) evaluated the antioxidant activity of Jordanian medicinal plant *Crataegus azarolus* stem and leaves. The level of antioxidant activity was determined by DPPH and ABTS assay in relation to total phenolic contents of the medicinally used part. They concluded that *Crataegus azarolus* stem and leaves have moderate antioxidant capacity (DPPH-TEAC range 20-80mg/g).

Crataegus monogyna **L. Family: Rosaceae**

English Name: Hawthorn

Rakotoarison *et al.* (1996) invstigated extraction, chemical analysis and evaluation of antioxidant properties of polyphenolics from callus and cell suspension cultures of *Crataegus monogyna*. Proportions in individual substances identified vary according to the kind of extract. Whatever the origin of each extract, they exhibit antioxidant activities against the tested reactive oxygen species. The flavanol type substances (-epicatechin and B2 dimer) appeared to be the most efficient scavengers of the reactive oxygen species, and the yield of this calss of molecules is the highest for cell suspension extracts. The compared results of tissue culture extracts and fresh flower extracts show quite analogous activities.

Kiselova *et al.* (2006) studied the antioxidant activity of *Crataegus monogyna*. Antioxidant activity was measurd by ABTS cation radical decoloration assay and the total polyphenol content was measured according to the Folin-Ciocalteu method. *Crataegus monogyna* extract exhibited higher antioxidant activity than mate. A positive correlation between antioxidant activity and polyphenol content was found, suggesting that the antioxidant capacity of the aqueous plant extract was due to a greater extent to their polyphenols. Leskovac *et al.* (2007) reported radioprotective and antioxidant properties of *Crataegus monogyna*. Treatment with 0.1mg/ml showed reduced levels of lipid peroxidation products (13.18 per cent).

Demiray *et al.* (2009) evaluated phenolic profiles and antioxidant activities of Turkish medicinal plant *Crataegus monogyna*. The antioxidant activity of the extracts was determined by ABTS radical cation scavenging activity. The Folin Ciocalteu procedure was used to assess the total phenolic concentrations of the extracts as gallic acid equivalents. The catechin content in the Hawthorn leaves ranged from 284 to 10631 in µg/g of dry weight of leaves. The highest catechin content was found in 70 per cent acetone extract. In acetone and methanol extracts, protocatechuic acid was the second prominent phenolic compound. Acetone extract had the highest phenolic content (36.6 mg GAE/g dw leaves). Water extract of Hawthorn leaves exhibited the highest antioxidant activity (11.5mg AAE/g dw leaves), followed by acetone extract (8.7 mg AAE/g dw leaves). Water extract had the highest content of protocatechuic, ferulic, coumaric, caffeic and gallic acids.

Crataegus oxyacantha L. Family: Rosaceae

The antioxidant capacity of extracts of *Crataegus oxyacantha*, plant native to North America, which have long been used in herbal medicine for the treatment of cardiac and circulatory functions has been investigated by da Silva *et al.* (2000). The total antioxidant potential conferred by all hydrogen donating antioxidants present in these extracts has been assessed by ABTS assay and the relative order of antioxidant potential has been established.

Crataegus pinnatifida Bge. Family: Rosaceae

Cai *et al.* (2004) evaluated the antioxidant activity and phenolic compounds in traditional Chinese medicinal plants associated with anticancer. The improved ABTS.+ method was used to systematically assess the total antioxidant capacity (Trolox equivalent antioxidant capacity, TEAC) of the medicinal extracts. The TEAC values and total phenolic content for methanolic extract of fruit of *Crataegus pinnatifida* were 884.3 µmol Trolox equivalent/100g dry weight (DW), and 4.28 g of gallic acid equivalent/100g DW respectively. A positive significant linear relationship between antioxidant activity and total phenolic content showed that phenolic compounds were the dominant antioxidant components in the tested herbs. Major types of phenolic compounds in *Crataegus pinnatifida* include flavonoids (epicatechin, quercetin, vitexin, quercitrin), phenolic acids (chlorogenic acid).

The aqueous extracts of 13 oriental medicinal plants were examined for their reducing power, scavenging ability towards superoxide and hydroxyl radicals and their inhibitory effect on lipid peroxidation. All extracts tested were found to be highly active on scavenging of superoxide radicals (Nam and Kang, 2004).

Crataeva nurvala Buch.-Ham. Family: Capparaceae

The cytoprotectiv action of lupeol isolated fom *Crataeva nurvala* stem bark against free radical toxicity has been investigated by Baskar *et al.* (1996) in experimental urolithiasis. Glycolate, an inducer of stone formation, significantly raised the renal tissue levels of calcium, oxalate, phosphorus and magnesium in calcculogenic rats and brought about a remarkable decrease in kidney oxalate level. The increase in lipid peroxidation and superoxide dismutase (SOD) activity, associated with decreased catalase activity and glutathione (GSH) level are the salient features observed in tissues of stone forming rats. Lupeol administration induced a remarkable decrease in kidney oxlate level and also was effective in counteracting the free radical toxicity by bringing about a significant decrease in peroxidative levels and increase in antioxidant status. These observations highlighten the antioxidant property of lupeol and its cytoprotective action against free radical toxicity.

Sunitha *et al.* (2001) investigated hepatoprotective effect of lupeol and lupeol lineolate on tissue antioxidant defence system in cadmium –induced hepatotoxicity in rats.

Effect of alcohol extract of *Crataeva nurvala* was evaluated on cisplatin-induced dysfunction model of renal proximal tubule cells by oxidative stress by Shirwaikar *et al.* (2004). The results suggest that the plant extract (250 and 500 mg/kg) was effective in significantly altering the indices of cisplatin induced dysfunction of renal proximal tubule cells under oxidative stress by decreasing the concentration of blood urea nitrogen, creatinine and lipid peroxidation. The increased glutathione and catalase activity are indicative of the antioxidant properties of stem bark extract (Shirwaikar *et al.*, 2004).

Cratoxylum cochinchinense (Lour.) Blume Family: Clusiaceae (Guttiferae)

Tang *et al.* (2004) carried out characterization of antioxidant and antiglycation properties and isolation of active ingredients from traditional Chinese medicines. Extract of *Cratoxylon cochinchinense* showed highest antioxidant activity. They attempted to isolate the active ingredient and obtained an extract containing at least 90 per cent mangiferin.

Two xanthones and two caged-prenylated xanthones, named cochinchinones A-D, respectively, and a synthetically known caged-prenylated xanthone, together with seven compounds were isolated by Mahabusarakam *et al.* (2006) from the roots of *Cratoxylum cochinchinense*. Some of the compounds exhibited effective antioxidative properties.

A new xanthone named cratoxylumxanthone A, together with five known compounds: dulcisxanthone B, α-mangotin, β-mangostin, 2-geranyl-1,3,7-trihydroxy-4-(3-methylbut-2-enyl)xanthone and tectochrystin were isolated from *Cratoxylum cochinchinense* by Phuwapraisirisan *et al.* (2006). The isolated compounds showed free radical scavenging against DPPH and lipid peroxidation inhibition.

Cratoxylum formosum Dyer Family: Clusiaceae (Guttiferae)

English Name: Teaw.

Antioxidant properties of *Cratoxylum formosum* extract in soybean oil and emulsions were investigated by Maisuthisakul *et al.* (2006).

Maisuthisakul *et al.* (2008) reported that extracts of *Cratoxylum formosum* possessed strong antioxidant activity, high total phenolic and flavonoid contents. They are of the opinion that plants with high antioxidant activities also have high total phenolic and flavonoid content.

Cremastra variabilis (Bl.) Nakai Family: Orchidaceae

Li *et al.* (2008) evaluated the antioxidant properties of 45 medicinal plants including *Cremastra variabilis* using FRAP and TEAC assays and the total phenolic contents of these plants were measured. *Cremastra variabilis* showed low antioxidant activity with 6.97 µmol Fe(II)/g FRAP value, 6.55µmol Trolox/g of TEAC value and 1.17 mg GAE/g of phenolic content.

Crepidiastrum lanceolatum Nakai Family: Asteraceae

Japanese name: Hosobawadan

Maeda *et al.* (2006) examined the phenolic contents of *Crepidiastrum lanceolatum* and their radical scavenging activities by using DPPH assay. High contents of phenolic compounds and strong DPPH scavenging activities were observed in extracts from this species. They also showed radical scavenging activities for the superoxide radical and tert-butylhydroperoxyl radical (t-BuOO·). They found five antioxidative phenolic compounds, D(+)-chicoric acid, chlorogenic acid, caffeic acid, luteolin7-O-β-D-glucopyranoside, and luteolin 7-O-β-D-glucuronide, in this species.

Crinum ornatum (Ait.) Bury Family: Amaryllidaceae

The antioxidant properties of the methanolic extract of the bulb of *Crinum ornatum* was investigated by Oloyede and Farombi (2010). The extract showed marked activity as a radical scavenger in the experiment using DPPH. The antioxidative activity of this compound towards hydrogen peroxide was also reported. A 92.7 per cent inhibition of hydrogen peroxide was observed when compared with control experiment at a concentration of 0.1 mg/ml. The extract also showed marked inhibitory effect on peroxidation of tissue lipids in the malonaldehyde (MDA) test. Administration of the crude extract decreased the concentration of thiobarbituric acid reactive substances of rat liver homogenate. At the highest concentration of 0.1mg/ml, 85.5 per cent inhibition of malonaldehyde was observed.

Crithmum maritimum L. Family: Apiaceae (Umbelliferae)

English Name: Marine Fennel.

The antioxidant activity of the essential oil obtained from *Crithmum maritimum* was evaluated by Ruberto *et al.* (2000) in two lipid model systems: A modified TBARS assay and a spectrophotometric detection of hydroperoxydienes from linoleic acid in a micellar system. The oil demonstrated antioxidant capacitity comparable in some cases to that of α-tocopherol and BHT, used as reference antioxidants.

The evaluation of the antioxidant activities of essential oils from *Crithmium maritimum* from Dalmatia (Croatia) by different methods showed that using TBARS method, the oils presented at 1 g/l (the highest concentration tested) lower ablity to inhibit lipid peroxidation than BHA, but close to that of BHT (Kulisic-Bilusic *et al.*,

2010). They also found that the activity was dose-dependent. The essential oil was predominantly constituted by sabinene and lomonene.

Crocus sativus **L. Family: Iridaceae**

English Name: Saffron.

Protocatechuic acid, kaempferol and kaempferol-7-O-β-D-glucopyranoside isolated from the methanol extract of the petals were more effective in scavenging DPPH radicals than α-tocopherol (Li *et al.*, 2004).

Assimopoulou *et al.* (2005) investigated the radical scavenging activity of *Crocus sativus* extract and its bioactive constituents. It was shown that a methanol extract of *Crocus sativus* exhibited high antioxidant activity, although it contains several active and inactive constituents. In trying to approximate a structure-activity relationship, two bioactive constituents of saffron extract were identified, namely crocin and safranal. Crocin showed high radical scavenging activity (50 per cent and 65 per cent for 500 and 1000ppm solution respectively), followed by safranal (34 per cent for 500 ppm solution). All the tested samples showed high radical scavenging activity due to the ability to donate hydrogen atom to the DPPH radical (Assimopoulou *et al.*, 2005)

Crude extract from *Crocus sativus* was screened by Sengul *et al.* (2009) for its *in vitro* antioxidant and antimicrobial properties. Total phenolic content of extract of this plant was also determined. β-carotene bleaching assay and Folin-Ciocalteu reagent were used to determine total antioxidant activity and total phenols of plant extract. Antioxidant activity was 63.15 per cent and the total phenolic content was 42.29 mgGAE/g DW).

Crossopteryx febrifuga **(Afzel) Benth. Family: Rubiaceae**

Maiga *et al.* (2006) reported antioxidant and lipoxygenase inhibitory activity of *Crossopteryx febrifuga.*

Crotalaria aegyptiaca **Benth. Family: Fabaceae (Papilionaceae)**

The chloroform extract of *Crotalaria aegyptiaca* was investigated for antioxidant, anti-inflammatory activities. The antioxidant activity assays *in vitro* using the ability of the extract to scavenge DPPH radicals, also β-carotene lineolate bleaching method and erythrocyte hemolysis mediated peroxyl free radicals method were used. Antioxidant assays *in vivo* including determination of hepatic antioxidant enzymes SOD, CAT and GSH-px activities as well as lipid peroxidation product as expressed by TBARS. The results revealed that the chloroform extract possesses a significant and dose-dependent antioxidant, anti-inflammatory and analgesic activities. Administration of chloroform extract of *C. aegyptiaca* prior to CCl_4 were shown to exhibit significant inhibition of TBARS production and cause a significant elevation of free radical scavenging enzyme activities such as SOD, CAT and GSH-px in dose dependent manner. The antioxidant and anti-inflammatory activities of the extract may be associated with the potential activity of flavonoids constituents (Ali *et al.*, 2004).

Crotalaria burhia Buch.-Ham. ex Benth. Family: Fabaceae (Papilionaceae)

Alang *et al.* (2010) evaluatd the free radical scavenging potential of the roots of *Crotalaria burhia*. The methanolic extract at 1280μg showed significant scavenging of the radical cation: DPPH, superoxide anion and lipid peroxidation potential. The antioxidant activity was found to be concentration dependent.

Crotalaria mucronata Desv. Family: Fabaceae (Papilionaceae)

Jiang *et al.* (2010) evaluated the antioxidant activity of ethanol extracts from *Crotalaria mucronata*. Ultrasonic-assisted ethanol extraction method was adopted to extract the antioxidant active components. Chemilumiscence analysis in H_2O_2-Luminol system was used to determine efficiency of scavenging hydroxyl free radical. The ethanol extract showed good scavenging effects of hydroxyl free radical.

Crotalaria sessiliflora L. Family: Fabaceae (Papilionaceae)

Seven antioxidative compounds were isolated from the EtOAc extract of the aerial part of *C. sessiliflora* by activity-guided fractionation with DPPH. Among the isolated compounds, hydroxyeucomic acid showed the strongest free radical-scavenging activity, which was almost identical to that of epigallocatechin gallate, against DPPH. Orientin and isoorientin showed strong anti-peroxidative activities toward linoleic acid and protective effects against the bactericidal action of the tert-butyl peroxyl radical. Their activities were nearly equal to that of epigallocatechin gallate (Munim *et al.*, 2003).

Woo *et al.* (2005) investigated the antioxidative activities of solvent extracts of leaf, stem and root of *Crotalaria sessiliflora*. Antioxidative activities decreased in the order BHT>leaf extract>stem extract>root extract> control.

Croton andinus Muell.-Arg. Family: Euphorbiaceae

The aerial parts of 17 Bolivian plants including *Croton andinus* were screened by Rosas-Romero and Saavedra (2005) to determine antioxidant activity. A methanol extract of each plant was prepared and portioned sequentially with hexane, chloroform, and ethyl acetate, leaving an aqueous solution. All extracts and their 5 fractions for a total of 102 samples, were evaluated using two techniques: an adaptation of the β-carotene bleaching technique using an emulsion of linoleic acid in water as the oxidizable substrate, and the DPPH free radical trapping technique. The results of the β-carotene bleaching technique were more discriminating and better related to the rancidity process under normal conditions; with this assay, 11 species provided at least one fraction with highly promising antioxidant activity. All species gave good results under the DPPH technique, and in most cases they performed better than BHA, which was used as a reference antioxidant (Rosas-Romero and Saavedra, 2005).

Croton cajucara Benth. Family: Euphorbiacee

Extract from *Croton cajucara* showed radical scavenging activity against the stable radical DPPH (75 per cent, 43 per cent and 25 per cent of the standard trolox at 1, 10 and 100 mg/ml, respectively; IC^{50} 218mg/ml) (Tieppo *et al.*, 2006).

Croton celtidifolius Baill. Family: Euphorbiaceae

Crude extracts, fractions and isolated compounds from *Croton celtidifolius* showed *in vitro* antioxidant properties through the superoxide scavenger capacity method by the nitro blue tetrazolium (NBT) reduction assay (Nardi *et al.*, 2003). The bark showed abundance of phenolic compounds such as catechin, gallocatechin and proanthocyanidins as phytochemical constituents, which are responsible for the antioxidant properties of this species (Nardi *et al.*, 2003).

Croton emporiorum Croizat Family: Euphorbiaceae

The aerial parts of 17 Bolivian plants including *Croton emporiorum* were screened by Rosas-Romero and Saavedra (2005) to determine antioxidant activity. A methanol extract of each plant was prepared and portioned sequentially with hexane, chloroform, and ethyl acetate, leaving an aqueous solution. All extracts and their 5 fractions for a total of 102 samples, were evaluated using two techniques: an adaptation of the β-carotene bleaching technique using an emulsion of linoleic acid in water as the oxidizable substrate, and the DPPH free radical trapping technique. The results of the β-carotene bleaching technique were more discriminating and better related to the rancidity process under normal conditions; with this assay, 11 species provided at least one fraction with highly promising antioxidant activity. All species gave good results under the DPPH technique, and in most cases they performed better than BHA, which was used as a reference antioxidant (Rosas-Romero and Saavedra, 2005)

Croton lechleri Muell.-Arg. Family: Euphorbiaceae

Activity of latex from *Croton lechleri* was determined by Risco *et al.* (2003) by *in vitro* assay in human serum. Free radical scavenging activity was evaluated using DPPH assay. The latex showed free radical scavenging, antioxidant, phagocytosis inhibitory and anti-inflammatory activities.

Croton lobatus L. Family: Euphorbiaceae

Lagnika *et al.* (2011) evaluated the antioxidant activity of *Croton lobatus* using DPPH radical scavenging assay. The extract showed good radical scavenging activity with IC_{50} value of 1.96μg/ml while the standard L-Ascorbic acid had IC_{50} value of 1.1μg/ml.

Croton mortibensis Family: Euphorbiaceae

In a study with 32 methanolic extracts, from Brazilian Caatinga plants, through DPPH radical scavenging assay showed that the species *Croton mortibensis* displayed good antioxidant activity with IC_{50} = 5.5 mg/l (David *et al.*, 2007).

Croton sparsiflorus Morong Family: Euphorbiaceae

Ahmad *et al.* (2010) examined xanthine oxidase inhibitory potential, antioxidant activity (DPPH and phosphomolybdenum assays) and total phenolics of *Croton sparsiflorus*. Dichloromethane, chloroform (CSC), ethyl acetate (CSEA), acetone (CSA), ethanol (CSE) and n-butanol extracts inhibited xanthine oxidase more than 50 per cent at 100 μg/ml concentration with IC_{50} values of 23.82, 17.34, 9.31, 11.35, 43.25

and 87.50 μg/ml, respectively. CSEA, CSM (methanolic extracts) and CSE extract exhibited significant DPPH free radical scavenging potential. Total phenolics, estimated as gallic acid equivalent (mg/g of the extract); ranged from 28 to 483 (CSC, CSEA and CSA more than 400 mg/g). A weak correlation of total phenolics with XOI and phosphomolybdenum assays proved phenolics are not the only bioactive compounds responsible for these activities. Strong correlation between total phenolics and DPPH free radical scavenging marked phenolics as the major free radical scavengers (Ahmad *et al.*, 2010).

***Cruciata taurica* (Pall. ex Willd.) Ehrend. Family: Rubiaceae**

Antioxidant and radical scavenging activities, reducing powers and the amount of total phenolic compounds of aqueous and methanolic extracts of *Cruciata taurica* were studied by Mavi *et al.* (2004). The aqueous extract showed highest peroxidation inhibition (IC_{50} : 0.00022 mg/l).

***Cryptotaenia japonica* L. Family: Apiceae (Umbelliferae)**

English Name: Japanese honewort; Japanese: Mitsuba.

Maeda *et al.* (2006) examined the phenolic contents of leaf of *Cryptotaenia japonica* and its radical scavenging activities for DPPH radical. The polyphenol content was 21.7 mg-gallic acid equivalent/100 g) and DPPH scavenging activity was 98.9 μmol-Trolox eq./100 g).

***Cucumis melo* L. Family: Cucurbitaceae**

English Name: Cantaloupe.

Ismail *et al.* (2010) determined the phenolic content and antioxidant activity of methanolic extracts from different parts of cantaloupe (leaf, stem, skin, seed and flesh). The flesh extract afforded the highest yield (89.6 per cent) whilst the lowest yield was obtained from the seed (13.7 per cent). The leaf extract showed the highest total phenolic content (26.4mg GAE/g extract) and total flavonoid content (69.7 μg RE/g extract) accompanied with best antioxidant activity through all antioxidant assays ($p<0.05$). In addition, the stem extract also exhibited good antioxidant activity.

Thirty-eight types of fruits commonly consumed in Singapore were systematically analyzed by Isabelle *et al.* (2010) for their hydrophilic oxygen radical absorbance capacity (H-ORAC), total phenolic content (TPC), ascorbic acid (AA) and various lipophilic antioxidants. Cantaloupe had the highest β-cryptoxanthin, lycopene and β-carotene per gram fresh weight.

***Cucumis sativus* L. Family: Cucurbitaceae**

English Name: Cucumber.

The antioxidative activity of phenolic extracts from leaves of cucumber was examined by autoxidation of methyl linoleate by Kahkonen *et al.* (1999). The total phenolics were low (3.8mg/g GAE) and the antioxidant activity was moderate, *i.e.*32 per cent at 500 ppm level.

Cai *et al.* (2004) evaluated the antioxidant activity and phenolic compounds in traditional Chinese medicinal plants and vegetables associated with anticancer. The improved ABTS.+ method was used to systematically assess the total antioxidant capacity (TEAC) of the medicinal extracts. The TEAC values and total phenolic content for methanolic extract of fruit of cucumber were 28.7 μmol Trolox equivalent/100g dry weight (DW), and 0.16 g of gallic acid equivalent/100g DW respectively.

Maeda *et al.* (2006) examined the phenolic contents of fruit of cucumber and its radical scavenging activities for DPPH radical. The polyphenol content was 13.1 mg-gallic acid equivalent/100 g) and DPPH scavenging activity was 29.9 μmol-Trolox eq./100 g).

Souri *et al.* (2008) evaluated the antioxidant activity of seed against linoleic acid peroxidation using 1,3-diethyl-2-thiobarbituric acid as reagent. Antioxidant activity (IC_{50}) against peroxidation of linoleic acid (2mg/ml) was 1.25 and phenolic content was 27.79 mg/100g dry weight.

Nema *et al.* (2011) screened the antioxidant activity of the lyophilized juice of *Cucumis sativus* fruit (CSLJ) by DPPH and superoxide radical scavenging assay in reference to BHT. The CSLJ exhibited DPPH-free radical and superoxide radical scavenging activity, IC_{50} at a concentration of 14.73 and 35.29 μg/ml respectively. CSLJ is the rich source or ascorbic acid and this study thereby rationalizes the use of *C. sativus* as potential anti-wrinkle agent in cosmetic-products.

Cucurbita maxima Duch. Family: Cucurbitaceae

English Name: Pumpkin.

Nara *et al.* (2009) evaluated the antioxidant activity of a water soluble polysaccharide fraction (WSP) from pumpkin fruits. In the WSP, DPPH radical scavenging and superoxide dismutase-like activity increased depending on the total sugar content. Furthermore, the WSP can serve as an inhibitor of ascorbic acid oxidation. The efficacy was also affected by the total sugar content.

Antioxidant potential of selected vegetables commonly used in diet in Asian subcontinent, including *Cucurbita maxima* fruits has been evaluated by Gacche *et al.* (2010). The DPPH radical scavenging activity was 40.7 per cent, inhibition of lipid peroxidation was 40.3 per cent, FICA was 22.3 per cent, Vitamin C content was 15.36 mg/100g and total phenols was 13.30 mg/g.

Cucurbita moschata Duchesne ex Poiret Family: Cucurbitaceae

English Name: Pumpkin.

Chanwitheesuk *et al.* (2005) investigated the antioxidant activity of *Cucurbita moschata* using a β-carotene bleaching method. The contents of plant chemicals such as vitamin C, Vitamin E, carotenoids, tannin and total phenolics, were also determined. Methanolic extract showed antioxidant activity. *Cucurbita moschata* has antioxidant index of 0.91, Vitamin C 27.8 mg per cent, Vitamin E 0.0045mg per cent, Total carotenes 1.92 mg per cent, Total xanthophylls 1.59mg per cent, Tannins 4.48mg per cent and Total phenolics of 87.8 mg per cent.

Chemical composition and antioxidant activity of an acidic polysaccharide extracted from *Cucurbita moschata* was investigated by Yang *et al.* (2007). The pumpkin polysaccharide was demonstrated to effectively inhibit the H_2O_2 caused decrease of cell viability, lactate dehydrogenase leakage, and malonaldehyde formation, and also reduced the H_2O_2-caused decline of superoxide dismutase activity and glutathione depletion in cultured mouse peritoneal macrophages.

Cucurbita pepo L. Family Cucurbitaceae

English Name: Pumpkin.

Nkosi *et al.* (2006) studied the effects of pumpkin seed protein isolate on the plasma levels of catalase (CA), superoxide dismutase (SOD), Glutathione peroxidase (GSHpx) and total antioxidant capacity (TAC) as well as glucose-6-phosphatase (G6Pase) in liver homogenates and lipid peroxidation (LPO-malondialdehyde-MDA) levels in liver homogenates and liver microsomal fractions on CCl_4 induced acute liver injury low-protein fed Sprague-Dawley rats were investigated. From the results Nkosi *et al.* (2006) concluded that pumpkin seed protein isolate administration was effective in alleviating the detrimental effects associated with protein malnutrition and CCl_4-induced intoxication. It is therefore apparent that pumpkin seed protein isolate has components that have antiperoxidative properties.

Pumpkin seeds are a popular snack food in several countries. They are consumed either raw or roasted (salted or not) and used in cooking and baking as an ingredient of bread, cereals, salads and cakes. Moreover, pumpkin seed oil now-a-days gains wide acceptance not only as edible oil but as a nutraceutical too. Pumpkin seeds have been implicated in providing health benefits. Xanthopoulou Marianna *et al.* (2009) treated four commercially available pumpkin seeds with two different extraction methodologies in order to obtain fractions with different content. These extracts were screened for their antioxidant activity using 2,2-diphenyl-1-picrylhydrazyl (DPPH) free radical scavenging assay and for their inhibitory activity against lipid peroxidation catalyzed by soybean lipoxygenase. Most extracts tested have demonstrated radical scavenging activity, which depends on their total phenolic content, with fractions rich in phenolics showing the strongest activity. On the other hand, the phenolic content of extracts does not determine their activity against lipoxygenase, as acetone and polar lipid fractions are its strongest inhibitors. The presence of molecules being able to scavenge radicals and inhibit lipoxygenase in pumpkin seeds may in part explain the health benefits attributed to them.

Culcitum reflexum H.B.K. Family: Asteraceae (Compositae)

Aquino *et al.* (2002) evaluated the antioxidant and photoprotective activity of a ethanolic extract of *Culcitum reflexum* leaves and their major flavonoids. The antioxidant activity of the extract was assessed by means of three different *in vitro* tests: bleaching of the stable DPPH radical, peroxidation induced by the water soluble radical initiator 2,2′-azobis(2-amidinopropane) hydrochloride, on mixed dipalmitoylphosphatidylcholine/linoleic acid unilamellar vesicles, UV radiation-induced peroxidation in phosphatidylcholine multilamellar vesicles. In all *in vitro* tests used the extract showed a significant antioxidant effect. Flvonols rutin, quercetin-3-O-β-D-galactopyranoside-4′-O- β-D-glucopyranoside, quercetin-3-O-β-D-glucopyranoside, isorhamnetin–3-O-β-D-galactopyranoside, quercetin and kaempferol were isolated from the extract.

Cuminum cyminum L. Family: Apiaceae

English Name: Cumin

Souri *et al.* (2008) evaluated the antioxidant activity against linoleic acid peroxidation using 1,3-diethyl-2-thiobarbituric acid as reagent. Antioxidant activity (IC_{50}) against peroxidation of linoleic acid (2mg/ml) was 5.76 and phenolic content was 241.41 mg/100g dry weight. The results of this study showed that there was no significant correlation between antioxidant activity and phenolic content of the studied plant materials and phenolic content could not be a good indicator of antioxidant capacity.

Mariutti *et al.* (2008) evaluated the antioxidant activity of Cumin employing ABTS and DPPH radical assay. DPPH EC_{50} value was 81g/kg while DPPH-TEAC value was 3.2mM/g. ABTS-TEAC value was 6.0mM/g.

Antioxidant activities of ethanol extracts from seven Apiaceae fruits have been studied by Nickavar and Abolhasani (2009) by the DPPH radical scavenging test. All the studied extracts, including *Cuminum cyminum*, showed antioxidant capability. For *Cuminum cyminum*IC_{50} value of the DPPH scavenging activity was 149.40μg/ml and total flavonoid content was 56.92μg/mg. A positive correlation was found between the antioxidant potency and flavonoid content of the fractions.

El-Ghorab *et al.* (2010) evaluated the chemical composition and antioxidant activity of cumin. The DPPH method showed the highest antioxidant activity for cumin essential oil (85.44 per cent). The FRAP of essential oils showed almost comparable results with DPPH. Cumin essential oil was found best in reducing Fe^{3+} ions.

Cuphea carthagenesis (Jacq.) Macbr. Family: Lythraceae

Schuldt *et al.* (2004) investigated the superoxide anion and hydroxyl radical scavenger properties, as well as the inhibition of lipid peroxidation by the crude hydroalcoholic extract (CE) and the butanolic (BF) and ethyl acetate fractions (EAF) of *Cuphea carthagenesis* leaves. The study suggest that extract (CE) and fractions (EAF and BF) from leaves are significant sources of phenolic compounds with antioxidant activity *in vitro* and may have important health effects, for example, in cardiovascular disease.

Curculigo orchioides Gaertn. Family: Hypoxidaceae

Antioxidant activity of methanol extract of rhizome of *Curculigo orchioides* (MEC) was investigated using CCl_4-intoxicated rat liver as the experimental model. Elevated level of glutathione transferase (GTS) observed in hepatotoxic rats showed sign of returning towards normalacy in MEC co-administered animals, thus corroborating the antioxidant efficacy of MEC (Venukumar and Latha, 2002a).

Nagesh *et al.* (2008) evaluated the antioxidant activity of *in vitro* derived callus and *in vivo* plant using DPPH assay, reducing power and liposome peroxidation assay. Dose dependent DPPH radical scavenging activity was observed in rhizome callus and *in vitro* rhizome with an EC_{50} of 0.145 mg and 0.04 mg GAE phenol respectively.

Curcuma alismatifolia Gagnep. Family: Zingiberaceae

Akter *et al.* (2008) evaluated the analgesic and antioxidant potential of leaves of *Curcuma alismatifolia*. The 80 per cent methanol extract showed significant antioxidant activities in lipid peroxidation assay compared to the standard antioxidant in a dose-dependent manner. In lipid peroxidation assay, the IC_{50} value was found to be 122.431g/ml, while IC_{50} value of reference ascorbic acid was 147.871 g/ml. Moreover, *Curcuma alismatifolia* extract showed strong total antioxidant capacity and good reducing power.

Hydromethanol extracts of 15 Bangladeshi medicinal plants, including *Curcuma alismatifolia* were evaluated by Hasan *et al.* (2009) for antioxidant potential using DPPH radical scavenging assay. *Curcuma alismatifolia* leaves exhibited radical scavenging activity with an IC_{50} value of 18.72 µg/ml compared to the IC_{50} value of 5.15µg/ml as shown by the reference antioxidant ascorbic acid, in a dose dependent fashion.

Curcuma amada Roxb. Family: Zingiberaceae

English Name: Mango ginger.

Policegoudra and Aradhya (2007) investigated biochemical changes and antioxidant activity of Mango ginger (*Curcuma amada*).The antioxidant activity was measure by DPPH scavenging activity.

Mango ginger free phenolics and Mango ginger bound phenolics exhibited free radical scavenging reducing power abilities (193-104 units/g), inhibition of lipid peroxidation and DNA protection (80 per cent at 4 microg), indicating strong antioxidant properties (Siddaraju and Dharmesh, 2007). Leelapornpisid *et al.* (2007) reported antioxidant activity of the volatile oil of *Curcumaa amada*.

Curcuma aromatica Salisb. Family: Zingiberaceae

Methanol aqueous extract of *Curcuma aromatica* was screened by Kim *et al.* (1997) for anti-oxidative activity using Fenton's reagent/ethyl lineolate system. The extract showed antioxidant activity.

The antioxidant activities of the essential oil and various extracts of *Curcuma aromatica* were evaluated by Al-Reza *et al.* (2010) by using DPPH and superoxide radical scavenging assays. The oil and the methanol extract showed potent DPPH radical scavenging activities (IC_{50}=14.45 and 16.58 µg/ml, respectively), which were higher than butylated hydroxyanisole (IC_{50}=18.27µg/ml). These extracts also exhibited remarkable superoxide radical scavenging activities (IC_{50}=22.26-45.27µg/ml) and the activity of the methanol extract was superior to all other extracts (IC_{50}=22.26µg/ml). Furthermore, the amount of total phenolic compounds was determined and its content in ethyl acetate extract was the highest as compared to other extracts.

Gan *et al.* (2010) evaluated the antioxidant activities of 40 medicinal plants associated with prevention and treatment of cardiovascular and cerebrovascular diseases using FRAP and TEAC assays and their total phenolic contents were measured by the Folin-Ciocalteu methods. The antioxidant activity of *Curcuma aromatica* was 11.43μmol Fe(II)/g in FRAP assay, 7.66μmol Trolox/g and Total phenolic content was 0.38 mg GAE/g. A high correlation between antioxidant capacity and total phenolic content indicated that phenolic compounds could be the main contributor of the antioxidant activity of these plants.

Curcuma comosa Roxb. Family: Zingiberaceae

The ethanol extract of *C. comosa* and its isolated diarylheptanoid compound exhibited radical scavenging activities (Jariyawat *et al.*, 2009). Cisplatin raised BUN, creatinine and kidney lipid peroxidation levels and lowered kidney GSH content and levels of GPx, SOD, and CAT activities, all of which (except SOD and CAT) could be restored to normal values by pretreatment with 200 mg/kg BW of *C. comosa* ethanol extract.

Povichit *et al.* (2010) investigated phenolic content and free radical scavenging effect of *Curcuma comosa* by DPPH and ABTS assays, anti-lipid peroxidation activity by TBARS and for antiglycation activity. The results revealed that the total phenolic content showed good correlation with free radical scavenging by ABTS and anti-lipid peroxidation by TBARS, but showed no correlation with antiglycation. The extract of *Curcuma comosa* showed moderate antioxidant effect and also showed a promising antiglycation effect. The IC_{50} were 0.50mg/ml for the DPPH method, TEAC value of 0.09 mg/Trolox/mg sample for the ABTS method, IC_{50} of 2.02 mg/ml for the TBARS method and IC_{50} of 12.30 for the antiglycation method.

Curcuma domestica Valeton Family: Zingiberaceae

Leelapornpisid *et al.* (2007) reported antioxidant activity of the volatile oil of *Curcumaa domestica.*

Curcuma kwangsiensis S.K.Lee et C.F.Liang Family: Zingiberaceae

The sesquiterpenoids are one of the major groups of antioxidants in *Curcuma* besides curcuminoids. Zho *et al.* (2010) determined the antioxidant activity of sesquiterpenoids in four species of *Curcuma* and two essential oils from *Curcuma* genus. Their antioxidant capacities were quantitatively evaluated using densitometry with detection at 530 nm using vitamin C as reference. The results showed that *Curcuma kwangsiensis* rhizomes had higher antioxidant capacity.

Curcuma longa L. Family: Zingiberaceae

English Name: Turmeric.

Curcuma longa rhizome have been reported to possess antioxidative activity (Toda *et al.*, 1985b), better than α-tocopherol.

Turmeric has several compounds with immunomodulatory and antioxidant properties. Selvam *et al.* (1995) have isolated turmeric antioxidant-protein (TAP) from aqueous extract of turmeric. The protein showed a concentration-dependent inhibitory effect on lipid peroxdation. Ca^{2+}-ATPase of rat brain homogenate was protected to nearly 50 per cent of the initial activity from the lipid peroxidant induced inactivation by this protein. This protection of Ca^{2+}-ATPase activity was found to be associated with the prevention of loss of –SH groups.

Turmeric contains several small molecular weight components with antioxidant, medicinal and immunomodulatory activities (Ruby *et al.*, 1995). Natural curcuminoids, isolated from turmeric, were compared for their cytotoxic, tumor reducing and antioxidant activities. These compounds were also checked for their potential use as anti-promoters. The curcuminoids inhibited lipid peroxidation besides the production of superoxides and hydroxyl radical.

Methanol aqueous extract of *Curcuma longa* was screened by Kim *et al.* (1997) for anti-oxidative activity using Fenton's reagent/ethyl lineolate system. The extract showed antioxidant activity.

Curcumin, an antioxidant present in turmeric, has been shown to inhibit chemical carcinogenesis in animal models and has been shown to be an anti-inflammatory agent. It modulates glutathione (GSH)-linked detoxification mechanisms in rats. Singhal *et al.* (1999) have examined the effects of curcumin on GSH-linked enzymes in K562 human erythroleukemia cells. These results suggest that glutathione transferases (GST) play a major role in detoxification of lipid peroxidation products in K562 cells, and that these enzymes are modulated by curcumin. Scartezzini and Speroni (2000) are of the opinion that this herb contains antioxidant principles that can explain and justify their use in traditional medicine in the past as well as the present.

Turmeric oil and its fractions were tested for antioxidant activity using the β-carotene-linolate model system and the phosphomolybdenum method (Jayaprakasha *et al.*, 2002). The fraction III showed maximum antioxidant capacity. Patricia *et al.* (2003) reported the antioxidant, anticancer and antimicrobial activities of extracts from turmeric. The extracts were obtained using super critical CO_2 with and without ethanol or isopropyl alcohol as co solvent.

Cai *et al.* (2004, 2006) evaluated the antioxidant activity and phenolic compounds in traditional Chinese medicinal plants associated with anticancer. The improved ABTS.+ method was used to systematically assess the total antioxidant capacity (TEAC) of the medicinal extracts. The TEAC values and total phenolic content for methanolic extract of rhizome of *Curcuma longa* were 132.1 μmol Trolox equivalent/100g dry weight (DW), and 0.80 g of gallic acid equivalent/100g DW respectively. Major types of phenolic compounds in *Curcuma longa* include curcuminoids (curcumin, bisdemethoxycurcumin, demethoxycurcumin) phenolic volatile oils.

Odukoya *et al.* (2005) evaluated the antioxidant activity of Nigerian dietary spices including *Curcuma longa*. The antioxidant activity (expressed as per cent inhibition of oxidation) was highest in turmeric 82.51 per cent, reducing power was 19.65 per cent and Total phenolic content was 176.25 mg/100g Tannic acid Equivalent. Antioxidant activity correlated significantly and positively with total phenolics while there was no linear correlation between total antioxidant activity and reducing power neither between reducing power and total phenolic content.

Phytochemical flavonols, carotenoids and the antioxidant capacity of rhizome of *Curcuma longa* was evaluated by Lako *et al.* (2006). *Curcuma longa* rhizome have the highest Total antioxidant capacity (360 mg/100g TEAC) and are rich in Total polyphenol content (TPP) (320mg/100g), fisetin (64mg/100g), quercetin (41mg/100g) and myricetin (17mg/100g).

Curcumin found in the turmeric prevents free radical formation, premature ageing and possesses anticancerous properties (Mishra *et al.*, 2007). Dai *et al.* (2009) evaluated the antioxidative effects of curcumin and its analogues against the free-radical-induced peroxidation of linoleic in micelles.

Curcumin from *Curcuma longa* shows protective effects during selenium induced toxicity on dehydrogenases in hepatic tissue and cataract (Padmaja and Raju, 2004, 2005).

Wojcikowski *et al.* (2007) evaluated the antioxidant activity of aerial parts of *Curcuma longa*. Oxygen radical absorbance capacity of the ethyl extract, methanol extract and aqueous extracts were 466.81, 61.13 and 29.45µmol Trolox equivalent per g of dried starting material, respectively. Phansawan and Poungbangpho (2007) reported antioxidant capacities of *Curcuma longa* by ABTS method. Ethanol was the best solvent for extraction to give the highest antioxidant capacity, followed by acetone, methanol, distilled water and acetic acid.

Aftab and Vieira (2009) in their study compared the antioxidant activity of curcumin with those of reservatrol, a polyphenol present in some dietary plants such as *Vitis vinifera* and *Arachis hypogaea* and many other, non-dietary plants. Curcumin exhibited antioxidant effects at all time points. Curcumin and reservatrol together resulted in a synergistic antioxidant effect 15.5 per cent greater than an average of individual activities. This synergy was significantly greater (about 4 fold) the curcumin together with the flavonol quercetin.

Gan *et al.* (2010) evaluated the antioxidant activities of 40 medicinal plants associated with prevention and treatment of cardiovascular and cerebrovascular diseases using FRAP and TEAC assays and their total phenolic contents were measured by the Folin-Ciocalteu methods. The antioxidant activity of *Curcuma longa* was 122.03µmol Fe(II)/g in FRAP assay, 242.78µmol Trolox/g and Total phenolic content was 15.33 mg GAE/g. A high correlation between antioxidant capacity and total phenolic content indicated that phenolic compounds could be the main contributor of the antioxidant activity of these plants.

The sesquiterpenoids are one of the major groups of antioxidants in *Curcuma* besides curcuminoids. Zhao *et al.* (2010) determined the antioxidant activity of sesquiterpenoids in four species of *Curcuma* and two essential oils from *Curcuma* genus. Their antioxidant capacities were quantitatively evaluated using densitometry with detection at 530 nm using vitamin C as reference. The results showed that *Curcuma longa* rhizomes had highest antioxidant capacity. The major sesquiterpenoids with antioxidant activity identified from the *Curcuma longa* include α-turmerone, β-turmerone and β-sesquiphellandrene. The essential oil of *C.longa* (373.1mg/g vs. 10.6-13.1 mg/g) showed much higher antioxidant activities than their related raw materials which suggested that essential oil with antioxidant sesquiterpenoids were, at least, one of their major antioxidant compounds.

Curcuma mangga Valeton Family: Zingiberaceae

Abbas *et al.* (2006) evaluated the antioxidant activityof *Curcuma mangga* using DPPH assay. Methanolic extract showed DPPH radical scavenging activity. Leelapornpisid *et al.* (2007) reported antioxidant activity of the volatile oil of *Curcuma mangga*.

Curcuma phaeocaulis Val. Family: Zingiberaceae

The sesquiterpenoids are one of the major groups of antioxidants in *Curcuma* besides curcuminoids. Zho *et al.* (2010) determined the antioxidant activity of sesquiterpenoids in four species of *Curcuma* and two essential oils from *Curcuma* genus. Their antioxidant capacities were quantitatively evaluated using densitometry with detection at 530 nm using vitamin C as reference. The results showed that *Curcuma phaeocaulis* rhizomes had lowest antioxidant capacity.

Curcuma wenyujin Y.H.Chen Family: Amaranthaceae

Gan *et al.* (2010) evaluated the antioxidant activities of 40 medicinal plants associated with prevention and treatment of cardiovascular and cerebrovascular diseases using FRAP and TEAC assays and their total phenolic contents were measured by the Folin-Ciocalteu methods. The antioxidant activity of *Curcuma wenyujin* was 12.15µmol Fe(II)/g in FRAP assay, 136.05µmol Trolox/g and Total phenolic content was 3.67 mg GAE/g. A high correlation between antioxidant capacity and total phenolic content indicated that phenolic compounds could be the main contributor of the antioxidant activity of these plants.

The sesquiterpenoids are one of the major groups of antioxidants in *Curcuma* besides curcuminoids. Zhao *et al.* (2010) determined the antioxidant activity of sesquiterpenoids in four species of *Curcuma* and two essential oils from *Curcuma* genus. Their antioxidant capacities were quantitatively evaluated using densitometry with detection at 530 nm using vitamin C as reference. The results showed that *Curcuma wenyujin* rhizomes had higher antioxidant capacity. The major sesquiterpenoids with antioxidant activity identified from the *Curcuma wenyujin* include kurzerene and furanodiene. The essential oil of *C.wenyujin* (759.3mg/g vs. 7.5-15.8 mg/g) showed much higher antioxidant activities than their related raw materials which suggested that essential oil with antioxidant sesquiterpenoids were, at least, one of their major antioxidant compounds.

Curcuma xanthorrhiza Roxb. Family: Zingiberaceae

A new curcumin analogue has been isolated by Masuda *et al.* (1992) from the rhizomes of *Curcuma xanthorrhiza* along with four known curcuminoids, and its structure has been determined as 1-(4-hydroxy-3,5-dimethoxyphenyl)-7-(4-hydroxy-3-methoxy phenyl)-(1E, 6E)-1,6-heptadiene-3,4-dione by spectral data. The new compound showed potent antioxidant activity as against autoxidation of linoleic acid in water-alcohol system.

Chanwitheesuk *et al.* (2005) investigated the antioxidant activity of rhizome of *Curcuma xanthorrhiza* using a β-carotene bleaching method. The contents of plant chemicals such as vitamin C, Vitamin E, carotenoids, tannin and total phenolics, were also determined. Methanolic extract showed antioxidant activity. *Curcuma xanthorrhiza* has antioxidant index of 0.69, Vitamin C 7.69 mg per cent, Vitamin E 0.0085mg per cent, Total carotenes 1.28 mg per cent, Total xanthophylls 0.65mg per cent, Tannins 14.4mg per cent and Total phenolics of 112mg per cent.

Curcuma zedoaria Rosc. Family: Zingiberaceae

English Name: Zedoary

Souri *et al.* (2008) investigated the antioxidant activity and radical scavenging activity of methanolic extract of rhizøme. The antioxidant activity of this plant extract against linoleic acid peroxidation expressed as IC_{50} was 3.77ng/ml and radical scavenging activity against DPPH expressed as IC_{50} was 10.60 µg/ml.

Paramapojn and Gritsanapan (2009) determined the free radical scavenging activity of *Curcuma zedoaria* extracts and curcuminoids using DPPH assay. Values of EC_{50} of all extracts were found in the range of 18.29 to 40.33µg/ml. Free radical scavenging activity of the separated pure compounds was found in descending order of curcumin>demethoxycurcumin>bisdemethoxycurcumin.

Gan *et al.* (2010) evaluated the antioxidant activities of 40 medicinal plants associated with prevention and treatment of cardiovascular and cerebrovascular diseases using FRAP and TEAC assays and their total phenolic contents were measured by the Folin-Ciocalteu methods. The antioxidant activity of *Curcuma zedoaria* was 213.72µmol Fe(II)/g in FRAP assay, 245.03µmol Trolox/g and Total phenolic content was 13.73 mg GAE/g. A high correlation between antioxidant capacity and total phenolic content indicated that phenolic compounds could be the main contributor of the antioxidant activity of these plants.

Cuscuta chinensis Lam. Family: Convolvulaceae

Cai *et al.* (2004, 2006) evaluated the antioxidant activity and phenolic compounds in traditional Chinese medicinal plants associated with anticancer. The improved ABTS.+ method was used to systematically assess the total antioxidant capacity (Trolox equivalent antioxidant capacity, TEAC) of the medicinal extracts. The TEAC values and total phenolic content for methanolic extract of seed of *Cuscuta chinensis* were 141.4 µmol Trolox equivalent/100g dry weight (DW), and 1.32 g of gallic acid equivalent/100g DW respectively. A positive significant linear relationship between antioxidant activity and total phenolic content showed that phenolic compounds were the dominant antioxidant components in the tested herbs. Major types of phenolic compounds in *Cuscuta chinensis* include flavonols (quercetin, astragalin), furofuran lignans, coumarins, tannins.

Ethanolic extract of *Cuscuta chinensis* exhibited a significant antioxidant effect (p less than 0.05) by increasing levels of superoxide dismutase (SOD), catalase (CAT), and glutathione peroxidase (GPx), and by reducing malondialdehyde (MDA) levels (Yen *et al.*, 2007).

Cuscuta reflexa Roxb. Family: Cuscutaceae

Yadav and Tripathi (2004) studied the antioxidant activity of *Cuscuta reflexa*.The ethanolic extract has more antioxidant activity than the benzene and chloroform elutent. The ethyl acetate fraction chromatographed on silica gel column eluted with chloroform-methanol different mixture yielded three flavonoid glycosides *viz.*, apigenin 7-O-glucoside, kaempferol 3-O-alpha-rhamnoside and myricetin3-O-rhamnoside. Myricetin 3-O-rhamnoside is more active than the kaempferol 3-O-α-rhamnoside and apigenin 7-O-glucoside.

Cussonia bancoensis Aubrev. et Pellegr. Family: Araliaceae

A new caffeoyl derivative *p*-caffeoyl phenol has been isolated from the n-butanol-soluble fraction of stem bark of *Cussonia bancoensis*, together with the known liriodendrin. This compound was screened for its potential antioxidant properties by using three different assays: DPPH radical assay, non-enzymatic superoxide anion generating system NADH/PMS and towards inhibition of xanthine oxidase (XO) enzyme, and revealed potent antioxidant activities (Tapondjou *et al.*, 2005).

Cussonia barteri Seem. Family: Araliaceae

Diallo *et al.* (2001) demonstrated with DPPH assay that *Cussonia barteri* possessed potent antioxidant activity. Biapa *et al.* (2007) investigated the phytochemical and antioxidant properties of four medicinal plants used in Cameroon, including *Cussona barteri*. Four extracts (methanol, hydro-ethanolic, aqueous and hydrolysed) from each of the plants were prepared and analysed. Folin Ciocaltee, FRAP and DPPH assays were used to determine the antioxidant potential. The highest antioxidant activites were observed in the hydrolysed extracts of each plant, while the aqueous extract showed the least activity irrespective of the method used.

Cyamopsis tetragonaloba L. Family: Fabaceae (Leguminosae)

English Name: Cluster bean

Antioxidant potential of selected vegetables commonly used in diet in Asian subcontinent, including *Cyamopsis tetragonaloba* fruits has been evaluated by Gacche *et al.* (2010). The DPPH radical scavenging activity was 51.6 per cent, inhibition of lipid peroxidation was 45.3 per cent, Vitamin C content was 15.36 mg/100g and total phenols was 5.79 mg/g.

Cyathula officinalis Kuan Family: Amaranthaceae

Gan *et al.* (2010) evaluated the antioxidant activities of 40 medicinal plants associated with prevention and treatment of cardiovascular and cerebrovascular diseases using FRAP and TEAC assays and their total phenolic contents were measured by the Folin-Ciocalteu methods. The antioxidant activity of *Cyathula officinalis* was very low, 0.24μmol Fe(II)/g in FRAP assay, 23.36μmol Trolox/g in TEAC assay and Total phenolic content was 3.73 mg GAE/g.

Cyclea peltata Lam. Family: Menispermaceae

The administration of methanolic leaf extract of *Cyclea peltata* showed a significant enhancement of antioxidants in gastric mucosa against ethanol induced gastric ulceration (Fijesh and Padikkala, 2008).

Cyclobalanopsis stenophylla (Makino) Liao var. stenophylloides (Hayata) Liao Family: Fagaceae

Hou *et al.* (2003) evaluated the antioxidant activity of 70 per cent aqueous acetone extract of *Cyclobalanopsis stenophylla* var. *stenophylloides* by various assays including DPPH, hydroxyl radicals and reducing power assay. The extract exhibited stronger activity against DPPH radicals with IC_{50} value of 5.4 µg/ml and inhibited formation of OH generated in the Fenton reaction system with IC_{50} value of 0.2 µg/ml). The amount of phenolics was 33.5 mg of GAE/g.

Cyclocarya paliurus **(Batal.) Iljinskaja Family: Juglandaaceae**

Xie *et al.* (2010) evaluated the antioxidant activity of a water soluble polysaccharide from *Cyclocarya paliurus.* The polysaccharide, named as CPP-1, exerted significant scavenging effects on DPPH radicals with a value of around 91.4 per cent compared to the reference controls of BHT (91.2 per cent) and ascorbic acid (98.9 per cent) at a concentration of 400µg/ml, and with EC_{50} value of 52.3µg/ml.

Cyclocodon lancifolia **(Roxb.) Kurz. subsp.** ***lancifolia*** **Family: Campanulaceae**

Chen *et al.* (2009) studied the antioxidant activity of aerial parts extract towards xanthine oxidase, lipoxygenase and tyrosinase activities. The inhibitory effect of acetone extract on xanthine oxidase activity was about 10 per cent, tyrosine activity was about 10 per cent and lipoxygenase activity was more than 20 per cent.

Cyclopia genistoides **R.Br. Family: Fabaceae (Leguminosae)**

The *in vitro* antioxidant activity of aqueous extracts prepared from four *Cyclopia* spp. (unfermented and fermented) was assessed by Joubert *et al.* (2008) using radical (ABTS) scavenging ferric ion reduction and inhibition of Fe^{2+}-induced microsomal lipid peroxidation as criteria. Qualitative and quantitative differences in penolic composition were demonstrated for the *Cyclopia* spp. The xanthone glycoside, a.k.a. mangiferin, was the major monomeric polyphenol present in the *Cyclopia* extracts, with both unfermented and fermented *C. genistoides* extracts containing the highest quantities. Fermentation resulted in a significant reduction in extract yields and their total polyphenolic and individual polyphenol contents. Unfermented plant material should preferabaly be used for preparation of extracts, as fermentation significantly lowered antioxidant activity of all species, except in the case of *C. genistoides,* where the ability to inhibit lipid peroxidation was not affected.

Cyclopia intermedia **E. Mey Family: Fabaceae (Leguminosae)**

English Name: Honeybush

Honeybush is a popular tisane in its native South Africa and have a worldwide market. Herbal tea is used traditionally for medicinal purpose and is rich in polyphenols. The principal polyphenols in honeybush include Xanthone mangiferin and the flavones hesperin and isokuranetin. McKay and Blumberg (2006) in animal model studies reported potent antioxidant and antimutagenic activities *in vitro*. The herbal teas possess a potent antioxidant, immune-modulating and chemopreventive actions. No adverse effects of honeybush consumption as tisane have been reported.

Cyclotrichium niveum (Boiss.) Manden et Scheng. Family: Lamiaceae (Labiatae)

The radical scavenging activity of essential oil was evaluated *in vitro* by DPPH method by Cetinus *et al.* (2007). From the overall results they concluded that *C. niveum* essential oil showed radical scavenging activity.

The dichloromethane, ethyl acetate, ethanol and aqueous extracts of *Cyclotrichium niveum* along with the essential oils were assessed by Orhan *et al.* (2009) for their antioxidant activities. Antioxidant activity was evaluated by DPPH radical scavenging and ferrous iron-chelating tests. Ferrous-reducing power (FRAP) was also tested. The water extracts displayed the highest ferrous iron-chelating effect.

The aerial parts of *Cyclotrichum niveum* were screened by Gursoy *et al.* (2009) for their possible antioxidant and antimicrobial activities in addition to their phenolic contents. Antioxidant activity was employed by two complementary test systems: DPPH free radical scavenging and β-carotene/linoleic acid.

Cyclotrichium origanifolium (Labill.) Manden et Scheng. Family: Lamiaceae (Labiatae)

The *in vitro* antimicrobial and antioxidant activities of the essential oil of *C. origanifolium* have been reported (Tepe *et al.*, 2005).

Cydonia oblonga Miller Family: Rosaceae

Methanolic extract was fractionated into a phenolic fraction and an organic acid fraction and was analyzed by HPLC/diode array detection and HPLC/UV, respectively. Free radical scavenging activities of the extracts and fractions were evaluated by a microassay using DPPH. The phenolic fraction always exhibited a stronger antioxdiant activity than the whole methanolic extract. Organic acid extracts were always the weakest in terms of antiradical activity, which seems to indicate that the phenolic fraction gives a higher contribution for the antioxidant potential for quince fruit and jam. The IC_{50} values of quince pulp, peel, and jam extracts were correlated with caffeoylquinic acids total content. Among the phenolic fractions, the seed extract was the one that exhibited the strongest antioxidant activity (Silva *et al.*, 2004).

Aslan *et al.* (2010) studied the antidiabetic and antioxidant activities of the ethanol extracts of leaves of *Cydonia oblonga*. TBARS and GSH livels in liver, kidney and heart tissues were measured by using spectrophotometric methods for antioxidant assay. Extracts induced significant alleviation on only hear tissue TBARS levels (45.7 per cent). None of the extracts restored GSH levels in kidney, liver and heart tissues of diabetic rats.

Cydonia vulgaris Pers. Family: Rosaceae

English Name: Quince

The antioxidant activities of water, ethanol, and ether extracts of leaves of *Cydonia vulgaris* were determined by Yildirin *et al.* (2001) by the thiocyanate method. The antioxidant activity of the water extract increased with the increasing amount of extract (200μg – 1000 μg) added to the linoleic acid emulsion. Ether extract was the most effective antioxidant among the extracts. Like antioxidant activity, the reducing power of water extract was concentration dependent. However, ethanol extract has the highest reducing power, and ether extract has the lowest. They suggested that although the reducing power of a substance may be an indicator of its potential antioxidant activity, there is not necessarily a linear correlation, between these two activities.

Cymbopogon citratus Stapf Family: Poaceae (Graminae)

Kruawan and Kangsadalampai (2006) investigated the antioxidant activity and phenolic compound contents of water extract of *Cymbopogon citratus*. The extract exhibited strong scavenging activity against DPPH radicals (81.69 per cent) and their FRAP value was 794.31 µmol/g. The content of phenolic compound in the extracts was determined using the Folin-Ciocalteu reagent and calculated as gallic acid equivalent (GAE). The herbal extract gave a high phenolic compound content (GAE 83.83 mg/g).

Leelapornpisid *et al.* (2007) reported antioxidant activity of the volatile oil of *Cymbopogon citratus*.

Cymbopogon flexuosus (Nees ex Steud.) Watson Family: Poaceae (Graminae)

It is found to have a refreshing effect on the body and possesses immune modulatory properties. β-sitosterol, myrcene and selenium in lemongrass are responsible for its antioxidative behaviour (Mishra *et al.*, 2007).

Cymbopogon schoenanthus (L.) Spreng Family: Poaceae (Graminae)

Khadri *et al.* (2008) evaluated the chemical composition and antioxidant activities of the essential oils from fresh leaves, dried leaves and roots of *Cymbopogon schoenanthus* collected from three different locations in Southern Tunisia. Antioxidant activity was assayed using DPPH radical scavenging and β-carotene-linoleic acid bleaching method. The best results (IC_{50} = 0.47mg/ml) were obtained with fresh leaves of the plants collected in the desert region.

Aqueous extract, proanthocyanidin rich extract and organic extracts of *Cymbopogon schoenanthus* shoots from three different locations in South Tunisia were screened by Khadri *et al.* (2008) for their antioxidant activity. In addition the contents of flavonoids and phenolic compounds were determined. Antioxidant activity measured by DPPH assay showed that the proanthocyanidins extract exhibited higher antioxidant activity than aaqueous extract. Extract concentration providing 50 per cent inhibition (IC_{50}) ranged from 16.4µg/ml to 26.4µg/ml. The antioxidant activity was also determined using the β-carotene/linoleic acid bleaching test. The best results (IC_{50} = 0.11mg/ml) were obtained with the proanthocyanidin extract of the plants collected from the desert region (Dhibat).

Cynanchum atratum Bge. Family: Asclepiadaceae

Li *et al.* (2008) evaluated the antioxidant properties of 45 medicinal plants including *Cynanchum atratum* using FRAP and TEAC assays and the total phenolic contents of these plants were measured. *Cynanchum atratum* showed 29.54 µmol Fe(II)/g FRAP value, 17.87µmol Trolox/g of TEAC value and 7.75 mg GAE/g of Phenolic content. A high correlation between antioxidant capacities and their phenolic contents indicated that phenolic compounds were a major contributor of antioxidant activity of these plants.

Cynanchum paniculatum (Bge.) Khadri et al., Family: Asclepiadaceae

Gan *et al.* (2010) evaluated the antioxidant activity of *Cynanchum paniculatum* using the FRAP and TEAC assays, and its total phenolic content was measured by the Folin-Ciocalteu method. For FRAP assay the value was 35.93 μmol Fe(II)/g dry weight. For TEAC assay, the value was 39.45μmol Trolox/g dry weight. Total phenolic content was 4.35 mg gallic acid equivalent/g dry weight of plant material.

Cynanchum stauntonii (Decne) Schltr. Family: Asclepiadeaceae

Antioxidant capacities of 56 selected Chinese medicinal plants, including *Cynanchum stauntonii* were evaluated by Song *et al.* (2010) using the TEAC and FRAP assays, and their total phenolic content was measured by the Folin-Ciocaleteu method. The TEAC value was 14.24μmol Trolox/g, FRAP value was 9.77μmol Fe^{2+}/g and phenolic content was 1.40mg GAE/g.

Cynanthillium cinerum (Less.) H.Robs. Family: Asteraceae (Compositae)

Guha *et al.* (2009b) investigated the antioxidant activity of polar (methanolic and aqueous) and non polar (hexane and chloroform) crude extracts of whole plant of *Cynanthillium cinerum*. Several parameters including free radicals (DPPH, ABTS, H_2O_2 and OH) scavenging, reducing power and protection of DNA against oxidative damage, cytotoxicity, inhibition of oxidative hemolysis in erythrocytes, total phenolic content and inhibition of lipid peroxidation were examined. All the free-radical generating assay methods demonstrated positive scavenging efficiency with different but considerable magnitude for the four extracts. However only the hexane extract showed significant H_2O_2 scavenging effect. Lipid peroxidation was estimated by the TBA-MDA reaction and high degree of inhibition was shown by all the extracts. Reducing power of the polar extracts was higher than the non-polar ones.

Cynara cardunculus L. Family: Asteraceae (Compositae)

English Name: Artichoke, cardoon.

Valentão *et al.* (2002b) examined the antioxidant activity of cardoon lyophilized infusion against superoxide radical, hydroxyl radical and hypochlorous acid. It contained considerable amounts of polyphenolic compounds.

Hypercholesterolemia and lipid peroxidation play a complementary role in atherosclerosis. Artichoke is rich in antioxidants and has a cholesterol-reducing effect. Kusku-Kiraz *et al.* (2009) investigated the effect of Artichoke leaf extract (ALE) on lipid levels and peroxidation in experimental hypoercholesterolemic condition. Serum cholesterol and triglyceride levels and the ration of cholesterol to HDL-cholesterol decreased due to ALE treatment in rats fed with fat diet. Significant decreases in serum MDA and DC levels and increases in plasma AOA were detected in serum in ALE treated hyperocholesterolemic rats. Endogenous diene conjugate and copper-induced malondialdehyde levels were also lower in LDL+VLDL due to ALE treatment in hyperocholesterolemic rats. Their results indicate that ALE may be useful for prevention of hypercholesterolemia and pro-oxidant state in LDL+VLDL fraction and the reduction of increased serum cholesterol and triglyceride levels.

Cynara scolymus L. Family: Asteraceae (Compositae)

English Name: Artichoke.

Fintelman (1996) reported antioxidant activity of leaf extract of artichoke. Artichoke extract showed strong antioxidant properties identified via inhibition of tert, hydroperoxide-induced lipid peroxidation (Gebhardt, 1996). Antioxidant and protective properties of extracts from the leaves of the artichoke against hyperoxide-induced oxidative stress in cultured rat hepatocytes were reported by Gebhardt (1997).

In vitro antioxidant activities of edible artichoke and effect on biomarkers of antioxidant in rats were investigated by Jimenez-Escrig *et al.* (2003). The antioxidant activity of aqueous-organic extracts were determined using three methods: DPPH scavenging, FRAP and inhibition of Copper(II)-catalyzed *in vitro* human low-density lipoprotein (LDL) oxidation. One gram (dry matter) had a DPPH activity and a FRAP value *in vitro* equivalent to those of 29.2 and 62.6 mg of vitamin C and those of 77.9 and 159 mg of vitamin E, respectively. Artichoke extracts showed good efficiency in the inhibition *in vitro* of LDL oxidation. Among different antioxidant enzymes measured (superoxide dismutase, glutathione peroxidase, glutathione reductase and catalase) in erythrocytes, only glutathione peroxidase activity was elevated in the artichoke group compared to the control group.

Cynodon dactylon (L.) Pers. Family: Poaceae (Graminae)

This grass grows throughout India and in almost all parts of the world. Auddy *et al.* (2003) have studied and demonstrated the antioxidant properties of *C. dactylon* using ABTS assay.

Valamarthi and Harathi (2009) evaluated the antioxidant activity in Daltons Lymphoma Ascite (DLA) transplanted Swiss albino male mice treated with protein fraction of *Cynodon dactylon*. Administration of protein fraction to the DLA induced mice showed enhanced level of enzymatic antioxidants and non-enzymatic antioxidants which proved their protective action against the free radical damage caused by DLA tumor cells and maintains normal function of the organ and confimrmed their antioxidant activity.

Cynomorium songaricum Rupr. Family: Cynomoriaceae - Balanophoraceae

In traditional Chinese medicine a number of herbs are used to alleviate the symptoms of aging, among them the stems of *Cynomorium songaricum*. Lu *et al.* (2008) evaluated the protective effect of different extracts of *C. songaricum* on staurosporin induced apoptotic cell death in SK-N-SH neuroblastoma cells. Staurosporine (100nm) reduced cell viability to about 55 per cent. The ethyl acetate fraction of *C. songaricum* significantly attenuated staurosporine-induced cell death at concentrations of 100 and 10 µg/ml. On the other hand, the dichloromethane as well as water fractions showed no protective effects. In order to further analyse the protective action, the superoxide anion scavenging activity of two extracts was evaluated in a xanthine/xanthinoxidase system. Ethyl acetate extract showed a good scavenging activity (IC_{50}

value 2.9µg/ml) without inhibition of xanthine oxidase. In conclusion their results proved the neuroprotective activity of *C. songaricum* extracts *in vitro*, thus supporting its traditional use (Lu *et al.*, 2008).

Cyperus rotundus L. Family: Cyperaceae

C. rotundus is a pantropic weed. *C. rotundus* has been found to exhibit significant antioxidant and antimutagenic activities. The preliminary chemical study of the tested extracts of *C. rotundus* revealed the presence of important quantities of flavonoids, tannins and coumarins in total oligomers flavonoids (TOF) and methanol extracts. Ethyl acetate extract has shown the presence of sterols and flavonoids, while antioxidant nature was proved by DPPH assay (Kilani *et al.*, 2005).

The *in vitro* antioxidant activity of roots and rhizomes of *Cyperus rotundus* has been investigated by Pal and Dutta (2006) by estimating degree of non-enzymatic haemoglobin glycosylation. The ethanol extract of the roots and rhizomes of *Cyperus rotundus* showed higher activity than other extracts of it. The antioxidant activity of the extracts are close and identical in magnitude and comparable to that of standard antioxidant compounds.

The antioxidant potential of hydroalcoholic extract of *Cyperus rotundus* was evaluated by various antioxidant assays, including antioxidant capacity by the phosphomolybdenum method, total antioxidant activity in linoleic acid emulsion systems, DPPH, superoxide, hydroxyl radicals and nitric oxide scavenging. They further evaluated the reducing potential of the extract as well as Fe(2+)/ascorbate-induced lipid peroxidation in rat liver homogenate. The extract exhibited high reduction capability and powerful free radical scavenging, especially against DPPH and superoxide anions as well as a moderate effect on NO. The extract also inhibited lipid peroxidation in rat liver homogenate induced by Fe(2+)/ascorbate and prevented deoxyribose degradation in both non-site-specific and site-specific assays showing the hydroxyl radical scavenging and metal chelating activity of the hydroalcoholic extract. Moreover, the peroxidation inhibiting activity of CRE was demonstrated in the linoleic acid emulsion system (Yazdanparast and Ardestani, 2007).

Nagulendran *et al.* (2007) evaluated the antioxidant activity of *Cyperus rotundus* rhizomes extract (CRRE) in a series of *in vitro* assay involving free radicals and reactive oxygen species. CRRE exhibited its scavenging effect in concentration dependent manner on superoxide anion radicals, hydroxyl radicals, nitric oxide radical, hydrogen peroxide and property of metal chelating and reducing power. The extract was also studied for lipid peroxidation assay by TBARS using young and aged rat brain mitochondria. The extract was also effective in preventing mitochondrial lipid peroxidation induced by $FeSO_4$/ascorbate in concentration dependent manner.

Cyperus scariosus L. Family: Cyperaceae

English Name: Umbrella sedge.

Kalim *et al.* (2010) evaluated the antioxidant activity of Methanol (50 per cent) extract of *Cyperus scariosus*. IC_{50} values for DPPH, ABTS, NO, OH, O_2 and ONOO were 11.10, 6.27, 18.23, 240.31, 28.85 and 590.23µg/ml. The extract contained significant amount of polyphenols.

Cytisus scoparius L. Family: Fabaceae (Leguminosae)

Raja *et al.* (2006) investigated the antioxidant activity of *Cytisus scoparius* on CCl_4 treated oxidative stress in wistar albino rats. Pretreatment of rats with different doses of plant extract (250 and 500mg/kg) significantly lowered SGOT, SGPT, LDH and TBARS levels against CCl_4 treated rats. GSH and hepatic enzymes like SOD, CAT, GPx, GRD and GSR were significantly increased by treatment with the plant extract, against CCl_4 treated rats. The activity of extract at the dose of 500mg/kg was comparable to the standard drug, silymarin (25mg/kg).

Dalbergia odorifera T.C.Chen. Family: Fabaceae (Papilionaceae)

Cheng *et al.* (1998) reported antioxidant properties of butein isolated from *Dalbergia odorifera*. Gan *et al.* (2010) evaluated the antioxidant activities of 40 medicinal plants associated with prevention and treatment of cardiovascular and cerebrovascular diseases using FRAP and TEAC assays and their total phenolic contents were measured by the Folin-Ciocalteu methods. The antioxidant activity of *Dalbergia odorifera* was 162.82μmol Fe(II)/g in FRAP assay, 307.39μmol Trolox/g TEAC assay and Total phenolic content was 7.76 mg GAE/g.

Dalbergia saxatilis Hook.f. Family: Fabaceae (Papilionaceae)

Sofidya *et al.* (2009) evaluated the DPPH radical scavenging, total antioxidant activities, reducing power and contents of phenolic compounds in methanolic leaf extracts of *Dalbergia saxatilis*. The extract showed a dose-dependent effect both on the free radical scavenging DPPH and also on Fe^{3+} reducing power. Total antioxidant activities correlated with total phenols. A linear positive relationship existed between the reducing power and total phenolics.

Dalbergia sissoo Roxb. Family: Fabacae (Papilionaceae)

Vadlamudi and Naidu (2010) evaluated the antioxidant activity of ethanolic extracts of *Dalbergia sissoo* leaves. SOD activity was 1.1U/mg, Catalase activity was 2.4U/mg while vitamin C content was 300 mg/100g.

Daniella oliveri Rolfe (Syn.: *D. thurifera*) Family: Fabaceae (Leguminosae)

Muanda *et al.* (2011) evaluated the antioxidant activity of three Malian medicinal plants including *Daniella oliveri* employing phosphomolybdenum assay (PPM), DPPH and ABTS tests. The content of TPC and TFC were higher in the stem barks. The PPM assay showed that the highest value was 606.0mg/g dw for the root bark, followed by stem bark (586 mg/g dw) and leaves (526mg/g dw). The values of ABTS were 127.5, 124.1 and 109.2 mg/g d w for stem bark, root bark and leaves, respectively. The values for DPPH were 193.7, 196.3 and 210.3 mg/g dw, for stem bark, root bark and leaves respectively.

Daphne giraldii Nitsche Family: Thymelaeaceae

Two new biflavonoids, daphnogirins A and B were obtained from the roots of *Daphne giraldii* by Zhou *et al.* (2007). Oxygen radical scavenging assay has indicated that they have significant antioxidant activity.

Daphne gnidium L. Family: Thymelaeaceae

Dessi *et al.* (2001) carried out studies on the effect of methanol extract and essential oil on lipid peroxidation. Methanol extract showed antioxidant activity.

Daphne oleoides Schreb. Family: Thymeleaceae

Demo *et al.* (1998) reported antioxidant α-tocopherol from the Hexane extract of *Daphne oleoides.*

Daphniphyllum calycinum Benth. Family: Daphniphyllaceae

A novel flavonoid diglycoside, 5,6,7,4′-tetrahydroxyflavonol 3-O-rutinoside (1), and a previously known compound, kaempferol 3-O-neohesperidoside (2), were isolated from an ethyl acetate extract of *Daphniphyllum calycinum* leaves that showed significant activity in a DPPH free-radical assay (Gamez *et al.*, 1998). Compounds 1 and 2 were found to be moderately active as antioxidants in the DPPH assay.

Datura metel L. Family: Solanaceae

Vadlamudi and Naidu (2010) evaluated the antioxidant activity of ethanolic extracts of *Datura metel* leaves. SOD activity was 1.6U/mg, Catalase activity was 3.7U/mg while vitamin C content was 60 mg/100g.

Datura stramonium L. Family: Solanaceae

Suresh Kumar *et al.* (2008) carried out a study to determine the antioxidant activity of *Datura stramonium*. The total alkaloid content was 31.4mg/g while flavonoid content was 23.063 mg/g. The inhibition of lipid peroxidation was 70 per cent which was lesser than Butylated hydroxyl toluene (BHT) (93.65).

Daucus carota L. Family: Apiaceae (Umbelliferae)

English Name: Carrot.

The antioxidative activity of phenolic extracts from flesh and peel of root tubers and leaf of carrot was examined by autoxidation of methyl linoleate by Kahkonen *et al.* (1999). The total phenolics were very low, 0.6 mg/g GAE in flesh, 7.4 mg/g GAE in leaf and 6.6 mg/g GAE in peel. The antioxidant activity was moderate, 34 per cent in flesh, 22 per cent in leaf and 52 per cent in peel.

Cai *et al.* (2004) evaluated the antioxidant activity and phenolic compounds in traditional Chinese medicinal plants and vegetables associated with anticancer. The improved ABTS.+ method was used to systematically assess the total antioxidant capacity (TEAC) of the medicinal extracts. The TEAC values and total phenolic content for methanolic extract of carrot were 38.7 µmol Trolox equivalent/100g dry weight (DW), and 0.22 g of gallic acid equivalent/100g DW respectively.

The β-carotene found in the carrot functions as antioxidant. It prevents accumulation of toxins and is effective in treatment of arthritis, edema and rheumatism. Alanine, α-tocopherol, ascorbic acid, camphene, eugenol, γ-terpinene, histidine are the main constituents in carrot that possess antioxidant properties (Mishra *et al.*, 2007).

A study was carried out by Raghu *et al.* (2010) to determine the antioxidant activity of aqueous extracts of ten selected common vegetables. For *Daucus carota* the IC_{50} value for DPPH scavenging was 1.70mg/ml while standard was 2.45µg/ml. Reducing power of the extract was moderate.

Decalepis hamiltonii **Wight et Arn. Family: Asclepiadaceae**

English Name: Swallow root.

Naik *et al.* (2007) studied the gastroprotective effect and antioxidantive mechanism of swallow root extract. Aqueous extract possessed reducing power, free radical ability with an IC_{50} of 0.17µg/ml gallic acid equivalent (GAE), comparable to that of BHA (IC_{50} – 0.08 µg/ml). Depletion in antioxidant enzymes and increased TBARS were observed in ulcer-induced rats while swallow root aqueous extract fed rats showed normal levels.

Srivastava and Sivanandappa (2010) studied the hepatoprotective activity of the aqueous extract of the roots of *Decalepis hamiltonii* with known antioxidant constituents against carbon tetrachloride (CCl_4)-induced oxidative stress and liver injury in rats. The extract prevented CCl_4-induced hepatic damage and also prevented oxidative stress in the rat liver by inhibiting lipid peroxidation and protein carbonylation, and restoring the levels of antioxidant enzymes (SOD, CAT, GPx, GR and GST) and glutathione. Protective effect of the aqueous extract of the roots of *Decalepis hamiltonii* against CCl_4-induced acute hepatotoxicity could be attributed to the antioxidant constituents.

Surveswaran *et al.* (2010) surveyed antioxidant properties and total phenolic and flavonoid contents of 15 samples, representing 12 Indian medicinal plant species from these subfamilies. Total antioxidant assay was performed using ABTS and FRAP methods. The highest antioxidant capacity and high levels of total phenolics and flavonoids were found in the leaves of *Decalepis hamiltonii*. The highly significant and positive correlations between total antioxidant capacity parameters and total phenolic content indicated that the phenolic compounds contributed significantly to the antioxidant activity of the tested plant samples.

Delonix elata **(L.) Gamble Family: Caesalpiniaceae**

Wijayasiriwardena *et al.* (2009) investigated the anti-inflammatory effect of *Delonix elata*. Mathanol extract of leaf at 300 mg/kg showed significant reduction in area under curve of rat paw edema in the later phases of inflammation (45.83 per cent) which was comparable to that of standard. The extract showed nitric scavenging activity IC_{50} 157.08µg/ml. At 300mg/kg dose it was found to have potent antioxidant capacity, which may play its role in reducing inflammation.

Delonix regia **Rafin. Family: Caesalpiniaceae**

Bade *et al.* (2009) investigated the *in vitro* antioxidant and free radical scavenging activity of ethyl acetate extract of flowers of *Delonix regia* by different *in vitro* antioxidant models. The ethyl acetate extrat showed better antioxidant activity as compared to the standard one by DPPH radical scavenging and phenylhydrazine induced haemolysis of erythrocyte methods. IC_{50} values for DPPH radical scavenging and phenylhydrazine induced haemolysis of erythrocyte were 29.69 and 14.586 respectively. Total polyphenolic and total flavonoids were 40.83 and 30.32 per cent w/w, respectively. Ethyl acetate extract contains quercetin.

Delphinium linearilobum (Trautv.) N. Busch. Family: Ranunculaceae

From the roots of *Delphinium linearilobum* two new norditerpene alkaloids linearilobin and linearilin, and the known alkaloids lycoctonine, 14-acetyltalatizamine, browniine, cammakonine, talatizamine and cochlearenine were isolated by Kolak *et al.* (2006). Antioxidant activity was performed by DPPH and metal chelating activity assays.

Dendrobium candidum Wall. Family: Orchidaceae

Dendrocandin C, Dendrocandin D, and Dendrocandin E, isolated from the stems of *Dendrobium candidum* exhibited potent antioxidant activity in the DPPH radical scavenging test with IC_{50} values of 34.2, 34.5 and 15.6 microM respectively (Li *et al.*, 2009). Vitamin C was used as positive control with IC_{50} 23.2 microM.

Four new bisbenzyl derivatives namely, dendrocandins F-I were isolated by Li *et al.* (2009b) from the stems of *Dendrobium candidum*. Compounds F-I were examined for antioxidant activity by DPPH free radical scavenging assay. Compound H and I showed significant activity with IC_{50} values 19.8 and 21.3 μM, respectively while F and G exhibited moderate activity with IC_{50} 55.8 and 32.4 μM, respectively.

Dendrobium denneanum Kerr. Family: Orchidaceae

The water-soluble crude polysaccharide obtained from the aqueous extracts of the stem of *Dendrobium denneanum* through hot water extraction followed by ethanol precipitation, was found to have antioxidant activity both *in vitro* and *in vivo* (Fan *et al.*, 2009; Luo *et al.*, 2011).

Dendrobium linawianum Reichb. f. Family: Orchidaceae

Lo *et al.* (2004b) investigated the DPPH radical scavenging activity of the methanolic extract of *in vitro* propagated plants of *Dendrobium linawianum*. The extract scavenged DPPH radical significantly in a concentration dependent manner.

Dendrobium moniliforme (L.) Sw. Family: Orchidaceae

Lo *et al.* (2004) investigated the free radical scavenging active components from *in vitro* propagated medicinal plant *Dendrobium moniliforme* using DPPH radical antioxidative assay. Extracts were prepared from plants grown in the green house for a period of six months. Methanolic extract scavenged DPPH at 83.4 per cent at a concentration of 0.4 mg/ml. Bioguided fractionation yielded components alkyl ferulates.

Dendrobium nobile Lindl. Family: Orchidaceae

Cai *et al.* (2004) evaluated the antioxidant activity and phenolic compounds in traditional Chinese medicinal plants associated with anticancer. The improved ABTS.+ method was used to systematically assess the total antioxidant capacity (Trolox equivalent antioxidant capacity, TEAC) of the medicinal extracts. The TEAC values and total phenolic content for methanolic extract of stem of *Dendrobium nobile* were 490.2 µmol Trolox equivalent/100g dry weight (DW), and 2.22 g of gallic acid equivalent/100g DW respectively. A positive significant linear relationship between antioxidant activity and total phenolic content showed that phenolic compounds were the dominant antioxidant components in the tested herbs. Major types of phenolic compounds in *Dendrobium nobile* include phenanthraquinones (denbinobin).

Luo *et al.* (2009, 2010) reported that DNP 4-2, a polysaccharide purified from the crude water extract of the stem of *Dendrobium nobile* has good potential for scavenging activity of ABTS radical and higher scavenging activity of hydroxyl radical and DPPH radical than the other polysaccharides.

Dendrobium officinale K.Kimura et Migo Family: Orchidaceae

Chen *et al.* (2009) studied the antioxidant activity of leaves extract towards xanthine oxidase, lipoxygenase and tyrosinase activities. The inhibitory effect of acetone extract on xanthine oxidase activity was less than 20 per cent, tyrosine activity was almost nil and lipoxygenase activity was about 40 per cent.

Dendrobium sonia Family: Orchidaceae

Radical scavenging activities of extract and solvent-solvent partition fractions from the orchid flower, *Dendrobium sonia* "Red Bom" were evaluated by Shafazila *et al.* (2010). Flower of *Dendrobium sonia* "Red Bom" was extracted with acidified methanol. The residue was redissolved in water which was then subjected to solvent-solvent partition with organic solvents. Crude extract (CEDS) and three different fractions namely hexane layer (HXL), ethyl acetate layer (EAL) and water layer (WL) were evaluated for the total phenolic and anthocyanin contents and 2,2-diphenyl-1-picrylhydrazyl (DPPH) radical scavenging activities. The DPPH radical scavenging activity was found to be the highest in WL. The direct relationship between the total anthocyanins content with the radical scavenging activity indicates that anthocyanins are the main contributors for the observed radical scavenging activities for CEDS and WL.

Dendrobium tosaense Makino Family: Orchidaceae

Lo *et al.* (2004a) investigated the free radical scavenging active components from *in vitro* propagated medicinal plant *Dendrobium tosaense* using DPPH radical antioxidative assay. Extracts were prepared from plants grown in the green house for a period of six months. Methanolic extract scavenged DPPH at 95.9 per cent at a concentration of 0.4 mg/ml. Bioguided fractionation yielded components quercetin. Methanolic extracts of *in vitro* propagated plants also showed DPPH radical scavenging activity, and it was concentration dependent (Lo *et al.*, 2004b).

Dendrocnide sinuata (Blume) Chew Family: Urticaceae

Tanti *et al.* (2010) evaluated the antimicrobial and antioxidant activities of *Dendrocnide sinuata*. The methanol and aqueous extracts were studied for their antioxidant activity using DPPH assay. The methanol extract of the investigated plant showed a high effective free radical scavenging in the DPPH assay. When methanol extract of the investigated plant was tested for DPPH radical scavenging activity, it was found that 75μg/ml and 100μg/ml of the extract lowered the DPPH radical levels above 64 and 74 per cent respectively. Inhibition of DPPH radicals above 50 per cent is considered as significant for antioxidant properties of any compound (Sanchez-Moreno *et al.*, 1998).

Dendrophthoe falcata (L.f.) Ettingsh. Family: Loranthaceae

Pattanayak and Sunitha (2008) evaluated the antioxidant potential of *Dendrophthoe falcata*. The results indicated that *Dendrophthoe falcata* extract possessed potent antioxidant activity in inhibiting lipid peroxidation, reduced glutathione, superoxide dismutase levels and increased the catalase activity.

Methanolic and aqueous extracts of *Dendrophthoe falcata* leaves were evaluated through DPPH, antilipid peroxidation and nitric oxide scavenging methods to assess the antioxidant activity (Patil *et al.*, 2011). It was found that the methanolic extract of *Dendrophthoe falcata* leaves demonstrates potent antioxidant activity as compared to aqueous extraction of *Dendrophthoe falcata* leaves for DPPH radical scavenging, antilipid peroxidation and nitric oxide scavenging ability respectively (having IC_{50} value 77.8, 79.36 and 86.2, 144, 87, 104).

Dendrophthoe trigona (Wight et Arn.) Danser ex Santapau Family: Loranthaceae

Dendrophthoe trigona parasitized on *Mangifera* was collected from Western Ghats of India and air-dried stem bark was used for extraction of secondary metabolites using hot-methanol in soxlet. The crude extract was rich in chlorophyll, quercetin, long chain ester and other compounds. Quercetin – a powerful antioxidant and potent anti-cancer compound and chlorophyll have been purified by column chromatography (Singh *et al.*, 2004).

Derris scandens (Roxb.) Benth. Family: Fabaceae (Papilionaceae)

Genistin and isoprenylated compounds, isolated from the stem extract, showed antioxidant activity (Laupattarksem *et al.*, 2004).

Descureania sophia (L.) Webb. et Berth. Family: Brassicaceae (Cruciferae)

English Name: Flix weed.

Souri *et al.* (2008) evaluated the antioxidant activity against linoleic acid peroxidation using 1,3-diethyl-2-thiobarbituric acid as reagent. Antioxidant activity (IC_{50}) against peroxidation of linoleic acid (2mg/ml) was 23.27 and phenolic content was 265.22 mg/100g dry weight. The results of this study showed that there is no significant correlation between antioxidant activity and phenolic content of the studied plant materials and phenolic content could not be a good indicator of antioxidant capacity.

Desmodium ascendes DC. Family: Fabaceae (Papilionaceae)

Mamyrbekova-Bekro *et al.* (2008) evaluated the phytocompounds of the extract and assessed their antioxidant potential of *Desmodium ascendes* leaves and root. The extracts showed the presence of polyphenols, flavonoids, tannins and coumarins.

The extracts showed DPPH scavenging activity. This activity is probably due to the presence of the phenolic compounds, flavonoids and tannins.

***Desmodium gangeticum* (L.) DC. Family: Fabaceae (Papilionaceae)**

Govindarajan *et al.* (2003) evaluated the antioxidant activities of *Desmodium gangeticum* using DPPH, nitric oxide, ferryl-bipyridyl and hypochlorous acid scavenging activity along with lipid peroxidation. The results indicate that *Desmodium gangeticum* extract has potent antioxidant activity.

Niranjan and Tewari (2008) studied the phytochemical composition and antioxidant potential of root and aerial parts of *Desmodium gangeticum*. The total phenolic content and antioxidant activity was marginally higher in roots than aerial parts. Both the parts of the plant showed medium antioxidant activity as compared to quercetin (74 per cent). The 5.0µg/ml root and aerial part extracts showed significant protection of DNA damage from oxidative stress (Niranjan and Tewari, 2008).

Kurian *et al.* (2009) investigated the *in vitro* and *in vivo* free radical scavenging effect of aqueous extract of *Desmodium gangeticum* root in different antioxidant models and experimentally induced ischemic reperfusion in an isolated rat heart. The results suggest that the aqueous extract of the *Desmodium gangeticum* root exhibit potential free radical scavenging effect.

The methanolic crude extracts of *Desmodium gangeticum* were screened by Veeru *et al.* (2009) for their free radical scavenging properties using ascorbic acid as standard antioxidant. Free radical scavenging activity was evaluated using DPPH free radical. The overall antioxidant activity of *Desmodium gangeticum* was found to be strongest with IC_{50} value of 0.19 mg/l. The ascorbic acid level was 21.33 mg/100g and the carotenoids content were 24 mg/100g in plant extracts. The total phenol content was 37.33 mg/g.

***Desmodium triflorum* (L.) DC Family: Fabaceae (Papilionaceae)**

Lai *et al.* (2010) evaluated the antioxidant and antiproliferative activities of the crude extract and fractions of *Desmodium triflorum*. The total phenolic content, DPPH free radical scavenging activity, TEAC, reducing power, total flavonoid content were evaluated for the exploration of its antioxidant activities. Among all fractions, ethyl acetate fraction was the most active in scavenging DPPH and TEAC radicals, of which 0.4 mg was equivalent to 186.6 µg and 82.5µg of α-tocopherol and trolox respectively. The total phenolic and flavonoid contents of the crude extract were equivalent to 36.60 mg catechin and 45.6 mg rutin per gram respectively. In the reducing power assay, 1.25 mg of crude extract was similar to 61.2µg of ascorbic acid.

***Desmos zeylanica* (Hook.f. et Thoms.) Safford Family: Annonaceae**

Puvanendran *et al.* (2007) evaluated the antioxidant activity of Dichloromethane (DCM) and MeOH extract of stem bark, leaves and seeds of *Desmos zeylanica* using DPPH assay. The extract showed moderate antioxidant activities. Percentage of inhibition of stem was 19.25 (MeOH extract); leaf extract was 9.70 (MeOH extract).

Detarium microcarpum Guill. et Perr. Family: Fabaceae (Leguminosae)

Lamien-Meda *et al.* (2008) analyzed the phenolic and flavonoid contents of wild edible fruits of 14 species of Burkina Faso and their antioxidant activities using DPPH, FRAP and ABTS methods. The data obtained show that the total phenolic and total flavonoid levels were significantly higher in the acetone than the methanol extracts. *Detarium microcarpum* fruit had the highest phenolic and flavonoid content. *Detarium microcarpum* fruit also showed the highest antioxidant activity using the three antioxidant assays. There was a strong correlation between total phenolic and flavonoid levels and antioxidant activity.

Dialium guineense Willd. Family: Fabaceae (Leguminosae)

Polyphenol content and antioxidant activity of fourteen wild edible fruits from Burkina Faso were investigated by Lamien-Meda *et al.* (2008). The data obtained show that the total phenolic and total flavonoid levels were significantly higher in the acetone than in the methanol extracts. *Dialium guinense* fruit had higher flavonoid content and higher antioxidant activity.

Dianthus superbus L. Family: Caryophyllaceae

Cai *et al.* (2004) evaluated the antioxidant activity and phenolic compounds in traditional Chinese medicinal plants associated with anticancer. The improved ABTS.+ method was used to systematically assess the total antioxidant capacity (Trolox equivalent antioxidant capacity, TEAC) of the medicinal extracts. The TEAC values and total phenolic content for methanolic extract of floral bud of aerial parts of *Dianthus caryophyllus* were 91.3 μmol Trolox equivalent/100g dry weight (DW), and 0.59 g of gallic acid equivalent/100g DW respectively. A positive significant linear relationship between antioxidant activity and total phenolic content showed that phenolic compounds were the dominant antioxidant components in the tested herbs.

Gan *et al.* (2010) evaluated the antioxidant activity of *Dianthus superbus* using the FRAP and TEAC assays, and its total phenolic content was measured by the Folin-Ciocalteu method. For FRAP assay the value was 68.57 μmol Fe(II)/g dry weight. For TEAC assay, the value was 37.12μmol Trolox/g dry weight. Total phenolic content was 5.00 mg gallic acid equivalent/g dry weight of plant material.

Dicentra eximia (Ker-Gawl.) Torr. Family: Fumariaceae

Borchardt *et al.* (2008) evaluated the antioxidant activity of methanol extracts of seeds of *Dicentra eximia* using DPPH antioxidant assay.The antioxidant value reported in μM Trolox/100g (TE) from DPPH radical scavenging activity of crude seeds extract was 55,145.

Dichrostachys cinerea Wight et Arn. Family: Mimosaceae

Alcoholic extract of roots showed marked nitric oxide scavenging effect and reducing power suggesting an antioxidant property (Sreedevi *et al.*, 2009). Lipid peroxidation was evaluated as malondialdehyde (MDA) production. Phytochemical investigation revealed the presence of n-octacosanol, β-sitosterol and β-amyrin acetate, pentacyclic triterpenoids such as β-amyrin, friedelan-3-one, frielelan-3 β-ol, friedelin and α-amyrin. Pentacyclic triterpenoids are known to exhibit several biological functions such as analgesic, anti-inflammatory, anti-ulcerogenic, hepatoprotective, anti-hyperglycemic activities.

Dictamnus dasycarpus Turcz. Family: Rutaceae

Li *et al.* (2008) evaluated the antioxidant properties of 45 medicinal plants including *Dictamnus dasycarpus* using FRAP and TEAC assays and the total phenolic contents of these plants were measured. *Dictamnus dasycarpus* showed low antioxidant activity with 10.09 μmol Fe(II)/g FRAP value, 5.49μmol Trolox/g of TEAC value and 2.92 mg GAE/g of Phenolic content.

Didymopanax venosum (Cham. & Schaldtl.) Marchal. Family: Araliaceae

Santos *et al.* (2009) evaluated the antioxidant activity of extracts from nine plant species belonging to the Brazilian flora acting on the free radicals DPPH and TEMPOL. Results showed that the extracts of *Didymopanax venosum* display moderate antioxidant activities with DPPH scavenging IC_{50} value of 0.375 mg/ml while Hydroxyl scavenging IC_{50} value was 9.567 mg/ml.

Dimocarpus longan Lour. Family: Sapindaceae

English Name: Longan.

Three different solvent extracts (methanol, ethyl acetate and n-Hexane) of longan (*Dimocarpus longan* Lour.) flowers were assayed with three different antioxidant capacity methods, namely the DPPH free radical scavenging effect, the ORAC assay and the inhibition of Cu(2+)-induced oxidation of human low-density lipoprotein (LDL) (Hsieh *et al.*, 2008). It was revealed that the methanol extract has the best antioxidative activity, followed by ethyl acetate and n-Hexane extracts. The ethyl acetate fraction was found to have the highest activtity of delaying LDL oxidation. Two major compounds – epicatechin and proanthocyanidin A2 were isolated from the fraction.

Pan *et al.* (2008) evaluated the antioxidant activity of microwave-assisted extract of *Dimocarpus longan* peel. The extract showed excellent antioxidant activity in all test systems compared to synthetic antioxidant BHT. The antioxidant activities of microwave-assisted extract were all superior to those of soxhlet extract longan peel.

Dioclea grandiflora Benth. Family: Fabaceae (Leguminosae)

Dioclein (150μM) isolated from the plant *Dioclea grandiflora* reduced the formation of reaction oxygen species observed during reperfusion in hearts (Vianna *et al.*, 2006)

Dioclea lasiophylla Mart. ex Benth. Family: Fabaceae (Leguminosae)

The chloroform extract of stems of *Dioclea lasiophylla* furnished a new compound namely 3-β-E-feruloyl oleanolic acid, in addition to 3-β-E-caffeoyl oleanolic, betulinic and oleanolic acids, vanillin, methyl vanillate, pinoresinol and syringaresinol. Antioxidant activities of isolates were measured using the DPPH free radical scavenging assay and the auto-oxidation of β-carotene in a linolenic acid suspension method. The 3-β-E-caffeoyl oleanolic, syringaresinol and the new compound 3-β-E-feruloyl oleanolic acid, exhibited antioxidant activities in both tests (David *et al.*, 2004).

Diodia apiculata (Willd. ex Roem. et Schult.) K.Schum. Family: Rubiaceae

In a study with 32 methanolic extracts, from Brazilian Caatinga plants, through DPPH radical scavenging assay showed that the species *Diodia apiculata* displayed good antioxidant activity with IC_{50} = 1.3 mg/l (David *et al.*, 2007).

Dioscorea alata L. Family: Dioscoreaceae

English Name: Greater Yam.

Mendoza *et al.* (1994) screened a number of greater yam varieties and found high antioxidant activities in some of them. A steroid yam extract clled dioscorea was reported to exhibit antioxidant activity which can modify serum lipid levels (Araghiniknam *et al.*, 1996).

Antioxidative effects of *Dioscorea alata* were investigated by Chang *et al.* (2004) in hyperhomocysteinemia (HHcy) rats. *D. alata* feeding exhibited its antioxidative effects in HHcy including alleviating PA, lipid peroxidation and oxidative stress but did not induce activity of antioxidant enzymes which had already adaptively increased by HHcy.

Chen *et al.* (2004) reported the free radical scavenging ability of compounds from Taiwan yams extracted by a continuous hot pressurized solvent extraction process. Lin *et al.* (2005) reported the antioxidant activities of crude and partially purified mucilages of three different cultivars of yam (*D.alata*).

Aqueous methanolic (50 per cent MeOH) extracts of the tubers (peel and flesh) of nine cultivars of greater yam (*Dioscorea alata*) were determined to have relatively high antioxidant activities among which two cultivars (Ubing upo, purple; LAO96, white) had activities as high as those of α-tocopherol and BHA (Lubag *et al.*, 2008).

A study was carried out by Raghu *et al.* (2010) to determine the antioxidant activity of aqueous extracts of ten selected common vegetables. For *Dioscorea alata* the IC_{50} value for DPPH scavenging was 1.70mg/ml while standard was 2.45μg/ml. Reducing power of the extract was moderate.

Dioscorea batatas Decne Family: Dioscoreaceae

English Name: Yam

Dioscorin, the storage protein isolated from yam was shown to have scavenging activity against DPPH radical in a pH-dependent manner (Hou *et al.*, 2001). There was a positive correlation between scavenging effects against DPPH (8-46 per cent) and amounts of 32kDa dioscorin (5.97-47.80 nmol) added in Tris-HCl buffer (pH 7.9) more than in distilled water compared to controls.

The yam tuber mucilage (YTM) was extracted and partially purified by Hou *et al.* (2002) SDS and heating treatments. This purified YTM exhibited antioxidant activities in a series of *in vitro* tests, including DPPH radical (half-inhibition concentration, IC_{50} was 0.86 mg/ml) and hydroxyl radical (IC_{50} was 22 μg/ml) scavenging activity assays, reducing power test, anti-lipid peroxidation and anti-human low density lipoprotein peroxidase tests (IC_{50} was 145.46 μg/ml) using BHT, reduced glutathione, or ascorbic acid for comparisons. With electron paramagnetic resonance (EPR) spectrometry for DPPH radical detection, the intensities of the EPR signals were decreased by the increased amounts of YTM added (IC_{50} was 1.62 mg/ml).

Dioscorea bulbifera **L. Family: Dioscoreaceae**

Antioxidant capacities of 56 selected Chinese medicinal plants, including *Dioscorea bulbifera* were evaluated by Song *et al.* (2010) using the TEAC and FRAP assays, and their total phenolic content was measured by the Folin-Ciocaleteu method. *Dioscorea bulbifera* showed the highest antioxidant activity with TEAC value of 708.73µmol Trolox/g, FRAP value of 856.92µmol Fe^{2+}/g and phenolic content was 59.43mg GAE/g. The strong correlation between TEAC value and FRAP value suggested that the antioxidants in these plants possess free radical scavenging activity and oxidant reducing power, and the high positive correlation between antioxidant capacities and total phenolic content implied that phenolic compounds are a major contributor to the antioxidant activity of these plants.

Dioscorea collettii **Hook.f. Family: Dioscoreaceae**

Gan *et al.* (2010) evaluated the antioxidant activity of *Dioscorea collettii* using the FRAP and TEAC assays, and its total phenolic content was measured by the Folin-Ciocalteu method. For FRAP assay the value was 16.05 µmol Fe(II)/g dry weight. For TEAC assay, the value was 10.82µmol Trolox/g dry weight. Total phenolic content was 1.31 mg gallic acid equivalent/g dry weight of plant material.

Dioscorea esculenta **(Lour.) Burkill Family: Dioscoreaceae**

Lako *et al.* (2006) analysed the total antioxidant capacity, total polyphenol content, total anthocyanin content as well as the major flavonol and carotenoid profiles of Fijian fruit and vegetables. For *Dioscorea esculenta* Total Antioxidant Capacity for var. red was 27mg/100g (TEAC) and var. white was 10 mg/100g (TEAC). Total polyphenol content was 38 mg GAE/100 for var. red and 17mg/GAE/100g for var. white.

Dioscorea glandulosa **Klotzsch ex Kunth Family: Dioscoreaceae**

Menezes *et al.* (2004) investigated the antioxidant activity of hydroalcoholic extracts of leaves of *Dioscorea glandulosa* by DPPH assay and phosphomolybdenum method. The relative activity in phosphomolybdenum complex was 0.46 while scavenging activity against DPPH in IC_{50} (µg/ml) was 260.88.

Dioscorea nipponica **Makino. Family: Dioscoreaceae**

Gan *et al.* (2010) evaluated the antioxidant activity of *Dioscorea nipponica* using the FRAP and TEAC assays, and its total phenolic content was measured by the Folin-Ciocalteu method. For FRAP assay the value was 39.61 µmol Fe(II)/g dry weight. For TEAC assay, the value was 54.86µmol Trolox/g dry weight. Total phenolic content was 4.82 mg gallic acid equivalent/g dry weight of plant material.

Dioscorea nummularia **L. Family: Dioscoreaceae**

Lako *et al.* (2006) analysed the total antioxidant capacity, total polyphenol content, total anthocyanin content as well as the major flavonol and carotenoid profiles of Fijian fruit and vegetables. For *Dioscorea nummularia* Total Antioxidant Capacity was 26mg/100g (TEAC). Total polyphenol content was 47 mg GAE/100.

Dioscorea villosa L. Family: Dioscoreaceae

Wojcikowski *et al.* (2007) evaluated the antioxidant activity of aerial parts of *Dioscorea villosa.* Oxygen radical absorbance capacity of the ethyl extract, methanol extract and aqueous extracts were 14.48, 171.57 and 196.08μmol Trolox equivalent per g of dried starting material, respectively.

Borchardt *et al.* (2008) evaluated the antioxidant activity of methanol extracts of seeds of *Dioscorea villosa* using DPPH antioxidant assay.The antioxidant value reported in μM Trolox/100g (TE) from DPPH radical scavenging activity of crude seeds extract was 45,196.

Diospyros abyssinica (Hiern) F. White Family: Ebenaceae

Maiga *et al.* (2006) reported antioxidant and lipoxygenase inhibitory activity of *Diospyros abyssinica.* Semipola extracts (80 per cent aqueous ethanol and methanol) showed the highest activity both as radical scavengers and lipoxygenase inhibitors and also gave the highest yield. Lipophilic extracts were not active as radical scavengers.

Diospyros discolor Willd. Family: Ebenaceae

Lee *et al.* (2006) investigated the antioxidant activity of ethanol extract of leaves using DPPH, hydroxyl and superoxide radical scavenging and reducing power activities as well as the induction of heme oxygenase-1 (HO-1). The IC_{50} values against DPPH, radical were 4.4μg/ml, OH radical 0.24μg, superoxide radical 16.6μg/ml and total phenolics was 36.8 mg of GAE/g.

Diospyros kaki L.f. Family: Ebenaceae

Katsube *et al.* (2004) screened aqueous ethanol extract of *Diospyros kaki* for antioxidant activity using DPPH radical scavenging activity and LDL oxidation induced by copper ion. Total phenolic content was also measured for comparisons with antioxidant activity in LDL. The study revealed high levels of LDL antioxidant activity in *Diospyros kaki.*

Diospyros malabarica Kostel Family: Ebenaceae

Defatted methanol extract of *Diospyros malabarica* bark was studied for its radical scavenging property on different *in vitro* models, *e.g.*, DPPH, nitric oxide superoxide, hydroxyl radical and lipid peroxide radical model (Mondal *et al.*, 2006). The extract showed good dose-dependent free radical scavenging activity in all the models except in hydroxyl radical inhibition assay. IC_{50} values were found to be 9.16, 13.21, 25.27 and 17.33mewg/ml, respectively in DPPH, nitric oxide, superoxide and lipid peroxidation inhibition assays. In hydroxyl radical inhibition assay 1000mewg/ml extract showed only 10 per cent inhibition with respect to the control. Measurement of total phenolic compounds indicated that 1 g of the extract contained 120.7 mewg equivalent of pyrocatechol. The results indicate that the antioxidant property of the extract may be due to the high content of phenol.

Diospyros mespiliformis Hochst. Family: Ebenaceae

Lamien-Meda *et al.* (2008) evaluated the antioxidant activity of fourteen wild edible fruits of Burkina Faso using DPPH, FRAP and ABTS methods. The data obtained show that the total phenolics and total flavonoid levels were significantly higher in the acetone than in the methanol extracts. In *Diospyros mespiliformis* Total phenolic content was 336.33mg GAE/100g in MeOH extract and 591.67 mgGAE/100g in acetone extract. Total flavonoids were 22.40 mg/100g in MeOH extract and 27.10 in Acetone extract. *Diospyros mespiliformis* showed moderate antioxidant activity.

Dipsacus asper **Wall. Family: Dipsacaceae**

The methanol extract from *Dipsacus asper* was found to show antioxidant activity against free radical and Cu(2+)-mediated LDL oxidation (Huang *et al.*, 2006). Active constituents isolated were 3,4-di-O-caffeoylquinic acid (1), methyl 3,4-di- O-caffeoylquinate (2), 3,5-di-O-caffeoylquinic acid (3), methyl 3,5-di- O-caffeoylquinate (4), 4,5-di-O-caffeoylquinic acid (5) and methyl 4,5-di- O-caffeoylquinate (6). These isolated compounds, 1-6, were found to be potent scavengers of the free radical DPPH and are more potent than BHT used as a positive control. These compounds 1-6 also inhibited Cu(2+)-mediated low-density lipoprotein (LDL) oxidation. They increased the lag time of conjugated dienes formation and inhibited the generation of TBARS in a dose dependent manner.

Dodecadenia grandiflora **Nees Family: Lauraceae**

Joshi *et al.* (2010) evaluated the antioxidant activity of seven Himalayan Lauraceae species. The essential oil of *Dedecadenia grandiflora* was able to inhibit linoleic acid oxidation. Furanodiene and germacrene D predominated in the oil.

Dodonaea viscosa **(L.) Jacq. Family: Sapindaceae**

Hepatoprotective activity of aqueous extract of *Dodonaea viscosa* against paracetamol and CCl_4 induced hepatic damage in albino rats was observed by Pawar *et al.* (2009) and the action may be due to antioxidant activity.

Methanolic extracts of *Dodonaea viscosa* showed high free radical scavenging activity (Mothana *et al.*, 2010).

The chloroform and methanolic extracts of 124 Egyptian plant species belonging to 56 families were investigated by Moussa *et al.* (2011) and compared their antioxidant activity by DPPH scavenging assay. Among the 124 plant species tested 18 exhibited extremely high antiradical activity (>80 per cent inhibition). The IC_{50} value of DPPH radical scavenging of *Dodonea viscosa* was 21.05µg/ml, while total phenolic and flavonoid contents were 113.66 mg Tannic acid equivalent/g extract and 208.26 mg Rutin equivalent/g extract, respectively. Correlation coefficient between DPPH radical scavenging activity and the total phenolic and flavonoid contents suggest that phenolics and flavonoids in the extracts were partly responsible for the antiradical activities.

Dorstenia ciliata **Engl. Family: Moraceae**

Kansci *et al.* (2003) observed that Camerounian plant *Dorstenia ciliata* used as both food ingredient and recipe in traditional medicine possessed strong anti-radical activity when evaluated with the DPPH test. They suggested that the antiradical potency should be the basis for its therapeutic efficacy in traditional medicine.

Dorstenia foetida Schweinf. Family: Moraceae

Marwah *et al.* (2007) investigated the antioxidant activity of wound-healing herbs using *in vitro* DPPH and phosphomolybdenum assay methods. Out of the 19 plants screened, the aqueous ethanol extract of *Dorstenia foetida* showed very low DPPH scavenging activities with per cent IP value 21.1 per cent, after 15 min of incubation at an effective test concentration of 50μg/ml. The Total Antioxidant Capacity as gallic acid equivalent of 269mg/g of ethanol extract was obtained in the phosphomolybdenum assay. Total phenolic content was 26.5 mg GAE/g of ethanol extract.

Dorstenia mannii Hook.f. Family: Moraceae

Duffall *et al.* (2003) demonstrated the antioxidant activity of prenylated flavonoids from *Dorstenia mannii,* a central African plant harvested from Cameroun. They attributed the medicinal action of the plant to the high concentration of the potent antioxidant prenylated flavonoids which inhibited Cu^{2+}-mediated oxidation of human low density lipoprotein (LDL), scavenged the free radical DPPH, and was more potent than butylated toluene (BHT), a common antioxidant used as a food additive.

Dorstenia multiformis Miquel Family: Moraceae

Balestrin *et al.* (2009) evaluated the phytochemistry and antioxidant activity of *Dorstenia multiformis.* Isolated substances, from ethanol extract and hexane, chloroform and ethyl acetate fractions have been determined. All the testes samples showed antioxidant activity, and the chloroform fraction was the one that presented the largest antioxidant activity.

Dorstenia psilurus Welw. Family: Moraceae

Kansci *et al.* (2003) observed that Camerounian plant *Dorstenia psilurus* used as both food ingredient and recipe in traditional medicine possessed strong anti-radical activity when evaluated with the DPPH test. They suggested that the antiradical potency should be the basis for its therapeutic efficacy in traditional medicine.

Dorycnium hirsutum (L.) Ser. Family: Fabaceae (Leguminosae)

Seven flavonol glycosides were isolated and identified from the aerial parts of *Dorycnium hirsutum* together with catechin, D-pinitol, β-sitosterol, and stearic acid (Pistelli *et al.*, 2006). The extracts and the isolated flavonol glycosides were evaluated for their antioxidant activity, using the DPPH test (radical scavenging) and the lipoxygenase assay (lipid peroxidation).

Dorystoechas hastata Boiss. et Heldr. Family: Lamiaceae (Labiatae)

A monotypic plant endemic to Antalya province of Turkey.

The antioxidant activities of the extracts were tested by Karagozler (2008) by Ferric thiocyanate (FTC) and DPPH free radical scavenging methods. Diethyl ether extract exhibited the greatest antioxidant activity with FTC method, whereas hot water extract exhibited the lowest IC_{50} value (6.17 μg/ml) for DPPH radical scavenging activity. Total phenolic contents of the extracts were estimated by Folin-Ciocalteu method and hot water extract was found to contain highest amount (554.17 mg GAE/g extract) of phenolics. Ethanol extract contained highest flavonoid content and there was a inverse linear correlation between IC_{50} values for DPPH radical scavenging activity and flavonoid contents of all extracts. Hot water extract was found to possess higher reducing power than equivalent amount of ascorbic acid at 20 and 25 µg/ml concentrations.

Draba nemerosa L. Family: Brassicaceae (Cruciferae)

Rahman *et al.* (2007) isolated antioxidant two new polyphenol glycosides, 1-O-(sinapoyl)-glucitol and 1,3-disinapyoylgentiobiose along with five known phenolic glycosides from *Draba nemerosa.*

Dracocephalum moldavica L. Family: Lamiaceae (Labiatae)

English Name: Dragonhead

Khalil *et al.* (2007) studied the growth, phenolic compounds and antioxidant activity of dragonhead grown under organic farming condition. Antioxidant activity of ethanol extract (inhibition per cent) was 88.74 at 200µl extract.

Dracocephalum surmandicum Family: Lamiaceae (Labiatae)

Sonboli *et al.* (2010) reported antioxidant activity of essential oil from the aerial parts of *Dracocephalum surmandicum.*

Dregea volubilis (L.f.) Benth. ex Hook.f. Family: Asclepiadaceae

Chanwitheesuk *et al.* (2005) investigated the antioxidant activity of *Dregea volubilis* using a β-carotene bleaching method. The contents of plant chemicals such as vitamin C, Vitamin E, carotenoids, tannin and total phenolics, were also determined. Methanolic extract showed antioxidant activity. *Dregea volubilis* has antioxidant index of 7.20, Vitamin C 20.0 mg per cent, Vitamin E 0.0015mg per cent, Total carotenes 6.14 mg per cent, Total xanthophylls 1.07mg per cent, Tannins 17.7mg per cent and Total phenolics of 100 mg per cent.

Droogmansia godefroyana (Kuntze) Schindl. Family: Fabaceae (Papilionaceae)

Patnibul *et al.* (2008) screened 18 vegetables for antioxidant activity using silica gel Thin Layer chromatography followed by spraying with DPPH. *Droogmansia godefroyana* exhibited strong antioxidant activity.

Drosera burmannii Vahl. Family: Lardizabalaceae

Gan *et al.* (2010) evaluated the antioxidant activity of *Drosera burmannii* using the FRAP and TEAC assays, and its total phenolic content was measured by the Folin-Ciocalteu method. For FRAP assay the value was 99.80 µmol Fe(II)/g dry weight. For TEAC assay, the value was 79.76µmol Trolox/g dry weight. Total phenolic content was 4.76 mg gallic acid equivalent/g dry weight of plant material.

Drymaria diandra Blume Family: Caryophyllaceae

Mandal *et al.* (2009) evaluated the phytochemical constituents like total phenol, ortho-dihydric phenol, flavonoids, tannins and antioxidant activity of leaves and stems of *Drymaria diandra*. DPPH, superoxide anion, hydroxyl radical, nitric oxide radical and anti-lipid peroxidation activity was studied. Strong antioxidant scavenging activities were observed in leaf and stem. Anti-oxidative efficiency to inhibit anti-lipid peroxidation of these plant extracts in the liver was investigated. Stem showed moderate class of anti-lipid peroxidation against thiobarbituric acid but the leaves have high anti-lipid peroxidation activity.

Durio zibethinus **L. Family: Bombacaceae**

English Name: Durian.

Thirty-eight types of fruits commonly consumed in Singapore were systematically analyzed by Isabelle *et al.* (2010) for their hydrophilic oxygen radical absorbance capacity (H-ORAC), total phenolic content (TPC), ascorbic acid (AA) and various lipophilic antioxidants. Durian was high in tocopherols.

Ebenus pinnata **Ait. Family: Fabaceae (Leguminosae)**

Abreu *et al.* (2007) isolated antioxidant compounds from *Ebenus pinnata*. Activity-guided fractionation of the methanol extract of aerial parts led to the isolation of ombuoside (1), kaempferol 3-O-rutinoside (2), rutin (3), catechin (4) and picein (5), along with β-sitosterol and β-sitosterol glucoside. Compounds 1-4 showed significant activity in the DPPH and TEAC, reducing power assays.

Echinacea angustifolia **DC. Family: Asteraceae (Compositae)**

English Name: Echinacea.

Methanol extracts of freeze-dried *Echinacea angustifolia* roots were examined for free radical savenging capacities and antioxidant activities. Root extracts were capable of scavenging hydroxyl radical, DPPH radical and ABTS radical. Echinacea root extracts suppressed the oxidation of human low density lipoprotein, as evaluated by reduced agarose electropheoretic mobility following oxidative modification by Cu(2+) (Hu and Kitts, 2000).

Chemical components and antioxidant capacity of roots and leaves of *Echinacea angustifolia* were analysed by Sloley *et al.* (2001). Cynarine and dodeca-2E,4E, 8Z, 10Z/E-tetraenoic acid isobutylamide predominated in extracts of *Echinacea angustifolia* root. Exracts of the roots and leaves were found to have antioxidant properties in a free radical scavenging assay and in a lipid peroxidation assay.

The radical scavenging activity of *Echinacea* methanolic extracts was evaluated *in vitro* with a spectrophotometric method based on the reduction of an alcoholic DPPH radical solution at 517 nm in the presence of a hydrogen donating antioxidant. As for pure compounds, echinoside had the highest capacity to quench DPPH radicals (EC_{50} = 6.6 microM), while caftaric acid had the lowest (EC_{50} = 20.5 microM). The average EC_{50} value of *E.angustifolia* was 231 microg/ml. The radical scavenging activity of *Echinacea* root extracts reflected their phenolic composition (Pellati *et al.*, 2004, 2005).

Ethanolic extracts of roots of Echinacea were prepared and evaluated by Amarowicz *et al.* (2004) for their free-radical scavenging capacity and their antioxidant activity, by a number of chemical assays. Assays employed included β-carotene linoleic acid (lineolate) model system, reducing power, scavenging effect on the DPPH free radical and capacity to scavenge hydroxyl free radical (HO), by use of electron paramagnetic resonance (EPR) spectroscopy.

Echinacea atrorubens Nutt. Family: Asteraceae (Compositae)

The radical scavenging activities of methanolic extracts of roots of *Echinacea* species was evaluated *in vitro* using the DPPH radical scavenging method. The EC_{50} values of the samples ranged from 122 to 1223 microg/ml. The radical scavenging activities of the root extracts were correlated with the content of phenolic compounds (Pellati *et al.*, 2005).

Echinacea laevigata (Boynton et Beadle) Blake Family: Asteraceae (Compositae)

The radical scavenging activities of methanolic extracts of roots of *Echinacea* species were evaluated *in vitro* using the DPPH radical scavenging method. The EC_{50} values of the samples ranged from 122 to 1223 microg/ml. The radical scavenging activities of the root extracts were correlated with the content of phenolic compounds (Pellati *et al.*, 2005).

Echinacea pallida Nutt. Family: Asteraceae (Compositae)

Methanol extracts of freeze-dried *Echinacea pallida* roots were examined for free radical savenging capacities and antioxidant activities. Root extracts were capable of scavenging hydroxyl radical, DPPH radical and ABTS radical. Echinacea root extracts suppressed the oxidation of human low density lipoprotein, as evaluated by reduced agarose electropheoretic mobility following oxidative modification by Cu(2+) (Hu and Kitts, 2000).

Chemical components and antioxidant capacity of roots and leaves of *Echinacea pallida* were analysed by Sloley *et al.* (2001). Echinacoside and 6-O-caffeoylechinacoside predominated in extracts of *Echinacea pallida* root. Exracts of the roots and leaves were found to have antioxidant properties in a free radical scavenging assay and in a lipid peroxidation assay.

The radical scavenging activity of Echinacea methanolic extracts was evaluated *in vitro* with a spectrophotometric method based on the reduction of an alcoholic DPPH radical solution at 517 nm in the presence of a hydrogen donating antioxidant. As for pure compounds, echinoside had the highest capacity to quench DPPH radicals (EC_{50} = 6.6 microM), while caftaric acid had the lowest (EC_{50} = 20.5 microM). The average EC_{50} value of *E.pallida* was 167 microg/ml. The radical scavenging activity of *Echinacea* root extracts reflected their phenolic composition (Pellati *et al.*, 2004, 2005).

The dichloromethane, ethyl acetate, ethanol and aqueous extracts of *Echinacea pallida* along with the essential oils were assessed by Orhan *et al.* (2009) for their antioxidant activities. Antioxidant activity was evaluated by DPPH radical scavenging and ferrous iron-chelating tests.

Echinacea purpurea (L.) Mienchs. Family: Asteraceae (Compositae)

English Name: Purple conflower.

Purple conflower is a well-known medicinal plant possessing bactericidal, antiviral and antifungal effects against some microbial cultures. *Echinacea purpurea* extracts were tested for antioxidant activity using ABTS (generated by metmyoglobin and H_2O_2 in the presence of peroxidase) scavenging method and it was found to be rather ineffective (Pietta *et al.*, 1998).

Methanol extracts of freeze-dried *Echinacea angustifolia* roots were examined for free radical scavenging capacities and antioxidant activities. Root extracts were capable of scavenging hydroxyl radical, DPPH radical and ABTS radical. Echinacea root extracts suppressed the oxidation of human low density lipoprotein, as evaluated by reduced agarose electropheoretic mobility following oxidative modification by Cu(2+) (Hu and Kitts, 2000).

Chemical components and antioxidant capacity of roots and leaves of *Echinacea purpurea* were analysed by Sloley *et al.* (2001). Cichoric acid and verbascoside predominated in extracts of *Echinacea purpurea* root. Exracts of the roots and leaves were found to have antioxidant properties in a free radical scavenging assay and in a lipid peroxidation assay.

Miliuskas *et al.* (2004) also investigated the radical scavenging activity using DPPH and ABTS assays. The acetone extract of *E. purpurea* was more effective than its methanolic extract, however, all the extracts from this plant were very weak radical scavengers.

The radical scavenging activity of Echinacea methanolic extracts was evaluated *in vitro* with a spectrophotometric method based on the reduction of an alcoholic DPPH radical solution at 517 nm in the presence of a hydrogen donating antioxidant. As for pure compounds, echinoside had the highest capacity to quench DPPH radicals (EC_{50} = 6.6 microM), while caftaric acid had the lowest (EC_{50} = 20.5 microM). The average EC_{50} value of *E. purpurea* was 134 microg/ml. The radical scavenging activity of Echinacea root extracts reflected their phenolic composition (Pellati *et al.*, 2004, 2005).

Effect of *Echinacea purpurea* was examined by Mishima *et al.* (2004) by evaluating changes in the peripheral blood cell count and peripheral blood antioxidant activity. *Echinacea* administration had a suppressive effect on radiation-induced leucopenia, especially on lymphocytes and monocytes, and resulted in faster recovery of blood cell counts. Mouse peripheral blood antioxidant activity was increased by *E. purpurea* and a relationship between the suppressive effect was suggested.

The dichloromethane, ethyl acetate, ethanol and aqueous extracts of *Echinacea purpurea* along with the essential oils were assessed by Orhan *et al.* (2009) for their antioxidant activities. Antioxidant activity was evaluated by DPPH radical scavenging and ferrous iron-chelating tests. Chloroform extract of the aerial parts of *Echinacea purpurea* displayed the highest ferrous iron-chelating effect.

Echinacea sanguinea Nuttal Family: Asteraceae (Compositae)

The radical scavenging activities of methanolic extracts of roots of *Echinacea* species was evaluated *in vitro* using the DPPH radical scavenging method. The EC_{50} values of the samples ranged from 122 to 1223 microg/ml. The radical scavenging activities of the root extracts were correlated with the content of phenolic compounds (Pellati *et al.*, 2005).

Echinacea simulata Mc Gregor Family: Asteraceae (Compositae)

The radical scavenging activities of methanolic extracts of roots of *Echinacea* species was evaluated *in vitro* using the DPPH radical scavenging method. The EC_{50} values of the samples ranged from 122 to 1223 microg/ml. The radical scavenging activities of the root extracts were correlated with the content of phenolic compounds (Pellati *et al.*, 2005).

Echinacea tennesseensis (Beadle) Small Family: Asteraceae (Compositae)

The radical scavenging activities of methanolic extracts of roots of *Echinacea* species was evaluated *in vitro* using the DPPH radical scavenging method. The EC_{50} values of the samples ranged from 122 to 1223 microg/ml. The radical scavenging activities of the root extracts were correlated with the content of phenolic compounds (Pellati *et al.*, 2005).

Echium amoenum Fisch et C.A.Mey Family: Boraginaceae

Ranjbar *et al.* (2006) evaluated the antioxidant activity of Iranium *Echium amoenum* flower decoction in humans. Blood samples before and after entering the study were measured for lipid peroxidation level (LPO), total antioxidant capacity (TAC) and total thio (SH) molecules. A significant reduction of blood LPO 24.65 versus 19.05 was observed after 14 days of *E. amoenum* consumption. Blood TAC(1.46 versus 1.70) and total thiol molecules (0.49 versus 0.56) increased after 14 days of *E. amoenum* consumption.

Echium rubrum Forssk. Family: Boraginaceae

Niiforovi *et al.* (2010) examined the antioxidant activity of *Echium rubrum* including total antioxidant capacity, DPPH free-radical scavenging, the inhibitory activity toward lipid peroxidation, Fe(3+)-reducing power, Fe(2+)-chelating ability and hydroxyl scavenging activity. Total phenolic and flavonoid contents were also determined for each alcoholic extract. *Echium rubrum* showed greatest hydroxyl radical scavenging activity, as well as ferrous ion chelating ability.

Echium vulgare L. Family: Boraginaceae

Niiforovi *et al.* (2010) examined the antioxidant activity of *Echium vulgare* including total antioxidant capacity, DPPH free-radical scavenging, the inhibitory activity toward lipid peroxidation, Fe(3+)-reducing power, Fe(2+)-chelating ability and hydroxyl scavenging activity. Total phenolic and flavonoid contents were also determined for each alcoholic extract. Methanol extract of *Echium vulgare* showed highest proportion of flavonoids (105 mg RU/g). *Echium vulgare* showed greatest hydroxyl radical scavenging activity, as well as ferrous ion chelating ability.

Eclipta prostrata L. (Syn.: _Eclipta alba_ (L.) Hassk.) Family: Asteraceae (Compositae)

Sabu and Kuttan (2003) investigated the antioxidant activity of Indian herbal drugs, including *Eclipta prostrata*, in rats with alloxan-induced diabetes. The extract of *Eclipta prostrata* showed antidiabetic and antioxidant activity.

Cai *et al.* (2004) evaluated the antioxidant activity and phenolic compounds in traditional Chinese medicinal plants associated with anticancer. The improved ABTS.+ method was used to systematically assess the total antioxidant capacity (Trolox equivalent antioxidant capacity, TEAC) of the medicinal extracts. The TEAC values and total phenolic content for methanolic extract of aerial parts of *Eclipta prostrata* were 54.3 µmol Trolox equivalent/100g dry weight (DW), and 0.29 g of gallic acid equivalent/100g DW respectively. A positive significant linear relationship between antioxidant activity and total phenolic content showed that phenolic compounds were the dominant antioxidant components in the tested herbs. Major types of phenolic compounds in *Eclipta prostrata* include coumarins (wedelolactone), tannins.

The *in vitro* antioxidant activity of ethanolic extract of leaves of *Eclipta prostrata* (Syn.: *Eclipta alba*) was investigated bt DPPH free radical and nitric oxide scavenging method (Sharma *et al.*, 2007). The ethanolic extract showed good antioxidant activity against syntheric free radicals like DPPH and nitric oxide.

The methanolic crude extracts of *Eclipta alba* were screened by Veeru *et al.* (2009) for their free radical scavenging properties using ascorbic acid as standard antioxidant. Free radical scavenging activity was evaluated using DPPH free radical. The overall antioxidant activity of *Eclipta alba* was found to be stronger with IC_{50} value of 0.08 mg/l. The ascorbic acid level was 9.83 mg/100g and the carotenoids content were 9 mg/100g in plant extracts. The total phenol content was 30.4 mg/g.

Ekebergia senegalensis A. Juss. Family: Meliaceae

Sofidya *et al.* (2006) evaluated the DPPH radical scavenging, total antioxidant activities, reducing power and contents of phenolic compounds in methanolic leaf extracts of *Ekebergia senegalensis.* The extract showed a dose-dependent effect both on the free radical scavenging DPPH and also on Fe^{3+} reducing power. *Ekebergia senegalensis* possess very high radical scavenging activity in both assays. Total antioxidant activities correlated with total phenols. A linear positive relationship existed between the reducing power and total phenolics.

Elais guineensis Jacq. Family: Arecaceae (Palmae)

English Name: Oil Palm.

Balasundram *et al.* (2005) determined the phenolic content of crude and ethanolic extracts of oil palm fruits and to evaluate the antioxidant properties of these extracts. The total phenolics content of the crude ethanol extracts as determined by the Folin-Ciocalteu method were found to be 40.3 and 49.6 mg GAE/g extract (dry basis), respectively. The radical scavenging activity of the extracts determined using DPPH indicated that both crude and ethanol extracts exhibited hydrogen donating capacity, and have antiradical power (ARP) comparable to ascorbic acid. The DPPH radical scavenging activity of the extracts were less than that of gallic acid, but the time-course variations of the curves suggest that the extracts acted by a mechanism similar to that of gallic acid. The reducing power of crude and ethanol extracts at 1mM GAE were found to be comparable to that of 0.3mM gallic acid. The extracts indicated complete scavenging of hydrogen peroxide at concentrations above 0.4mM GAE.

The total phenolic content (TPC) of the oil palm fruit extracts was determined using the Folin-Ciocalteu method and found to range from 5.03 to 9.04 g/l per g of dried weight (DW). The antioxidant activities of oil palm phenolic extracts were analyzed using free radical scavenging assay and results showed that oil palm phenolic extracts contained antioxidant activities in the order of insoluble bound > Esterified > soluble free. Ferulic acid, *p*-hydroxybenzoic and *p*-coumaric acid were the dominant phenolic acids found in oil palm fruit extracts and ranged from 55 to 376 µg/g od DW (Neo *et al.*, 2010).

***Elettaria cardamomum* (L.) Maton Family: Zingiberaceae**

English Name: Cardamon.

Badei *et al.* (2002) showed that powdered cardamom and its essential oil both protect cookie lipids against peroxidation.

Hydrodistillated extracts from cardamon were assessed for their total phenol content, and antioxidant (iron(III) reduction, inhibition of linoleic acid peroxidation, iron (II) chelation, DPPH radical-scavenging and inhibition of hydroxyl radical-mediated 2-deoxy-D-ribose degradation, site and nonsite-specific) activities by Hinneburg *et al.* (2005). The extracts showed significant antioxidant activity.

Omankutty (2007) reported that extract obtained from cardamom have shown potent antioxidant property.

Khalaf *et al.* (2008) reported weak antioxidant activity in the methanolic crude extracts of cardamon using DPPH assay. Souri *et al.* (2008) evaluated the antioxidant activity against linoleic acid peroxidation using 1,3-diethyl-2-thiobarbituric acid as reagent. Antioxidant activity (IC_{50}) against peroxidation of linoleic acid (2mg/ml) was 167.29 and phenolic content was 84.19 mg/100g dry weight. The results of this study showed that there is no significant correlation between antioxidant activity and phenolic content of the studied plant materials and phenolic content could not be a good indicator of antioxidant capacity.

Mariutti *et al.* (2008) evaluated the antioxidant activity of cardamon employing ABTS and DPPH radical assay. DPPH EC_{50} value was 89g/kg while DPPH-TEAC value was 2.8mM/g. ABTS-TEAC value was 3.9mM/g.

Bhatti *et al.* (2010) evaluated the antioxidant potential of green cardamom. Green cardamom showed the total phenolic contents from 0.317 g/100g; total flavonoids from 11.33 g/100g. Antioxidant activities of the extracts were determined using thiocyanate method and percentage inhibition of peroxidation was 84.2-90 per cent. Significant differences were observed in the antioxidant efficacy of green cardamom extracts within the different solvent concentration.

Eleusine coracana (L.) Gaertn. Family: Poaceae (Graminae)

English Name: Finger millet, Hindi: Ragi.

Sreeramulu *et al.* (2009) evaluated the antioxidant activity of finger millet by DPPH scavenging assay, FRAP assay and reducing power. Phenolic content was 373.15 mg/100g. DPPH radical scavenging activity expressed as trolox equivalent, FRAP (μmol/g) and reducing power (mg/g) were 1.73, 472 and 4.54, respectively in finger millet.

Eleutherococcus gracilistylus (W.W. Sm.) S.Y. Hu Family: Araliaceae

Wojcikowski *et al.* (2007) evaluated the antioxidant activity of aerial parts of *Eleutherococcus gracilistylus.* Oxygen radical absorbance capacity of the methanol extract and aqueous extracts were 51.96 and 31.85μmol Trolox equivalent per g of dried starting material, respectively.

Elsholtzia splendens Nakai et F. Maek. Family: Lamiaceae (Labiatae)

Antioxidant capacities of 56 selected Chinese medicinal plants, including *Elsholtzia splendens* were evaluated by Song *et al.* (2010) using the TEAC and FRAP assays, and their total phenolic content was measured by the Folin-Ciocaleteu method. *Elsholtzia splendens* showed the highest antioxidant activity with TEAC value of 59.84μmol Trolox/g, FRAP value of 52.55μmol Fe^{2+}/g and phenolic content was 7.71mg GAE/g.

Elymus repens (L.) Gould Family: Poaceae (Graminae)

Wojcikowski *et al.* (2007) evaluated the antioxidant activity of aerial parts of *Elymus repens.* Oxygen radical absorbance capacity of the ethyl extract, methanol extract and aqueous extracts were 8.37, 119.43 and 11.09μmol Trolox equivalent per g of dried starting material, respectively.

Embelia ribes Burm.f. Family: Myrsinaceae

Free radical scavenging reactions and antioxidant activity of embelin (2,5-dihydroxy-3-undecyl-1,4-benzoquinone) from *Embelia ribes* have been studied by Joshi *et al.* (2007). It has been found to scavenge DPPH radical and inhibit hydroxyl radical induced deoxyribose degradation. It has also been found to inhibit lipid peroxidation and restore impaired Mn-superoxide dismutase in rat liver mitochondria.

Emilia sonchifolia (L.) DC. Family: Asteraceae (Compositae)

Fresh juice and methanolic extract of leaves were found to be potent inhibitors of hydroxyl radical formation and superoxide radical generation *in vitro* (Shylesh and Padikkala, 1999). Phytochemical analysis of methanolic extract and juce of leaves showed the presence of flavonoids which may account for the observed antioxidant and anti-inflammatory properties.

Patnibul *et al.* (2008) screened 18 vegetables for antioxidant activity using silica gel Thin Layer chromatography followed by spraying with DPPH. *Emilia sonchifolia* exhibited strong antioxidant activity.

Eminium spiculatum Kuntze Family: Araceae

Janakat and Al-Thanibat (2008) evaluated the antilipoperoxidative activity of the aqueous, boiled and non boiled and methanolic extracts of leaves of *Eminium spiculatum* against lipid peroxidation (LPO) induced by a rat liver homogenate model. Nonboiled aqueous extracts exhibited the highest LPO inhibition percent followed by methanolic and boiled aqueous extracts. Moreover, LPO inhibition percent was dose dependent, where 1.65 x 10-2 mg/ml caused the highest inhibition percent in all aqueous extracts.

Empetrum nigrum L. Family: Empetraceae

English Name: Crowberry.

The antioxidative activity of phenolic extracts from fruits of *Empetrum nigrum* was examined by autoxidation of methyl linoleate by Kahkonen *et al.* (1999). Remarkable high antioxidant activity (98 per cent) and high total phenolic content (50.8 mg/g GAE) were found in crowberry. In addition, berry wines made of berries showed stronger antioxidative activity in the oxidation of MeLo.

Enicosanthum acuminata (Thwaites) Airy Shaw Family: Annonaceae

Puvanendran *et al.* (2007) evaluated the antioxidant activity of Dichloromethane (DCM) and MeOH extract of stem, stem bark and leaves of *Enicosanthum acuminata* using DPPH assay. The extract showed high antioxidant activities. Percentage of inhibition of stem bark was 45.59 (MeOH extract); leaf extract was 36.52 (MeOH extract) while that of stem was 13.52 (DCM extract).

Enicostemma axillare (Lam.) Raynal (Syn.: _Enicostemma littorale_ Bl.)

Family: Gentianaceae

It contains phenols, tannins, flavonoids, glycosides, anthraquinones, sterols (Vasu *et al.*, 2005), gentianine and swertiamarin (Govindachari *et al.*, 1966; Viswakarma *et al.*, 2004). Studies have also confirmed its hypoglycemic, antioxidant and hypolipidaemic potential (Vasu *et al.*, 2005). The *in vitro* antioxidant activity of aqueous, hydroalcoholic, methanolic, and chloroform extracts of this plant was investigated by Deore *et al.* (2008) by using two different model covering nitric oxide and DPPH method. The results indicated efficacy of extracts for antioxidant activity in the following sequence: methanol > hydroalcoholic > aqueous >chloroform.

Administration of methanol extract (2.5g/kg body weight/day) to diabetic rats for 20 days reduced blood glucose levels from 466 to 237 (Maroo *et al.*, 2003). *E. littorale* also increased the serum insulin levels of diabetic rats and improved the antioxidant studs of diabetic rats.

Enicostemma littorale aqueous extract (1.5 g/100g body weight/day, p.o.) was administered to rats along with hypercholesterolaemic diet for 6 weeks and the hypolipidaemic and antioxidant effect was evaluated by Vasu *et al.* (2005). *Enicostemma littorale* treatment increased HDL levels and decreased serum cholesterol, triglyceride, LDL, VLDL, LDL/HDL ratio and a decrease in activities of erythrocyte catalase, superoxide dismutase and lipid peroxidation levels, with an increase in reduced glutathione levels as compared to cholesterol fed untreated rats (Vasu *et al.*, 2005).

Jaishree *et al.* (2008) investigated the antioxidant activity of *Enicostemma axillare*. Antioxidant and hepatoprotective effect of the ethyl acetate extract of *Enicostemma axillare* against CCl_4-induced liver injury in rats was evaluated by Jaishree *et al.* (2010).

Swertiamarin, a secoiridoid isolated from the ethyl acetate extract of *Enicostemma axillare*, was screened for *in vitro* antioxidant activity using seven different methods: good activity was observed in hydroxyl radical by deoxyribose and lipid peroxidation methods, with IC_{50} values of 2.83, 5.70, 52.56 and 78.33µg/ml, respectively. The total antioxidant capacity was found to be 4.51mM of ascorbic acid per gram of swertiamarin (Jaishree and Badami, 2009, 2010).

Entada africana Guill. et Perr. Family: Mimosaceae

Using Trolox assay, Cook *et al.* (1998) estimated the antioxidant activity of 17 wild edible plants of Niger Republic used for food and traditional medicine. They observed that *Entada africana* possessed strong antioxidant activity. Extract treatment to the diabetic rats significantly increased reduced glutatihione levels and decreased erythrocyte catalase activity and lipid peroxidation.

Tibri *et al.* (2010) evaluated the antioxidant activity, total phenolic and flavonoid contents of *Entada africana*. Aqueous, chloroform, ethyl acetate and methanol extracts of leaves, stem bark and roots wre used for this investigation. The radical scavenging activity was studied using DPPH assay and phenolic content by Folin-Ciocalteu method. The extracts showed high radical scavenging activity with EC_{50} values ranging from 3.39 to 16.72 µg/ml

Epilobium angustifolium L. Family: Onagraceae

English Name: Willow herb.

The antioxidative activity of phenolic extracts from willow herb was examined by autoxidation of methyl linoleate by Kahkonen *et al.* (1999). The extracts of the herb showed high antioxidant activity (90 per cent inhibition at 500ppm level). The amount of total phenolics was 32.2 mg/g GAE dry material. Several monoglycosides of kaempferol, quercetin, and myricetin have been isolated from willow herb (Duckrey *et al.*, 1995). It also contains a flavonol, glucuronide, that has very strong anti-inflammatory properties (Hiermann *et al.*, 1991).

Five commercially available water-soluble extracts prepared from the aerial parts of *Epilobium angustifolium* were screened by Shikov *et al.* (2006) for antioxidant-related properties in six *in vitro* assays. The extracts demonstrated varying degrees of efficacy in each screen. Two extracts, denoted as nonfermented and Tver, were the most effective toward reducing iron (III), scavenging DPPH free radicals, inhibiting hydroxyl radical-catalyzed bovine-brain-derived phospholipids degradation, and non-site and site-specific hydroxyl radical-catalyzed 2-deoxy-D-ribose degradation. The activity profile of the extracts changed, however, when their iron (II) chelating ability was assessed. The nonfermented and Tver extracts were not as effective iron(II) chelators as the extract denoted as Lotos. All the extracts contained Folin-Ciocalteu-reactive substances which was confirmed by the presence of predominently polar phenolic analytes (hydroxylated benzoic acid derivatives and flavonoids).

Štajner *et al.* (2007) investigated the antioxidative properties of root, leaf and stalk of *Epilobium angustifolium*. The activities of antioxidant enzymes (superoxide dismutase, catalase, peroxidase), quantities of malonyldialdehyde, superoxide and hydroxyl radicals and reduced glutathione and also the content of total flavonoids, chlorophylls a and b, carotenoids, vitamin C and soluble proteins were determined. The results indicate that extracts from all plant organs exhibited antioxidant activity. The highest antioxidant activity was observed in the leaves. Radical scavenging capacity (RSC), determined by DPPH assay, was highest in the leaves as well as total reducing power determined by FRAP assay.

Borchardt *et al.* (2008) evaluated the antioxidant activity of methanol extracts of seeds of *Epilobium angustifolium* using DPPH antioxidant assay.The antioxidant value reported in µM Trolox/100g (TE) from DPPH radical scavenging activity of crude seeds extract was 70,940.

Hevesi Toth *et al.* (2009a,b) investigated the phytochemical and antioxidant activity of *Epilobium angustifolium* by ABTS method and DPPH radical assay. The extract showed high radical-scavenging activity, comparable to that of well-known antioxidants, Trolox and ascorbic acid. Flavonoid composition was 0.69g/100g dried herb, Total polyphenol was 22.3 g/100g dried herb, and Tannin (pyrogallol) was 18.7 g/100 g.

Epilobium hirsutum L. Family: Onagraceae

Wojdylo *et al.* (2007) investigated Total equivalent antioxidant capacities (TEAC) and phenolic contents of 32 species extracts from 21 botanical families grown in Poland. The total antioxidant capacity was estimated by ABTS, DPPH and FRAP expressed as TEAC. The herbs with highest TEAC values were *Syzygium aromaticum* and *Epilobium hirsutum*.

Epilobium montanum L. Family: Onagraceae

Hevesi Toth *et al.* (2009a,b) investigated the phytochemical and antioxidant activity of *Epilobium montanum* by ABTS method and DPPH radical assay. The extract showed high radical-scavenging activity, comparable to that of well-known antioxidants, Trolox and ascorbic acid. Flavonoid composition was 0.80g/100g dried herb, Total polyphenol was 30.2 g/100g dried herb, and Tannin (pyrogallol) was 25.2 g/100 g.

Epilobium parviflorum Schreb. Family: Onagraceae

The aqueous acetone extract of *Epilobium parviflorum* showed higher antioxidant effect in the DPPH assay (IC_{50}=2.37mg/ml). In concentration of 0.2 µg/ml the extract possessed a protective effect, comparable to catalase (250 iu/ml), against oxidative damage, generated in fibroblast cells. In the COX inhibition assay *E. parviflorum* decreased the PGE2 release which showed inhibition of the COX enzyme (IC_{50}=1.4 µg/ml) (Balazs *et al.*, 2003).

Crude water and ethanolic extracts of *Epilobium parviflorum* scavenged hydroxyl radicals (Steenkamp *et al.*, 2006). Arrendondo *et al.* (2004) reported antioxidant activity of *Epilobium parviflorum* infusion.

Hevesi Toth *et al.* (2009a,b) investigated the phytochemical and antioxidant activity of five species of *Epilobium* including *Epilobium parviflorum* by ABTS method and DPPH radical assay. The extract showed high radical-scavenging activity, comparable to that of well-known antioxidants, Trolox and ascorbic acid. Among the species examined, extract of *Epilobium parviflorum* possessed the highest antioxidant capacity (EC_{50} = 1.71µg/ml. Flavonoid composition was 0.83g/100g dried herb, Total polyphenol was 34.8 g/100g dried herb, and Tannin (pyrogallol) was 25.8 g/100 g.

Epilobium roseum L. Family: Onagraceae

Hevesi Toth *et al.* (2009a,b) investigated the phytochemical and antioxidant activity of *Epilobium montanum* by ABTS method and DPPH radical assay. The extract showed high radical-scavenging activity, comparable to that of well-known antioxidants, Trolox and ascorbic acid. Flavonoid composition was 0.82g/100g dried herb, Total polyphenol was 28.6 g/100g dried herb, and Tannin (pyrogallol) was 21.4 g/100 g.

Epilobium tetragonum L. Family: Onagraceae

Hevesi Toth *et al.* (2009a,b) investigated the phytochemical and antioxidant activity of *Epilobium tetragonum* by ABTS method and DPPH radical assay. The extract showed high radical-scavenging activity, comparable to that of well-known antioxidants, Trolox and ascorbic acid. Flavonoid composition was 0.73g/100g dried herb, Total polyphenol was 25.8 g/100g dried herb, and Tannin (pyrogallol) was 18.7 g/100 g.

Epimedium brevicornum Maxim. Family: Berberidaceae

Cai *et al.* (2004) evaluated the antioxidant activity and phenolic compounds in traditional Chinese medicinal plants associated with anticancer. The improved ABTS.+method was used to systematically assess the total antioxidant capacity (Trolox equivalent antioxidant capacity, TEAC) of the medicinal extracts. The TEAC values and total phenolic content for methanolic extract of *Epimedium brevicornum* were 420.6 µmol Trolox equivalent/100g dry weight (DW), and 2.78 g of gallic acid equivalent/100g DW respectively. A positive significant linear relationship between antioxidant activity and total phenolic content showed that phenolic compounds were the dominant antioxidant components in the tested herbs. Major types of phenolic compounds in *Epimedium brevicornum* include flavonols (epimedins, quercetin and their glycosides) and lignans.

Epipremnum pinnatum (L.) Engl. (Syn. _Rhapidophora pinnatifida_ L.) Family: Araceae

Linnert *et al.* (2010) evaluated anti-inflammatory, analgesic and anti-lipid peroxidative effects of aerial parts of *Epipremnum pinnatum*. The ethanol extract showed significant anti-lipid peroxidant effects *in vitro* in rat liver homogenate. There was a significant increase of malondialdehyde (MDA) in $FeCl_2$-AA treated rat liver homogenate, compared to normal control.

Erechtites valerianaefolia (Wolf.) DC. Family: Asteraceae (Compositae)

The aerial parts of 17 Bolivian plants including *Erechtites valerianaefolia* were screened by Rosas-Romero and Saavedra (2005) to determine antioxidant activity. A methanol extract of each plant was prepared and portioned sequentially with hexane, chloroform, and ethyl acetate, leaving an aqueous solution. All extracts and their 5 fractions for a total of 102 samples, were evaluated using two techniques: an adaptation of the β-carotene bleaching technique using an emulsion of linoleic acid in water as the oxidizable substrate, and the DPPH free radical trapping technique. The results of the β-carotene bleaching technique were more discriminating and better related to the rancidity process under normal conditions; with this assay, 11 species provided at least one fraction with highly promising antioxidant activity. All species gave good results under the DPPH technique, and in most cases they performed better than BHA, which was used as a reference antioxidant (Rosas-Romero and Saavedra, 2005).

Eremostachys adenantha Jaub et Spach Family: Lamiaceae (Labiatae)

Firuzi *et al.* (2010) evaluated the antioxidant activity and total phenolic content of 24 Lamiaceae species growing in Iran. *Eremostachys adenantha* showed low antioxidant activity with 11.4μM/g DW FRAP value and IC_{50} = 988.6μg/ml of DPPH radical scavenging activity. It also showed lowest phenolic content of 3.6 mg CE/g.

Eremostachys laciniata (L.)Bunge Family: Lamiaceae (Labiatae)

Erdemoglu *et al.* (2006) evaluated the antioxidant activity of *Eremostachys laciniata* extract by using DPPH as well as by flow-injection analysis-luminol chemiluminiscence (FIA-CL). Extracts were shown to possess a significant scavenger activity against DPPH free radical and an inhibitory effect on H_2O_2- or HOCl-luminol chemiluminiscence. The results concluded that the extracts have a potential source of antioxidants of natural origin.

Erica arborea L. Family: Ericaceae

The antioxidant properties of the methanol extract of leaves and flowers of *Erica arborea* and the ethyl acetate, butanol and water soluble fractions were investigated by Ay *et al.* (2007). The ethyl acetate extract was found to be the richest for phenolic and flavonoid content which showed the highest antioxidant activity.

Nazemiyeh *et al.* (2008) investigated the antioxidant activity of the phenolic compounds isolated from the methanol extraxct of leaves of *Erica arborea.*

Erigeron acris L. Family: Asteraceae (Compositae)

6-Caffeoylerigeroside, isolated from *Erigeron acris,* showed antioxidant activity (Nalewajo-Sielwoniuk *et al.*, 2008).

Erigeron annuus (L.) Pers. Family: Asteraceae (Compositae)

Antioxidant activities of the fractions from *Erigeron annuus* leaf were examined by Jeong *et al.* (2011) with ABTS and FRAP assays. The study demonstrated that butanol fraction had the highest antioxidant activity among all solvent fractions from methanolic extract *of E. annuus* leaf. Butanol fraction had the highest total phenolic contents (396.49 mg of GAE/g). Caffeic acid, an isolated active compound from butanol fraction, showed dose-dependent *in vitro* antioxidant activity. Moreover, neuronal cell protection against oxidative stress induced cytotoxicity was also demonstrated.

Erigeron breviscapus (Vant.) Hand.-Mazz. Family: Asteraceae (Compositae)

Antioxidative activity of the ethanol extract of *Erigeron breviscapus* were studied by Liu *et al.* (2005) by using neuron oxidative injury model induced by superoxide radical. The extract at the dose of 10-140μg/ml reduced significantly the lipid peroxidation levels and lactic dehydrogenase (LDH) release from neuron exposed to superoxide radical (generated by xanthine and xanthine oxidase). These results suggest that neuroprotective actions of the *Erigeron breviscapus* extract may be due to its antioxidant or radical scavenging activity.

Eriobotrya deflexa (Hemsl) Nakai Family: Rosaceae

Hou *et al.* (2003) evaluated the antioxidant activity of 70 per cent aqueous acetone extract of *Eriobotrya deflexa* by various assays including DPPH, hydroxyl radicals and reducing power assay. The extract exhibited moderate activity against DPPH radicals with IC_{50} value of 16.94 μg/ml and inhibited formation of OH generated in the Fenton reaction system with IC_{50} value of 0.8 μg/ml). The amount of phenolics was 29.6 mg of GAE/g.

Eriobotrya japonica (Thunb.) Lindl. Family: Rosaceae

A 70 per cent ethanol extract of *Eriobotrya japonica* seeds was prepared and administered the extract to rats with renal disorder induced by a single administration of 7 mg/kg body weight adriamycin, and investigated the usefulness of the extract (Hamada *et al.*, 2004. Increases in indices of renal function, plasma urea nitrogen, were significantly inhibited in rats treated with the *Eriobotrya* extract compared to rats treated with tap water. In addition, the renal tissue level of reduced glutathione was significantly high in rats that ingested the extract, while the lipid peroxide levels in plasma and renal tissue were significantly low. However, no effect on renal tissue antioxidative enzymes was observed suggesting that seed extract has direct antioxidant activity. Based on these findings, seed extract may be effective in reducing the oxidative stress of adriamycin-induced renal disorder. Therefore, ingestion of seed extract may contribute to a reduction of the adverse effects of adriamycin (Hamada *et al.*, 2004).

Extracts of *Eriobotrya japonica* seed were prepared using various solvents n-hexane, ethyl acetate, n-butanol, methanol and aqueous and the antioxidative activity of the solvent fractions and components was evaluated based on the scavenging of various radicals (DPPH and O^{2-}) measured by the ESR method and the inhibition of Fe^{3+}-ADP induced NADPH dependent lipid peroxidation in rat liver microsomes. The radical scavenging activities and inhibitory activities on lipid peroxidation differed among the solvent fractions and components. In the n-butanol, MeoH and aqueous fractions, radical scavenging activity and inhibitory activity on lipid peroxidation were high. In addition, these fractions contained polyphenols, and the radical scavenging activity increased with the polyphenol content. In the low-polar n-hexane and EtOAc fractions, the radical scavenging activity was low, but the lipid peroxidation inhibition activity was high. These fractions contained β-sitosterol, and the inhibitory activity on lipid peroxidation was high. Based on these findings, the antioxidant activity of seed extract may be derived from many components involved in a complex mechanism, resulting in high activity (Yokota *et al.*, 2006).

Antioxidant capacities of 56 selected Chinese medicinal plants, including *Eriobotrya japonica* were evaluated by Song *et al.* (2010) using the TEAC and FRAP assays, and their total phenolic content was measured by the Folin-Ciocaleteu method. *Eriobotrya japonica* showed high antioxidant activity with TEAC value of 326.87µmol Trolox/g, FRAP value of 437.4µmol Fe^{2+}/g and phenolic content was 31.47mg GAE/g. The strong correlation between TEAC value and FRAP value suggested that the antioxidants in these plants possess free radical scavenging activity and oxidant reducing power, and the high positive correlation between antioxidant capacities and total phenolic content implied that phenolic compounds are a major contributor to the antioxidant activity of these plants.

Erodium cicutarium L'Herit. Family: Geraniaceae

Phytochemical analysis and immunotropic activity of polyphenolic fraction from *Erodium cicutarium* was carried out by Fecka *et al.* (1997). Polyphenolic content was found to be 3.41 per cent, flavonoids as quercetin 0.45 per cent, tannins 0.78 per cent. Methanol extract contains quercetin, kampferol, myricetin, their glycosidic derivatives and free polyphenolic acids. Low concentration of extract stimulated while the high concentration inhibited free radical activity of human granulocytes *in vitro*. Antioxidant properties of the herb have been explained (Fecka *et al.*, 1997).

***Erodium oxyrhynchum* M.B. subsp. *oxyrhynchum* Family: Geraniaceae**

Souri *et al.* (2004) evaluated the antioxidative activity of *Erodium oxyrhynchum* subsp. *oxyrhynchum* by linoleic acid peroxidation test using 1,3-diethyl-2-thiobarbituric acid as the reagent. Methanolic extract showed 71.00 per cent inhibition at 40µg concentration with IC_{50} value of 6.57µg.

Eruca sativa Mill. Family: Brassicaceae (Cruciferae)

English named: Rocket salad.

The antioxidant activity of 4-(β-D-glucopyranosyldisulfanyl)butyl desulfo-glucosinolate isolated from leaves of *Eruca sativa* was measured by Kim *et al.* (2004). The antioxidant activity (2.089 unit/g) was relatively higher than that of dimeric 4-mercaptobutyl desulfo-glucosinolate (1.227).

Alam *et al.* (2007) reported that *Eruca sativa* seeds possess antioxidant activity and exert a protective effect on mercuric chloride induced renal toxicity. The ethanolic extract of the seeds was found to possess a potent antioxidant effect, with a large amount of polyphenols and a high reducing ability. *E. sativa* extract significantly scavenged several reactive oxygen species (ROS) and reactive nitrogen species (RNS).

Erucaria microcarpa Boiss. Family: Brassicaceae (Cruciferae)

A chromone has been isolated from the aerial parts of *Erucaria microcarpa* (Hashim, 2007). Testing this chromone for electron spin resonance, it did not give any free radical along, but with singlet oxygen, it gave one free radical, which rapidly combined with each other to give a stable compound. This chromone can be, thus considered a singlet oxyben scavenger which changes to ordinary oxygen.

Erycibe obstusifolia Benth. Family: Convolvulaceae

Hsu *et al.* (1999) evaluated the antioxidant activity of *Erycibe obtusifolia* assessed by the measurements of hepatic, renal and spleenic peroxides (measured as malondialdehyde) after treatments. The results suggest that a high dose of *Erycibe obtusifolia* administration may contribute a little antioxidant activity in both liver and kidney.

Eryngium caucasicum Trautv. Family: Apiaceae (Umbelliferaae)

Methanol extracts of leaves and inflorescence of *Eryngium caucasicum* at flowering stage were investigated by Ebrhimzadeh *et al.* (2009) for their antioxidant activities employing six *in vitro* assay systems, *i.e.* DPPH and nitric oxide radical scavenging, reducing power, linoleic acid and iron ion chelating power. IC_{50} for DPPH radical-scavenging activity was 0.15 for leaves and 0.39 mg ml^{-1} for inflorescence. Reducing powers of both extracts increased with the increase of their concentrations. Leaves extract showed better activity than Vitamin C. Extracts showed weak nitric oxide-scavenging activity. Leaves extract exhibited better Fe^{2+} chelating ability (IC_{50} =0.25 mg ml^{-1}) that was comparable with EDTA. (IC_{50}=18 µg ml^{-1}). Inflorescence extracts had shown a very weak activity. Extracts showed very good scavenging activity of H_2O_2. IC_{50} was 25.5 for leaves and 177.2 mg/ml for inflorescence, respectively. No antioxidant activity exhibited in linoleic acid test. Extracts exhibited different levels of antioxidant activity in all the models studied.

Eryngium foetidum L. Family: Apiaceae (Umbelliferae)

English Name: False coriander.

Chanwitheesuk *et al.* (2005) investigated the antioxidant activity of leaves of *Eryngium foetidum* using a β-carotene bleaching method. The contents of plant chemicals such as vitamin C, Vitamin E, carotenoids, tannin and total phenolics, were also determined. Methanolic extract showed antioxidant activity. *Eryngium foetidum* has antioxidant index of 5.65, Vitamin C 11.4 mg per cent, Vitamin E 0.0069mg per cent, Total carotenes 1.92 mg per cent, Total xanthophylls 1.60mg per cent, Tannins 17.7mg per cent and Total phenolics of 98.4mg per cent.

Wong *et al.* (2006) investigated the antioxidant properties of 25 edible tropical plants including *Eryngium foetidum* using DPPH and FRAP assays. Total phenolic content was also estimated. *Eryngium foetidum* showed moderate antioxidant activity.

Erythrina abyssinica Lam. Family: Fabaceae (Papilionaceae)

Sigmoidin A isolated from the stem bark of *Erythrina abyssinica* showed prooxidant effect and cytotoxicity (Habterman and Dagne, 2010).

Erythrina crista Galli. Family: Fabaceae (Papilionaceae)

Maisuthisakul *et al.* (2008) reported that extracts of *Erythrina crista* possessed strong antioxidant activity, high total phenolic and flavonoid contents. They are of the opinion that plants with high antioxidant activities also have high total phenolic and flavonoid content.

Erythrina latissima E. Meyer Family: Fabaceae (Papilionaceae)

From the stem wood of *Erythrina latissima* two isoflavones and a flavanone were isolated and characterized in addition to 10 known flavonoids (Chacha *et al.*, 2005). The isolated compounds exhibited weak radical scavenging properties towards DPPH radical.

Erythrina mildbraedii Harms Family: Fabaceae (Papilionaceae)

Erycristagallin, a pterocarpene isolated from *Erythrina mildbraedii* was tested *in vitro* for its antioxidant properties as also, was tested on different experimental models of inflammation. *In vitro*, erycristallin showed antioxidant activity in DPPH test (Najmen *et al.*, 2003).

Erythrina sigmoidea Hua Family: Fabaceae (Papilionaceae)

Njamen *et al.* (2004) isolated two prenylated flavanones from *Erythrina sigmoidea* (sigmoidin A and B) and studied their ability to inhibit the stable DPPH free radical and arachidonic acid metabolism. Both sigmoidins A and B proved to be potent scavengers of the DPPH radical.

Erythrina stricta Roxb. Family: Fabaceae (Papilionaceae)

Ashok Kumar *et al.* (2008) evaluated the antioxidant activities of *Erythrina stricta* using various *in vitro* and *ex vivo* models. The leaves were extracted with ethanol (70 per cent) and used for the evaluation of various *in vitro* antioxidant assays which include H-donor activity, nitric oxide scavenging, total phenolic content, total flavonoid content, total antioxidant activity by thiocyanate and phosphomolybdenum method, metal chelating, β-carotene bleaching, total peroxyl radical assays. *Ex vivo* models like lipid peroxidation and erythrocyte haemolysis were also used to study the antioxidant property of the extract. The various antioxidant activities were compared with suitable standard antioxidants such as ascorbic acid, BHT, α-tocopherol, curcumin, quercetin and Trolox. The generation of free radicals *viz.* O_2, OH, H_2O_2, NO and peroxyl radicals were effectively scavenged by the ethanol extract of *Erythrina stricta.* In all the methods, the extract offered strong antioxidant activity in a concentration dependent manner.

Erythrina variegata (Thunb.) Koidz. Family: Lardizabalaceae

Gan *et al.* (2010) evaluated the antioxidant activity of *Akebia trifoliata* using the FRAP and TEAC assays, and its total phenolic content was measured by the Folin-Ciocalteu method. For FRAP assay the value was 185.91 μmol Fe(II)/g dry weight. For TEAC assay, the value was 124.31μmol Trolox/g dry weight. Total phenolic content was 9.12 mg gallic acid equivalent/g dry weight of plant material.

Etlingera elatior (Jack) R.M.Sm. Family: Zingiberaceae

Diarylheptanoids 1-3 isolated from the rhizomes of *Etlingera elatior* were found to inhibit lipid peroxidation in a more potent manner than α-tocopherol (Mohamad *et al.*, 2005).

Extracts from 11 vegetables of Indonesian origin, including *Etlingera elatior* were screened for flavonoid content, total phenolics, and antioxidant activity by Andarwulan *et al.* (2010).

Eucalyptus camaldulensis Dehn. Family: Myrtaceae

Souri *et al.* (2004) evaluated the antioxidative activity of *Eucalyptus camaldulensis* by linoleic acid peroxidation test using 1,3-diethyl-2-thiobarbituric acid as the reagent. Methanolic extract showed 92.88 per cent inhibition at 40µg concentration with IC_{50} value of 2.16µg.

Barra *et al.* (2010) found high antioxidant activities of essential oils of *Eucalyptus camaldulensis* growing wild in different localities of Sardinia (Italy). The activities changed according to the chemical composition, nevertheless generally constituted by *p*-cymene, 1,8-cineole, β-phellandrene, spathulenol and cryptone in different proportions according to the harvesting place and period. The antioxidant activity assessed by DPPH-test and expressed as Trolox equivalent antioxidant capacity, showed values ranging between 0.5 and 5.8 mmol/l.

Eucalyptus camaldulensis Dehn. var. _brevirostris_ Family: Myrtaceae

El-Ghorab *et al.* (2003) reported on the promising antioxidative activities of ethanol extract from the leaves of *Eucalyptus camaldulensis* var. *brevirostris* harvested from Egypt which contained gallic and ellagic acid as the major components. The essential oil obtained from Pakistani *Eucalyptus camaldulensis* var. *brevirostris* exhibited clear antioxidant activity with dose-response in DPPH radical scavenging and β-carotene bleaching assays (El-Ghorab *et al.*, 2009)

Eucalyptus globulus Labill. Family: Myrtaceae

Dessi *et al.* (2001) carried out studies on the effect of methanol extract and essential oil on lipid peroxidation *in vitro* systems. Essential oil showed antioxidant activity. Amakura *et al.* (2009) studied the marker constituents of the natural antioxidant Eucalyptus leaf extract for the evaluation of food additives.

In order to find their potential sources, the crude extracts of different parts of some selected food and medicinal plants, including *Eucalyptus globulus*, were studied by Bajpai *et al.* (2005) for total phenolic contents (TPC) and antioxidant activity (AOA). AOA was assayed by auto-oxidation of β-carotene and linoleic acid and expressed as the per cent of inhibition relative to control. TPC in leaves and bark of *Eucalyptus globulus* was 40.1 and 42.9mg/g (GAE) respectively while antioxidant activity was 85.6 per cent and 74.5 per cent respectively.

Eucalyptus grandis L. Family: Myrtaceae

The effectiveness of subcritical liquid water processing for extraction of antioxidants from the leaves of *Eucalyptus grandis* biomass was evaluated by Kulkarni *et al.* (2007) by determining the capability of the processed samples to scavenge peroxynitrite radicals *in vitro*, as compared with the extracts obtained by conventional extraction. Pyrogallol, obtained from the subcritical water extract, exhibited stronger antioxidant activity than BHT, which was used as a standard for the antioxidant activity measurement.

Eucalyptus robusta Sm. Family: Myrtaceae

Fu *et al.* (2010) evaluated the antioxidant capacity and total phenolic contents of 56 wild fruits from South China using FRAP and TEAC assays. A high correlation between antioxidant capacity and phenolic compounds could be the main contributor to the measured antioxidant activity. The results showed that fruits of *Eucalyptus robusta* has possessed the highest antioxidant capacity and total phenolic contents.

Eucalyptus rostrata Schlecht. Family: Myrtaceae

Antioxidant activity of an extract from *Eucalyptus rostrata* was the highest among 16 *Eucalyptus* species tested (Okamura *et al.*, 1993). Antioxidant activity of tannins and acylated flavonol glycosides, all with galloyl groups, was much higher than that of a synthetic antioxidant. Two of the compounds, spiraeoside and quercetin 4'-glucosylgallate, showed high superoxide dismutase activity (Okamura *et al.*, 1993).

The chloroform and methanolic extracts of 124 Egyptian plant species belonging to 56 families were investigated by Moussa *et al.* (2011) and compared their antioxidant activity by DPPH scavenging assay. Among the 124 plant species tested 18 exhibited extremely high antiradical activity (>80 per cent inhibition). The IC_{50} value of DPPH radical scavenging of *Eucalyptus rostrata* was 20.73µg/ml, while total phenolic and flavonoid contents were 67.76 mg Tannic acid equivalent/g extract and 220.26 mg Rutin equivalent/g extract, respectively. Correlation coefficient between DPPH radical scavenging activity and the total phenolic and flavonoid contents suggest that phenolics and flavonoids in the extracts were partly responsible for the antiradical activities.

Eucalyptus tereticornis Sm. Family: Myrtaceae

The essential oils hydrodistilled from fresh and decaying leaves were assayed by Singh *et al.* (2009) for antioxidant activity in terms of scavenging DPPH and hydroxyl radical, and superoxide anion. Both essential oils exhibited a strong radical scavenging activity against DPPH radical with IC_{50} values of 110 and 139.8µg/ml for fresh and decaying leaf oil, respectively (IC_{50} of BHT = 164.2µg/ml). Further essential oils also exhibited OH (56-62 per cent) and O_2(-*) (65-68 per cent) scavenging activity parallel to the commercial antioxidant BHT/ascorbic acid. However, unlike the essential oils, the major monoterpene constituents exhibited significantly less scavenging acitity (<35 per cent DPPH or OH(*); at 400µg/ml).

Eucommia ulmoides Oliv. Family: Eucommiaceae

Chinese name: Du-zhong

Hsieh and Yen (2000) investigated the antioxidant effect of water extracts of Du-zhong (WEDZ) on oxidative damage in biomolecules such as deoxyribose, DNA, and 2'-deoxyguanosine (2'-dG) as induced by Fenton reaction. The WEDZ used included leaves, raw cortex, and roasted cortex. All of the WEDZ inhibited the oxidation of deoxyribose induced by Fe(3+)-EDTA/H_2O_2/ascorbic acid in a concentration dependent manner. At a concentration of 1.14mg/ml, the inhibitory effect of the extracts of leaves, roasted cortex and raw cortex was 85.2 per cent, 68.0 per cent and 49.3 per cent, respectively. The extract of leaves inhibited the strand breaking of DNA induced by the Fenton reaction at a concentrations of 5 and 10 µg/ml. This inhibitory effect was similar to mannitol whereas the extracts of raw cortex and roasted cortex had no inhibitory effect at all. WEDZ also inhibited the oxidation of 2'dG to 8-OH-2'-dG induced by Fe(3+)-EDTA/H_2O_2/ascorbic acid. Gallic acid had a prooxidant effect, but trolox and mannitol had an antioxidant effect. The leaf extract had a marked inhibitory effect on Fenton reaction induced oxidative damage on biomolecules. The extract of roasted cortex exhibited modest inhibition while the extract of raw cortex had the least inhibitory effect on oxidative damage in biomolecules (Hsieh and Yen, 2000).

Enzyme-assisted water extracts (EWEDL) and ethanol extracts of Du-zhong (EEDL) were evaluated by Luo *et al.* (2009) for their antioxidant activities using DPPH radical-scavenging assay, Fe(2+)-chelating assay, and inhibition ability of the linoleic acid peroxidation assay. In general, the antioxidant activity of Du-zhong leaf extracts increased with increasing concentration. Based on the two extracting methods with different antioxidative reactions, it was shown that the enzyme-assisted water extracting method was more effective for antioxidant extraction from Du-zhong leaves. The main phenolic compounds in both the extracts are geniposidic acid, epicatechin and chlorogenic acid. The total phenolic contents of EWEDL and EEDL were 13.84 and 14.72 mg chlorogenic acid equivalents in each gram extract, respectively. However, there was no positive correlation between total phenolic content and antioxidant activities of EWEDL and EEDL measured by the three different assays.

The ethanol extracts from leaf, roasted cortex, and seed of Du-zhong were examined by Xu *et al.* (2010) for total phenolics content (TPC) and total flavonoids content (TFC). The antioxidant activity of the extracts was evaluated by measuring the radical scavenging activity of DPPH, FRAP, and lipid peroxidation capacity in beta-carotene/linoleic acid system. Du-zhong leaf extract was found to have the highest TPC content (94.46 mg of gallic acid equivalent/g of solid extrat) and TFC content (61.36 mg of catechin equiv./g of solid extract. In the above three antioxidant assay systems, Du-zhong leaf extract also exhibited the strongest antioxidant capacities followed by roasted cortex extract and seed extract. The effects of Du-zhong extracts (leaf, roasted cortex and seed) on lipid peroxidation, meat color, and metmyoglobin formation in raw pork patties were investigated and compared with that of BHT during refrigerated storage at 4°C for 8 days. The results indicated that the addition of leaf extract at 0.1 per cent (w/w), roasted cortex extract at 0.1 per cent (w/w) and BHT at 0.01 per cent (w/w) decreased day 8 TBARS values by 35, 20 and 37 per cent, respectively.

Eugenia caryophyllata (Spreng.) Bullock et Harrison Family: Myrtaceae

English Name: Cloves.

Methanol aqueous extract of *Eugenia caryophyllata* was screened by Kim *et al.* (1997) for anti-oxidative activity using Fenton's reagent/ethyl lineolate system. The extract showed antioxidant activity.

Khalaf *et al.* (2008) reported antioxidant activity in the methanolic crude extracts of cloves using DPPH assay.The inflorescence of clove is effective against treatment of arthritis, rheumatism and sprains. Acetyl-eugenol, ascorbic acid, β-carotene, β-sitosterol, caryophyllene-oxide, eugenol and isoeugenol are the main constituents in clove that possess antioxidative property.

Jayathilakan *et al.* (2007) evaluated the antioxidant potential of synthetic and natural antioxidants and its effect on warmed-over-flavour in different species of meat. Cloves showed good antioxidant activity.

Twenty-five essential oils, including Clove leaf, were tested by Wei and Shibamoto (2010) for antioxidant activities using a conjugated diene assay, the aldehyde/ carboxylic acid assay, the DPPH free radical scavenging assay, and the malonaldehyde/gas chromatography (MA/GC) assay. They were also tested for lipoxygenase inhibitory activities using the lipoxygenase inhibitor-screening assay. Clove leaf oil showed good antioxidant activity (53-100 per cent).

Wei *et al.* (2010) reported that Essential oil of *Eugenia caryophyllata* had appreciable antioxidant activities comparable to that of α-tocopherol, the reference chosen by the authors. In the clove oil eugenol and β-caryophyllene predominated.

Antioxidant activity of essential oils of five spice plants, including *Eugenia caryophyllata* used in the Mediterranean diet was evaluated by Viuda-Martos *et al.* (2010). All the essential oils tested had antioxidant activity. Politeo *et al.* (2010 reported the chemical composition and antioxidant activity of clove glycodically bound volatile compounds in comparison with their essential oil. DPPH radical scavenging method and FRAP assay were followed. Both the methods showed that the clove volatile aglycones possess lower antioxidant properties than its essential oil.

Eugenia jambolana Lam. Family: Myrtaceae

Administration of *Eugenia jambolana* seed kernel to diabetic rats significantly decreased the levels of blood glucose, glycosylated hemoglobin and increased body weight gain, plasma insulin and hemoglobin. The diabetic rats showed the low activities of superoxide dismutase, catalase, glutathione peroxidase and reduced glutathione content in liver and kidney, which were restored to near normal levels by treatment with the seed kernel extract. The increased levels of lipid peroxidation and hydroperoxides in diabetic rats were reverted back to near normal levels after the treatment with seed kernel extract (Ravi *et al.*, 2004).

Eugenia polyantha Wight Family: Myrtaceae

Wong *et al.* (2006) investigated the antioxidant properties of 25 edible tropical plants including *Eugenia polyantha* using DPPH and FRAP assays. Total phenolic content was also estimated. *Eugenia polyantha* showed very high antioxidant activity. A strong correlation between TEAC values obtained for the DPPH assay and those for FRAP assay implied that compounds in the extracts were capable of scavenging DPPH radical and reducing ferric ions. A satisfactory correlation of TPC with TEAC suggested that polyphenols in the extracts were partly responsible for the antioxidant activities while its correlation with CCA was poor, indicating that polyphenols might not be the main cupric ion chelators.

Antioxidative activities of three bark extracts (methanol, methanol-water and water) from *Eugenia polyantha* grown in Indonesia were evaluated using various *in vitro* assays: DPPH, hydrogen peroxide-scavenging and β-carotene bleaching assays. From the assays, *Eugenia polyantha* bark extracts were found to be potential antioxidative activities. The methanol-water extract showed the highest level of free radical-scavenging activity (Effective dose (ED_{50}) = 0.18 mg/ml) and protection from β-carotene bleaching (85.7 per cent at 100µg/ml). The methanol-water extracts showed the highest total phenolic content (856 mg GAE/g and 161 mg/catechin equivalent (CE)/g) and total antioxidative capacity (449 ascorbic acid equivalent (AAE)/g). Furthermore it exhibited dose-dependent antioxidative activities. A relationship between total antioxidative capacity and total phenolic content was recognized in the three extracts from *Eugenia polyantha* bark (Leleno *et al.*, 2009).

Eugenia siamensis Craib. Family: Myrtaceae

Maisuthisakul *et al.* (2008) reported that extracts of *Eugenia siamensis* possessed strong antioxidant activity, total phenolic and flavonoid contents.

Eugenia uniflora L. Family: Myrtaceae

The antioxidant property of *Eugenia uniflora* was studied using free radical-generating systems (Velazquez *et al.*, 2003). The methanol extract protected against enzymatic and non-enzymatic lipid peroxidation in microsomal membranes of rat. The extract showed the highest scavenging activity on the superoxide and DPPH radicals.

Euonymus alatus (Thunb.) Siebold. Family: Celastraceae

Kim *et al.* (2008) investigated the effects of *Euonymus alatus* on cultured hepatocyte cell system, H_2O_2-induced cytotoxicity and antioxidative enzyme activities including catalase, superoxide dismutase, glutathione peroxidase and glutathione S-transferase in treatment conditions. *E. alatus* treatment had significant protective or elevating activities on the oxidative enzyme activities compared to a normal group. The results indicate that *E. alatus* provides a strong antioxidant protection of cells against H_2O_2 induced oxidative stress.

Chon *et al.* (2009) evaluated the antioxidant activities and total phenolics level and cytotoxicity of young sprouts of some traditional Korean salad plants. Methanol extract of *Euonymus alatus* at 63 mg/kg exhibited the highest dose-dependent DPPH radical scavenging activity by 91.2 per cent and highest amount of phenolics (235.7mg/kg).

Euonymus latifolius (L.) Mill. subsp. *latifolius* Family: Celastraceae

Serteser *et al.* (2009) studied the antioxidant activity of 50 per cent aqueous methanol extracts by various antioxidant assays, including free radical scavenging, hydrogen peroxide scavenging and metal (Fe^{2+}) chelating activities. The DPPH radical scavenging effects of leaves extract was 0.413 while fruit extract was 0.608. Fe^{2+} chelating activity (per cent) of leaves extracts was 30.26 while fruit extracts showed 38.65 per cent inhibition. Hydrogen peroxide inhibition activity of leaves extracts was 38.03 per cent and fruit extracts was 36.53 per cent.

Eupatorium articulatum Sch.-Bip. Family: Asteraceae (Compositae)

Ethanolic extracts from 15 plant species used in traditional medicine in Ecuador were evaluated by De las Heras *et al.* (1998) for anti-inflammatory and antioxidant activities. *Eupatorium articulatum* extract was active as antioxidant.

Eupatorium bunifolium **Hook. et Arn. var.** ***bunifolium*** **Family: Asteraceae (Compositae)**

Borneo *et al.* (2008) evaluated the antioxidant activity of ethanolic extract of *Eupatorium bunifolium* var. *bunifolium*. The FRAP value was 319.5μmol of Fe(II)/g and DPPH radical scavenging activity was EC_{50} = 270.0.

Eupatorium fortunei **Turcz. Family: Asteraceae (Compositae)**

Gan *et al.* (2010) evaluated the antioxidant activity of *Eupatorium fortunei* using the FRAP and TEAC assays, and its total phenolic content was measured by the Folin-Ciocalteu method. For FRAP assay the value was 111.80 μmol Fe(II)/g dry weight. For TEAC assay, the value was 71.93μmol Trolox/g dry weight. Total phenolic content was 8.65 mg gallic acid equivalent/g dry weight of plant material.

Eupatorium odoratum **L. Family: Asteraceae (Compositae)**

Chomnawang *et al.* (2007) reported that *Eupatorium odoratum* had a moderate antioxidant activity. Antioxidant activity was determined by DPPH scavenging and NBT reduction assay.

Euphorbia acanthothamnos **Heldr. et Sart. ex Boiss. Family: Euphorbiaceae**

Barla *et al.* (2007) investigated the antioxidant activity of extracts of *Euphorbia acanthothamnos* by β-carotene bleaching method, DPPH assay and ferrous metal chelating activity. The ethanol textract was found to be most active in all tests. Kallimanis and Philiianos (1980) have reported polyphenolic compounds in this species.

Euphorbia antiquorum **L. Family: Euphorbiaceae**

Veeresh (2009) evaluated the protective effect of *Euphorbia antiquorum*, a potent antioxidant, on haloperidol induced catalepsy in mice. The results suggested that administration of methanolic extract has protective effect against haperidol – induced catalepsy.

Euphorbia formosana **Hayata Family: Euphorbiaceae**

Chen *et al.* (2009) studied the antioxidant activity of leaves extract towards xanthine oxidase, lipoxygenase and tyrosinase activities. The inhibitory effect of acetone extract on xanthine oxidase activity was less than 40 per cent, tyrosine activity was more than 40 per cent and lipoxygenase activity was more than 50 per cent.

Euphorbia geniculata **Ortega Family: Euphorbiaceae**

Chougale *et al.* (2011) investigated the antibacterial and antioxidant activity of latex of some plants. Ethanol extract of *Euphorbia geniculata* plant latex has shown almost 93.33 per cent inhibition.

Euphorbia hirta **L. Family: Euphorbiacae**

Kumar *et al.* (2010) evaluated the antidiabetic and *in vitro* free radical scavenging effects of flower extract of *Euphorbia hirta*. Free radicals scavenging effect of ethanolic extract was evaluated by various antioxidant assays, including DPPH free radical scavenging activity, superoxide anion radical scavenging, nitric oxide scavenging and reducing power assy. All the extracts showed antioxidant activity in all the tested methods.

Leaves, stems, flowers and roots from *Euphorbia hirta* were tested for phenolic content and flavonoids content and *in vitro* antioxidant activity by DPPH scavenging activity of 72.96 per cent followed by the flowers, roots and stems whose scavenging activities were 52.45, 48.59 and 44.42 respectively. The standard BHT was 75.13. The IC_{50} for leaves, flowers, roots, stems and BHT were 0.803, 0.972, 0.989, 1.358 and 0.794mg/ml, respectively. The reducing power of the leaves extract was comparable with that of ascorbic acid and found to be dose dependent. Leaves extract had the highest total phenolic content 206.17 mg GAE/g, followed by flowers, roots and stems extracts which were 117.08, 83.15 and 65.70 mg GAE/g respectively. On the other hand, total flavonoids content also from leaves had the highest value (37.970 mg CEQ/g) followed by flowers, roots and stems extracts which were 35.2, 24.35, 24.12 mg CEQ/g, respectively (Basma *et al.*, 2011).

Habila *et al.* (2011) evaluated the phytochemicals and antioxidant activity of four Nigerian medicinal plants, including *Euphorbia hirta*. The percentage of antioxidant activity (AA per cent) of the plants extract showed a dose dependent increase. The results of reducing potential and Total Phenolic contents expressed in terms of Gallic acid equivalent showed that *Euphorbia hirta* possessed moderate reducing potential (0.451nm, 7.00mg) when compared with Gallic acid standard (1.268nm).

Euphorbia macroclada Boiss. Family: Euphorbiaceae

Barla *et al.* (2007) investigated the antioxidant activity of extracts of *Euphorbia macroclada* by β-carotene bleaching method, DPPH assay and ferrous metal chelating activity. All tested extracts of *E. macroclada* showed higher antioxidant activity than BHT and α-tocopherol. The acetone extract demonstrated the highest flavonoid content.

Euphorbia rigida Bieb. Family: Euphorbiaceae

Barla *et al.* (2007) investigated the antioxidant activity of extracts of *Euphorbia macroclada* by β-carotene bleaching method, DPPH assay and ferrous metal chelating activity. Acetone extract showed low antioxidant activity.

Euphorbia royleana Boiss. Family: Euphorbiaceae

Antioxidant activity of different extracts of *Euphorbia royleana* were evaluated by Shahwar *et al.* (2010) through DPPH radical scavenging activity, phosphomolybdate and ferric thiocyanate methods. The DPPH radical scavenging activity of ethyl acetate extract was 34.7 per cent, total antioxidant activity was 0.115 and inhibition of lipid peroxidation was 18.2 per cent while total phenols was 98.0 mg GAE/g of crude extract.

Euphorbia thymifolia L. Family: Euphorbiaceae

Methanol, chloroform, ethyl acetate, n-butanol and aqueous fractions as well as pure compounds isolated from *Euphorbia thymifolia* displayed antioxidant activities as was evidenced by the cytochrome C reduction method (Lin *et al.*, 2002).

Extracts of *Euphorbia thymifolia* leaf showed remarkable free radical scavenging activity with IC_{50} value of 11.0μg/ml and significant inhibitory effects on lipid peroxidation with 79.7 per cent inhibition at the concentration of 50ng/ml (Thoung *et al.*, 2006).

Habila *et al.* (2011) evaluated the phytochemicals and antioxidant activity of four Nigerian medicinal plants, including *Euphorbia thymifolia*. The percentage of antioxidant activity (AA per cent) of the plants extract showed a dose dependent increase. The results of reducing potential and Total Phenolic contents expressed in terms of Gallic acid equivalent showed that *Euphorbia thymifolia* possessed moderate reducing potential (0.482nm, 7.46mg) when compared with Gallic acid standard (1.268nm).

Euphrasia officinalis L. Family: Orobanchaceae

Katalinic *et al.* (2006) reported antioxidant activity and presence of phenolic compounds in the herb *Euphrasia officinalis*.

Eurya nitida Korth Family: Theaceae

Fu *et al.* (2010) evaluated the antioxidant capacity and total phenolic contents of 56 wild fruits from South China using FRAP and TEAC assays. A high correlation between antioxidant capacity and phenolic compounds could be the main contributor to the measured antioxidant activity. The results showed that fruits of *Eurya nitida* has possessed the highest antioxidant capacity and total phenolic contents.

Euryale ferox Salisb. Family: Nymphaeaceae

Various aspects of antioxidant activity were evaluated by Lee *et al.* (2002) in total extracts and fractions derived from *Euryale ferox*. Total extracts (IC_{50} 5.6 μg/ml) showed relatively high level radical scavenging activity toward DPPH and also enhanced viability of Chinese hamster lung fibroblast (V79-4) cells under exposure to oxidative agents. Upon further fractionation, the highest levels of DPPH radical scavenging and lipid peroxidation inhibitory activities were found in the ethyl acetate and butanol fractions. The ethyl acetate fractions, the butanol fractions, and total extracts also dose-dependently enhanced the activities of superoxide dismutase, catalase and glutathione peroxidase in V79-4 cells.

Evodia daniellii (Benn.) Hemsl. Family: Rutaceae

Study was conducted by Thuong *et al.* (2010) to evaluate the structure-activity relationships of the antioxidative of natural coumarins isolated from *Evodia daniellii*. The free radical scavenging and lipid peroxidation assays revealed that five phenolic coumarins, scopoletin (1), aesculetin (2), fraxetin (3), umbelliferone (18), and daphnetin (19), possessed considerable antioxidant activities. The coumarins having a catechol group, 2,3 and 19, showed significant free radical scavenging activity and inhibitory effects on lipid peroxidation indicating that the catechol group contributed to the antioxidant activities of coumarins. In contrast, the sugar moiety markedly reduced the activities of coumarin glycosides. The results also demonstrate that the alpha-pyrone ring of coumarins significantly enhanced the capacity of inhibiting oxidative reactions of coumarins (Thuong *et al.*, 2010).

Evodia officinalis Dode Family: Rutaceae

Methanol aqueous extract of *Evodia officinalis* was screened by Kim *et al.* (1997) for anti-oxidative activity using Fenton's reagent/ethyl lineolate system. The extract showed antioxidant activity.

Evolvulus alsinoides L. Family: Convolvulaceae

Both ethanolic extract and water infusion of *E. alsinoides* have been found to have antioxidant effects, shown by the inhibition of lipid peroxidation and ABTS assay (Auddav *et al.*, 2003).

Alcoholic extract of *Evolvulus alsinoides* showed significant free radical scavenging activity *in vitro* (Achliya *et al.*, 2003). The phytochemical screening revealed presence of triterpenoids and glycosides.

Jain and Agrawal (2009) reported antioxidant activity of methanolic extract of *Evolvulus alsinoides.*

Excoecaria agallocha L. Family: Euphorbiaceae

The hydroalcoholic extract of the dried and ground bark of *Exocoecaria agalloca* was assessed by Subhan *et al.* (2008) for antioxidant activity using a series of well-established assays including DPPH, the lipid peroxidation by the TBA, the reducing power, the nitric oxide, and the hydrogen peroxide scavenging assays. In the DPPH, the NO and the H_2O_2 scavenging assays, the extract displayed significant antioxidant activities with IC_{50} values of 179.16, 120.24 and 134.29µg/ml, respectively. The reducing power of the extract increased dose-dependently, and the extract reduced the most Fe^{3+} ions to the extent less than BHT. In the lipid peroxidation assay, the extract showed significant inhibition of peroxidation effect at all concentrations, with an IC_{50} value of 189.27µg/ml.

Vadlapudi and Naidu (2009) evaluated the antioxidant potential of *Excoecaria agallocha*, a mangrove plant. The SOD was 0.72U/mg, Catalase was 0.55U/mg, Ascorbic acid was 300mg/100g, DPPH radical inhibition was 53 per cent and FRAP units were 1290.

Patra *et al.* (2009) reported that aqueous extract of *Excoecaria agallocha* showed DPPH radical scavenging activity, reducing power and hydrogen peroxide scavenging activity, which increased with increasing concentration.

Excoecaria kawakamii Hayata Family: Euphorbiaceae

Lee *et al.* (2006) investigated the antioxidant activity of ethanol extract of leaves using DPPH, hydroxyl and superoxide radical scavenging and reducing power activities as well as the induction of heme oxygenase-1 (HO-1). The IC_{50} values against DPPH radical were 8.7µg/ml, OH radical 0.32µg, superoxide radical 86.7µg/ml and total phenolics was 23.2 mg of GAE/g.

Eysenhardtia platycarpa Pennell et Saff. Family: Fabaceae (Leguminosae)

Methanolic extracts from *Eysenhardtia platycarpa* protected the rat pancreatic homogenate from AAPH induced damage (Narvez-Mastache *et al.*, 2007). The inhibitory effect was dose-dependent at concentrations of 10-1000 ppm and correlated with DPPH radical scavenging capacity. 3-O-Acetyl-11α, 12 α-epoxy-oleanan-28, 13β-olide, (+)catechin, and (+)-catechin 3-O- β-D-galactopyranoside constituents isolated from the branches of *E.platycarpa* displayed significant antioxidant activity. Compounds 1 and 2 significantly increased the pancreatic glutathione (GSH) concentration alone and in combination with AAPH treatment. Compound 1, at a dose of 100mg/kg b.wt. (i.p) for 5d, also significantly increased the activities of glutathione peroxidase (GSPx) and catalase. Simultaneous treatment of 1 (100 mg.kg b. wt.) and STZ significantly reduced the pancreatic thiobarbituric acid reactive substances concentration together with a significant increase in the activities of GSPx and CAT, preventing hyperglycemia induced by STZ after 5d treatment. Study demonstrated antioxidant and anihyperglycemic activities of compound 1 isolated from *Eysenhardtia* species.

Eysenhardtia punctata Pennell Family: Fabaceae (Leguminosae)

Methanolic extracts from *Eysenhardtia punctata* protected the rat pancreatic homogenate from AAPH induced damage (Narvez-Mastache *et al.*, 2007). The inhibitory effect was dose-dependent at concentrations of 10-1000 ppm and correlated with DPPH radical scavenging capacity.

Eysenhardtia subcoriacea Pennell Family: Fabaceae (Leguminosae)

Methanolic extracts from *Eysenhardtia subcoriacea* protected the rat pancreatic homogenate from AAPH induced damage (Narvez-Mastache *et al.*, 2007). The inhibitory effect was dose-dependent at concentrations of 10-1000 ppm and correlated with DPPH radical scavenging capacity.

Fagaropsis angolensis (Engl.) Dale Family: Rutaceae

Kuglerova *et al.* (2008) investigated antioxidant activity of seven Ugandan tree species. The crude ethanolic extract of bark of *Fagaropsis angolensis* exhibited DPPH scavenging activity with IC_{50} value 175μg/ml.

Fagonia cretica L. Family: Zygophyllaceae

It has been reported to contain a wide variety of antioxidants and have been in use in the Eastern system of medicine for various disorders. *Fagonia cretica* exhibits strong free radical scavenging properties against reactive oxygen and nitrogen species as revealed by electron paramagnetic resonance spectroscopy, diminishing the expression of iNOS gene. Therefore, it attenuate oxidative stress mediated cell injury during oxygen glucose deprivation (OGD) and exert the above effects at both the cytosolic as well as at gene expression levels and may be effective therapeutic tool against ischemic brain damage. Further protective ability may be attributed to the GSH and vitamin C content of *F. cretica* (Rawal *et al.*, 2004).

Fagopyrum cymosum (Trev.) Meisn. Family: Polygonaceae

Cai *et al.* (2004) evaluated the antioxidant activity and phenolic compounds in traditional Chinese medicinal plants associated with anticancer. The improved ABTS.+ method was used to systematically assess the total antioxidant capacity (Trolox equivalent antioxidant capacity, TEAC) of the medicinal extracts. The TEAC values and total phenolic content for methanolic extract of seed of *Fagopyrum cymosum* were 1628.7 µmol Trolox equivalent/100g dry weight (DW), and 6.67 g of gallic acid equivalent/100g DW respectively. A positive significant linear relationship between antioxidant activity and total phenolic content showed that phenolic compounds were the dominant antioxidant components in the tested herbs. Major types of phenolic compounds in *Fagopyrum cymosum* include phenolic acids (*p*-coumaric acid, ferulic acid), flavonols (rutin), tannins.

Fagopyrum esculentum Moench Family: Polygonaceae

English Name: Buckwheat.

Wojcicki *et al.* (2006) investigated the effect of buckwheat extract on free radical generation in rabbits administered with high fat diet. Quite distinct superiority of buckwheat extract, compared with rutin, was shown.

Liu *et al.* (2007) reported that trace element water improves the antioxidant activity of Buckwheat sprouts.

Wojcikowski *et al.* (2007) evaluated the antioxidant activity of aerial parts of *Fagopyrum esculentum.* Oxygen radical absorbance capacity of the ethyl extract, methanol extract and aqueous extracts were 0.22, 13.70 and 2.71µmol Trolox equivalent per g of dried starting material, respectively.

Liu *et al.* (2008) compared the differences of two types of buckwheat sprouts, namely, common buckwheat (*Fagopyrum esculentum)* and tartary buckwheat (*Fagopyrum tataricum*), in general composition, functional components, and antioxidant capacity. The ethanol extracts of tartary buckwheat sprouts had higher reducing power, free radical scavenging activity, and superoxide anion scavenging activity than those of common buckwheat sprouts. As for chelating effects on ferrous ions, common buckwheat had higher values than tartary buckwheat. Rutin was the major flavonoid found in these two types of buckwheat sprouts, and tartary buckwheat sprouts was 5 fold higher in rutin than common buckwheat sprouts.

Alvarez-Jubete *et al.* (2010) examined the polyphenol composition and antioxidant properties of methanolic extracts from buckwheat, and evaluated how these properties were affected following two types of processing: sprouting and baking. The total phenol content amongst the seed extracts were significantly higher in buckwheat (323.4mg GAE/100g). Antioxidant capacity, measured by the DPPH scavenging capacity and the ferric ion reducing antioxidant power assays was also highest for buckwheat seed extract. The total phenol content and antioxidant activity was generally found to increase with sprouting and derease in levels was observed following bread making. Analysis by liquid chromatography coupled with diode array detector revealed the presence of phenolic acids, catechins, flavonol, flavone and flavonol glycosides.

Buckwheat is a source of rutin and other antioxidant phytochemicals. Rutin is highest during the flowering stage of crop growth. A longitudinal study was performed by Metzger *et al.* (2010) in the spring of 2007 to characterize the expression of other antioxidant phytochemcals from sprout to seed development on a large scale farming operation. Carotenoids decreased during the growth of the plant, while the rutin and anthocyanin concentrations peaked between weeks 4-5 before declining. Total phenolics increased and were maximal between weeks 4-7. There was no change in vitamin levels. Antioxidant activity against iron induced oxidation was not different during the growing period; however, the antioxidant activity against a peroxyl radical increased over the experiment and was maximal 7 weeks after sowing. Based on summary of the data, the optimal harvest time to enhance antioxidant and phytochemical content of buckwheat plants is between 4-5 weeks after sowing.

Fagopyrum tataricum (L.) Gaertn. Family: Polygonaceae

English Name: Tartary buckwheat.

Liu *et al.* (2008) compared the differences of two types of buckwheat sprouts, namely, common buckwheat (*Fagopyrum esculentum)* and tartary buckwheat (*Fagopyrum tataricum*), in general composition, functional components, and antioxidant capacity. The ethanol extracts of tartary buckwheat sprouts had higher reducing power, free radical scavenging activity, and superoxide anion scavenging activity than those of common buckwheat sprouts. As for chelating effects on ferrous ions, Common buckwheat had higher values than tartary buckwheat. Rutin was the major flavonoid found in these two types of buckwheat sprouts, and tartary buckwheat sprouts was 5 fold higher in rutin that common buckwheat sprouts.

Fagraea blumei G. Don Family: Loganiaceae

Blumeosides A and D isolated from the methanolic stem-bark-extract of *Fagraea blumei* demonstrated scavenging properties towards DPPH radical in TLC autographic and spectrophtometric assays (Cuendet *et al.*, 1997).

Falcaria vulgaris Bernh. Family: Apiaceae (Umbelliferae)

Antioxidant activity of the essential oils from leaf, flower and stem of *Falcaria vulgaris* were investigated by Shafaghat (2010) using DPPH radical scavenging method.The essential oils from different parts and localities of *Falcaria vulgaris* showed free radical scavenging activity.

Fallopia convolvulus (L.) A.Love Family: Polygonaceae

The antioxidative activity of phenolic extracts from *Fallopia convolvulus* was examined by autoxidation of methyl linoleate by Kahkonen *et al.* (1999). The total phenolics were very low 2.8 mg/g GAE and the antioxidant activity was also very low (13 per cent at 500 ppm level and 2 per cent at 5000 ppm level).

Feronia elephantum Correa Family: Rutaceae

Antioxidant activity of the fruit extract was assessed by Nithya and Saraswathi (2010) by estimating total phenols, flavonoids, tannins, total antioxidant capacity and iron chelating activity. The results suggested that the extract exhibited dose dependent antioxidant ability which might possibly be due to the presence of phenolic compounds. The total phenolic content of the plant extract was found to be 10.17 mg of pyrocatechol/g plant tissue. The flavonoid content of plant extract was recorded to be 2.3 mg of catechin/g plant tissue. Tannin content of the plant extract was calculated as 0.045 mg of tannic acid/g plant tissue. Ferric Reducing power of ascorbate and plant extract as indicated by its absorbance was found to be 1.91 and 1.25 at 1000 µg/nl, respectively. This assay suggests that the reducing ability of the plant extract was comparable to that of ascorbate. The total antioxidant capacity was measured as 610 mg of ascorbate/g plant tissue by phosphomolybdic acid method which is based on the ability of plant extract to form phosphomolybednum complex by the transformation of Mo (VI) to Mo (V).

Feronia limonia L. Family: Rutaceae

Hydromethanol extracts of 15 Bangladeshi medicinal plants, including *Feronia limonia*, were evaluated by Hasan *et al.* (2009) for antioxidant potential using DPPH radical scavenging assay. *Feronia limonia* leaves exhibited radical scavenging activity with an IC_{50} value of 17.60 µg/ml compared to the IC_{50} value of 5.15µg/ml as shown by the reference antioxidant ascorbic acid, in a dose dependent fashion.

Ferula hermonis Boiss. Family: Apiaceae (Umbelliferae)

Suboh *et al.* (2004) investigated the antioxidant activity effects of various concentrations of methanol extracts of *Ferula hermonis* and six other medicinal plants. They reported that the extract significantly reduced 10mM hydrogen peroxide-induced lipid peroxidation caused by H_2O_2 in human erythrocytes.

Ferula orientalis L. Family: Apiaceae (Umbelliferae)

Kartal *et al.* (2005) examined the *in vitro* antioxidant properties of the essential oil and various extracts prepared from the herbal parts of *Ferula orientalis.* The highest DPPH radical-scavenging activity was found in the polar extract, *e.g.*, methanol-water (1:1), obtained from non-deodorised materials with IC_{50} values at 99.1µg/ml. In the β-carotene/linoleic acid assay, the deodorized acetone extract exhibited stronger activity than the polar one. The relative antioxidant activities of the extracts ranged from 10.1 per cent to 76.1 per cent, respectively. Extraction with methano-water (1:1) mixture was concluded to be the most appropriate method in terms of higher extract yield, as well as effectiveness, observed in both assays. Although the essential oil showed antioxidative potential, it was not as strong as that of positive control (BHT).

Ferulago campestris (Besser) Grecescu Family: Apiaceae (Umbelliferae)

Basile *et al.* (2009) investigated the antimicrobial and antioxidant activities of coumarins from the roots of *Ferulago campestris* by their effects on human blood leucocytes (WB) and on isolated polymorphonucleate (PMN), chemiluminiscence (CL), PMA stimulated resting.

Fibraurea tinctoria Lour. Family: Menispermaceae

Keawpradub *et al.* (2005) evaluated the antioxidant activities of *Fibraurea tinctoria* employing DPPH radical scavenging assay. Methanol extract exhibited moderate antioxidant activity with EC_{50} values of 25-55µg/ml.

Ficus asperifolia Miq. Family: Moraceeae

Annan and Houghton (2008) investigated the antioxidant activity of *Ficus asperifolia.* Aqueous extracts significantly protected fibroblast cells against oxidative damage a doses up to 50µg/ml.

Ficus benghalensis L. Family: Moraceae

English Name: Banyan tree.

The antioxidant effect of aqueous extract of the bark of Banyan tree has been evaluated by Shukla *et al.* (2004) in hypocholestrolaemic rabbits. Treatment with this extract led to a decrease in lipid peroxidation as evidenced by fall in thiobarbituric acid reactive substances with corresponding increase in blood glutathione content. Further, there was a significant increase in the activities of antioxidant enzymes, superoxide dismutase, catalase, glutathione peroxidase. The water extract of the bark of Banyan tree has significant antioxidant effect, in addition to hypolipidaemic effect.

The stem bark and fruits of *Ficus benghalensis* are used in India for the treatment of diabetes and a number of other diseases. Since these effects may be correlated with the presence of antioxidant compounds, methanol and 70 per cent acetone extracts of aerial root were evaluated by Manian *et al.* (2008) for their antioxidant activity and radical scavenging capacity in comparison with *Camellia sinensis* (green tea). Methanol extracts of green tea and *Ficus benghalensis* contained relatively higher levels of total phenolics than the other extracts. The antioxidant potential of the extracts was assessed by employing different *in vitro* assays. Though all the extracts exhibited dose dependent reducing power activity, methanol extracts of all the samples were found to have more hydrogen donating ability. Similar line of dose dependent activity has been maintained in all the samples in DPPH and OH scavenging systems. All the extracts exhibited antioxidant activity against the linoleic acid emulsion system (34-38 per cent). The potential of multiple antioxidant activity was evident as it possessed antihemolytic activity and metal ion chelating potency.

Thakare *et al.* (2010) reported that stem bark extraction of *Ficus benghalensis* exhibited anti-inflammatory and analgesic acivity in animal models and they attributed this largely to its antioxidant activity.

Patel *et al.* (2010) evaluated the antioxidant activity of *Ficus benghlensis* leaf. Alcoholic extract had significant free radical scavenging activity. In Hydrogen peroxide scavenging aassay, IC_{50} value of alcoholic extract and ascorbic acid was 399.5µg/ml and 276µg/ml, respectively. In DPPH assay, IC_{50} value of alcoholic extract and ascorbic acid was 158.2µg/ml and 104.2µg/ml respectively. In nitric oxide scavenging assay IC_{50} value of alcoholic extract and ascorbic acid was 373µg/ml and 264.5µg/ml, respectively. Flavonoid content of alcoholic extract was 0.18 per cent catechin equivalent/100 g of fresh mass. Phenolic content of alcoholic and aqueous extract was 0.45 per cent and 0.09 per cent of GAE/100g fresh mass respectively. The reducing power was also comparable to ascorbic acid.

Free radical scavenging activity of extracts of young prop roots of *Ficus benghalensis* was evaluated by Nidhiya *et al.* (2010) employing DPPH assay, ABTS radical scavenging assay and iron chelating activity. In all the methods the alcoholic and aqueous extracts showed concentration dependent radcal scavenging activity and IC_{50} values were 50.87 and 95.7µg/ml for DPPH assay, 49.23 and 103.08µg/ml for iron chelating activity, 19.89 and 392µg/ml for ABTS scavenging. The alcoholic extract exhibited promising antioxidant activity compared to aqueous extracts and the results were comparable with standard antioxidant ascorbic acid.

Ficus capensis (Thunb.) Forssk. Family: Moraceae

Muanda *et al.* (2011) evaluated the antioxidant activity of three Malian medicinal plants including *Ficus capensis* employing phosphomolybdenum assay (PPM), DPPH and ABTS tests. The content of TPC and TFC were higher in the stem barks. The PPM assay showed that the highest value was 280.0mg/g dw for the stem bark, followed by root bark (60 mg/g dw). The values of ABTS were 120.8 and 122.5 mg/g d w for stem bark and root bark, respectively. The values for DPPH were 195.8 and 91.3 mg/g dw, for stem bark and root bark respectively.

Ficus carica L. subsp. _carica_ Family: Moraceae

Souri *et al.* (2004) evaluated the antioxidative activity of *Ficus carica* subsp. *carica* by linoleic acid peroxidation test using 1,3-diethyl-2-thiobarbituric acid as the reagent. Methanolic extract showed 97.05 per cent inhibition at 40µg concentration with IC_{50} value of 3.18µg.

Serteser *et al.* (2009) studied the antioxidant activity of 50 per cent aqueous methanol extracts by various antioxidant assays, including free radical scavenging, hydrogen peroxide scavenging and metal (Fe^{2+}) chelating activities. The methanolic extract examined in the assay showed the strongest activities. The DPPH radical scavenging effects of leaves extract was 0.549 while fruit extract was 0.640. Fe^{2+} chelating activity (per cent) of leaves extracts is 32.86 while fruit extracts showed 40.69 per cent inhibition. Hydrogen peroxide inhibition activity of leaves extracts was 39.62 per cent and fruit extracts was 47.18 per cent.

Ficus cotinifolia H.B.& K. Family: Moraceae

Ruiz-Terán *et al.* (2008) evaluated the antioxidant activity of aerial parts of *Ficus cotinifolia* using DPPH and β-carotene-linoleic acid bleaching assay. The extract displayed antioxidant activity comparable to the commercial antioxidant BHA. For the methanolic extract analysed, a clear relation between the total phenolic content of the extracts and their antioxidant activity was found.

Ficus deltoidea Jack Family: Moraceae

Hakiman and Maziah (2009) investigated the antioxidant activities in aqueous extract of different accestion of *Ficus deltoidea*. They found that female leaf extracts scavenge more free radicals compared to the male leaf extracts. They also investigated the flavonoid content in the leaves.

Ficus involuta Miq. Family: Moraceae

Ruiz-Terán *et al.* (2008) evaluated the antioxidant activity of aerial parts of *Ficus involuta* using DPPH and β-carotene-linoleic acid bleaching assay. The extract displayed antioxidant activity comparable to the commercial antioxidant BHA. For the methanolic extract analysed, a clear relation between the total phenolic content of the extracts and their antioxidant activity was found.

Ficus lutea Vahl Family: Moraceae

Marwah *et al.* (2007) investigated the antioxidant activity of wound-healing herbs using *in vitro* DPPH and phosphomolybdenum assay methods. Out of the 19 plants screened, the aqueous ethanol extract of *Ficus lutea* showed the best DPPH scavenging activities with per cent IP value 84.6 per cent, after 15 min of incubation at an effective test concentration of 50µg/ml. The Total Antioxidant Capacity as gallic acid equivalent of 335mg/g of ethanol extract was obtained in the phosphomolybdenum assay. Total phenolic content was 224 mg GAE/g of ethanol extract.

Ficus microcarpa L. Family: Moraceae

Ao *et al.* (2008) evaluated the antioxidant and antibacterial activities of extracts of *Ficus microcarpa*. Results showed that methanol extracts of bark, fruits and leaves exhibited excellent antioxidant activities. Ethyl acetate fraction of bark extract exerted strong antioxidant antibacterial effects and contained high amount of total phenolics (436 GAE mg/g extract). EC_{50} values of bark extract were 4.83, 1.62 and 63.2 µg/ml in DPPH, ABTS, superoxide radicals scavenging methods, respectively. Twelve phenolic compounds were identified in bark extract.

Ficus racemosa L. (Syn.: *Ficus glomerata* Roxb.) Family: Moraceae

Veerapur *et al.* (2007) also reported that ethanol extract exhibited potent antioxidant activity against DPPH, ABTS, hydroxyl radical, superoxide radical scavenging and inhibited lipid peroxidation in concentration dependent doses. The methanol extract of stem bark has shown potent *in vitro* antioxidant activity when compared to the methanol extract of its roots (Channabasavaraj *et al.*, 2008). Ahmed *et al.* (2010) also reported radical scavenging activity of cold and hot aqueous extracts of *Ficus racemosa* stem bark. Radical scavenging activity was determined using DPPH method. Both the extracts exhibited excellent radical scavenging activity. Hot aqueous extract showed significantly higher radical scavenging activity than cold aqueous extract.

The stem bark and fruits of *Ficus racemosa* are used in India for the treatment of diabetes and a number of other diseases. Since these effects may be correlated with the presence of antioxidant compounds, methanol and 70 per cent acetone extracts of stem bark were evaluated by Manian *et al.* (2008) for their antioxidant activity and radical scavenging capacity in comparison with *Camellia sinensis* (green tea). Methanol extracts of green tea and 70 per cent acetone extract of *Ficus racemosa* contained relatively higher levels of total phenolics than the other extracts. The antioxidant potential of the extracts was assessed by employing different *in vitro* assays. Though all the extracts exhibited dose dependent reducing power activity, methanol extracts of all the samples were found to have more hydrogen donating ability. Similar line of dose dependent activity has been maintained in all the samples in DPPH and OH scavenging systems. All the extracts exhibited antioxidant activity against the linoleic acid emulsion system (34-38 per cent). The potential of multiple antioxidant activity was evident as it possessed antihemolytic activity and metal ion chelating potency.

Hypoglycemic and antioxidant activities of *Ficus racemosa* fruits was investigated by Ara Jahan *et al.* (2009). The 1-BuOH soluble part of the ethanol extract exhibited significant antioxidant activity in DPPH free radical scavenging assay. 3-I-(E)-Caffeoyl quinate was isolated for the first time from this plant, which also showed significant antioxidant activity.

Ahmed and Urooj (2010) investigated hepatoprotective effects of *Ficus racemosa* stem bark against carbon tetrachloride-induced hepatic damage in albino rats. The petroleum ether and methanol extracts improved the antioxidant status considerably as reflected by low TBARS and high GSH values.

Patel *et al.* (2010) evaluated the antioxidant activity of *Ficus racemosa* leaf. Alcoholic extract had significant free radical scavenging activity. In Hydrogen peroxide scavenging assay, IC_{50} value of alcoholic extract and ascorbic acid was 461µg/ml and 276µg/ml, respectively. In DPPH assay, IC_{50} value of alcoholic extract and ascorbic acid was 122µg/ml and 104.2µg/ml respectively. In nitric oxide scavenging assay IC_{50} value of alcoholic extract and ascorbic acid was 343.4µg/ml and 264.5µg/ml, respectively. Flavonoid content of alcoholic extract was 0.14 per cent catechin equivalent/100 g of fresh mass. Phenolic content of alcoholic and aqueous extract was 0.37 per cent and 0.04 per cent of GAE/100g fresh mass respectively. The reducing power was also comparable to ascorbic acid.

Ficus religiosa L. Family: Moraceae

Rathee *et al.* (2010) assessed the *in vitro* potential of methanolic extract of *Ficus religiosa* as a natural antioxidant. The DPPH activity of the *Ficus religiosa* fruit extract increased in a dose dependent manner which was found in the range of 43.5-73.92 per cent as compared to ascorbic acid 55.25 – 85.79 per cent. There was a significant increase in reducing power of the fruit extract with increasing concentration. Fruit extract of *Ficus religiosa* demonstrated the ability to inhibit the formation of TBARS in a concentration dependent manner. Phenolic content was 0.2 per cent w/w.

Filipendula hexapetala Gilib. ex Maxim. Family: Rosaceae

The antioxidant activity of the methanol extract of *Filipendula hexapetala* flowers was assessed by the assay for FRAP, the assay for DPPH free radical scavenging ability and the assay for the influence of lipid peroxidation in liposomes, induced by Fe^{2+}/ascorbate system and measured by the TBA test. The activity of the investigated extract in all test-systems was found to be significant. The principal constituent responsible for the observed effects was isolated and identified as spiraeoside (Maksimovi *et al.*, 2007).

Filipendula ulmaria (L.) Maxim. Family: Rosaceae

English Name: Meadowsweet

The antioxidative activity of phenolic extracts from fruits of Meadowsweet was examined by autoxidation of methyl linoleate by Kahkonen *et al.* (1999). Remarkable high antioxidant activity (95 per cent inhibition at 500 ppm level and 87 per cent at 5000 ppm level) and high total phenolic content (26.8 mg/g GAE) were found in meadowsweet. Two of the main polyphenolic constituents of meadowsweet were the tannins tellimagrandin II and rugosin D (Haslam, 1996). Spiraeoside, hyperoside, kaempferol-4-glucoside and quercetin have been identified in the flowers and flowering tops (Lamaison *et al.*, 1991).

Nikolova and Dzhurmanski (2009) evaluated the free radical scavenging activity of total methanol extract of *Filipendula ulmaria*. The extract exhibited strong antioxidant activity and its IC_{50} value for DPPH radical was 9.04 µg/ml.